# PLANT BIOTECHNOLOGY

*Comprehensive Biotechnology*
*Second Supplement*

**PERGAMON MAJOR REFERENCE WORKS**

Comprehensive Inorganic Chemistry (1973)

Comprehensive Organic Chemistry (1979)

Comprehensive Organometallic Chemistry (1982)

Comprehensive Heterocyclic Chemistry (1984)

International Encyclopedia of Education (1985)

Comprehensive Insect Physiology, Biochemistry & Pharmacology (1985)

Comprehensive Biotechnology (1985)

Physics in Medicine & Biology Encyclopedia (1986)

Encyclopedia of Materials Science & Engineering (1986)

World Encyclopedia of Peace (1986)

Systems & Control Encyclopedia (1987)

Comprehensive Coordination Chemistry (1987)

Comprehensive Polymer Science (1989)

Comprehensive Electrocardiology (1989)

Comprehensive Medicinal Chemistry (1990)

Comprehensive Organic Synthesis (1991)

Comprehensive Rock Engineering (1993)

# PLANT BIOTECHNOLOGY

*Comprehensive Biotechnology*
*Second Supplement*

VOLUME EDITORS

MICHAEL W. FOWLER

*University of Sheffield, UK*

&

GRAHAM S. WARREN

*University of Sheffield, UK*

EDITOR-IN-CHIEF

MURRAY MOO-YOUNG

*University of Waterloo, Ontario, Canada*

**PERGAMON PRESS**

**OXFORD · NEW YORK · SEOUL · TOKYO**

| | |
|---|---|
| U.K. | Pergamon Press plc, Headington Hill Hall, Oxford OX3 0BW, England |
| U.S.A. | Pergamon Press, Inc., 395 Saw Mill River Road, Elmsford, New York 10523, USA |
| KOREA | Pergamon Press Korea, KPO Box 315, Seoul 110-603, Korea |
| JAPAN | Pergamon Press Japan, Tsunashima Building Annex, 3-20-12 Yushima, Bunkyo-ku, Tokyo 113, Japan |



First edition 1992

**Library of Congress Cataloging in Publication Data**

Plant biotechnology: Comprehensive biotechnology. Second supplement volume editors, Michael W. Fowler & Graham S. Warren; editor-in-chief, Murray Moo-Young.
p. cm.
Includes bibliographical references and index.
1. Plant biotechnology. I. Fowler, Michael W. II. Warren, Graham S., 1952– . III. Moo-Young, Murray. IV. Comprehensive biotechnology.
TP248.27.P55P56 1991 91–16681
631—dc20

**British Library Cataloguing in Publication Data**

Fowler, Michael W.
Plant biotechnology. – (Comprehensive biotechnology supplement series; no. 2) I. Title II. Warren, Graham S. III. Series
660.6

ISBN 0-08-034731-2

∞™The paper used in this publication meets the minimum requirements of the American National Standard for Information Sciences—Permanence of Paper for Printed Library Materials, ANSI Z39.48-1984.

Printed and bound in Great Britain by BPCC Wheatons Ltd, Exeter

# Contents


Foreword vii
Preface ix
Contributors xiii
Contents of All Volumes xv

**Biology of Plant Cells**

1 The Cell Biology of Plant Culture Systems 1
G. S. Warren, *University of Sheffield, UK*

2 The Molecular Biology of Plant Cells and Cultures 19
C. A. Cullis, *Case Western Reserve University, Cleveland, OH, USA*

**Systems for the Exploitation of Cell Cultures**

3 Cell Selection 33
W. H.-T. Loh, *DNA Plant Technology Corporation, Cinnaminson, NJ, USA*

4 Bioreactors for the Mass Cultivation of Plant Cells 45
A. H. Scragg, *Bristol Polytechnic, UK*

5 Immobilized Plant Cells 63
P. D. Williams and F. Mavituna, *University of Manchester Institute of Science and Technology, UK*

6 Plant Cell Culture, Process Systems and Product Synthesis 79
M. W. Fowler and A. M. Stafford, *University of Sheffield, UK*

**Regeneration and Propagation Systems**

7 Plant Regeneration from Cultured Protoplasts of Higher Plants 99
S. J. Ochatt, *INRA, Angers, France* and J. B. Power, *University of Nottingham, UK*

8 Micropropagation: Principles and Commercial Practice 129
R. D. Rice, *London, UK*, P. G. Alderson, *Nottingham University School of Agriculture, Loughborough, UK*, J. Hall, *Leicester Polytechnic School of Life Sciences, UK* and A. Ranchhod, *Sheffield City Polytechnic, UK*

**Genetic Manipulation of Plant Cells**

9 Plant Genetic Transformation 151
A. G. Day, *University of Cambridge, UK* and C. P. Lichtenstein, *Imperial College of Science, Technology and Medicine, London, UK*

10 Expression of Plant Genes in Yeast and Bacteria 183
N. Overbeeke and C. T. Verrips, *Unilever Research Laboratorium Vlaardingen, The Netherlands*

11 Protoplast Fusion 199
H. Morikawa and Y. Yamada, *Kyoto University, Japan*

12 Genetic Engineering of Plants and Cultures 223
G. Ooms, *AFRC Institute of Arable Crops Research, Herts, UK*

**Specific Applications in Plant Biotechnology**

13 Application of Unconventional Techniques in Classical Plant Production 259
G. Wenzel, *Institute for Resistance Genetics, Grünbach, Germany*

14 Cell Culture and Recombinant DNA Technology in Plant Pathology 283
G. S. Warren, *University of Sheffield, UK*

15 Exploitation of Chloroplast Systems in Biotechnology: Stabilization and Regulation of Photosynthesis 293
C. H. Foyer, *INRA, Versailles, France* and R. T. Furbank, *CSIRO, Canberra, Australia*

**Subject Index** 317

# Foreword

In 1985, *Comprehensive Biotechnology* was published as a major reference work (four volumes; 3764 pages) for effective 'one stop shopping' in a diverse, multidisciplinary field which had previously been treated only in specialist publications. It was well received by the international audience, as shown by citations from *Nature*, the American Chemical Society, the Society for Industrial Microbiology and the Institution of Chemical Engineers, among others.

Since 1985, the biotechnology field has grown significantly, especially in the agriculture-related aspects. To address the changes, the supplement *Animal Biotechnology* was published in 1989 and this year (1992), a complementary supplement *Plant Biotechnology* is being released. As with the first supplement, the editors of *Plant Biotechnology* (M. Fowler and G. Warren) undertook the difficult task of capturing, in an authoritative way, the current status and future trends in this important aspect of biotechnology. We owe them both a word of appreciation for their dedication.

In the future, Pergamon intends to keep the four volume foundation work of *Comprehensive Biotechnology* updated on an ongoing basis with review articles in its quarterly journal, *Biotechnology Advances.*

MURRAY MOO-YOUNG
*Waterloo, Canada*
*June 1991*

# Preface

### Introduction

Some 10 to 15 years ago when the first stirrings of what might be termed 'modern' biotechnology began to take shape, and organizations such as the European Federation of Biotechnology were formed to nurture its development, great doubts were expressed as to the applicability of what might be loosely termed plant sciences to biotechnology. What a difference a decade has made. Today it is generally accepted that one of the key areas of biotechnology for the next century will be in plant-based biotechnology, with implications ranging from improved crops and food provision, to alternative bioremediation systems and novel high value chemicals, including enzymes. Tremendous progress has been made in all aspects of plant physiology, biochemistry and molecular biology over the last decade, much of it driven from a biotechnological standpoint.

The application of novel techniques aimed at improving the performance of plants or plant cells has been a particular growth area in modern biotechnology. Such growth has been directed largely by the needs of agriculture and fuelled by the rapid development of the central techniques of regeneration from plant cell and tissue cultures, gene vector design and plant transformation. Other factors have also been key, notably the interest of many multinational companies who have been quick to see the commercial potential of engineered crop species. Equally, organizations such as the European Commission and World Bank have seen the wider impact upon farming economies and Third World development, and have been instrumental in assembling multinational research programmes focusing on plant biotechnology. In parallel, plant biotechnologists have also recognized that recently unreachable goals have almost overnight come within their grasp, enabling them to reach towards opportunities both scientific and commercial.

Like every branch of biotechnology, plant biotechnology is an ever-broadening subject that is difficult to characterize. We have attempted for ease of presentation to divide this volume into five subject areas.

### Biology of Plant Cells

Much that is fundamental to the development of plant biotechnology in the future lies in the structure, composition and functionality of plant cells and genes. It is on this area that the first section focuses, with particular emphasis on the cellular and molecular biology of plants and cultured cells. An appreciation of these basic aspects is important in the recognition of new possibilities and the limitations of many of the individual techniques currently in use. The pursuit of applied aims without due attention to the underlying mechanisms involved is likely to be a self-limiting and inefficient process. This is seen in molecular biology where, while protocols for cloning genes have become almost routine, the question 'which gene shall we clone?' has become commonplace. Plant biotechnology at present is greatly limited by the lack of basic understanding of most of the useful characters we wish to manipulate.

The state of knowledge of plant systems has always lagged far behind that of microorganisms and animals. This to some extent reflects the smaller number of researchers working with plants, but also reflects the technical problems associated with experimentation on plant systems. The new techniques, especially recombinant DNA methods, linked to cell and tissue culture offer great scope to relieve this situation, leading to a deeper and wider appreciation of plant cell and molecular biology.

### Systems for the Exploitation of Cell Cultures

This section is concerned with the direct exploitation of cell cultures for the production of useful substances. In principle, this area of research should be perhaps the most immediately fruitful because engineered or selected cells can be produced with characteristics that do not need to be compatible with plantlet regeneration or plant fertility, constraints that have limited progress in certain agricultural applications. However, other obstacles have been encountered, in particular the failure of many cell lines in culture to express those genes coding for secondary metabolite synthesis, genetic instability of cells maintained in culture and the lack of basic information about the biochemistry of biosynthetic pathways. Largely for these reasons, progress in the development of plant cell cultures for industrial use has generally been disappointing. In addition to more basic research, the more widespread adoption of gene transfer methods for the manipulation of pathways may alter this situation. Most likely, plant cell culture will prove to be a feasible route to the production of specific substances in particular, favourable circumstances, for example novel products unique to cultured cells, and the formation of protein products from foreign genes. The alternative strategy of the transfer of plant genes to microorganisms for heterologous expression is also being actively pursued.

## Regeneration and Propagation Systems

Central to the application of novel techniques to the improvement of plants is the ability to regenerate whole, fertile plants from individual cells and protoplasts. Although this process was first described some 40 years ago, it has long been restricted to a relatively small range of species. Extensive research has resulted in an improvement in this situation to the point that at least low levels of regeneration are now possible from most of the important crop species. However, much of the progress has been empirically based, and plant regeneration is still devoid of a satisfactory theoretical foundation. The desired applications of cell fusion and transformation techniques have placed extra demands on regeneration systems. The requirement that newly developed plant varieties must be compatible with breeding programmes necessitates high fertility of the regenerants. This situation is usually the exception rather than the norm. It has been found that, although transformation and regeneration can often be readily achieved separately, transformation of regeneration competent cells is much rarer, and therefore the frequency of transgenic plants recovered can be very low. New developments are occurring rapidly in this area, however, and the recent reports of fertile, transgenic rice and maize perhaps indicate that these problems are near to a general solution.

Plant regeneration has long been exploited commercially for the micropropagation of ornamentals and disease-free stock. However, tissue culture techniques are labour intensive and time consuming and careful planning is required to ensure that a micropropagation scheme is economically viable. There is currently increasing interest in the mechanization of certain culture manipulations with a view to reducing the cost of the process.

## Genetic Manipulation of Plant Cells

The next section considers the increasingly central area of genetic manipulation of plant cell systems. For a number of years, manipulation of the genetic make-up of plants by protoplast fusion, mutagenesis and culture-induced variation has resulted in steady but relatively slow progress towards improved plant varieties. Now specific gene transfer has moved to centre stage. The ability to add just the desired genes to a plant, without the near certainty of downgrading previously optimized characters, gives special appeal to this approach. However, the other novel methods will continue to be useful in specific instances, especially in cases in which there is little understanding of the molecular mechanisms governing the desired characters. Traditional methods will probably remain central to crop improvement but will be increasingly complemented by the new technology. Biotechnology is all about application and commercialization. The technique that achieves the desired aim most cheaply will be adopted and not necessarily that which is novel, or technically most elegant or sophisticated.

Genetic engineering has already been responsible for the production of plants with enhancement in a range of desirable traits, notably disease resistance, insect resistance, ripening properties and nutritional and commercial value. Progress is also being made towards longer term aims such as improvement of the efficiency of photosynthesis and other polygenic mechanisms. Enthusiasm for transgenesis is currently high. The bounds of the possible have been radically widened. However, this enthusiasm is being tempered by the growing debate on the potential hazards of releasing transgenic plants, the ethical and emotional concerns associated with changing our flora, and whether there is an actual need to create certain new plant varieties.

This book is a survey of these various facets of plant biotechnology. The individual chapters and the follow-up literature cited should allow a relatively easy access to the various subject areas and hopefully stimulate interest in these rapidly moving and exciting fields of research.

MICHAEL W. FOWLER
*Sheffield, UK*
*June 1991*

GRAHAM S. WARREN
*Sheffield, UK*
*June 1991*

# Contributors

Dr P. G. Alderson
Nottingham University School of Agriculture, Sutton Bonington, Loughborough, Leicestershire LE12 5RD, UK

Dr C. A. Cullis
Department of Biology, Case Western Reserve University, Cleveland, OH 44106, USA

Dr A. G. Day
MRC Unit for Protein Function and Design, Cambridge IRC for Protein Engineering, Department of Chemistry, University of Cambridge, Lensfield Road, Cambridge CB2 1EW, UK

Professor M. W. Fowler
Department of Molecular Biology & Biotechnology, University of Sheffield, Sheffield S10 2TN, UK

Dr C. H. Foyer
Laboratoire de Métabolisme, INRA, Route de St Cyr, F-78026 Versailles Cedex, France

Dr R. T. Furbank
Division of Plant Industry, CSIRO, PO Box 1600, Canberra, 2601 ACT, Australia

Dr J. Hall
Leicester Polytechnic School of Life Sciences, Scraptoft Campus, Scraptoft, Leicester LE7 9SU, UK

Dr C. P. Lichtenstein
Centre for Biotechnology, Imperial College of Science, Technology and Medicine, London SW7 2AZ, UK

Dr W. H.-T. Loh
Applied Molecular Genetics & Oilseeds, DNA Plant Technology Corporation, Cinnaminson, NJ 08077, USA

Dr F. Mavituna
Department of Chemical Engineering, University of Manchester Institute of Science and Technology, Sackville Street, PO Box 88, Manchester M60 1QD, UK

Dr H. Morikawa
Research Centre for Cell & Tissue Culture, Kyoto University, Kyoto 606, Japan

Dr S. J. Ochatt
INRA, Station d'Amelioration des Espèces Fruitières et Ornamentales, Domaine de Bois l'Abbe, Beaucouze, F-49000 Angers, France

Dr G. Ooms
AFRC Institute of Arable Crops Research, Rothamsted Experimental Station, Harpenden, Herts AL5 2JQ, UK

Dr N. Overbeeke
Unilever Research Laboratorium Vlaardingen, PO Box 114, 3130 AC Vlaardingen, The Netherlands

Dr J. B. Power
Department of Botany, Plant Genetic Manipulation Group, University of Nottingham, University Park, Nottingham NG7 2RD, UK

Mr A. Ranchhod
Management Centre, Southampton Institute, East Park Terrace, Southampton SO9 4WW, UK

Dr R. D. Rice
11 Cloysters Green, St Katherine by the Tower, London E1 9LU, UK

Dr A. H. Scragg
Bristol Polytechnic, Coldharbour Lane, Frenchay, Bristol BS16 1QY, UK

Dr A. M. Stafford
Department of Molecular Biology & Biotechnology, University of Sheffield, Sheffield S10 2TN, UK

Dr C. T. Verrips
Unilever Research Laboratorium Vlaardingen, PO Box 114, 3130 AC Vlaardingen, The Netherlands

Dr G. S. Warren
Department of Molecular Biology & Biotechnology, University of Sheffield, Sheffield S10 2TN, UK

Professor G. Wenzel
Federal Biology Research Centre for Agriculture & Forestry, Institute for Resistance Genetics, D-8059 Grünbach, Germany

Dr P. D. Williams
Department of Chemical Engineering, University of Manchester Institute of Science and Technology, Sackville Street, PO Box 88, Manchester M60 1QD, UK

Professor Y. Yamada
Research Centre for Cell & Tissue Culture, Kyoto University, Kyoto 606, Japan

# Contents of All Volumes

## COMPREHENSIVE BIOTECHNOLOGY

### Volume 1

### Section 1: Genetic and Biological Fundamentals



### Section 2: Chemical and Biochemical Fundamentals

## Volume 2

### Section 1: Bioreactor Design, Operation and Control



### Section 2: Upstream and Downstream Processing



## Volume 3

### Section 1: Healthcare Products

# 1

# The Cell Biology of Plant Cell Culture Systems

GRAHAM S. WARREN
*University of Sheffield, UK*

## 1.1 INTRODUCTION

The major difference between what might be termed 'traditional' and the emerging 'novel' plant biotechnologies is that the former are essentially organism based, whereas the latter are largely cell based. The technique of plant cell culture gives access to individual plant cells, and thus makes possible a range of manipulations that have both academic and biotechnological importance. Plant cell culture occupies a central role in plant biotechnology. Subsequent to gene manipulation or cell selection experiments, the resulting cells must be induced to regenerate into plants by control of the relevant culture parameters. The culture process itself induces genetic variation that can be exploited as a mutagenic treatment for the production of novel genotypes. Finally, cultured cells can be utilized directly for the production of useful metabolites and enzymes. The aim of this chapter is to discuss aspects of the structure and growth of cultured plant cells that have direct relevance to biotechnological processes. Features in common with cells in the intact plant will not be discussed in detail, and the following is not intended to be an exhaustive review of plant cell biology, for which the reader is directed elsewhere (Alberts *et al.*, 1989). For reviews of the technical aspects of plant cell culture and experimental protocols, see Dixon (1985), Evans *et al.* (1983), Stafford and Warren (1991) and Street (1977). Aspects of the culture process specific to cell differentiation and organogenesis, as well as the special requirements of protoplasts, are not considered in this review (see Stafford and Warren, 1991 and Ochatt and Power, Chapter 7).

The majority of the structural and metabolic features of the plant cell are common to all eukaryotic cells. The differences that do exist are generally differences in detail and not in basic molecular mechanisms. However, three major characteristics do distinguish plant cells from those of other eukaryotes and these are: (i) the presence of a rigid cell wall; (ii) the ability to fix $CO_2$ in the chloroplasts; and (iii) the presence of large, membrane-bounded vacuoles that can occupy up to 95% of the cytoplasmic volume, depending on cell type. The initiation of plant cell cultures involves the surface sterilization of suitable plant material (the explant) and the incubation of this in the appropriate solid or liquid nutrient medium. When established, plant cell cultures resemble microbial colonies or suspensions and can be handled in a similar fashion. Perhaps the overriding characteristic of plant cells in culture is the absence of (largely unknown) control systems operating in the whole plant. Cell cultures can therefore behave in a relatively uncontrolled fashion and exhibit a variability with respect to structural, metabolic and genetic features not seen at the plant level. An understanding of the nature, extent and possible control of this variability is crucial to the application of plant cell culture technology.

## 1.2 THE MAJOR CELL TYPES

### 1.2.1 Meristem

These are small (*ca.* 20 μm) near-isodiametric cells characterized by thin walls, high metabolic activity and the ability to divide sustainedly. Due to their division potential, explants for the initiation of cultures are usually chosen so as to contain a high population of meristem cells. Clusters of small, meristem-like cells in culture are usually indicative of high morphogenic potential. Because of their thin walls, meristematic cells often form protoplasts easily.

### 1.2.2 Mesophyll

These are chloroplast-containing cells, usually about 50 $\mu$m in length, and often able to divide. Mesophyll (leaf) tissue is commonly selected for protoplasting because it is available in large quantities, is relatively homogeneous and the cells have thin walls and division potential.

### 1.2.3 Phloem and Xylem

These are elongated cells of the vascular system. Phloem cells have lost their nuclei but are still metabolically active. Xylem cells are dead and heavily lignified, and have characteristic ribbed or spiral appearance. The presence of vascular cells in cultures is indicative of differentiation.

### 1.2.4 Epidermal

These are specialized cells on the external surfaces of the plant, often organized into monolayers. These cells secrete waxes onto the external faces of the tissue, giving rise to a cuticle which is highly impermeable (Kolattukudy, 1980). This cuticle is poorly formed or absent in plants recently regenerated from cultures. Slow acclimatization of such plants to lower humidities induces cuticle maturation, and is a necessary procedure prior to soil planting.

This list covers the cell types with most relevance to the culture process. A more detailed discussion of cell types is contained in Alberts *et al.* (1989).

## 1.3 CELL TYPES IN CULTURE

In principle, all the cell types present in the explant used to initiate a culture can be represented in that culture. However, not all cell types enter a culture with equal ease, and consequently some degree of selection may occur. It has been reported that a rich culture medium favours the culture of a wider range of cell types from an explant than a more basic formulation, and this situation probably reflects the range of biosynthetic capabilities of the starting cells (see Burgess, 1985). Microscopic examination of a suspension culture will reveal an enormous range of cell morphology. It is not possible (at present) to identify different cell types in a sophisticated fashion, such as is routinely possible with animal cell cultures. However, certain broad classes of cell can be recognized. Cells that are 'poised' for differentiation, or that have a high morphogenic potential, are usually small (*ca.* 20 $\mu$m), have dense cytoplasm, and are highly aggregated. These cells are usually considered to be equivalent to meristematic cells in the plant. Greatly elongated cells on the other hand are frequently of low morphogenic potential and are often (but not always) polyploid. Elongation can be medium dependent. Other types of cell can sometimes be recognized by the presence of morphological markers, *e.g.* starch grains or coloured secondary products. Such markers are usually not associated with a characteristic cell size or shape. It is important to realize, however, that these features may not be indicative of truly different cell types (stable phenotypes) but rather may occur in the same cells at different stages of the growth cycle, because cell suspension cultures are highly asynchronous in division (see Section 1.13).

Cell typing of plant cells, as can be achieved with animal cells using such probes as lectins and antibodies, has only developed to a very rudimentary level. The major difficulty has been the presence of the cell wall, which prevents direct access to the membrane where cell identity markers (determinants) may reside. However, in a few cases, tissue-specific antigens and lectins have been found to be expressed in culture, although they are frequently lost with increasing time in culture (Raff *et al.*, 1979). This result illustrates a general principle of plant cell cultures, namely that the culture conditions, in particular the growth regulator (hormone) regime, is a far more important determining factor of cell behaviour than is tissue origin. This situation has implications for the design of culture protocols. For example, in experiments to produce secondary products from plant cell cultures, it is not usually necessary to use the tissue in which the product is naturally synthesized as an explant for culture initiation. In a few cases, however, longer-lived determinants of cell identity have been characterized, and some of these are potentially useful in the exploitation of cultured cells. Both wall-located fucose residues (Wernicke and Gorst, 1987) and certain peroxidase isozymes (Joersbo *et al.*, 1989) have been shown to be specific for certain embryogenic or organogenic cell types. Such markers could be used to select for regenerative material during plant propagation or improvement schemes. Stable, wall-located *Agrobacterium* binding sites have been demonstrated in cells of dicots and hormone-habituated tissue cultures, but not in monocot cells, embryonic cells or crown gall tissue (Lippincott and Lippincott, 1978). Plant regeneration from cell cultures occurs from clusters of meristematic-like cells which may constitute a relatively small proportion of the total cell population. In certain cases, these cells appear to exist more or less independently from the other cells present, and consequently may represent a stable cell type, originating from the parent explant. However, in other cases, meristem-like cells and morphogenetically dormant cells can be interconverted by the appropriate hormonal stimuli (Jones, 1974).

## 1.4 SEPARATION OF CELL TYPES

Cell separations have only rarely proved possible for cultured plant cells. Large, vacuolated cells can often be separated from small, meristematic cells simply by allowing cultures to settle out.

Filtration and centrifugation methods exist for the separation of somatic embryos of various stages (Warren and Fowler, 1977). Protoplasts from certain cell types have also been isolated in relatively pure preparations. For example, enzymic conditions can be formulated that favour the release of protoplasts from a particular cell type. Alternatively, density gradient centrifugation has been used to isolate one type of protoplast from a mixture (Galun, 1981). Other separation methods that have been used include adsorption to lectin-coated beads (Warren and Fallon, 1984), isoelectric focusing (Griffing *et al.*, 1985) and fluorescence-activated cell sorting (Alfonso *et al.*, 1985).

## 1.5 FRACTIONATION OF PLANT CELLS

The disruption of plant cells for the isolation of subcellular organelles, enzymes, *etc.* poses some special problems (Price, 1974). Specifically, sophisticated cell disruption methods such as bead homogenization and sonication frequently fail to give acceptable results. The traditional mortar and pestle is often the method of choice. An additional problem is the deleterious effect of phenolics released when vacuoles are ruptured (Loomis and Battaile, 1966).

## 1.6 SUBCELLULAR ORGANELLES

The properties of a range of cellular organelles and structures are now discussed in relation to the culture process. The transfer of organelles has been proposed as a means of genetic manipulation of cells (McDaniel, 1989).

### 1.6.1 The Cell Wall

The cell wall provides the structural rigidity required for the plant or cell to withstand a range of environmental stresses. Despite this fact, cultured plant cells can be relatively sensitive to shear forces, such as those generated in turbine bioreactors. Often air-lift vessels have provided a gentler form of culture, although selection of shear-resistant cells occurs quite readily. The plant cell primary wall consists of an array of cellulose and hemicellulose components, cemented together with pectic substances. Protoplast isolation therefore, requires the action of cellulase, hemicellulase and pectinase enzymes. Treatment with pectinase alone often yields individual, walled cells that can be useful in cell-cloning experiments. The walls of cultured cells may differ in structure from those of intact plant cells of the same species (Asamizu and Nishi, 1980). In addition, the walls of cultured cells change in thickness during the growth cycle of the culture. Exponential phase cells usually have very thin walls and are a good source of protoplasts, whereas stationary phase cells have much thicker walls. The high strength of the cell wall imposes certain constraints on the experimental manipulation of cultured plant cells. For instance, considerable force is required to rupture the wall for liberation of the cell contents. Often this amount of force disrupts cellular organelles, preventing their isolation in an intact state. In these cases, isolation by gentle osmotic rupture of protoplasts is often employed. The cell wall tends to break or deflect microinjection needles (*e.g.* as used in the direct injection of DNA). This problem can be overcome by the use of protoplasts, or alternatively through the use of laser, electrical injection, or projectile methods. The plant cell wall has a high hydroxyl and ionic group content, and consequently many proteins and other cellular metabolites can bind tightly to wall components. As a result of this behaviour, such molecules may be incompletely extracted from plant tissue, and wall sequestration should be considered in experiments in which total product extraction is important. The highly cross-linked nature of the cell wall means that its porosity is severely limited. A current estimate of pore size is about 4 nm (Carpita, 1982). This size of pore makes the trans-wall diffusion of molecules above 15 000 Da (a small protein) extremely slow. Consequently most enzymes, cellular probes, such as lectins and antibodies, and DNA cannot enter the cell directly. Cultured cells may have altered wall porosity, possibly higher than in the whole plant in the case of rapidly-dividing exponential phase, or meristem-like, cells, and probably lower in the case of staionary phase cells. Cell lines with a lower porosity have been selected by growth in poly(ethylene glycol), a toxic polymer, in which case the synthesis of a less porous wall was suspected.

For further information on the biosynthesis and structure of the plant cell wall see Delmer (1987) and Varner and Lin (1989).

### 1.6.2 Protoplasts

The physical barrier imposed by the cell wall is an impediment to many biotechnological manipulations, *e.g.* direct DNA delivery and cell fusion, as well as many fundamental studies on plant cells. Consequently the isolation of wall-free protoplasts was developed using mixtures of wall-degrading enzymes. Isolated protoplasts are fully viable cells. They can begin to reform a cell wall (which may be abnormal in structure) within a few hours of the removal of the protoplasting enzymes, divide, and eventually regenerate into fertile plants. Protoplasts, having exposed membranes are often used as experimental models for membrane function. For a discussion of the biotechnological exploitation of protoplasts see Chapter 11.

### 1.6.3 The Plasma Membrane (Plasmalemma)

Structurally, plant membranes are essentially similar to those of animal and microorganism cells. In common with virtually all biomembranes, $Ca^{2+}$ is essential for physical stability. The $Ca^{2+}$ content of culture media is therefore important in experimental manipulations of protoplasts. Isolation methods for plant cell membranes are available, but, because suitable marker enzyme systems have not been fully developed, assessment of purity can be difficult (Hodges and Leonard, 1973; Polonenko and Maclachan, 1984). It is believed that the plant cell membrane is constantly being recycled (in common with that of animal cells) to compensate for vesicle-mediated excretion of wall components (Steer, 1988). It is generally assumed, however, that endocytosis is severely limited by the osmotic compression of the membrane against the wall, although there are reports of endocytosis in protoplasts (Suzuki *et al.*, 1977; Wright and Oparka, 1989). However, endocytosis in the intact plant cell cannot be ruled out because coated pits, which have a function in endocytosis, are commonly observed on the membrane surface (Coleman *et al.*, 1988). It is an open question as to the extent of enzyme or product excretion from plant cells by exocytosis (Robinson, 1984). Plant plasma membranes have binding sites for a number of lectins, indicating the presence of exposed glycoconjugates. Lectin-binding studies have been performed with a view to distinguishing between different cell types and probing the structure of the membranes (Larkin, 1978). The properties of the plant cell endomembrane systems have been reviewed by Harris (1986).

### 1.6.4 The Vacuole

Vacuoles are organelles, bounded by a single membrane (the tonoplast), which are present in all plant cells. Vacuoles may be specialized to perform a range of cellular functions, and often several specialized types of vacuole are present in the same cell. Usually a meristematic cell contains a number of small vacuoles that fuse as the cell matures to form a single large structure that can occupy up to 95% of the cell volume. The mechanisms controlling vacuolar fusion are unknown. Vacuoles are thought to originate from vesicles budding off from the endoplasmic reticulum and Golgi (Boller and Wiemken, 1986). Types of vacuole specialization include the possession of hydrolytic enzymes indicating a degradative function (closely related to lysosomes in animal cells) (Moriyasu and Tazawa, 1988), and storage functions, both reversible and irreversible (*e.g.* waste disposal) for such products as proteins, organic acids and secondary metabolites. Useful secondary products are usually locked in the vacuole and this situation at present limits the effectiveness of immobilized cell systems for secondary metabolite production. The vacuole has a lower pH (*ca.* pH 5) than the cytosol (*ca.* pH 7.0). This can result in sequestration of certain ionizable compounds in the vacuole by 'ion trapping', although the occurrence of this mechanism has been questioned and specific tonoplast transport systems have been invoked (Deus-Neumann and Zenk, 1985). In mature cells the vacuole is often so large that the cytoplasm is confined to a thin ribbon between the vacuole and the plasma membrane. In this situation, the vacuole has a space-filling function, allowing the cell to expand osmotically without the need to maintain a large volume of cytoplasm. Diffusion problems are likewise minimized. Most vacuolar proteins are glycosylated but no vacuole-specific sugars or sugar sequences have yet been identified. Methods based on gentle osmotic rupture of protoplasts are available for vacuole isolation (Kreis and Reinhard, 1985; Wagner and Siegelman, 1975). However, the study of isolated vacuoles has proved to be difficult because of their fragility, and because vacuole–cytosol cooperative mechanisms are suspected.

### 1.6.5 Plastids

Plastids are another group of membrane-bounded organelles that can specialize in several ways, for example into chloroplasts or amyloplasts. All types of plastid appear to arise from small (*ca.* 1 μm dia.) vesicles (proplastids) present in meristematic cells. They possess their own (partial) genomes and can divide.

Functional chloroplasts are rarely present in cultured cells grown under normal laboratory conditions; such cultures therefore conduct little photosynthesis, and require an external carbon source in their growth medium. Cells have been selected on the basis of their green (chlorophyll) pigmentation, but it has been shown that this characteristic correlates only weakly with photosynthetic activity (Sato *et al.*, 1979). Highly photosynthetic cultures can, however, be selected under the appropriate conditions. $CO_2$ levels of around 1% (usually produced by the $CO_2$ buffer system $K_2CO_3 + KHCO_3$) are required together with a high light intensity (10 000 lux). Carbohydrates are usually omitted from the culture medium, indeed sucrose above 0.25% has been found to inhibit chlorophyll synthesis and photosynthetic oxygen evolution (Dalton and Street, 1977). The choice of auxin is also important. 2,4-D (2,4-dichlorophenoxyacetic acid) can inhibit chlorophyll synthesis and IAA (3-indoleacetic acid) is inactivated by light; consequently NAA (1-naphthylacetic acid) is frequently used. Even under these conditions, not all clones or cell lines from a particular explant appear to have the potential to become photoautotrophic. It has been reported that selection should proceed immediately after culture initiation, established cultures being much less likely to develop photoautotrophy (Yamada and Sato, 1978). For biotechnological purposes, the main interest in photoautotrophic cultures is in the saving of carbohydrate (the most expensive medium component) or in enhancing the biosynthetic potential of the cells. However, in the majority of cases studied secondary metabolite production is greatly suppressed in photoautotrophic cells (*e.g.* Hagimori *et al.*, 1984). An exception to this finding is the biosynthesis of products related to photosynthesis, *e.g.* plastoquinone, which can be produced in greater amounts in photoautotrophic cultures (Husemann *et al.*, 1989). Although chloroplasts possess their own genomes, the majority of chloroplastic proteins are encoded on nuclear genes. Amino acid sequences (transit peptides) have been determined that direct cytoplasmically produced proteins to the chloroplast compartment (Smeekens *et al.*, 1990), and these can be functional when incorporated in chimeric proteins. This and similar mechanisms will be important in the correct spatial expression of engineered DNA (see Chapter 12).

Amyloplasts, specialized organelles of starch synthesis and storage, also have relevance in the biotechnological exploitation of cultured plant cells. Starch represents stored energy, and it will often accumulate in cultured cells exposed to a plentiful carbon supply. Media analyses may indicate carbohydrate depletion, whereas the cells may continue to grow actively by starch utilization. There can thus be a discrepancy between the nutritional environment imposed by the media, and that perceived by the metabolic machinery of the cell, and such effects can affect the interpretation of experimental results. Similar effects can occur with other nutrients (*e.g.* phosphate). Continuous or semicontinuous application of small amounts of such nutrients can be a more economical method of supply, and may minimize catabolite repression of product synthesis in cases in which this mechanism operates. The presence in the cell of dense starch grains can be detrimental to protoplast isolation, because the angular grains can rupture the protoplast membrane during centrifugation (washing) steps. The starch content of cultured cells is often reduced prior to protoplasting by a period of starvation or rapid growth (short subculture interval). The breakdown of starch and concomitant build up of soluble saccharides has been correlated with an increased osmotic tolerance of the derived protoplasts. Sometimes the presence of starch grains can be advantageous, for example they can provide a morphological marker to facilitate the selection of protoplast fusion products.

### 1.6.6 Plasmodesmata

Plasmodesmata are essentially cytoplasmic connections that occur between most cells in the plant. They have been observed in lower numbers in clusters of cultured plant cells. It has been shown that molecules larger than about 800 Da pass through plasmodesmata only slowly (Gunning and Robards, 1976). Ruptured plasmodesmata constitute a possible route into the cell for macromolecules. Vortexing of clusters of cultured plant cells in the presence of viruses results in uptake of the latter (Murakishi *et al.*, 1970). This type of system might have an application in the genetic transformation of cells.

## 1.7 HORMONAL EFFECTS ON CULTURED PLANT CELLS

Several classes of small, diffusable molecules are exploited by plants to coordinate cellular activities and development, both within the plant, and in response to environmental fluctuations. Generally speaking, each of these substances can take on a number of different regulatory roles, some of which are hormonal (*i.e.* they provoke a response remote from their site of synthesis). Three main classes of hormone or growth regulator have so far been identified, namely auxins, cytokinins and gibberellins. In addition, ABA (abscisic acid) and ethylene play important regulatory roles. The concentration of ethylene in the gas phase above cultures can affect cell growth (ethylene is generally inhibitory). Flask closures should therefore allow gas exchange to prevent the build up of ethylene. A uniformity of closure between vessels is necessary if growth data are to be compared. Many other substances (*e.g.* nicotine, salicylic acid, lipids) have been reported to affect plant growth, but in many cases growth regulation may be a secondary (or nonphysiological) activity. The activity of plant hormones or growth regulators shows biochemical differences when compared with the classical animal hormones, *e.g.* insulin. Animal hormones typically operate over a narrow concentration range with a sharply defined optimum, whereas plant hormones often exert an effect over a $10^4$–$10^5$-fold concentration range. Animal hormones have specific receptor molecules that are saturable. In contrast, few true receptors have been identified for plant growth substances, despite intense effort, although many hormone-binding proteins have been found (*e.g.* Shimomura *et al.*, 1988). The significance of hormone-binding activities is a matter of debate (Kende and Gardner, 1976). Furthermore, plant responses do not exhibit classical saturation kinetics. These considerations have led some workers to question whether the term 'hormone' is appropriate for these substances. However, some clear cut hormonal (action at a distance) effects do occur, and so this terminology will be continued here for convenience.

Each of the hormone types listed above mediates a bewildering range of physiological responses, which can vary depending on the species of plant, type of tissue, developmental state, and hormone concentration. Although the range of responses elicited by a given hormone can appear unrelated, it is possible that the primary cellular action of the hormone (as yet largely unknown) might be the same in each case. Sometimes, substances that induce the same behaviour bear little structural resemblance. Clearly these observations can make it difficult to assign characteristic roles to individual hormone types.

## 1.8 HORMONE BIOCHEMISTRY AND CELLULAR ROLES

The control of hormone levels and the nature of the responses elicited are subjects largely beyond the scope of this chapter. The reader is referred to recent reviews for a detailed picture (Creelman, 1989; Key, 1989; Sauerbrey *et al.*, 1987). However, plant hormones constitute by far the most important determinants of the behaviour of plant cells in culture. Biotechnologically important processes such as plantlet regeneration and secondary metabolite production are extremely sensitive to the hormone regime imposed. The effects of hormones in individual cases are so diverse that the original literature must always be consulted, and extrapolation from other systems should not be relied upon.

## 1.9 THE INTERPRETATION OF OBSERVATIONS ON THE EFFECTS OF HORMONES IN CELL CULTURES

A survey of the plant hormone literature will quickly reveal that contradictory results and species/tissue differences in response preclude a clear general view of the primary effects of these substances in plant cells, and make prediction of likely responses in culture experiments almost impossible. The definition of a given hormonal regime is usually based on the quantities of hormones added to the culture medium. However, there is unlikely to be a simple relationship between the hormone concentrations in the medium and those inside the cultured cells. It is possible that a large number of the apparent contradictions referred to above might be resolved when it is possible to reanalyze results with a better appreciation of all the factors involved. The important principles now recognized as influencing the relationship between medium and cellular hormone levels are discussed below. In the majority of cases precise mechanisms and specific details are lacking, but these general considerations at least provide conceptual tools for interpretation of the behaviour of cultured cells in response to hormones in the culture medium. These general principles are: (i) exogenous hormones can be processed (stored, modified or degraded) by the plant cells; (ii) there can be feedback control of synthesis of endogenous hormones by added substances; (iii) medium and cellular carry-over from previous treatment can occur; and (iv) cell clusters can create their own internal microenvironments. These factors will now be considered in more detail.

### 1.9.1 Hormone Processing

Conjugation to amino acids and sugars appears to be an important mechanism controlling the availability of hormones (especially auxins) in plant cells (Cohen and Bandurski, 1982). Some conjugates are hormonally inactive (*e.g.* auxin–sugars) whereas others retain hormonal activity (*e.g.* auxin–amino acids). Conjugates often show differences in transport characteristics. Species/tissue variations in conjugation activity can lead to markedly different fates of exogenous hormones, as considered below. As an example of the possible importance of hormone conjugation, it has been found that calli can adapt to high (toxic) concentrations of 2,4-D by induction of a 2,4-D-sugar (hormonally inactive) conjugation system (Davidonis *et al.*, 1982). In other cases 2,4-D is incorporated into lignin, which may also consititute a detoxification system (Scheel and Sandermann, 1981).

### 1.9.2 Feedback Control

Many biosynthetic pathways are under end product negative feedback control, *i.e.* the system is switched off by the product. Similar controls are thought likely in plant hormone biosynthesis. It is therefore possible that exogenous hormones may in fact inhibit endogenous hormone production, and in the extreme case may actually lower the cellular active hormone concentration. As an example, ABA synthesis is switched off by high levels of ABA.

### 1.9.3 Hormone Carryover

Hormones may be carried over from one medium to another simply by adherence to the cells. Thorough washing may eliminate this effect, but has been found to be relatively ineffective in depleting cellular pools of hormones. Such internal carryover effects must always be considered in evaluating changes in medium hormone regime.

A carryover effect resulting from hormone conjugation was put forward to explain the following, seemingly contradictory results. It was observed that the transfer of carrot and soybean cells from a high to a low auxin medium resulted in morphogenesis in the carrot, but not in the soybean. High auxin levels characteristically inhibit morphogenesis and so the soybean result appeared anomalous. The carrot cells were subsequently found to have a low auxin–amino acid conjugation activity, and consequently most cellular auxin was present in the free form which could efflux from the cells fairly rapidly after transfer to the low auxin medium. In contrast, the soybean cells rapidly converted auxin to amino acid conjugates (still hormonally active) which, because of their charged nature, were effectively locked in the cell and continued to influence cellular behaviour in the low auxin medium (Montague *et al.*, 1981). IAA conjugates have in fact been added to culture media to act as 'slow release' sources of the hormone (Hangarter *et al.*, 1979).

### 1.9.4 Tissue Microenvironments

There is some evidence that cell clusters and tissue fragments may regulate their own intracellular hormonal environment to give conditions essentially independent of the external medium. This concept can be extended in principle to include isolated cells. The idea is illustrated in the following example. Soybean root cultures, and calli showing morphogenesis, rapidly regulated their internal free 2,4-D level to a constant value that was independent of the concentration of the hormone in the medium. This regulation was achieved by conjugation of 2,4-D with amino acids (the conjugates were still hormonally active). Young, disorganized calluses performed a similar regulation (to a higher final 2,4-D concentration) much more slowly (Davidonis *et al.*, 1978). This type of observation clearly has important implications in the consideration of the differences between the composition of the medium and the environment actually perceived by the cells. It remains to be seen whether whole cell hormone concentration is a valid criterion of hormone activity.

## 1.10 HORMONE–HORMONE INTERACTIONS

An additional level of complexity is imposed on the evaluation of effects of plant hormones by the fact that the various hormone classes do not act independently. Many cases of synergistic and antagonistic interactions have been documented. Hormones of one type can also affect the availability of other classes by effects on synthesis or modification. The following examples illustrate some of the more general principles that have emerged.

### 1.10.1 Interactions at the Physiological Level

#### *1.10.1.1 Auxin:cytokinin ratios*

One of the most useful and widely (but not universally) applicable concepts to emerge from investigations of hormone effects on differentiation in tissue cultures is that of the auxin:cytokinin ratio. In many cases the balance between these growth regulators in the medium influences the developmental pathway followed. Usually a high auxin:cytokinin ratio promotes disorganized growth or root formation, whereas a low auxin:cytokinin ratio induces shoot development. The 'all-or-nothing' nature of this switch suggests permissive/inhibitory hormonal control, rather than a specific primary function for the hormones in cell differentiation at the gene level. Recent work on the *Agrobacterium* Ti plasmid, in which the auxin:cytokinin ratio imposed on the host plant tissue can be manipulated, supports the above scheme.

In rare cases, the absolute amounts of auxins and cytokinins dictate the course of development, *e.g.* in tomato cultures 2 mg $L^{-1}$ indoleacetic acid (IAA) and 2 mg $L^{-1}$ kinetin promote root formation, and 4 mg $L^{-1}$ IAA and 4 mg $L^{-1}$ kinetin induce shoot formation (Burgess, 1985).

#### *1.10.1.2 Cell division*

Cell division requires the presence (either endogenous or externally supplied) of an auxin and cytokinin. Therefore, the large majority of tissue culture media contain a member of each of these growth regulator classes.

### 1.10.2 Interactions at the Molecular Level

Many molecular interactions between the various pathways of hormone synthesis and transport have been documented. Some of the most widely recognized mechanisms are: (i) ethylene inhibits the polar transport of auxin; (ii) ethylene inhibits auxin synthesis; and (iii) high auxin levels stimulate ethylene synthesis. For a general survey of the properties of plant growth regulators see Wareing and Phillips (1981).

## 1.11 HORMONE HABITUATION

The capacity of cultured plant cells to synthesize endogenous hormones frequently increases with increasing time in culture. This hormone 'habituation' is being studied as an interesting example of possible stable epigenetic variation (Meins, 1989). Changes in endogenous hormone production frequently adversely affect a cell's abililty to regenerate and produce secondary metabolites.

## 1.12 OTHER CELLULAR PROCESSES IN CULTURED PLANT CELLS

Most cellular mechanisms have been more thoroughly studied in animal and microorganism systems than in plants. However, research with plant cells is proceeding at an ever increasing rate and many processes originally elucidated in other cell types are now being identified in plants and cultured plant cells. These processes include phosphoinositide signalling (Das *et al.*, 1987; Ettlinger and Lehle, 1988), cyclic AMP control (Brown and Newton, 1981), and calcium/calmodulin systems (Timmers *et al.*, 1989). Although the study of these mechanisms is generally still at a rudimentary level, work in these areas may have biotechnological relevance in the future by leading to a deeper understanding of the control of a variety of cellular functions.

## 1.13 THE GROWTH OF PLANT CELLS IN SUSPENSION CULTURE

Plant cells may be grown in solid phase (callus) cultures, or as suspensions. Callus cultures are slow growing ($t_D$ typically 3–5 d) and growth can be heterogeneous, mainly due to the vectorial nutrient supply. Critical experiments are rarely done on callus, rather its usual function is in cell line maintenance. During solid-phase growth of many microorganisms, colony morphology is diagnostic for genotype and this situation has been of the utmost usefulness in the selection of desirable mutants. In contrast, plant cell callus growth is very amorphous, highly dependent on the conditions of culture and has rarely proved to be useful as a diagnostic tool. The ease of culture initiation is often highly genotype dependent and, unless results from a specific variety are required, it is advantageous to screen for the most amenable related genotype.

The growth cycle of a typical suspension culture is shown in Figure 1. The growth kinetics are essentially similar to those of microorganisms, although the timescale is much longer. The lag, exponential, stationary and death phases can be interpreted as adaption, rapid growth, nutrient limitation, and cell lysis respectively. Mathematical models are available for the description and prediction of plant cell growth kinetics (de Gunst *et al.*, 1990). As shown in Figure 1, fresh weight accumulation often continues to rise after dry weight increase has ceased. Because of this effect, the apparent productivity of cell cultures can vary depending on the basis on which data are expressed. This effect results from the phenomenon of expansion growth of plant cells, as seen in the whole plant most visibly in roots and shoots. The expansion phase of plant cell growth, which is due to water uptake, requires that the rigid crosslinking of the wall components is loosened to allow elongation. According to the currently accepted model, on receipt of the appropriate signal (often auxin), protons are pumped out through the membrane and cause acidification of the wall region. The resulting drop in pH may directly destabilize the crosslinking of wall polymers, and may activate glucanase and glucan synthase enzymes (Heyn, 1981). This 'acid growth' model has implications for plant cell culture. Firstly, cell expansion may be sensitive to buffers in culture media. Effects of buffering on growth and secondary metabolism have been reported (Banthorpe and Brown, 1990; Parfitt *et al.*, 1988). Secondly, the uncontrolled exposure of plant cells to auxins, as happens in culture, may lead to abnormal wall growth. This can take the form of the synthesis of a wall with an altered structure, behaviour that is relevant to the performance of enzyme systems used for protoplast isolation. Over-production of wall components can also occur, resulting in the appearance of wall polymers in the culture medium, which in turn can contribute to the blocking of bioreactors if stable foams are formed. Cell expansion itself, which is frequently unidirectional, increases the viscosity of the culture, which may be undesirable in large-scale growth. Elevated osmotic potentials have been reported to limit cell expansion in culture (Ammirato and Steward, 1971).

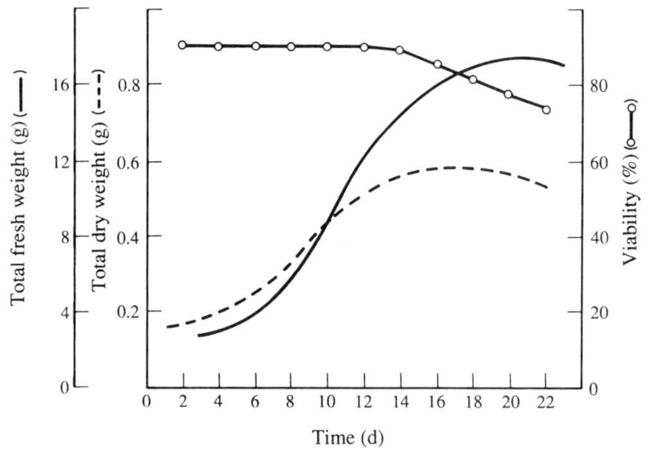


**Figure 1** Growth curve for a typical plant cell suspension culture

Plant cells in culture will not divide at cell concentrations below the critical inoculation density. A typical value for this density is $10^4$ cells $mL^{-1}$. This factor means that cell- and protoplast-cloning experiments require special conditions, such as the use of microdroplets and feeder cultures (Schweiger *et al.*, 1987). The critical minimum density requirement is thought to result from an initial efflux of metabolites from the cells, and an equilibration with the fresh growth medium.

Substances involved in this 'conditioning' of the growth medium include amino acids and hormones (Stuart and Street, 1969, 1971). Conditioning factors appear to be largely non-species-specific (Benbadis, 1968). Critical minimum inoculation density is an important factor in the planning of production schedules for the inocula for large bioreactors. Cells in larger clusters frequently divide faster than those in smaller clusters, or free cells. This may be due to a nursing effect (provision of complex nutrients) by the viable, but nondividing, cells in the cluster (often only the outer cells divide), or it may be the case that nutrient loss into the medium from larger clusters is reduced. The minimum inoculation density is often lower for cell clusters than for free cells, probably for the same reasons. Single cells are usually capable of cell division, but in some cases can represent a terminal stage in the development of the culture, and, therefore, increase in number towards the stationary phase of the culture growth cycle. Cell division in plant cell cultures is typically highly asynchronous, even for the first division after subculture. However, methods for the synchronization of division (*e.g.* starvation and cold treatment) have been developed (Okamura *et al.*, 1973), although these appear to have been rarely exploited. Using synchronized cultures, metabolic changes related to cell cycle phase have been identified (Amino *et al.*, 1985).

## 1.14 CELLULAR COHESIVENESS IN CULTURES

Plant cell cultures usually exist as cell clusters (commonly 10–100 units) rather than individual cells. The cluster size is presumably governed by the degree of cohesiveness of the cells. Cell cohesiveness is an important variable in several aspects of the culture process and its biotechnological application. The precise molecular nature of plant cell cohesiveness is poorly understood, but is certainly a function of the composition of the cell wall, most probably the pectic components, and is possibly related to the degree of differentiation of the wall. Ease of initiation of suspension cultures is highly dependent on the nature of the starting callus. If the latter is soft and friable, then suspensions usually form very easily. In contrast, if the callus is hard, suspensions can be difficult to obtain. Calli from certain species (*e.g.* cereals, brassicas, soybean) are highly organized and consist of masses of embryo- or meristem-like structures. It can be very difficult, or impossible (at present), to initiate suspensions from certain genotypes in these situations. Cells in the whole plant are clearly organized and exist in a highly cohesive state. It is often found that cultured cells in tight clusters also show some degree of morphological organization (*e.g.* cell alignments, vascularization, *etc.*) or exhibit specialized functions (*e.g.* chlorophyll synthesis). These observations have led to the model that cell cluster friability is a consequence of a loss of organization. This model is probably a gross over-simplification, but does provide some basis for the prediction of factors likely to increase cell separation. For example, conditions known to suppress differentiation, *e.g.* high auxin levels, have been reported to result in finer suspensions. The size of cell clusters varies throughout the growth cycle in some cultures, and therefore the timing of subculture may affect the cluster size distribution. Several experimental procedures have been employed to increase cell separation. These treatments include the settling of vessels prior to subculture to remove the larger clusters, sieving through nylon mesh, the use of dilute solutions of cellulase enzymes (Street, 1977), and cell immobilization in calcium alginate (Morris and Fowler, 1981). It is technically difficult to determine the size distribution in a plant cell culture. A continuous spectrum of cell and cluster sizes is usually present, and the available methods of sizing, often filtration through a range of mesh sizes, are imprecise and laborious due to the high viscosity of the cultures. Cluster size is an important determinant of culture behaviour. Cells in large and small clusters often perform differently, possibly because of the different microenvironments to which the cells are exposed, or because such cells possibly represent true variant phenotypes (or genotypes) with altered cohesiveness as one of the variant properties. A range of biochemical parameters have been found to depend on cell cluster size; these include cytochrome oxidase, peroxidase and catalase activities, rate of protein synthesis, pool size of free amino acids and secondary product accumulation (Hulst *et al.*, 1989; Street, 1977). In some cases it may be necessary to select for a given cluster size to amplify the activity under investigation. Subculture of suspension cultures is often performed using a syringe or automatic pipette. Care must be exercised to ensure that the canula or tip bore is sufficiently wide to prevent exclusion of the larger cell clusters. Similar considerations apply to the design of sampling ports in bioreactors.

A phenomenon related to cell–cell cohesiveness is that of wall growth in flasks of bioreactors, which in extreme cases can result in complete blocking of the vessel. It may be possible to select for cell lines with less tendency towards wall growth. A relatively high degree of cohesiveness may be advantageous in cells to be immobilized, for example in polyurethane foams.

## 1.15 SELECTION, COOPERATION AND COMPETITION IN CELL CULTURES

Suspension cultures of the majority of microorganisms behave under most conditions like relatively homogeneous systems. Plant cells in suspension have to be thought of in a completely different way, because of their inherent genetic heterogeneity and instability. These characteristics are of course present in microorganism cultures, but to a much more limited extent. Genetic heterogeneity in plant cell suspension may reflect the mixed population of cells in the original explant, and may arise due to somaclonal variation (see Chapter 2). Cell selection can occur when cultures are initiated, because not all cell types will enter a culture with the same degree of ease. Rich culture media are thought to favour more heterogeneous starting cultures. The culture process itself tends to select for the fastest-growing cells. Rapid growth may not be compatible with desirable characteristics of cultured cells, for instance growth rate and secondary product formation are often inversely related.

The heterogeneity of plant cell suspension cultures has important consequences when considering such questions as: (i) the mechanism of response of the culture to changes in growth conditions; and (ii) the relationship between individual properties of the culture. By way of an illustration of the first of these problems, consider an experiment in which a culture with a certain desirable property is transferred to a new culture medium and that, after a two-week growth period, the desirable property is greatly reduced. One could not assume that the new medium directly suppresses the behaviour in question. An equally valid interpretation at this stage is that a previously minor cell type, lacking the desirable property, has become dominant in the new medium due to a selective advantage in growth. Indeed it is possible in the case of secondary product formation, that the new medium might actually stimulate product formation, but product accumulation might then inhibit cell division and allow the nonproducing cells present to take over the culture. In this type of situation, the misconception that the medium was inhibitory could lead to a completely inappropriate development strategy. Some actual instances of this type of competition between cell types in suspension have been documented. For example in a suspension culture of carrot it was found that one cell type was more efficient in the uptake of phosphate, and became dominant under conditions of phosphate-limited growth (Bayliss, 1977). Such cell types can come to constitute the large majority of the cells in a culture very rapidly, say over one or two subculture periods. Other changes in the behaviour of cultures, *e.g.* the progressive loss of morphogenetic potential (Smith and Street, 1974) are also thought to result in part from gradual changes in the proportions of various cell types in suspension cultures.

The opposite of competition, *i.e.* cell cooperativity, has also been observed, for example in early experiments in which attempts were made to regenerate colonies from cells engineered by the *Agrobacterium* Ti plasmid. In this case, transformants were selected by the ability to produce their own hormones and therefore grow on a medium lacking these substances. However, hormones from autotrophic cells were able to 'cross-feed' untransformed cells in the same culture, and regeneration was equally if not more likely to occur from these latter cells. For this reason, one could not assume on this evidence that regenerated colonies selected in this way were transformed.

An example of difficulties in interpreting the relationship between individual properties of a suspension culture is afforded by the case of certain profiles of secondary product accumulation by cells throughout their growth cycle. A representative case is shown in Figure 2. It might appear that product accumulation occurs during the period of maximum growth. However, an alternative interpretation, namely that a small subpopulation of nondividing cells is producing the product observed, is equally valid on the basis of this evidence. In the development of a large-scale process for secondary product formation, a correct evaluation of the relationship between cell division and product accumulation could be crucial. For instance it would influence the decision whether or not to adopt a continuous culture production system, in which the majority of the cells would be continuously dividing.

These examples of possible different interpretations of tissue culture experiments serve to illustrate the general consequences of mixed populations of cells in suspension cultures. It is clear that in these cases it would be necessary to apply appropriate additional criteria to specify precisely the mechanisms involved. It must also be borne in mind that suspension cultures are not synchronized and therefore cells at all stages of the cell cycle will be present at any one time. Although culture variability has so far been stressed, relatively stable (in terms of growth rate, morphogenetic potential, product formation, *etc.*) cultures have been isolated. Cultures often become less variable but possibly more divergent from the parent genotype with increasing time in culture. Presumably this reflects increasing adaptation to culture conditions such that new mutants or variants become less likely to possess a selective advantage. Newly initiated cultures (callus or suspension) should not be used

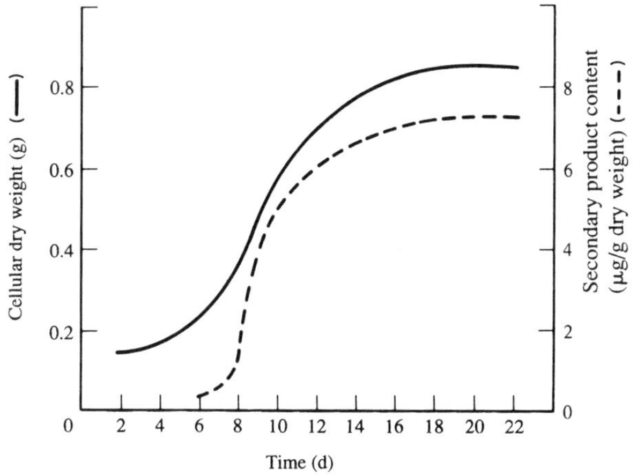


**Figure 2** Growth and secondary metabolite accumulation in a plant cell suspension culture

for critical quantitative experiments. It is often highly instructive to monitor the required parameters over successive subcultures from the time of initiation. The acquisition of relative stability can thus be determined, and some clues concerning controlling factors or covariant parameters might be obtained. In stable systems, the inoculum will be identical for each subculture. By definition, a varying culture will generate an inoculum different from that produced by the previous subculture, and so to some extent (ignoring possible homeostatic mechanisms) variability in a culture is self-propagating. This factor may be involved in the oscillations in parameters measured over successive subcultures that have frequently been observed.

## 1.16 ELECTRICAL EFFECTS ON PLANT CELL GROWTH

There is mounting evidence that exposure to electrical fields stimulates the growth of plant cells and cultures. The effect has been related to changes in the polar transport of auxin (Goldsworthy and Rathore, 1985). Apparatus for the treatment of cultures with electrical fields has been described (Rathore *et al.*, 1988). Electroporation-enhanced regeneration of plant protoplasts has been widely noted and exploited.

## 1.17 CELL LINES AND CLONES

It is commonly found that similar explants from the same individual plant can yield cultures with different properties, *e.g.* growth rate, appearance, degree of aggregation, biosynthetic potential, *etc.* Similarly, offshoots from the same suspension culture can behave differently (Endo *et al.*, 1988). For this reason it is accepted that any culture maintained in isolation constitutes an individual cell line, and data are not automatically applicable to different cell lines, unless further validation is provided. It is possible to clone individual plant cells (or single clusters assuming single cell origin), but, because extensive somaclonal variation rapidly takes place, similar cell line variation is usually seen. Stable cultures can be obtained, and usually result from long periods (years) in culture during which time highly adapted cell types can arise. In some cases stable 'clones' have been isolated from heterogeneous suspensions (Coutos-Thevenot *et al.*, 1990). The problem of cell line divergence (*i.e.* variation between independently maintained lines) must be addressed when considering a long-running programme of experimentation, or when designing a process requiring stable plant cell material. A practical solution is to operate a scheme of regular (*e.g.* once every 4–6 months) mixing of equivalent cell lines. The potential hazard with this approach is the high vulnerability of a single culture to loss by infection. In fact, the principal reason for maintaining multiple equivalent cell lines (on different shakers) is to protect against accidental contamination or mechanical breakdown. The risk associated with cell line mixing can be minimized by maintaining the original cell lines until it is proven that the new, mixed line is free of infection. It is usual to maintain callus material as a back-up in case the derived suspensions are accidentally lost. However, because of the much higher growth rate of suspension-cultured cells, the latter may rapidly diverge from the callus. Thus if suspensions have to be reinitiated from the callus, they may have very different properties to the originals. For this reason, it is prudent to back-initiate calli from suspensions periodically (*e.g.* every 6 months) to act as a suitable back-up material. Frequently, insufficient information about the cell lines used is presented in published reports, making full interpretation of the results impossible. Cell line documentation should include (in addition to the growth conditions), aggregate sizes and cell number, cell shape, morphogenic potential, viability and, if possible, degree of ploidy.

## 1.18 CELL LINE PRESERVATION

Longer term storage of cells has been achieved by cryopreservation (Withers, 1986), by maintenance of callus under oil at 4 °C, or by drying (Kaimori and Takahashi, 1989). It is possible that these treatments may select for cell subpopulations more resistant to the stresses involved. For this reason the assumption cannot be made that the recovered material has identical properties to the original.

## 1.19 NOVEL METHODS OF CULTURE

Many variations on the standard growth configurations of plant cells have been suggested, and these may prove worthy of investigation in specific situations. These procedures include: the growth of cells in dense slurries (Gibson and Christen, 1989); the use of customized flasks (Manasse, 1972); the growth of cells in aqueous two-phase polymer systems (Hooker and Lee, 1990), which can be used to aid the isolation of lipophilic substances (Beiderbeck, 1983); and coculture of different plant genotypes (Carlson, 1977), plant and microbial cells (DiCosmo and Tallevi, 1985), and plant and algal cells (Bradley, 1981). In this last case, the algae resided in the plant intercellular spaces and supplied nitrogen to the plant cells. Plant cells have also been cultured in a wide variety of bioreactors and immobilized systems. These configurations and the associated cellular growth characteristics are described in Chapters 4 (Scragg) and 5 (Williams and Mavituna).

## 1.20 AUTOMATION

The rate-limiting operations in plant cell culture, especially when applied to micropropagation, are the manual subculture and transfer steps, Automation could make a significant economic impact here, and various suggestions for automating certain processes have been made (Levin *et al.*, 1988). However, the situation is complicated by the fact that a high degree of judgement is required for the selection of the appropriate material for tansfer.

## 1.21 CONCLUSIONS

The techniques of plant cell culture have now reached a level of sophistication at which it is possible to get most plant species into culture. The degree of understanding of the cell biology of cultured cells, although low in comparison to some other systems, is also steadily increasing. However, two major impediments to a fuller exploitation of the culture process remain. Firstly, the genetic instability of plant cells in culture is still essentially uncontrollable. Advances are being made concerning the gene-level events responsible for this variation, but until control is possible, cell cultures are likely to remain a somewhat unpredictable material. Secondly, the processes involved in cell regeneration to the plantlet stage are extremely poorly understood. The range of species and genotypes for which regeneration is a routine process is still fairly small. Blocks in the regeneration pathway at present greatly limit the application of gene manipulation and cell selection techniques, and advances here will have a major impact on agricultural biotechnology. Unfortunately, progress towards an understanding of differentiation and development has been slow, even in other model systems, including those taken from among animals and microorganisms.

When considering the use of plant cell and tissue cultures for the study of basic plant cell biology one fundamental point should be borne in mind; cultured plant cells are not whole plants. As with any experimental system, the greatest rewards are to be gained by observing which features are exaggerated in, or uniquely expressed by, the system and studying these, rather than concentrating on mechanisms that might be more easily, and relevantly followed elsewhere.

## 1.22 REFERENCES


Alberts, A., D. Bray, J. Lewis, M. Raff, K. Roberts and J. D. Watson (eds.) (1989). *Molecular Biology of the Cell.* 2nd edn. Garland, New York.

Alfonso, C. I., K. R. Harkins, M. A. Thomas-Compton, A. E. Krejci and D. W. Galbraith (1985). Selection of somatic hybrid plants in *Nicotiana* through fluorescence-activated sorting of protoplasts. *Biotechnology*, **3**, 811–816.

Amino, S., Y. Takeuchi and A. Komamine (1985). Changes in synthetic activity of cell walls during the cell cycle in a synchronous culture of *Catharanthus roseus. Physiol. Plant.*, **64**, 202–206.

Ammirato, P. V. and F. C. Steward (1971). Some effects of environment on the development of embryos from cultured free cells. *Bot. Gaz. (Chicago)*, **132**, 149–158.

Asamizu, T. and A. Nishi (1980). Regenerated cell wall of carrot protoplasts isolated from suspension-cultured cells. *Physiol. Plant.*, **48**, 207–212.

Banthorpe, D. V. and G. D. Brown (1990). Growth and secondary metabolism in cell cultures of *Tanacetum mentha* and *Anethum* species in buffered media. *Plant Sci.*, **67**, 107–113.

Bayliss, M. W. (1977). The causes of competition between two cell lines of *Daucus carota. Protoplasma*, **92**, 117–125.

Beiderbeck, R. (1982). Two-phase culture — a method for the isolation of lipophilic substances from plant suspension cultures. *Z. Pflanzenphysiol.*, **108**, 27–30.

Benbadis, A. (1968). Culture des cellules isolees: le probleme du conditonment des milieux de culture. In *Les Cultures de Tissus de Plantes*, pp. 121–129. CNRS, Paris.

Boller, T. and A. Wiemken (1986). Dynamics of vacuolar compartmentation. *Annu. Rev. Plant Physiol.*, **37**, 137–164.

Bradley, P. M. (1981). Coculture of carrot cells and a green alga on a medium deficient in nitrogen. *Z. Pflanzenphysiol.*, **100**, 65–67.

Brown, E. G. and R. P. Newton (1981). Cyclic AMP and higher plants. *Phytochemistry*, **20**, 2453–2463.

Burgess, J. (1985). *An Introduction to Plant Cell Development.* Cambridge University Press, Cambridge.

Carlson, P. S. (1977). Novel cellular associations formed *in vitro.* In *Molecular Genetic Modification of Eukaryotes*, ed. I. Rubenstein, R. L. Phillips, C. E. Green and R. J Desnick, pp. 43–56. Academic Press, New York.

Carpita, N. C. (1982). Limiting diameters of pores and the surface structure of plant cell walls. *Science (Washington, D.C.)*, **218**, 813–814.

Cohen, J. D. and R. S. Bandurski (1982). Chemistry and physiology of bound auxins. *Annu. Rev. Plant Physiol.*, **33**, 403–430.

Coleman, T., D. Evans and C. Hawes (1988). Plant coated vesicles. *Plant Cell Environ.*, **11**, 669–684.

Coutos-Thevenot, P., J. P. Jouanneau, S. Brown, V. Petiard and J. Guern (1990). Embryogenic and nonembryogenic cell lines of *Daucus carota* cloned from meristematic cell clusters. Relation with cell ploidy determined by flow cytometry. *Plant Cell Rep.*, **8**, 605–608.

Creelman, R. A. (1989). Abscissic acid physiology and biosynthesis in higher plants. *Physiol. Plant.*, **75**, 131–136.

Dalton, C. C. and H. E. Street (1977). The influence of applied carbohydrates on the growth and greening of cultured spinach (*Spinacia oleracea* L.) cells. *Plant Sci. Lett.*, **10**, 157–164.

Das, R., S. Bagga and S. K. Sopory (1987). Involvement of phosphoinositides, calmodulin and glyoxalase-I in cell proliferation in callus cultures of *Amaranthus paniculatus. Plant Sci.*, **53**, 45–51.

Davidonis, G. H., R. H. Hamilton and R. O. Mumma (1978). Metabolism of 2,4-D in soybean root callus and differentiated soybean root cultures as a function of concentration and tissue age. *Plant Physiol.*, **62**, 80–83.

Davidonis, G. H., R. H. Hamilton and R. O. Mumma (1982). Metabolism of 2,4-dichlorophenoxyacetic acid in 2,4-dichlorophenoxyacetic acid-resistant soybean callus tissue. *Plant Physiol.*, **70**, 104–107.

de Gunst, M. C. M., P. A. A. Harkes, J. Val, W. R. van Zwet and K. R. Libbenga (1990). Modelling the growth of batch culture of plant cells: a corpuscular approach. *Enzyme Microb. Technol.*, **12**, 61–71.

Delmer, P. (1987). Cellulose biosynthesis. *Annu. Rev. Plant Physiol.*, **38**, 259–290.

Deus-Neumann, B. and M. H. Zenk (1985). Accumulation of alkaloids in plant vacuoles does not involve an ion-trap mechanism. *Planta*, **167**, 44–53.

DiCosmo, F. and S. G. Tallevi (1985). Plant cell cultures and microbial insult: interactions with biotechnological potential. *Trends Biotechnol.*, **3**, 110–111.

Dixon, R. A. (ed.) (1985). *Plant Cell Culture, A Practical Approach.* IRL Press, Oxford.

Endo, T., T. Komiya, M. Mino, F. Nakanishi, S. Fujita and Y. Yamada (1988). Genetic diversity among sublines originating from a single somatic hybrid cell of *Duboisia hopwoodii* x *Nicotiana tabacum. Theor. Appl. Genet.*, **76**, 641–646.

Ettlinger, C. and L. Lehle (1988). Auxin induces rapid changes in phosphatidylinositol metabolites. *Nature (London)*, **331**, 176–178.

Evans, D. A., W. R. Sharp, P. V. Ammirato and Y. Yamada (eds.) (1983). *Handbook of Plant Cell Culture*, vol. 1. Macmillan, New York.

Galun, E. (1981). Plant protoplasts as physiological tools. *Annu. Rev. Plant Physiol.*, **32**, 237–266.

Gibson, D. M. and A. A. Christen (1989). Slurry-producing agents provide buoyancy and allow growth of plant suspension cells in stationary culture. *Biotechnol. Tech.*, **3**, 305–308.

Goldsworthy, A. and K. S. Rathore (1985). The electrical control of growth in plant tissue cultures: the polar transport of auxin. *J. Exp. Bot.*, **36**, 1134–1141.

Griffing, L. R., A. J. Culter, P. D. Shargool and L. C. Fowke (1985). Isoelectric focusing of plant cell protoplasts. *Plant Physiol.*, **77**, 765–769.

Gunning, B. E. S. and A. W. Robards (eds.) (1976). *Intercellular Communication Studies on Plasmodesmata.* Springer, Berlin.

Hagimori, M., T. Matsumoto and Y. Mikami (1984). Photoautotrophic culture of undifferentiated cells and shoot-forming cultures of *Digitalis purpurea* L. *Plant Cell Physiol.*, **25**, 1099–1102.

Hangarter, R., R. Griesbach, B. Martin and N. Good (1979). The use of IAA conjugates in plant tissue culture. *Plant Physiol.*, **63**, (suppl. 5), 117.

Harris, N. (1986). Organisation of the endomembrane system. *Annu. Rev. Plant Physiol.*, **37**, 73–92.

Heyn, A. N. J. (1981). Molecular basis of auxin-regulated extension growth and role of dextranase. *Proc. Natl. Acad. Sci. USA*, **78**, 6608–6612.

Hodges, T. K. and R. T. Leonard (1973). Purification of a membrane-bound adenosine triphosphatase from plant roots. *Methods Enzymol.*, **32**, 392–406.
Hooker, B. S. and J. M. Lee (1990). Cultivation of plant cells in aqueous two-phase polymer systems. *Plant Cell Rep.*, **9**, 546–549.
Hulst, A. C., M. M. T. Meyer, H. Breteler and J. Tramper (1989). Effect of aggregate size in cell cultures of *Tagetes patula* on thiophene production and cell growth. *Appl. Microbiol. Biotechnol.*, **30**, 18–25.
Husemann, W., K. Fischer, I. Mittlebach, S. Hubner, G. Richter and W. Barz (1989). Photoautotrophic plant cell cultures for studies on primary and secondary metabolism. In *Primary and Secondary Metabolism of Plant Cell Cultures*, ed. W. G. W. Kurz, vol. 2. Springer, Berlin.
Joersbo, M., J. M. Andersen, F. T. Okkels and R. Rajogopal (1989). Isoperoxidases as markers of somatic embryogenesis in carrot cell suspension cultures. *Physiol. Plant.*, **76**, 10–16.
Jones, L. H. (1974). Factors influencing embryogenesis in carrot cultures. *Ann. Bot.* (*London*), **38**, 1077–1088.
Kaimori, N. and N. Takahashi (1989). Plant regeneration from dried callus of carrot (*Daucus Carota L.*). *Jpn. J. Breed.*, **39**, 379–382.
Kende, H. and G. Gardner (1976). Hormone binding in plants. *Annu. Rev. Plant Physiol.*, **27**, 267–290.
Key, J. L. (1989). Modulation of gene expression by auxin. *Bioessays*, **11**, 52–58.
Kolattukudy, P. E. (1980). Biopolyester membranes of plants: cutin and suberin. *Science* (*Washington, D.C.*), **208**, 990–1000.
Kreis, W. and E. Reinhard (1985). Rapid isolation of vacuoles from suspension-cultured *Digitalis lanata* cells. *J. Plant Physiol.*, **121**, 385–390.
Larkin, P. J. (1978). Plant protoplast agglutination by lectins. *Plant Physiol.*, **61**, 626–629.
Levin, R., V. Gaba, B. Tal, S. Hirsch. D. DeNola and I. K. Vasil (1988). Automated plant tissue culture for mass propagation. *Biotechnology*, **6**, 1035–1040.
Lippincott, J. A. and B. B. Lippincott (1978). Crown gall and embryonic cells lack *Agrobacterium* adherence sites. *Science* (*Washington, D.C.*), **199**, 1075–1078.
Loomis, W. D. and J. Battaile (1966). Plant phenolic compounds and the isolation of plant enzymes. *Phytochemistry*, **5**, 423–428.
Manasse, R. J. (1972). A new culture flask for plant tissue suspension cultures. *Experientia*, **28**, 723–724.
McDaniel, R. G. (1989). Plant genetic engineering *via* organelle transfer. *Plant Breed Rev.*, **2**, 283–302.
Meins, F. (1989). Habituation: heritable variation in the requirement of cultured plant cells for hormones. *Annu. Rev. Genet.*, **23**, 395–408.
Moriyasu, Y. and M. Tazawa (1988). Degradation of proteins artificially introduced into vacuoles of *Chara australis*. *Plant Physiol.*, **88**, 1092–1096.
Montague, M. J., R. K. Enns, N. R. Siegel and E. G. Jaworski (1981). A comparison of 2,4-dichlorophenoxyacetic acid metabolism in cultured soybean cells and in embryogenic carrot cells. *Plant physiol.*, **67**, 603–607.
Morris, P. and M. W. Fowler (1981). A new method for the production of fine plant cell suspension cultures. *Plant Cell, Tissue Organ Cult.*, **1**, 15–24.
Murakishi, H. H., J. X. Hartmann, L. E. Pelcher and R. N. Beachy (1970). Improved inoculation of cultured plant cells resulting in high virus titre and crystal formation. *Virology*, **41**, 365–367.
Okamura, S., K. Miyasaka and A. Nishi (1973). Synchronisation of carrot cell cultures by starvation and cold treatment. *Exp. Cell Res.*, **78**, 467–470.
Parfitt, D. E., A. A. Almehdi and L. N. Bloksberg (1988). Use of organic buffers on plant tissue-culture systems. *Sci. Horti.* (*Amsterdam*), **36**, 157–164.
Polonenko, D. R. and G. A. Maclachan (1984). Plasma membrane sheets from pea protoplasts. *J. Exp. Bot.*, **35**, 1342–1349.
Price, C. A. (1974). Plant cell fractionation. *Methods Enzymol.*, **31A**, 501–519.
Raff, J. W., J. F. Hutchinson, R. B. Knox and A. E. Clarke (1979). Cell recognition: antigenic determinants of plant organs and their cultured callus cells. *Differentiation* (*Berlin*), **12**, 179–186.
Rathore, K. S., T. K. Hodges and K. R. Robinson (1988). A refined technique to apply electrical currents to callus cultures. *Plant Physiol.*, **88**, 515–517.
Robinson, D. G. (1984). Membranes and secretion in higher plants. *Annu. Proc. Phytochem. Soc. Eur.*, **24**, 147–161.
Sato, F., K. Asada and Y. Yamada (1979). Photoautotrophy and the photosynthetic potential of chlorophyllous cells in mixotrophic cultures. *Plant Cell Physiol.*, **20**, 193–200.
Sauerbrey, E., K. Grossman and J. Jung (1987). Is ethylene involved in the regulation of growth of suspension cultured cells? *J. Plant Physiol.*, **127**, 471–479.
Scheel, D. and H. Sandermann (1981). Metabolism of 2,4-dichlorophenoxyacetic acid in cell suspension cultures of soybean (*Glycine max* L.) and wheat (*Triticum aestivum* L.) II. Evidence for incorporation onto lignin. *Planta*, **152** 253–258.
Schweiger, H. G., J. Dirk, H. V. Koop, E. Kranz, G. Neuhaus, G. Spangenberg and D. Wolf (1987). Individual selection, culture and manipulation of higher plant cells. *Theor. Appl. Genet.*, **73**, 769–783.
Shimomura, S., N. Inohara, T. Fukui and M. Futai (1988). Different properties at two types of auxin-binding sites in membranes from maize coleoptides. *Planta*, **175**, 558–566.
Smeekens, S., P. Weisbeek and C. Robinson (1990). Protein transport into and within chloroplasts. *Trends Biochem. Sci.* (*Pers. Ed.*), **15**, 73–76.
Smith, S. M. and H. E. Street (1974). The decline of embryogenic poteintial as callus and suspension cultures of carrot (*Daucus carota.* L.) are serially subcultured. *Ann. Bot.* (*London*), **38**, 223–241.
Stafford, A. and G. S. Warren (eds.) (1991). *Plant Cell and Tissue Culture*. Open University Press, Milton Keynes.
Steer, M. W. (1988). Plasma membrane turnover in plant cells. *J. Exp. Bot.*, **39**, 987–996.
Street, H. E. (ed.) (1977). *Plant Tissue and Cell Culture*, 2nd edn. Blackwell, Oxford.
Stuart, R. and H. E. Street (1969). Studies on the growth in culture of plant cells. 4. The initiation of division in suspensions of stationary-phase cells of *Acer pseudoplatanus*. *J. Exp. Bot.*, **20** 556–571.
Stuart, R. and H. E. Street (1971). Studies on the growth in culture of plant cells. 10. Further studies on the conditioning of culture media by suspensions of *Acer pseudoplatanus*. *J. Exp. Bot.*, **22**, 96–106.
Suzuki, M., I. Takebe, S. Kajita, Y. Honda and C. Matsui (1977). Endocytosis of polystyrene spheres by tobacco leaf protoplasts. *Exp. Cell Res.*, **105**, 127–135.

Timmers, A. C. J., S. C. de Vries and J. H. N. Schel (1989). Distribution of membrane-bound calcium and activated calmodulin during somatic embryogenesis of carrot (*Daucus carota* L.). *Protoplasma*, **153**, 24–29.

Varner, J. E. and L.-S. Lin (1989). Plant cell wall architecture. *Cell (Cambridge, Mass.)*, **56**, 231–239.

Wagner, G. J. and H. W. Siegelman (1975). Large-scale isolation of intact vacuoles and isolation of chloroplasts from protoplasts of mature plant tissues. *Science (Washington, D.C.)*, **190**, 1298–1299.

Wareing, P. F. and I. D. J. Phillips (1981). *The Control of Growth and Differentiation in Plants*, 3rd edn. Pergamon Press, Oxford.

Warren, G. S. and M. W. Fowler (1977). A physical method for the separation of various stages in the embryogenesis of carrot cell cultures. *Plant. Sci. Lett.*, **9**, 71–76.

Warren, G. S. and R. Fallon (1984). Reversible, lectin-mediated immobilization of plant protoplasts on agarose beads. *Planta*, **161**, 201–206.

Wernicke, W. and J. Gorst (1987). Fucose-specific lectins disclose perpetuation of root primordia in callus cultures of maize (*Zea mays* L.) . *Dev. Biol.*, **120**, 85–91.

Withers, L. A. (1986). Cryopreservation and genebanks. In *Plant Cell Culture Technology*, ed. M. M. Yeoman, pp. 96–140. Blackwell, Oxford.

Wright, K. M. and K. J. Oparka (1989). Uptake of lucifer yellow CH into plant cell protoplasts: a quantitative assessment of fluid-phase endocytosis. *Planta*, **179**, 257–264.

Yamada, Y. and F. Sato (1978). The photoautotrophic culture of chlorophyllous cells. *Plant Cell Physiol.*, **19**, 691–699.

# 2

# The Molecular Biology of Plant Cells and Cultures

CHRISTOPHER A. CULLIS
*Case Western Reserve University, Cleveland, OH, USA*



## 2.1 INTRODUCTION

An understanding of the molecular genetics of plants is important not only for the practical value in potential improvement of crops, but also for the opportunity to gain an insight into the basic processes unique to plants. There are two extreme paths that can be taken in describing the molecular basis of plant growth and development. One is where much is discovered about model systems in the expectation that the principles can be directly applied to an important crop plant. The other is where all work is done on the plant to which the results will eventually be applied. A position between these two extremes is obviously the most sensible approach, but if model systems are used, there must be some basis for the belief that the principles, and details discovered, will be applicable to some group of economically important plants.

The ease with which the recombinant DNA work can be done depends, in part, on the size of the genome. The smaller the genome the less difficulty is encountered in the isolation of specific genes. The genome size is not correlated with the complexity of the organism. However, the complexity of the genome affects the types of investigations which can be undertaken and the difficulty in developing important information. For example, the generation of a molecular linkage map at the same level of resolution is much simpler, and requires fewer markers, for a small genome than for a large one. Thus, starting *de novo*, it would be much easier to generate a map for *Arabidopsis thaliana* than for corn, wheat, or a conifer.

Another limiting factor in the ability to manipulate plants is the ease with which they can be transformed and regenerated. In this latter area, the use of model systems has led to some general principles, but for many new species the process has had to start right from the beginning each time. Even the use of closely related species, or even cultivars, has not ensured the immediate transferability of the results. Finally, the stability of the organism through the processes involved in the manipulation is important so that the desired event can be recognized against a background of similar phenotypic changes caused by mutations.

Here the role of genome size and alterations therein will be considered in relation to the ease of manipulation of higher plants, both in regard to their characterization by classical genetics as well as by cell culture techniques.

## 2.2 GENOME SIZE AND ORGANIZATION IN HIGHER PLANTS

Plants have large differences in their nuclear DNA content. Within the angiosperms there is nearly a 1000-fold range of variation and there does not appear to be any correlation between genome size and organismal complexity (Bennett and Smith, 1976). The haploid genome sizes of various flowering plants are shown in Table 1. The values have been calculated from the kinetic complexities except for *Pinus radiata*, which comes from spectrophotometric measurement.

**Table 1** Haploid Genome Size in Plants and the Number of Lambda Clones that must be Screened to have a 99% Chance of Isolating a Single Copy Sequence from these Genomes[a]

| *Plant* | *Haploid genome size (kilobase pairs)* | *Number of lambda clones in a complete library* |
|---|---|---|
| *Arabidopsis* | 70 000 | 16 000 |
| Flax[b] | 280 000 | 65 000 |
| Mung bean | 470 000 | 110 000 |
| Cotton | 780 000 | 180 000 |
| Tobacco | 1 600 000 | 370 000 |
| Soybean | 1 800 000 | 440 000 |
| Pea | 4 500 000 | 1 000 000 |
| Wheat | 5 900 000 | 1 400 000 |
| *Pinus radiata*[c] | 10 000 000 | 2 300 000 |

[a]The data are from Myerowitz and Pruitt (1985) unless otherwise stated. [b]Cullis (1981). [c]Miksche (1985).

In *Arabidopsis* the characterization of the genome by molecular analysis has indicated that the nuclear genome consists predominantly of unique sequences and that most of the nuclear repetitive DNA is ribosomal DNA (Myerowitz, 1987). The genes of *Arabidopsis* appear to be shorter than in other higher plants and there are obviously far fewer dispersed repetitive sequences in this very small genome. Taking the value of 75% of the total genome as comprising the unique fraction of the *Arabidopsis* genome suggests that the minimum information content required for a functional higher plant is 52 500 kb. Thus in all other higher plants shown in Table 1 the necessary information would comprise a much lower proportion of the total DNA. Examples of the amount of essential coding capacity would be 20% of the flax genome but only 0.2% of the pine genome.

The differences between the *Arabidopsis* genome and those larger genomes comprise three parts. Firstly, the genes in *Arabidopsis* appear to be smaller than those in other plants with shorter introns, therefore requiring fewer nucleotide pairs to encode them. Secondly there appear to be fewer members of the multigene families in *Arabidopsis*. For example, *Arabidopsis* has one gene for alcohol dehydrogenase (Chang and Meyerowitz, 1988), while maize has two, and three genes for the chlorophyll a/b binding protein, while petunia has 16, pea eight and wheat seven (Myerowitz, 1987). Thirdly, there are few dispersed repeats in the *Arabidopsis* genome, while in the maize genome this class can make up to 40% of the total. Finally, there are few repeated tandemly arrayed sequences apart from the ribosomal RNA genes, while this class of repeats makes up some 35% of the flux genome (Cullis and Cleary, 1986a).

The vast majority of the genome in most higher plants cannot be responsible for direct coding sequences. How has this wide range of variation in DNA content arisen? While a large genome size and heterogeneity seem to be indicators for evolutionary flexibility and progressivity, decline in total DNA content apparently accompanies evolutionary specialization and adaptation to certain ecological niches. As genome evolution is predominantly a molecular process governed by sequence amplification, divergence, dispersal and loss, the current status of the genome may be determined by a balance between these events.

The variation in nuclear DNA content per genome among both plant and animal taxa has been extensively documented (Bachmann *et al.*, 1972; Sparrow *et al.*, 1972; Bennett and Smith, 1976). This nuclear DNA content variation has not been correlated with any measure of evolutionary advancement or genetic complexity, with both increases and decreases in DNA content being observed in eukaryotic evolution (Price, 1976). The changes in nuclear DNA, apart from polyploidization and chromosome endoreduplication, affect the relative proportions of fractions of the genome. Thus amplifications or losses of DNA usually appear to occur in the repetitive fraction of the nuclear genome, the majority of which, although it may have no specific coding capacity or function, may still be important. The DNA content by virtue of its mere bulk can affect parameters such as nuclear volume, cell volume and mitotic and meiotic cycle times (Bennett, 1972; Price, 1976; Cavalier-Smith, 1978). At what level of difference in total nuclear DNA do differences become of significance to the organism? Price (1988) suggests that those in the range of 5–10% may be sufficient to be subject to selection.

The evidence is accumulating that rapid changes in nuclear DNA content can accumulate within a species. What is not yet clear is the mechanism(s) by which these variations arise, and their frequency. There is a latitudinal cline in the DNA content of crop plants, such that those with low DNA content are naturally adapted to equatorial latitudes, while those with higher DNA contents are found at higher latitudes (Bennett, 1976; Bennett and Smith, 1976). Increasing latitude and elevation may be associated with intraspecific increases in DNA in conifers (Miksche, 1967, 1968). The intraspecific variation in DNA content in the genus *Microseris* also appears to be related to environmental conditions (Price, 1988). The function and composition of the DNA fractions which alter is currently unknown, although the changes in heterochromatin are consistent with these alterations occurring in the highly repetitive sequence fractions.

In spite of the extensive variation between species, and also within a species, within an inbred line there is little evidence for quantitative variation in DNA content. One exception to this is in *Helianthus annuus* (Cavallini *et al.*, 1986, 1989). Here the nuclear DNA content in the progeny of single plants from an inbred line were compared for total DNA content by microdensitometry of Feulgen-stained material. Significant differences were observed between the progeny from a single plant. However, when these individuals were themselves inbred, the range did not continue to increase. The original distribution was recreated, so that the mean of the progeny from a high DNA plant was lower than the parental value, while the mean of the progeny from a low DNA parent was higher than the parental value.

### 2.2.1 Consequences of DNA Variation

DNA differences can exist between species which can be interbred. The combination of lines with such differences can lead to unexpected results. A specific example is in *Microseris* species. Here both interspecific and intraspecific crosses can result in progeny with a nuclear DNA content different from that expected from the parental values (Price, 1988). Another example concerns the fate of telomeric heterochromatin, present on the rye chromosomes in wheat–rye hybrids (triticale; Gustafson *et al.*, 1983). These rye telomeric heterochromatic segments are directly involved in abnormal bridge formation in triticale, resulting in abnormally polyploid nuclei in young endosperms, and are correlated with grain shrivelling and seed abortion at maturity. One possible explanation for this acitivity is a change in the time of replication of the heterochromatic sequences. As detailed later, one suggested cause of somaclonal variation is the disruption by late-replicating heterochromatic blocks. If this is so, then the alterations in DNA content seen in certain crosses may be a result of some form of differential cell cycle control, especially in respect to DNA replication, in the two parents. Since a series of genes would be expected to be involved in the control of DNA replication, it should be possible to mark how many of these genes are present and their relative importance from the interactive crosses.

The chromosome elimination in interspecific crosses between *Hordeum* species is another example of incompatibility between the cell cycle and the chromosome complement (Bennett *et al.*, 1976). This type of interaction must be related to chromosome structure. How the structure is determined with regard to the nucleotide composition and its interaction during the cell cycle is unknown. What proportion of the chromosome can be made up of sequences derived from *Hordeum bulbosum* and still be recognized as a *Hordeum vulgare* chromosome in an interspecific cross?

A contrast to chromosome elimination is the accumulation of supernumerary chromosomes in many plants. These are usually heterochromatic and have no demonstrable transcriptional activity. However, these chromosomes do appear to affect the cell cycle. One of the advances in yeast genetics which has had profound effects is the ability to construct artificial chromosomes (Blackburn, 1985). The isolation of centromeric and telomeric sequences with the addition of origins of replication has allowed such chromosomes to be constructed. These components have not yet all been assembled in higher plants, although a telomeric sequence (Richards and Ausebel, 1988) and possible origins of replication have been identified. In spite of the tolerance of higher plants to nuclear DNA variation, the fate of an artificial chromosome is not clear. Will an artificial chromosome be expressed, will it be recognized as a supernumerary chromosome? The ability to build a large fragment outside the cell and introduce it as a self-replicating entity would eliminate any difficulties in the control of introduced genes due to position effects.

The overall organization of the plant genome is well understood in terms of the types of sequences present and the way in which they are organized with respect to one another. However, the vast majority of the genome is still looking for a function. Is all the highly repetitive DNA (with the exception of the ribosomal RNA genes) just so much junk, or does it really have a sequence independent function? It is clear that this fraction can have effects on the cell cycle and, perhaps, on the stability of the genome overall. If the elimination/amplification of sets of repetitive sequences causes a general destabilization of the genome, then the compatibility of blocks of repetitive sequences within and between species becomes a significant consideration.

### 2.2.2 Transcriptional Specificities

Incompatibility can occur at the transcriptional level as well as at the level of replication. Two examples of incompatibility of the transcriptional signals come from the activity of the nucleolar regions in cereals, and the activity of genes transferred between monocots and dicots by genetic engineering techniques. In both cases the species specific nature of the control lends caution to an oversimplification of the ease by which genes can be moved around.

The use of chromosome replacement in the cereals has been a powerful tool in the genetic analysis and manipulation of these species. The addition of *Aegilops* chromosomes containing ribosomal RNA genes into a wheat background completely disrupted the normal functioning of the rDNA in the production of rRNA. These observations resulted in the construction of a hierarchy of nucleolar organizers in terms of their dominance (Flavell, 1986). This ordering did not prove to be on the basis of the number of genes present at a particular locus, but more on the structure of those genes. Thus, in a given background genotype, the performance of a gene is dependent on its structure, sequence and its position (the location of any heterochromatic blocks, for example).

A significant fraction of the signalling structure for the developmental and enviromental control of transcription appears to be conserved within higher plants. Thus, genes isolated from a range of dicots are appropriately regulated in heterologous transgenic systems at the RNA level. However, when the protein production is followed, then the accumulation and processing of the introduced gene in the heterologous systems is generally much less efficient than the homologous gene.

The conservation of the transcription signals between monocots and dicots is not as good as within the dicots. Thus maize genes transferred to tobacco cannot be transcribed and regulated efficiently solely by their own promoter. The addition of an enhancer sequence to increase transcription demonstrated that once the transcription can be initiated, the appropriate developmental and enviromental constraints are active. Because of the lack of good transformation in monocots, it is not known whether or not the transcription problems are reciprocal. With this type of incompatibility between monocots and dicots, the potential problems in engineering gymnosperms with the currently available vectors should not be underestimated. The constitutive promoters may well function efficiently enough for a general characterization but the development of regulated transformants may require much more information concerning the structure of gymnosperm genes themselves.

## 2.3 EFFECT OF GENOME SIZE ON THE MANIPULATION OF PLANT GENOMES

### 2.3.1 Construction of Genomic Libraries

The ease of isolation of a gene from an organism is greatly affected by the genome size of that organism. Thus a small virus can be cloned as a single insert in a plasmid vector. However the number of lambda recombinants needed to have a 99% probability of having a specific single copy gene present for a variety of plants is shown in Table 1 (calculations based on those in Maniatis *et al.*, 1982). The larger number of clones to be screened is not the only complication caused by a larger genome size. In general, the larger the genome size, the higher the extent of methylation of the plant genome. Most of the currently available cloning vectors do not handle methylated DNA as efficiently as nonmethylated DNA. Thus, the higher number of clones has to be obtained with a lower efficiency. The organization of the genome can also affect the representation of the total genome in the library. Special precautions must be taken to effectively clone tandemly repetitive sequences. This is not usually a problem since an increase in genome size is generally accompanied by an increase in the dispersed repetitive fraction, rather than the tandemly repeated families, but cannot be ignored.

The genes in higher animals are generally longer than those in higher plants. The introns in plant genes are relatively short so that the size of insert available in current vectors is generally able to accommodate both the complete gene as well as the sequences 3′ and 5′ to the coding region. The only exception to this identified so far is the rDNA in conifers. The repeat unit in *Pinus radiata* appears to be 27 kb (Cullis *et al.*, 1988) and so has to be cloned in a cosmid library in order to obtain the whole repeat unit. Since this species also has one of the largest genomes, the problem is further compounded. Insufficient genes have been isolated from gymnosperms to determine whether or not the genes in this group are significantly different from those in angiosperms, and therefore require special consideration.

### 2.3.2 Restriction Fragment Length Polymorphism Mapping

The term 'restriction fragment length polymorphism' (RFLP) was coined to denote the difference in molecular weight of homologous fragments of restriction enzyme digested genomic DNA sometimes observed between two genetically distinct individuals. These differences can be due to base pair changes at the site recognized by the restriction enzyme, to a rearrangement encompassing that site or to internal deletion/insertion events. These differences are due to differences in the genetic material of the individuals and so can be used as genetic markers to detect loci in a similar fashion to any morphological marker.

RFLPs posssess several advantages over conventional genetic markers. The potential number of markers is virtually unlimited since digestion of the DNA of a higher eukaryote by a restriction enzyme will generate, on the average, at least $10^6$ fragments and there are over 100 restriction enzyme specificities described to date. The RFLP marker is essentially silent and so there are no complicating interactions when a large number of them are evaluated in a single cross, as can happen with epistatic interference associated with the use of multiple morphological markers (Helentjaris *et al.*, 1985; Tanksley *et al.*, 1988, 1989). For the same reason the interpretation of RFLP data is independent of environmental influences. RFLP studies are now in progress in several plants including maize (Helentjaris, 1897), soybean, brassicas, pea, tomato, lettuce (Landry *et al.*, 1987), flax and *Arabidopsis* (Chang *et al.*, 1988). Clearly, the ease with which an RFLP map can be constructed depends on the size of the genome and the degree of variability within the genome among related lines. Thus it appears that while predominantly outcrossing species, such as maize and the brassicas, are extremely variable, the self-pollinating species, such as tomato, do not exhibit many polymorphisms. The size of the genome also plays a role in determining the number of RFLPs required to generate an informative map. *Arabidopsis*, with a genome size of 70 000 kb, would have a marker every 1000 kb with 70 evenly spaced RFLPs, while in pine the same coverage could only be achieved by the use of 1000 such markers. A second consideration is the status of the conventional genetic map in the species under investigation. In maize, the existence of a detailed genetic map with marker stocks has greatly facilitated the assignment of RFLP markers to specific chromosomes. However, in the absence of a preliminary map, the use of RFLPs should be the quickest and easiest method of generating the series of linkage groups, although it may not be possible to assign any one of these linkage groups to a specific chromosome.

What is the potential of an RFLP map once it has been generated? One use is the mapping of quantitative trait loci in maize, and the dissection of them into individual component loci (Lander and Botstein, 1989; Patterson *et al.*, 1988; Tanksley *et al.*, 1989). Thus the location of a quantitative trait locus can be determined, its relative effect on the overall variance for a complex trait evaluated and its gene action characterized as being either dominant/recessive or additive. Through the use of multiple molecular markers and recombinant inbred lines, it will be possible to examine the effects of the environment on the expression of individual loci involved in the determination of complex traits. This will lead to the determination of functionally just what a 'genotype-by-environment' interaction is. Such a dissection of the components of yield would provide significant new information to be incorporated into conventional breeding programs. Another trait which will be amenable to dissection by the use of multiple markers is the phenomenon of heterosis. A dissection of the loci associated with heterosis in certain combinations, by the isolation of probes closely linked to those loci, will open new avenues to the understanding of this phenomenon and, perhaps, allow its widespread manipulation and use in crop species.

It might be possible to use the existence of an RFLP marker closely linked to a locus of interest to clone that locus. The cloning may be done by chromosome walking using libraries constructed in any one of the following vectors: lambda phage vectors, cosmids or yeast artificial chromosomes. Once again the organization and size of the genome are important in the ease with which such a task can be completed. As mentioned above, to cover the genome to the same degree of resolution in *Arabidopsis* and pine would require 70 and 1000 markers respectively. Thus the mapping has to be done with a much larger number of clones in the plants with bigger genomes or the walking has to be performed over a longer distance. An additional complication arises in genomes larger than that of *Arabidopsis*, namely, the existence of repeated sequences. Both dispersed repeated sequences and tandem arrays have the potential to interfere with the isolation of loci by chromosome walking. If the locus in question is flanked by a long tandem array on both sides then it may not be possible to isolate identifiable fragments to move from the RFLP marker to the locus in question. Dispersed repeated sequences interfere with the selection of probes to identify overlapping clones and may make the process tedious, although they should not render it impossible.

The greatest advance in this technology is likely to come from the use of yeast artificial chromosomes (YACs; Blackburn, 1985) in conjunction with the transformation of plants with large ($>500$ kb) DNA fragments. When large pieces can be cloned, the number of steps in a walk from an RFLP to the locus in question will be minimized. Secondly, the ability to transform plant cells with large pieces will mean fewer transformants will be needed to demonstrate that the gene is present on a specific piece of DNA. Ideally, the identification of a gene by its function will be quicker and more direct than by sequence similarity to an analogous sequence.

### 2.3.3 Comparison of Related Species

The use of multiple RFLP markers can confirm conserved blocks of genes in related species. Thus, if a fragment is required to be transferred from a wild relative to a crop speices, its introgression can be followed using these markers (Young *et al.*, 1988). The size of the incorporated segment can be determined by the extent of the appearance of flanking markers. Thus the size of the alien fragment can be directly estimated. Blocks of RFLP markers can also be used to identify duplicate loci within a genome and provide evidence for their origin (Tanksley *et al.*, 1988; Bonierbale *et al.*, 1988). Maize is generally considered to be a true diploid but it has been repeatedly suggested that it originated as an amphidiploid hybrid and has an allopolyploid genomic structure. In the course of the construction of an RFLP map in maize a number of duplicate loci were observed. However, the chromosomal distribution of these loci indicated that maize is not a recent allotetraploid and that the distribution was also consistent with the duplicated segments having been generated through internal duplications.

Another use of RFLPs is in the identification of strains, especially in vegetatively propagated material. The use of bank of markers would make it possible to fingerprint an individual line. The use of a collection of RFLP markers can be simulated by a single sequence, the so-called hypervarible minisatellites (Jeffreys *et al.*, 1985). These have been used to fingerprint a variety of organisms and they appear to be ubiquitous. There are two current sources of these hypervariable probes, one from the bacteriophage M13 and one from an intron from a human gene. Both of these probes have been shown to hybridize to plant DNAs and to highlight polymorphisms (Dallas, 1988; Rogstad *et al.*, 1988; Zimmerman *et al.*, 1989). Their use is likely to be restricted to line identification and, perhaps, the characterization of the degree of diversity within populations. The hypervariable sequences are not especially useful for genetic mapping since they appear to be composed of tandem repetitive sequences which are prone to recombinational events, thereby varying between generations. In addition, the same size of band in two lines does not ensure the allelic nature of the band in those two lines. Additional characterization through a series of generations would be necessary to demonstrate their effectiveness as genetic markers.

### 2.3.4 Gene Isolation

What role does the genome size play in various scenarios involved in the isolation of desirable genes in plants? Already covered above is the use of RFLPs to locate the approximate position of a gene of interest and its subsequent cloning by chromosome walking. Three other possible routes are available for the isolation of specific genes. These are: the use of transposons as mutagens and specific tags, gene rescue by shotgun cloning, and the use of heterologous probes.

A number of approaches to gene cloning in plants are available. The simplest, and most widely used, is the use of a heterologous probe, *i.e.* an identical gene from a different organism is used to isolate the gene from the specific plant of interest. Probes derived from plants, animals and yeast have been used to isolate higher plant genes in specific cases. The feasibility of this approach, commonly known as clone by phone, is dependent on the evolutionary conservation of the gene in question. An example of how wrong this can be is the use of mammalian oncogenes to isolate their counterpart in higher plants. In general this approach has isolated genes for abundant cell wall proteins, with no known, or as yet demonstrated, function in growth control. However, this is not to imply that these plant genes are not interesting in themselves, but that their function is not related to that of the heterologous probe.

The recent technical advances in the transformation of plants have resulted in the ability to generate thousands of transformed cells which can be regenerated into whole plants (Caplan *et al.*, 1983). Thus a genomic library can be assembled in *Escherichia coli*, transferred to *Agrobacterium tumefaciens* and transformed *en masse* back into the plant. The use of a suitable, selectable or scorable phenotype associated with the gene of interest, makes it possible to rescue that gene from such a shotgun cloning experiment. The feasibility of the approach has been demonstrated for *Arabidopsis*, using an introduced antibiotic selection marker, namely the neomycin phosphotransferase gene, as a model system (Klee *et al.*, 1988). There are some major limitations to this approach. Although it can be accomplished for *Arabidopsis*, the necessity of screening large numbers of clones rules out the use of plants with larger genomes as gene sources for the shotgun approach. Many genes of interest will not confer a dominant selectable phenotype on the transformed plant. The labour involved in the phenotypic screening of transformed plants is probably not practical as the regeneration of 2000 independent tobacco transformants would be equivalent to one full man-year. Since tobacco is one of the most efficient transformation and regeneration systems currently available, any other plant species would require significantly more effort. Given the limitations of genome size, transformation frequency and ability to score for the presence of a gene in the transformed plant, the shotgun technique may be useful for the cloning of important genes which are not amenable to the present approaches used by plant molecular biologists.

A variation on the shotgun theme has been very successful in lower eukaryotes, particularly yeast. In this process the size of the insert contained in the vector is increased enormously by incorporating it into a chromosome. Since plant cells can be transformed directly, such an approach should be feasible in higher plants. In this case, the plant transformed could be scored for the altered phenotype directly and genes of agronomic importance, which are difficult to isolate in other ways, would become available. One potential snag in the use of such a plant artificial chromosome is the apparent inactivity of naturally occurring supernumerary chromosomes (B-chromosomes). If an introduced artificial chromosome suffered the same fate as a B-chromosome then direct testing for the gene would be rendered impossible.

## 2.4 TRANSPOSABLE ELEMENTS

Transposable elements are specific fragments of DNA which can move about the genome. Their presence has been recognized by the occurrence of unstable mutations caused by their insertion and excision from genes. Transposable elements were first genetically described in maize (McClintock, 1947), and described in molecular terms in *Escherichia coli*. Subsequently they have been found in a wide variety of prokaryotes and eukaryotes and are probably ubiquitous in nature. They are recognized as a major force in shaping the structure of the genome as well as being responsible for a major portion of mutational events. Studies on higher plant transposable element families has centered on those found in *Zea mays* and *Antirrhinum majus*. Representatives from different families of transposable elements from these species have been cloned, characterized and transferred to other plant species.

### 2.4.1 Transposable Elements from *Zea mays*

The collection of unstable mutations caused by transposable elements and the identification of these systems in maize far outnumber the identification of these elements in other eukaryotes. Many transposable element systems in maize have been described as acting as regulators of gene action. A single element is capable of simultaneous regulation of independent genes acting at different times during plant development (Dellaporta and Chomet, 1985).

Four categories of transposons have been described in maize. These are: (i) controlling elements, of which there have been eight families so far discovered; (ii) a retroviral-like element; (iii) Mu; and (iv) biologically uncharacterized insertions with structure similar to transposons (Lillis and Freeling, 1986). The controlling elements can be grouped into two classes. Firstly, there are autonomous elements which contain all the information required for their own transposition. In spite of being autonomous, their activity can be affected by the background genotype as well as environmental factors. The second class contains the nonautonomous elements, which can only be mobilized in the presence elsewhere in the genome of a related, active element. Thus the inactive, mobilizable fragment and the activator comprise a two-element system. The best-understood two-element system is Ac-Ds. Most Ds elements have been shown to be degenerate Ac elements, *i.e.* they are related to active Ac elements but have incurred mutations and/or deletions which have inactivated the transposition functions. However, many Ds elements only have homology with active Ac elements in the repeated termini which are necessary targets in the transposition process. Thus a non-autonomous member of a two-element family can be constructed from the appropriate end fragments, which define the ends of the fragment to be mobilized, separated by any sequence. The upper limit of the size of fragment which can be mobilized by maize transposons has not been determined. However, in *Drosophila melanogaster*, the elements can be hundreds of kilobases long.

The base sequence for integration of the transposable elements appears to be essentially random. However, the relationship between the original site of the element and the site to which it transposes is not necessarily random. Thus, Ac moves to positions close to the original site much more frequently than to positions further away, and rarely moves between chromosomes. Ac is usually present in a small number of copies in the genome so it is relatively simple to track the transpositions both genetically and in molecular terms. In fact, for Ac, the higher the number of copies present, the lower the transposition frequency. Robertson's mutator (Mu), on the other hand, appears to move randomly within the genome and not to be limited to short transpositions. However, this element is present in multiple copies and, since the specific one which has been mobilized to produce the new mutation is difficult to identify, the exact nature of the mobilization and distance moved is not clearly defined.

### 2.4.2 Manipulation of Transposable Elements

Transposable elements have been used to clone specific genes in both maize and *Antirrhinum majus* (Federoff *et al.*, 1984). The insertion of a transposable element into a gene disrupts the normal function of that gene. The subsequent isolation of a fragment containing the transposon and a comparison with the wild-type sequence has led to the isolation of genes in these two systems. Many of the genes isolated in both species have been involved in anthocyanin synthesis because of the history of characterization of unstable pigment mutants, and the availability of genetically characterized lines. However, more recently, in maize, Mu has been used to isolate the gene for knotted, a morphological mutant.

The success of transposons in the isolation of genes has led to their development as transposon tags in other species. This development requires a number of steps to be successful. Firstly, the transposon must be transferred to the material of choice. Secondly, the element must be able to be mobilized at a frequency which gives a reasonable chance for success in the isolation of the mutant of interest. Thirdly, the mutation must be stabilized so that the fragment containing the transposon, together with some surrounding sequence, can be isolated. Finally, a comparison of the cloned fragments with the normal line must be made to identity the gene itself. A number of these steps have been successfully carried out.

It has been reported that the maize element Ac is mobile in tobacco, *Arabidopsis thaliana* and *Daucus carota* (Baker *et al.*, 1986; Van Sluys *et al.*, 1987). In these experiments the Ac was introduced in *Agrobacterim tumefaciens* T-DNA vector and the fate of the Ac subsequently followed. The fate in tobacco indicated that the Ac element was excised from at least one copy of the T-DNA on which it was introduced in 25–75% of the transformed tobacco cells (Baker *et al.*, 1986). By following the fate of the Ac in transformed carrot roots, it was found that the element was excised early after transformation in 28% of the lines analyzed. These elements remained mobile as the roots continued to be cultured (Van Sluys *et al.*, 1987). The evidence for *Arabidopsis* was that more than half the transformed lines showed Ac excision and transposition (Van Sluys *et al.*, 1987). Thus it would appear that the maize elements are capable of being mobilized in heterologous systems. Since this transfer is between a monocot and a dicot, the potential problems in the control of gene expression have already been discussed above. The *Antirrhinum* transposons, on the other hand, when transferred by T-DNA, would be expected to function in other dicots. Thus the ability to introduce a transposable element into a new species is only limited by the ability to transform and regenerate that species.

What are the ideal characteristics of a transposon tag? Firstly, a high frequency of transposition is necessary so that the mutation of interest, caused by an insertion, can be identified among the spontaneous mutations. Secondly, the insertional mutation must be stabilized so that it can be recovered. These two properties are initially incompatible, but can be reconciled. One method is to use a two-element system, based on Ac-Ds. Here a Ds element, that is one that cannot transpose itself, but can be mobilized, is introduced into one line. Another Ds, which cannot be mobilized itself, but can mobilize other Ds elements is introduced into a second line. These two lines can be checked to ensure that the insertions are not on the same chromosome, although this is not an essential requirement. A hybrid between these two lines will contain an active transposable element system. Once the insertional mutation has occurred, the two elements can be separated through meiosis, so stabilizing the mutation.

A second method of reconciliation would be to use the property that some transposons are differentially sensitive to environmental conditions. The rate of transposition of the *Antirrhinum* element Tam3 is dependent on the temperature (Harrison and Fincham, 1964). Thus, the frequency of transposition can be three orders of magnitude higher at low temperatures than at high temperatures. Therefore, using this system, plants grown at low temperatures will have high rates of transposition. Once the desired mutation is obtained, raising the growth temperature will effectively stabilize the mutation.

A third possibility is to allow the plant to stabilize the mutation itself. Transposable elements are neither uniformly, nor always, active. The organism appears to be able to inactivate transposons, and for their reactivation to occur under a number of conditions, frequently in response to stress or 'genomic shocks' (McClintock, 1984). In Mu, especially, the inactivation of the transposable element is associated with modification of the DNA (Chandler *et al.*, 1986). Thus, the use of this element, which has a very high rate of mutation induction, would require the following of the desired mutation until it became stabilized (presumably, most frequently by an inactivation of the element rather than by some other means) and the subsequent isolation of the gene.

An alternative to the use of a transposon as a gene tag is the use of T-DNA as an insertional mutagen. The T-DNA inserts randomly into the genome and the regeneration of a large number of independent transformants may result in the identification of mutants produced by T-DNA insertion. However, there are two major drawbacks to this approach, one being the numbers to be produced and the other being the possibility of not getting a selectable insertion in the gene desired. A lack of a selectable insertion could be due to the target gene being in an inactive compartment of the genome at the transformation stage, so any insertions in there cannot be obtained as the introduced selectable marker cannot be expressed.

The genome size of the plant in question clearly affects this approach, and also applies to the transposon tagging described above. If the target size of the gene in question is 1 kb, then in a random model, one in 70 000 transpositions will be in the gene of interest. This is a reasonable value for endogenous transpositions, but the regeneration of these numbers of independent transformants is not feasible. However, as the genome size increases, the frequency of mutation must approach that of the insertional mutagen. In *Pinus*, the size of the genome, taking the same target size, would result in only one in $10^7$ insertions being in the gene of interest. Even with a very stable gene, this background of spontaneous mutations may be too high to allow this method of gene isolation to be efficient.

A method of improving the odds of isolation of the desired gene is to use the property of Ac to transpose most frequently to a neighbouring site. The strategy here would be to select a line in which the initial insertion of Ac mapped close to the gene of interest. The increase in rate would then raise the probability of an insertional mutation being caused by the introduced transposon. However, this approach would be limited to those species which are well characterized and already have a genetic map available.

Thus, the use of transposon mutagenesis is likely to be very productive in the isolation of genes. However, the system used may have to be tailored to the specific objective in order to recover the product. In addition, those species with extremely large genome sizes may not be amenable to this type of gene isolation by a random process, but have to be approached in a more directed fashion.

## 2.5 SOMACLONAL VARIATION

Efficient cell culture and regeneration of agriculturally important species is a relatively recent field of study. Phenotypic and genetic variants arising from this process were initially considered as 'artefacts of the culture process'. However, as the range of species exhibiting variation, and extent of variation arising, increased, the possibility that tissue culture could be a source of useful genetic variation—somaclonal variation—became widely investigated.

### 2.5.1 Genetic Nature of Somaclonal Variation

The analyses of somaclonal variation at the genetic, molecular and cytogenetic levels have revealed the extent of the variation. Mutants which arose could behave as classical Mendelian mutants, with changes in characters controlled by multigene families, and changes in polygenic and quantitative traits as well as the appearance of the activation of transposable elements. Many chromosomal abnormalities were also observed, covering the whole range of variants found to occur naturally (Scowcroft and Larkin, 1988).

The genetic analysis of plants regenerated from tissue cultures from many species including tobacco, tomato, celery, lettuce, rapeseed, wheat, rice and maize have demonstrated that classical point mutations have arisen during culture. One such example was that of an alcohol dehydrogenase variant in maize (Brettell *et al.*, 1986). The other types of variation are considered separately below.

### 2.5.2 Origin of Somaclonal Variation

What is the source of somaclonal variation? Were the variants isolated after culture preexisting, or were they generated during the culture process? Some of the variation arising in regenerated plants may reflect preexisting heterogeneity in the explant. This could occur from a heterogeneity of cell type in the explant material or from somatic changes which had taken place in the donor explant prior to being placed in culture, in spite of the explant being of uniform cell type. However, most of the variation would appear to occur during the culture process. This can be most obviously seen in the changes in the frequency of chromosomal abnormalities with time in culture. Meiotic analysis in maize showed that there were no chromosomal abnormalities in plants regenerated after three to four months in culture, while those regenerated after eight to nine months were cytologically abnormal (Lee and Phillips, 1987).

#### *2.5.2.1 Chromosomal abnormalities*

The most frequent cause of somaclonal variation is chromosomal rearrangement. The types of observable rearrangements are deletions, fusions and interchanges, as well as changes in ploidy (Larkin, 1987). The consequences of these types of chromosome abnormalities are frequently changes in the phenotypic expression of one or more genes. The genes involved may not actually be altered themselves, but by being placed in a new, and novel, chromosome environment they may be altered in the control of their expression. Precedents for such position effects occur in many systems, particularly *Drosophila melanogaster* (Hilliker and Appels, 1982) and mammals (for example X-chromosome inactivation).

One of the most dramatic examples of changes in chromosome size, number and DNA content is in *Scilla siberica* (Deumling and Clermont, 1989). When bulb tissue from this species (a triploid) was placed into culture massive changes in DNA content occurred, as well as changes in ploidy level and chromosome size. Different parts of the genome were differentially affected. Plants could only be regenerated from one type of callus, which had a massive reduction in DNA, but a disproportionate loss of satellite sequences. When these regenerated plants were grown for four years in normal conditions, the amount of satellite DNA was increased, but still not to the level of the starting material. The observation that all the regeneration occurred from this altered callus raises the question of DNA variation and regeneration. Is there a requirement for some form of chromatin diminution before the cells become competent to regenerate? Frequently this DNA change may be cryptic with regard to the phenotype. However, that part of the variation which does have phenotypic effects would be responsible for the high mutation rate seen in culture.

The cause of the high frequency of chromosomal abnormalities arising in culture has yet to be explained. One suggestion is that of chromatin diminution described above. Another theory is that chromosome breaks may be induced in culture by the late replication of heterochromatin (Johnson *et al.*, 1987; Less and Phillips, 1987). The replication of heterochromatin may extend into anaphase, which would result in the formation of chromosome bridges and subsequent chromosome breakage. The genetic consequences of a breakage–fusion–bridge cycle have been well documented, especially with regard to the activation of transposable elements. Support for this proposal comes from the apparent involvement of heterochromatin in many of the chromosome abnormalities observed. The breakpoints in maize regenerants were primarily on chromosome arms which contained large blocks of heterochromatin such as knobs (Lee and Phillips, 1987). Twelve of the 13 breakpoints involved in interchanges and deletions in wheat–rye hybrids were located in heterochromatin (Lapitan *et al.*, 1984).

The frequency of chromosomal abnormalities associated with heterochromatin raises the question of whether the structure of the chromosome region, *i.e.*, the fact that it is heterochromatic, or the nucleotide composition (most heterochromatic regions have been shown to contain tandemly arrayed repetitive sequences), is most important. Many plants do not have large blocks of recognizable heterochromatin, so would these be more stable in culture? This question is particularly relevant to forest trees, especially conifers. There are few, if any, heterochromatic blocks in these chromosomes, with few highly repetitive tandemly arrayed sequences. Since vegetative propagation and tissue culture systems offer real advantages in the bulking of superior selections, the stability of culture and vegetative propagation in conifers is an important consideration.

#### *2.5.2.2* ***Gene amplification and diminution***

The organization of the plant genome, with repetitive sequences being present as both tandem arrays and dispersed sequences, and low copy number sequences was described earlier. Changes can be found in this whole range of sequences during the process of culture and regeneration. These changes can be in both the number of copies of a sequence as well as in its state of modification, for example the state of methylation, which may affect its transcriptional activity.

The ribosomal RNA genes (rDNA) are an example of a tandemly repeated set of genes which is frequently altered in culture. Changes in this family have been demonstrated in flax (Cullis and Cleary, 1986b), triticale (Brettell *et al.*, 1986) and pea (Cullis and Creissen, 1987). In the case of both triticale and pea, the reduction in rDNA could be localized to particular nucleolar organizer regions. This apparent localization may indicate that all the portions of the genome are not equally accessible to the mechanisms by which this variation is generated. In flax, a whole range of repetitive sequence families, all organized as tandem arrays of a relatively short repetitive unit, were shown to vary through a cycle of tissue culture and regeneration. Unfortunately, flax has many blocks of heterochromatin which are the expected locations of these repetitive families, so the data are not useful in discriminating between the importance of chromosome structure and nucleotide sequence organization. However, the rDNA in flax does not lie in a distinct region of heterochromatin (Cullis and Creissen, 1987), but is still very variable.

Changes in the copy number of specific genes have also been selected in culture. Cells with an amplified number of genes have been selected using resistance to the herbicides phosphinotricin (Donn *et al.*, 1984) and glyphosate (Shah *et al.*, 1986; Ye *et al.*, 1987). These results are the exception rather than the rule as it has not been possible to demonstrate the amplification of many genes in cultured cells. Whether this is an absolute failure, or simply due to the application of inappropriate selection conditions is not known. However, it is clear that the ease with which different genes can be modified in culture is very variable. This observation is again consistent with the notion that a limited portion of the genome is accessible to somaclonal variation.

#### *2.5.2.3* ***Transposable elements***

One of the consequences of a chromosome breakage in maize crosses is the activation of transposable elements (McClintock, 1951). Since one of the proposed possible causes of somaclonal variation is the activation of transposable elements, studies to confirm this have been undertaken. Apparent activation of transposable elements have been found in studies with maize (Peschke *et al.*, 1987) and alfalfa (Groose and Bingham, 1986). The paucity of information about the involvement of transposition in the generation of variation through tissue culture leaves its general applicability in doubt. It is still not clear whether the activation of transposable elements is the major cause of somaclonal variation in systems other than maize and alfalfa, or even in maize and alflafa themselves.

#### 2.5.2.4 *Genetic control of variation*

Can the response of all plant materials to cycles of culture be generalized? It would appear that not all genotypes within a given species may respond to culture in the same way. The response of flax genotrophs to a cycle of culture and regeneration was very variable, both at the phenotypic level and at the level of genomic responses (Cullis and Cleary, 1986b). Thus some lines varied in a range of DNA families, while others appeared to be completely stable. The stable forms were also those which had the lowest DNA amount of the repetitive sequence families. Thus, if there is a relationship between heterochromatin and somaclonal variation then the genotype, in terms of blocks of repetitive DNA, will be important in determining the stability of the genome.

The occurrence of the repetitive blocks is not normally completely absent from a given line. Evidence has already been obtained that some parts of the plant genome are more labile than others. Thus the response to culture may be in a subset of the genome. The ability of the plant to recognize those subsets may also be genetically controlled. Thus the presence of a variable portion will be required in addition to the genes controlling the response to obtain extensive somaclonal variation. Conversely, the absence of either, or both, would be expected to reduce, or eliminate, somaclonal variation to a background mutational level. Once again, this may be most important in the forestry industry, where the numbers of individuals likely to be produced would still include mutants, even with a normal background level of mutation. An example of the scale of the forestry operation is that in the southern USA some 1.9 billion seedlings are planted each year, for just a single tree species.

## 2.6 GENERAL CONCLUSIONS

The plant genome is in a state of flux. The reorganization events which occur over a short timescale, as well as in an evolutionary framework, include amplifications, deletions and transpositions. The net result is a wide variety of DNA content and organization patterns in the higher plants. Increasing DNA content affects the ease with which plant cells can be manipulated. However, the characterization and manipulation of transposable elements offers real possibilities for generalized gene isolation. The increase in genome size, and particularly, the increase in heterochromatin, may be responsible for the destabilization of the genome in the process of culture and regeneration. However, a recognition of this effect may lead to the selection of certain types for stability, and others where variability is specifically desired. The concentration of changes in a subset of the genome would suggest that the range of variants arising from culture may only be a subset of those available by other mutational schemes, and this limitation must be recognized.

## 2.7 REFERENCES


Bachmann, K., O. B. Goin and C. J. Goin (1972). Nuclear DNA amounts in vertebrates. *Brookhaven Symp. Biol.*, **23**, 419–450.

Baker, B., J. Schell, H. Lortz and N. Federoff (1986). Transposition of the maize controlling element 'activator' in tobacco. *Proc. Natl. Acad. Sci. USA*, **83**, 4844–4848.

Bennett, M. D. (1972). Nuclear DNA content and minimum generation time in herbaceous plants. *Proc. R. Soc. London, Ser. B.* **181**, 109–135.

Bennett, M. D. (1976). DNA amount, latitude and crop plant distribution. In *Current Chromosome Research*, ed. K. Jones and P. E. Brandham, pp. 152–158. North Holland, Amsterdam.

Bennett, M. D., R. A. Finch and I. R. Barclay (1976). The time, rate and mechanism of chromosome elimination in Hordeum hybrids. *Chromosoma*, **54**, 175–200.

Bennett, M. D. and J. B. Smith (1976). Nuclear DNA amounts in angiosperms. *Philos. Trans. R. Soc. London, Ser. B.* **274**, 227–240.

Blackburn, E. H. (1985). Artificial chromosomes in yeast. *Trends Genetics*, **1**, 8–12.

Bonierbale, M. W., R. L. Plaisted and S. D. Tanksley (1988). RFLP maps based on a common set of clones reveal modes of evolution in potato and tomato. *Genetics*, **120**, 1095–1103.

Brettell, R. I. S., E. S. Dennis, W. R. Scowcroft and W. J. Peacock (1986). Molecular analysis of a somaclonal mutant of maize alcohol dehydrogenase. *Mol. Gen. Genet.*, **202**, 235–239.

Caplan, A., L. Herrera-Estrella, D. Inze, E. Van Haute and M. Van Montague (1983). Introduction of genetic material into plant cells. *Science* (*Washington, D.C.*), **222**, 815–821.

Cavalier-Smith, T. (1978). Nuclear volume control by nucleoskeletal DNA, selection for cell volume and cell growth rate, and the solution of the DNA C-value paradox, *J. Cell Sci.*, **34**, 247–278.

Cavallini, A., C. Zolfino, G. Cionini, R. Cremonini, L. Natali, O. Sassoli, and P. G. Cionini (1986). Nuclear DNA changes within *Helianthus annuus* L: cytophotometric, karyological and biochemical analyses. *Thero. Appl. Genet.*, **77**, 20 26.

Cavallini, A., C. Zolfino, L. Natali, G. Cionini and P. G. Cionini (1989). Nuclear DNA changes with *Helianthus annuus* L.: origin and control mechanism. *Theor. Appl. Genet.*, **77**, 12–16.

Chandler, V., C. Rivin and V. Walbot (1986). Stable non-Mutator stocks of maize have sequences homologous to the Mu 1 transposable element. *Genetics*, **114**, 1007–1021.
Chang, C., J. C. Bowman, A. W. DeJohn, E. S. Lander and E. M. Meyerowitz (1988). Restriction fragment length polymorphism linkage map for *Arabidopsis thaliana*. *Proc. Natl. Acad. Sci. USA*, **85**, 6856–6860.
Chang, C. and E. M. Meyerowitz (1988). Molecular cloning and DNA sequence of the *Arabidopsis thaliana* alcohol dehydrogenase gene. *Proc. Natl. Acad. Sci. USA*, **83**, 1208–1212.
Cullis, C. A. (1981). DNA sequence organization in the flax genome. *Biochim. Biophys. Acta*, **652**, 1–15.
Cullis, C. A. and W. Cleary (1986a). Rapidly varying DNA sequences in flax. *Can. J. Genet. Cytol.*, **28**, 252–259.
Cullis, C. A. and W. Cleary (1986b). DNA variation in flax tissue culture. *Can. J. Genet. Cytol.*, **28**, 247–251.
Cullis, C. A. and L. M Charlton (1981). The induction of ribosomal DNA changes in flax. *Plant. Sci. Lett.*, **20**, 213–217.
Cullis, C. A. and G. P. Creissen (1987). Genomic variation in plants, *Ann. Bot. (London)*, **60**, Suppl. 4, 103–113.
Cullis, C. A., G. P. Creissen, S. W. Gorman and R. D. Teasdale (1988). The 25S, 18S and 5S ribosomal RNA genes from *Pinus radiata D. Don.* In *Molecular Genetics of Forest Trees*, pp. 34–40, ed. W. M. Cheliak and A. C. Yapa, pp. 34–40. Petewawa National Forestry Institute, Canada.
Dallas, J. F. (1988). Detection of DNA fingerprints of cultivated rice with a human minisatellite DNA probe. *Proc. Natl. Acad. Sci. USA*, **85**, 6831–6835.
Dellaporta, S. L. and P. S. Chomet (1985). The activation of maize controlling elements. *Plant Gene Res.*, **2**, 169–216.
Deumling, B. and L. Clermont (1989). Changes in DNA content and chromosomal size during cell culture and plant regeneration of *Scilla siberica*: selective chromatin diminution in response to environmental conditions. *Chromosoma*, **97**, 439–448.
Donn, G., E. Tischer, J. A. Smith and H. M. Goodman (1984). Herbicide resistant alfalfa cells: an example of gene amplification in plants *J. Mol. Appl. Genet.*, **2**, 621–635.
Federoff, N. V., D. B. Furtek and O. E. Nelson, Jr. (1984). Cloning of the bronze locus in maize by a simple and generalizable procedure using the transposable controlling element Activator (Ac). *Proc. Natl. Acad. Sci. USA*, **81**, 3825–3829.
Flavell, R. B. (1986). The structure and control of expression of ribosomal RNA genes. *Oxford Surv. Plant Cell Mol. Biol.*, **3**, 251–274.
Groose, R. W. and E. T. Bingham (1986). An unstable anthocyanin mutation recovered from tissue culture of alfalfa (*Medicago sativa*). High frequency of reversion upon reculture. *Plant Cell Rep.*, **5**, 104–107.
Gustafson, J. P., A. J. Lukaszewski and M. D. Bennett (1983). Somatic deletion and redistribution of telomeric heterochromatin in the genus Secale and in *Triticale*. *Chromosoma*, **88**, 293–298.
Harrison, B. J. and J. R. S. Fincham (1964). Instability at the Pal locus in *Antirrhinum majus*. Effects of environment on frequencies of somatic and germinal mutation. *Heredity*, **19**, 237–258.
Helentjaris, T. (1987). A genetic linkage map for maize based on RFLPs. *Trends Genet.*, **3**, 217–221.
Helentjaris, T., G. King, M. Slocum, C. Siedenstrang and S. Wegman (1985). Restriction length polymorphisms as probes for plant diversity and their development as tools for applied plant breeding. *Plant Mol. Biol.*, **3**, 109–118.
Hilliker, A. J. and R. Appels (1982). Pleitropic effects associated with the deletion of heterochromatin surrounding rDNA on the X chromosome of *Drosophila*. *Chromosoma*, **86**, 469–490.
Jeffreys, A. J., V. Wilson and S. L. Thein (1985). Hypervariable 'minisatellite' regions in human DNA. *Nature (London)*, **314**, 67–73.
Johnson, S. S., R. L. Phillips and H.W. Rines (1987). Possible role of heterochromatin in chromosome breakage induced by tissue culture in oats (*Avena sativa* L.). *Genome*, **29**, 439–446.
Klee, H. J., M. B. Hayford and S. B. Rogers (1988). Gene rescue in plants: a model for 'shotgun' cloning by retransformation. *Mol. Gen. Genet.*, **210**, 282–287.
Lander, E. S. and D. Botstein (1989). Mapping mendelian factors underlying quantitative traits using RFLP linkage maps. *Genetics*, **121**, 185–199.
Landry, B. S., R. V. Kesseli, B. Farrara and R. W. Michelmore (1987). A genetic linkage map of lettuce (*Lactuca sativa* L.) with restriction fragment length polymorphism, isozyme, disease resistance and morphological markers. *Genetics*, **116**, 331–337.
Lapitan, N. L. V., R. G. Sears and B. S. Gill (1984). Translocations and other karyotypic structural changes in wheat–rye hybrids regenerated from tissue culture. *Theor. Appl. Genet.*, **68**, 547–554.
Larkin, P. J. (1987). Somaclonal variation: history, method and meaning. *Iowa State J. Res.*, **61**, 293–434.
Lee, M. and R. L. Phillips (1987). Genomic rearrangements in maize induced by tissue culture. *Genome*, **29**, 122–128.
Lillis, M. and M. Freeling (1986). Mu transposons in maize. *Trends Genet.*, **2**, 183–187.
Maniatis, T., E. F. Frisch and J. Sambrook (1982). *Molecular Cloning*. Cold Spring Harbor Laboratory, Cold Spring Harbor, NY.
McClintock, B. (1947). Cytogenetic studies of maize and *Neurospora*. *Carnegie Inst. Washington, Year Book*, **46**, 146–152.
McClintock, B. (1951). Chromosome organization and genetic expression. *Cold Spring Harbor Symp. Quant. Biol.*, **16**, 13–47.
McClintock, B. (1984). The significance of responses of the genome to challenge. *Science (Washington, D.C)*., **26**, 729–801.
Miksche, J. P. (1967). Variation in DNA content of several gymnosperms. *Can. J. Genet. Cytol.*, **9**, 717–722.
Miksche, J. P. (1968). Quantitative study of intraspecific variation of DNA per cell in *Picea glauca* and *Pinus banksiana*. *Can. J. Genet. Cytol.*, **10**, 590–600.
Miksche, J. P. (1985). Recent advances of biotechnology and forest trees. *For. Chron.*, **61**, 449–453.
Myerowitz, E. M. (1987). *Arabidopsis thaliana*. *Annu. Rev. Genet.*, **21**, 93–111.
Myerowitz, E. M. and R. E. Pruitt (1985). *Arabidopsis thaliana* and plant molecular genetics. *Science (Washington, D.C.)*, **229**, 1214–1218.
Patterson, A. H., E. S. Lander, J. D. Hewitt, S. Peterson, S. E. Lincoln and S. D. Tanksley (1988). Resolution of quantitative traits into Mendelian factors by using a complete RFLP linkage map. *Nature (London)*, **335**, 721–726.
Peschke, V. M., R. L. Phillips and B. G. Gegenbach (1987). Discovery of transposable element activity among progeny of tissue culture-derived maize plants. *Science (Washington, D.C.)*, **238**, 804–807.
Price, H. J. (1976). Evolution of DNA content in higher plants. *Bot. Rev.*, **42**, 27–52.
Price, H. J. (1988). Nuclear DNA content variation within angiosperm species. *Evol. Trends Plants*, **2**, 53–60.
Richards, E. J. and F. M. Ausubel (1988). Isolation of a higher eukaryotic telomere from *Arabidopsis thaliana*. *Cell*, **53**, 127–136.

Rogstad, S. H., J. C. Patton, II and B. A. Schaal (1988). M13 repeat probe detects DNA minisatellite-like sequences in gymnosperms and angiosperms. *Proc. Natl. Acad. Sci. USA*, **85**, 9176–9178.
Scowcroft, W. R. and P. J. Larkin (1988). Somaclonal variation. Applications of plant cell and tissue culture. In *Ciba Foundation Symposium 137*, pp. 21–35. Wiley, Chichester.
Shah, D. M., R. B. Horsch, H. J. Klee, G. M. Kishore, J. A. Winter, N. E. Tumer, C. M. Hironaka, P. R. Sanders, C. S. Gasser, S. Aykent, N. R. Siegel, S. G. Rogers and R. T. Fraley (1986). Engineering herbicide tolerance in transgenic plants. *Science (Washington D.C.)*, **233**, 478–481.
Sparrow, A. H., H. J. Price and A. G. Underbrink (1972). A survey of DNA content per cell and per chromosome of prokaryotic and eukaryotic organisms: some evolutionary considerations. *Brookhaven Symp. Biol.*, **23**, 451–494.
Tanksley, S. D., J. Miller, A. Paterson and R. Bernatzky (1988). Molecular mapping of plant chromosomes. In *Chromosome Structure and Function*, ed. J. P. Gustafson and R. Appels, pp. 157–173. Plenum Press, New York.
Tanksley, S. D., N. D. Young, A. H. Paterson and M. W. Bonierbale (1989). RFLP mapping in plant breeding: new tools for an old science. *Biotechnology*, **7**, 257–265.
Van Sluys, M. A., J. Tempe and N. Federoff (1987). Studies on the introduction and mobility of the maize activator element in *Arabidopsis thaliana* and *Daucus carota*. *EMBO J.*, **6**, 3881–3889.
Ye, J., R. M. Hauptman, A. G. Smith and J. M. Widholm (1987). Selection of a *Nicotiana plumbaginifolia* universal hybridizer and its use in intergenic somatic hybrid formation. *Mol. Gen. Genet.*, **208**, 474–480.
Young, N. D., D. Zamir, M. W. Ganal and S. D. Tanksley (1988). Use of isogenic lines and simultaneous probing to identify DNA markers tightly linked to the Tm-2a gene in tomato. *Genetics*, **120**, 579–585.
Zimmerman, P. A., N. Lang-Unnasch and C. A. Cullis (1989). Polymorphic regions in plant genomes detected by an M13 probe. *Genome*, **32**, 824–828.

# 3
# Cell Selection

WILLIE H.-T. LOH
*Intermountain Canola Company, Cinnaminson, NJ, USA*



## 3.1 INTRODUCTION

Microbiologists have combined mutagenesis and cell selection to develop mutant bacterial strains for biochemical, physiological and genetic studies. Later researchers have used cell selection to identify mutants for a range of commercial applications, *i.e.* producing valuable pharmaceuticals, degrading toxic compounds, increasing agricultural productivity, *etc.*

As the field of plant tissue culture developed, similar approaches were envisioned using plant cell selection. Through callus and cell suspension cultures, large numbers of cells can be routinely manipulated. Mutations can be induced by the application of chemical and/or physical mutagens, although cultured cells also exhibit somaclonal variation, a form of *in vitro* mutagenesis (Evans *et al.*, 1984; Larkin and Scowcroft, 1981). Since mutant cells can be selected and regenerated into plants, cell selection in tissue culture should give rise to commercially useful mutant crop plants.

The objective of this report is to evaluate the utility of this approach as a method of producing commercially useful mutants. A number of excellent reviews have been published recently on plant cell selection (Bourgin, 1986; Chaleff, 1983a,b; 1986; Chen and Gusta, 1986; Dix, 1986; Duncan and Widholm, 1986; Flick, 1983; Flick and Evans, 1984; Hughes, 1983; Jacobs *et al.*, 1987; Rains *et al.*, 1986; Sacristán, 1986; Tal, 1983; Yamada and Sato, 1983). These range from comprehensive summaries of all reports on cell selection to critical reviews of specific applications of this strategy, *i.e.* herbicide resistance, salt tolerance, disease tolerance, *etc.*

This review will assess the claims of various cell selection systems based on the recovery of mutant plants. Particular effort has been made to evaluate the commercial impact of these plants. The application of cell selection for biochemical and genetic analysis, *i.e.* resistance to 5-bromodeoxyuridine, methotrexate, hydroxyurea, amino acid auxotrophy and hormone habituated lines, are not reviewed. Similarly, the isolation of mutant donor lines for protoplast fusion is beyond the scope of this discussion.

## 3.2 SELECTION SYSTEMS

A typical cell selection scheme is illustrated in Figure 1. Callus cultures are induced from a variety of explant sources. Cell suspensions may be proliferated from callus explants, although this step is not essential for cell selection. With the exception of a few systems, *i.e.* carrot, caraway, *etc.*, plant recovery from single cell suspension is slower and less frequent. The range of genetic variability may be enhanced by exposure to chemical and/or physical mutagens. Genetic variability may also be enhanced by increasing culture time and/or cytokinin content of the culture medium; both factors have been shown to increase the level of somaclonal variation. Selection pressure is imposed after the mutagen is withdrawn. The cultured cells are subsequently transferred to a medium appropriate for plant regeneration. Selection may extend to after plant regeneration.

A number of alternative cell selection systems have been proposed and tested. Many researchers have selected at the protoplast level to ensure that mutants are derived from single cells, thereby reducing the possibility of recovering chimeric plants. This approach is restricted to those systems, commonly involving solanaceous plants, for which protoplast isolation is routine, and cell wall synthesis and plant regeneration can be induced at relatively high frequencies. Otherwise, the strategy differs little from the cell selection protocol described above.

Researchers have also advanced the concept of cell selection to haploid systems, notably *via* anther culture (Hoffmann *et al.*, 1982). In general, mass selection has been restricted to microspore culture systems where large populations can be generated. The genome of the haploid plants are subsequently doubled by colchicine treatment to create diploids with stable, heritable mutations. *Brassica* is the only genus to date in which microspore culture has been applied for cell selection.

Some very innovative researchers have proposed applying cell selection to pollen grains. Various papers have demonstrated high correlations between gametophytic and sporophytic generations in vigor (Ottaviano *et al.*, 1980; Windsor *et al.*, 1987), frost hardiness (Zamir *et al.*, 1982), salt tolerance (Sacher *et al.*, 1983), herbicide tolerance (Smith, 1985), heavy metal tolerance (Searcy and Mulcahy, 1985) and seed oil compostition (Evans *et al.*, 1988). A similar approach has been applied to select for herbicide tolerance in bryophytes, which have a distinct and complete gametophytic

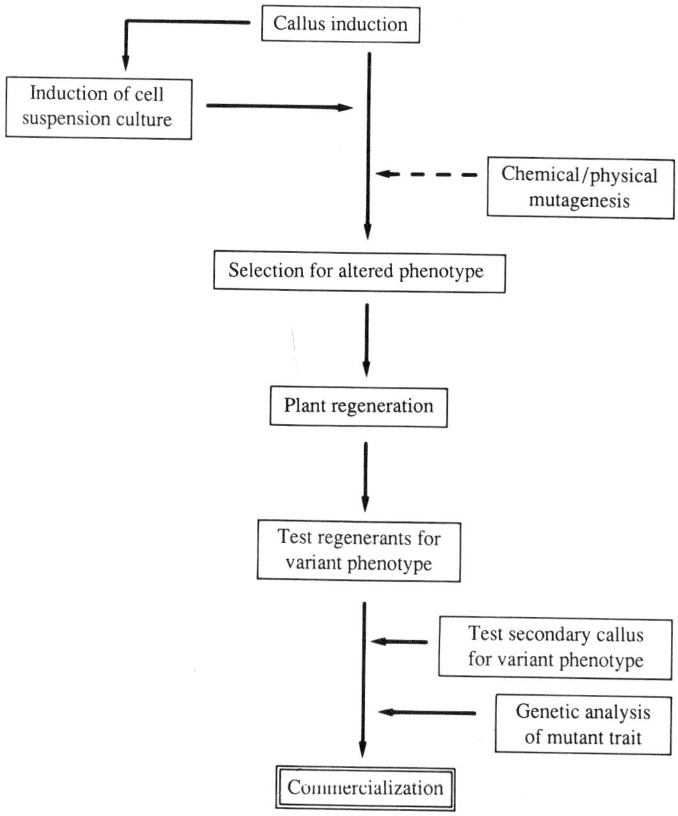


**Figure 1** Experimental scheme for developing useful mutants *via* cell selection

life cycle (Hickok and Schwarz, 1986). Physical stresses imposed on pollen, however, have surprisingly complex effects on the surviving progeny (Mulinix and Iezzoni, 1988).

## 3.3 THEORETICAL LIMITATIONS

Unlike microbes, plant cells normally function within an organized tissue system. In culture, these cells proliferate in an unorganized manner. Many plant metabolic functions, *e.g.* photosynthesis, are not present in cultured cells. Thus, single cell manipulations are limited to those traits which are expressed both in culture and in the intact plant. Traits such as plant habit and pest resistance are not approachable *via* cell selection.

Conversely, many traits expressed *in vitro, e.g.* auxotrophy or cytokinin habituation, are not expressed in the whole plant. Some traits are not expressed in the organized plant under strict developmental and/or temporal regulation. Thus, a mutant cell line may not express the selected trait when regenerated. The literature describes these as epigenetic changes, which are stable through mitosis but reversible through differentiation.

Further confounding the cell selection approach is the occurrence of somaclonal variation. In the absence of selection pressure and mutagenesis, regenerant plants exhibit a significant frequency of genetically stable mutations (Evans and Sharp, 1983). The culture process itself has been shown to induce mutations. In addition, somatic mutations in the donor explant may be rescued through tissue culture. Thus, genetic changes induced during cell selection may be masked by somaclonal variation.

Unlike microbial cultures, plant cell cultures represent highly variable populations of cells undergoing unorganized growth. If mutagenesis is imposed, the mutation effect is likely to be highly variable given the range of physiological states within the cell population. Quite often, only a subpopulation of cells express the trait to be selected against. Similarly, only a subpopulation of the selected cells are capable of cell division. Yet, only those mutations from within this population will be proliferated. Finally, only a subpopulation of these cells are capable of regeneration. Thus, not all mutations can be rescued. Examples of some which have been rescued are described in the following sections.

## 3.4 HERBICIDE TOLERANCE

Due to its economic importance and relatively simple genetics, herbicide resistance has been an extremely attractive target for biotechnology researchers. Significant resources have been devoted by both public and private institutions, utilizing cell selection and other approaches, to the development of plants resistant to specific herbicide classes.

Chaleff and Parsons (1978) were the first to report the recovery of herbicide tolerant plants *via* cell selection. Resistant tobacco lines were recovered from cell suspensions plated on a medium containing 500 $\mu$M picloram. Genetic analysis of progeny from the resistant plants identified both dominant and semidominant mutations for picloram tolerance (Chaleff, 1980, 1981). Genetic analysis defined four mutations into three distinct linkage groups.

Miller and Hughes (1980) selected for paraquat resistance by culturing tobacco lines on selective media. Herbicide tolerance was transmitted in $<2\%$ of the progency obtained by self-fertilization (Hughes, 1983). Thomas and Pratt (1982) isolated paraquat tolerant cell lines of a *Lycopersicon esculentum* × *L. pennellii* hybrid. The progeny of regenerated plants segregated for both herbicide tolerance and sensitivity. Paraquat tolerant mutants were also isolated by sowing haploid gametophytes of the fern *Ceratopteris richardii* on selective media (Hickok and Schwarz, 1986). Two mutants exhibited single nuclear genes which confer increased tolerance to the herbicide. However, these selections were conducted on 0.5 $\mu$M paraquat, compared to the 10–150 $\mu$M levels applied to tobacco and tomato.

Amitrole tolerant tobacco cells were isolated from suspension cultures plated on a selective medium (Singer and McDaniel, 1984). However, none of the lines were stably tolerant to amitrole in the absence of selective pressure. The selfed progeny of regenerated plants segregated for both herbicide tolerance and sensitivity, but at non-Mendelian ratios. The frequency of herbicide tolerant plants dropped in subsequent generations.

Chaleff and colleagues successfully applied cell selection to the development of sulfonylurea resistant tobacco (Chaleff and Ray, 1984). Two semidominant alleles confer tolerance to chlorsulfuron and sulfometuron methyl. An altered acetolactate synthase, the target of sulfonylurea herbicides, resulted in the resistant phenotype. Reselection on higher levels of sulfometuron methyl produced a second mutation which enhanced the level of herbicide resistance (Creason and Chaleff, 1988). Genetic analysis showed that the second mutation was also a single, semidominant, nuclear trait.

Polsoni *et al.* (1988) first proposed the use of microspore selection to develop herbicide tolerant rape, but no data were presented. Swanson *et al.* (1988) used this approach to select for sulfonylurea resistance in *Brassica napus*. Microspores and derived embryos were cultured on a medium containing the herbicide. Similarly, haploid protoplasts and derived calli were cultured on selective media. Plants regenerated from both microspore-derived embryos and protoplast-derived calli were doubled with colchicine and self-pollinated to produce seed for testing and genetic analysis. Plants recovered from both selection systems were 10–80 times more tolerant than the donor genotype to sulfonylurea application. Cell lines derived from these plants were 10–100 times more tolerant. Genetic analysis suggests that the mutation is inherited as a single nuclear, semidominant trait.

Swanson *et al.* (1989) also cultured mutagenized microspores of *B. napus* in media containing an imidazolinone herbicide. Like sulfonylurea herbicides, imidazolinones also target acetolactate synthase. Following embryogenesis, haploid plants were regenerated, doubled with colchicine and self-pollinated. Five lines produced selfed progeny with higher tolerance to the herbicide 'Pursuit'. Genetic analysis indicates that resistance was encoded by at least two nuclear, semidominant and complementary genes. The mutant acetolactate synthase was 500 times more tolerant to the imidazolinone herbicide. Imidazolinone resistant cell lines have also been developed in maize *via* cell selection (Shanner *et al.*, 1985).

Rain and Carlson (1978) reported a novel method for *in situ* cell selection on differentiated tissue. Tobacco plants were sprayed with one of two herbicides, bentazone and phenmedipham. Green islands of cells, which were visible in the otherwise necrotic tissue, were rescued. Eight bentazon resistant and two phenmedipham resistant clones were cultured and regenerated into plants. Both types of herbicide resistance were shown to be inherited as recessive Mendelian traits. This approach was used because photosynthetic herbicides have no effects on the callus.

Similarly, Cséplö *et al.* (1985) selected for resistance to a photosynthetic herbicide by imposing selection pressure on a photomixotrophic tobacco cell culture. Mesophyll protoplasts were isolated from *Nicotiana plumbaginifolia*, and induced to undergo cell division. Photomixotrophic growth was achieved by culturing the callus on a low sugar (0.3% sucrose) medium under high light intensity (1500 lux). Terbutryn, a triazine herbicide, was incorporated into the medium. Resistant calli were isolated as green colonies in an otherwise bleached cell population. Two resistant plants were regenerated. Hill reaction measurements of isolated chloroplasts and genetic analysis of the progeny localized the mutation to the chloroplast genome.

## 3.5 SALT TOLERANCE

High salt levels in the soil severely restrict agricultural production. Irrigation also leads to increased soil salinity. One of the first applications of cell selection to crop improvement was to select for salt tolerance by screening cell lines in media containing high salt concentrations. The development of salt tolerant crops would increase the range of agricultural production and permit the use of brackish for irrigation.

Nabors *et al.* (1980) selected for NaCl resistance in *N. tabacum* cell lines. Regenerated plants were then watered with a saline solution. NaCl resistance persisted in the intact plants for two generations beyond regeneration. However, plants regenerated on a low salt medium are less salt tolerant and $F_2$ plants are more salt tolerant if the parent $F_1$ plant was exposed to salt. The authors suggest that increased tolerance may have resulted from physiological adaptation, rather than a genetic change. Similar observations were made by Hasegawa *et al.* (1980), who selected for salt tolerance in *N. tabacum* suspension cultures. Although resistant lines were selected, the resistance was lost in the absence of selective pressure.

Nabors *et al.* (1982) also selected for salt resistance in oats by inducing callus proliferation from germinating roots directly on media containing NaCl. Plants were regenerated from surviving callus, but no tests for salt tolerance were reported.

McHughen and Swartz (1984) selected for salt tolerance in flax (*Linum usitatissimum*) by challenging callus with high levels of $Na^+$, $Ca^{2+}$, $Mg^{2+}$, $K^+$, $Cl^-$ and $SO_4^{2-}$. This salt mixture mirrors the ion profile of saline soils. Surviving pieces of green calli were rescued by transfer to a nonselective medium, from which plants were regenerated. Progeny from one of the two rescued plants was more salt tolerant, based on increased plant height when grown on saline soils. The authors reported that the trait was stably inherited, although no genetic data were presented.

McCoy (1987) selected for salt resistance in five alfalfa lines by culturing diploid callus on 0.5% NaCl, then on 1.0% NaCl. Stably resistant cell lines were regenerated and the plants were tested. All the regenerants were aneuploid and exhibited abnormal phenotypes. In contrast, plants regenerated from control cultures of the same age were normal or nearly normal. Those plants regenerated from the resistant cell lines, however, exhibited no increased salt tolerance in hydroponic testing.

The increase in variation of cultured cells under salt stress was also noted by Kononowicz *et al.* (1990a), who cultured *N. tabacum* on increasing concentrations of NaCl to select for salt tolerant cell lines. The surviving cell lines accumulated cells with a hexaploid chromosome number. However, the hexaploid cells are no more salt tolerant then their tetraploid progenitor. All plants regenerated from the selected cell line were hexaploid, while plants selected from unselected cell lines remained tetraploid (Kononowicz *et al.*, 1990b). No information was provided on salt tolerance in the intact plants.

Ye *et al.* (1987) enriched for salt tolerant progeny in a cross between salt sensitive and salt tolerant barley cultivars by culturing anthers of the $F_1$ hybrid in media containing sodium sulfate. Regenerated plants were doubled and self-pollinated to produce seed for testing. Seeds were germinated in varying salt solutions to test for tolerance. The results suggest that the screening procedure effectively eliminated salt sensitive microspores. Whether salt tolerance during germination is related to field tolerance is unknown.

Ben-Hayyim and Goffer (1989) screened cell lines of Shamouti orange for imposing NaCl stress during embryogenesis. Unlike the control treatment, embryos exposed to salt stress required kinetin prior to rooting on naphthalene acetic acid. The salt treatment also arrested further shoot development. Despite the recovery of normal-looking embryos, no further characterization was possible.

Rahman and Kaul (1989) screened tomato calli on media containing increasing concentrations of NaCl to select for salt tolerance. Shoot and root differentiation were initiated for the surviving callus, but no data were presented on salt tolerance from regenerated plants.

Freytag *et al.* (1990) induced shoot organogenesis from sugar beet petiole explants cultured directly on media containing multiple salts. Regenerated plants were stressed by watering with a saline solution. Surviving plants were vernalized and the $R_1$ generation was tested for salt tolerance by germination in a saline soil. Four lines maintained salt tolerance to the $R_1$ generation. However, no progeny data were presented for genetic analysis.

## 3.6 HEAVY METAL TOLERANCE

Cell selection for heavy metal resistance has been tested as a method of producing plants which can grow in soils contaminated with heavy metals, including aluminum, cadmium, copper, mercury and zinc. In mammalian cell cultures, heavy metal resistance is induced by accumulation of metallothionein, which sequesters metal ions away from the metabolic process. Theoretically, it may be possible to develop plants with increased metallothionein production that can grow in contaminated soils. The range of crop plants may be increased, if these plants do not accumulate heavy metals in their edible parts.

Bennetzen and Adams (1984) selected for cadmium resistance by stepwise exposure of a tomato (*L. peruvianum*) cell suspension. After each exposure, surviving cells were transferred to solid media to permit resistant lines to proliferate in the absence of cell lysate. Resistance to cadmium ions was increased from 300 $\mu$M in unselected cultures to 1500 $\mu$M in selected cultures. The cadmium resistant cell lines also exhibited tenfold greater resistance to copper ions. Enzyme and protein analyses suggest that resistance is related to increased metallothionein production. However, no plants were regenerated.

Jackson *et al.* (1984) selected for cadmium resistance by plating nonmutagenized *Datura innoxia* suspension cultures on selective media. Surviving colonies were used to reinitiate suspension cultures, which were plated on media containing higher cadmium levels. Selected cell lines grew in 250 $\mu$M of cadmium, 20-fold higher than unselected lines, and maintained this level of resistance in the absence of selective pressure for over 400 generations. No plants were regenerated in this study either.

Connor and Meredith (1985) selected for aluminum resistant plants by screening diploid *N. plumbaginifolia* cell cultures. A large number of variants were recovered, half of which maintained aluminum resistance after passage through nonselective media. All 40 aluminum resistant plants were tested for trait stability, and transmitted resistance as a single nuclear, dominant mutation.

## 3.7 DISEASE RESISTANCE

Cell selection was first applied to the development of disease resistant plants by Carlson (1973). The general approach has been to screen cell cultures with the phytotoxin produced by the bacterial or fungal pathogen. Spent culture filtrate, which may contain minor components associated with infection, have also been used as selective agents. The approach is only applicable to toxin-associated diseases.

Carlson selected for resistance to *Pseudomonas syringae* cv. *tabaci*, the causative agent of wildfire disease, in tobacco. Mutagenized haploid cells and protoplasts of tobacco were screened with methionine sulfoximine, an analog of the toxin produced by the bacterial pathogen. Plants regenerated from selection did not produce the characteristic leaf chlorosis induced by inoculation of sensitive plants with either the pathogen or methionine sulfoximine. However, small necrotic spots appeared on the resistant plants, resembling infection by a nonvirulent *Pseudomonas* which lacks the toxin.

Gengenbach *et al.* (1977) screened embryo-derived calli from corn varieties carrying the Texas male sterile cytoplasm to pathotoxin from *Helminthosporium maydis* race T. Maize genotypes with normal cytoplasms are resistant to *H. maydis*, while those carrying the cms-T cytoplasms are susceptible. Using a sublethal enrichment procedure, cell lines were screened five times with increasing selection pressure. Plants regenerated from cultures resistant to lethal levels of the toxin were also resistant to the pathotoxin, but were male fertile. Genetic analysis of the progeny showed that both the toxin resistance and male fertility traits were maternally inherited.

Brettell and Thomas (1980) also selected for resistance to the T-toxin in a maize callus culture. Plants regenerated from the resistant cultures retained resistance when T-toxin was applied to rapidly growing leaf tissue. Fertile and toxin resistant plants were also regenerated from unselected cultures. One regenerant, which was fertile and toxin sensitive, may have broken the linkage between toxin sensitivity and male sterility.

Rines and Luke (1985) produced resistance in oats to *H. victoriae* by selecting for pathotoxin insensitivity at the callus stage. Calli homozygous for the dominant sensitive allele *Vb*, which encodes for resistance to several races of crown rust, failed to survive the toxin treatment. Plants were recovered from calli heterozygous for this locus. Nine of 16 regenerated plants maintained toxin insensitivity, as did all the derived progeny. However, all pathotoxin resistant plants that were recovered were sensitive to crown rust.

Ling *et al.* (1985) screened embryo-derived calli of rice with phytotoxin from *H. oryzae* to select for plants with fungal resistance. After toxin treatment, the callus was transferred to a toxin-free medium for subculture and regeneration. Three of 98 regenerated plants were found to be resistant by leaf bioassay. Segregation in the selfed progeny suggests that resistance to brown spot disease in rice results from a dominant mutation.

Hartman *et al.* (1984) selected for resistance to *Fusarium oxysporum* f. sp. *medicaginis* in alfalfa by culturing calli on media containing the fungal filtrate. Surviving sectors were subcultured repeatedly on media containing the toxin and induced to undergo organogenesis. Resistance was maintained in the absence of selective pressure. Regenerated plants, all of which had higher ploidly, were resistant to the pathogen *in vivo*, and produced calli which were resistant to the fungal filtrate *in vitro*. No genetic analysis was reported.

Arcioni *et al.* (1987) also selected for resistance to the same pathogen by screening diploid alfalfa callus on a medium containing fungal filtrate. Plants regenerated from the resistant calli were recultured. Calli derived from the resistant plants showed no growth inhibition on the medium containing the filtrate. Leaf discs from regenerated plants exhibited altered membrane permeability, with reduced electrolyte loss, when challenged with the filtrate. The intact plants also expressed increased resistance when inoculated with the pathogen. However, no genetic crosses were made to identify the nature of the mutation.

Thanutong *et al.* (1983) screened protoplast-derived calli of *N. tabacum* cv. Samsun with the wildfire toxin, and also with the toxin from *Alternaria alternata* pathotype tobacco, which causes brown spot. The toxins were added to the solidified culture medium. Leaves from the regenerated plants were tested for resistance to wildfire by inoculation with the pathogen, and for resistance to brown spot by spray application of a spore suspension of the pathogen. Because of a high escape rate with the selection system, two successive selection cycles were imposed on the calli. Plants resistant to both toxins were recovered. Analysis of the selfed progeny produced complex segregation patterns for resistance to each pathogen.

Matern *et al.* (1978) screened Russet Burbank potato plants, regenerated from mesophyll protoplasts, with fungal filtrate to select for resistance to *A. solani*. A semipurified toxin preparation was shown to produce fungal lesions typical of early blight in leaf tissue. The size of the lesion produced was proportional to the amount of toxin applied and toxin sensitivity was strongly correlated with susceptibility to *A. solani*. Tubers from resistant plants were planted and leaf tissue from the second vegetative generation was retested. Both sensitive and resistant plants maintained their phenotypic reaction to the fungal preparation in the second generation. However, no evidence exists for sexual transmission of the trait.

Behnke (1979) cultured haploid potato calli on media containing filtrate of *Phytophthora infestans* to select for fungal resistance. Surviving sectors were selected and shown to maintain resistance in the absence of the toxin. Calli induced from the selected plants also demonstrated increased tolerance to the fungal toxin. Inoculation with the fungal pathogen showed that *P. infestans* grew much more slowly on leaves of resistant plants (Behnke, 1980).

Sacristán (1982) selected for resistance to *Phoma lingam* (blackleg disease) in *B. napus* using mutagenized calli, cell suspension and stem embryo cultures derived from somatic tissue of vegetatively propagated haploid plants. Calli and stem embryos were cultured in a liquid medium containing pycnidiospores, then plated on a solid medium, or transferred directly to agar containing a layer with spores. Cultures which were not covered by the developing mycelium were selected for regeneration. Alternatively, the toxic fungal filtrate was incorporated into the culture medium. Cultures which survive two passages on the toxic medium were selected. Regenerated plants and progeny were tested by two different leaf bioassays for resistance to fungal infection. Both resistant and tolerant lines were recovered, although the resistant plants were all male sterile. Preliminary progeny tests of the tolerant lines showed some segregation for improved tolerance.

Newsholme *et al.* (1989) screened secondary embroys *B. napus* ssp. *oleifera* with both culture filtrate from *Phoma lingam* and with sirodesmin PL, the blackleg toxin. Seed germination and embryo development under selective pressure was correlated with known resistance of the genotypes to the fungal pathogen. The proportion of resistant embryos increased with successive rounds of selection. However, regenerated plants selected for resistance to either sirodesmin PL or the culture filtrate were no more resistant than plants regenerated from control cultures. Similar results had also been obtained with selection by partially purified filtrate from *A. brassicicola* (MacDonald and Ingram, 1986).

Sjödin and Glimelius (1989) also screened protoplasts, cell aggregates, leaves and roots of *Brassica* accessions with sirodesmin PL. Protoplasts in culture were unable to tolerate any concentration of sirodesmin PL. However, the development of cell clusters was significantly correlated to the development of lesions on leaves and to field resistant to *P. lingam*. The authors concluded that it was possible to select for *Brassica* plants resistant to blackleg *in vitro* using sirodesmin PL.

## 3.8 AMINO ACID OVERPRODUCTION

Widholm (1972) had shown that cell selection may be used to select for resistance to amino acid analogs. Tobacco suspension cultures developed resistance to 5-methyltryptophan, a toxic analog of tryptophan, by an alteration in anthranilate synthase, the biosynthetic control enzyme for tryptophan. The enzyme became less sensitive to feedback inhibition by the end product, thereby overproducing tryptophan. A similar approach has been applied to alter the amino acid composition of edible plant parts for improved nutritional balance.

Hibberd *et al.* (1980) selected corn calli resistant to growth inhibitory levels of lysine and threonine. Calli proliferated from immature embryos were exposed to stepwise increases of the two amino acids for six cycles. One resistant cell line had increased levels of the following free amino acids: threonine ( × 8.8), free methionine ( × 3.8), lysine ( × 2) and isoleucine ( × 2). Enzyme analysis showed that the activity of aspartokinase was less sensitive to feedback inhibition from lysine. Resistance was maintained for 12 months in the absence of selective pressure. Although plants were regenerated, no seeds were produced for amino acid analysis or progeny testing.

Hibberd and Green (1982) cultured an embryo-derived callus culture of maize, which had been mutagenized with sodium azide, in the presence of inhibitory concentrations of lysine and threonine. A stable resistant line was selected and plants were regenerated. Resistance to these two amino acids was inherited as a single dominant mutation. Tissue cultures derived from resistant plants exhibited reduced sensitivity to high levels of lysine and threonine. Free threonine in kernels of regenerated plants increased 75–100 times, increasing the total threonine content by 33–59%. Other than the amino acid profile, regenerated plants had normal phenotypes.

Negrutiu *et al.* (1984) screened *N. sylvestris* protoplasts with *S*-(2-aminoethyl)-L-cysteine (AEC) to select for cell lines which overproduce lysine. Plants regenerated from resistant lines maintained AEC resistance when they were explanted to proliferate calli. AEC resistant lines contained significant increases in free lysine content in both calli (15-fold) and intact plants (28-fold), and a significant decrease in arginine content (2.5-fold). Dihydrodipicolinate synthase, one of two enzymes which regulate lysine synthesis, was more feedback insensitive in the resistant plants. Progeny analysis showed that the mutation was inherited as a single, dominant nuclear gene.

The same approach was used by Hibberd *et al.* (1986) to increase the level of tryptophan in maize. Calli proliferated from immature embryos were exposed to 5-methyltryptophan over several subcultures with increasing selective pressure. Plants regenerated from surviving calli were self- or cross-pollinated. Genetic analysis of the progeny indicates that resistance to 5-methyltryptophan is inherited as a single dominant trait. Free tryptophan levels were increased from 0.01 mg $g^{-1}$ dry seed weight to 2.2 mg $g^{-1}$ in seeds homozygous for the resistant trait.

Reisch *et al.* (1981) screened a diploid alfalfa (*Medicago sativa* L.) suspension culture with ethionine to select for methionine overproduction. Less than 15% of the selected cell lines maintained resistance to ethionine in the absence of selective pressure. Some ethionine resistant lines also expressed improved tolerance to inhibitory concentrations of lysine and threonine. One selected cell line had a 10-fold increase in soluble methionine. However, no data were presented on methionine content in intact plants regenerated from the resistant cell lines.

## 3.9 MISCELLANEOUS

Huitema *et al.* (1989) selected for low temperature tolerant (LTT) mutants of chrysanthemum by screening for plant regeneration at 6 °C from a suspension culture following X-irradiation. Regenerants were also recovered at 24 °C. Low temperature tolerance was measured by the ability to produce flowers at 12 °C. At the higher temperature, LTT mutants were selected from both irradiated and nonirradiated cultures. At the lower temperature, LTT mutants appeared at twice the frequency (3.6%) among the regenerants. However, no genetic analysis was presented for the LTT trait.

## 3.10 CONCLUSIONS

An extensive review of the literature suggests that the application of cell selection to crop improvement has had mixed success. Mutants have been selected in all the major categories that were reviewed. Both nuclear and cytoplasmic mutations have been recovered. However, the testing procedures employed to identify and document selected mutations often were not rigorous.

Many workers tested regenerants ($R_0$) selected from cell culture directly. However, $R_0$ plants often have aberrant physiology due to the effects of tissue culture. Testing for the selected traits will not be effective until the $R_1$ generation. For traits which are subject to maternal influence, screening should be delayed to $R_2$ or later generations.

All of the traits which were selected for contribute to yield. Thus, the mutations which were recovered should be evaluated under field conditions. For example, salt tolerant lines should be evaluated in direct comparison with the donor material in replicated field trials under saline conditions. However, field data were rarely included in these reports. Without such data, the efficacy of the approach is difficult to determine.

Many plants that were recovered exhibit additional genetic changes as a result of somaclonal variation. These changes may indirectly affect the expression of the selected trait. In such cases, comparisons to the donor material are difficult. The mutant trait must be backcrossed to the original donor to develop near-isogenic lines before conducting field trials.

Cell selection has produced herbicide resistant plants in a number of instances. The literature reports on resistance to picloram, paraquat, bentazon, phenmedipham, triazine, sulfonylurea and imidazolinone herbicides. In most cases, the mechanism of resistance has been characterized and mutant genes have been cloned. Despite the availability of molecular approaches, cell selection has demonstrated its efficacy in rapidly producing herbicide resistant plants.

The efficacy of cell selection for salt tolerance remains unproven. Despite extensive work, very few reports demonstrate the development of stable and heritable salt resistance from cell selection. McHughen and Swartz (1984) identified one flax line, and Freytag *et al.* (1990) identified four sugar beet lines, with increased salt tolerance. In saline soils, the selected flax line grew taller, while the selected sugar beet lines had increased germination. No genetic data were provided in either report.

A demonstration of improved salt tolerance would require yield comparisons of the donor and selected lines under salt stress conditions.

Cell selection has been successfully used to develop cell lines which accumulate metallothioneins. However, no plants have been regenerated in these studies. Therefore, it is not possible to determine whether plants which overproduce this enzyme would survive on soils contaminated with heavy metals. This approach has been surpassed by molecular methods to clone the metallothionein gene and plant transformation.

Connor and Meredith (1985) were able to develop *N. plumbaginifolia* plants with increased aluminum tolerance. Although no yield comparisons were made between selected plants and the aluminum sensitive donor line, their data demonstrated that this trait was stably maintained and transmitted to subsequent generations.

Cell selection has been applied to select for resistance to a wide array of bacterial and fungal pathogens, including *Helminthosporium maydis, H. victoriae, H. oryzae, Pseudomonas syringae, Alternaria alternata, A. solani, Fusarium oxysporum* f. sp. *medicaginis, Phytophthora infestans* and *Phoma lingam*, in many different plant species. The results have been mixed and reflect the inherent complexity of pathogenesis.

Carlson (1973) and Thanutong *et al.* (1983) developed tobacco plants resistant to the *P. syringae* toxin through cell selection. Potato lines resistant to the *P. infestans* toxin were reported by Behnke (1979, 1980). All the selected plants were susceptible to fungal infection, although pathogenesis was significantly reduced in severity. These studies demonstrate that the toxin was only a codeterminant of virulence. In both cases, the pathogen was able to invade toxin resistant plants.

Gengenbach *et al.* (1977) and Brettell and Thomas (1980) selected for corn plants resistant to *H. maydis* toxin, and Rines and Luke (1985) selected for oats resistant to *H. victoriae.* In each instance, the mutation was linked to other, less desirable, changes. In maize, resistance to the *H. maydis* toxin was linked to male fertility, rendering the selected line useless for hybrid seed production. In oats, resistance to *H. victoriae* toxin was linked to sensitivity to crown rust. Thus, none of these lines were suitable for commercialization.

Ling *et al.* (1985) showed that *Fusarium* resistance in rice cell lines selected against the toxin was stably inherited as a dominant mutation. Thanutong *et al.* (1983) also conducted genetic analysis for resistance to the wildfire and *A. alternata* toxins. Sacristán (1982) reported that *B. napus* lines selected for resistance to *P. lingam* showed segregation for improved tolerance. Transmission of these disease resistance traits supports the efficacy of the cell selection approach. However, disease resistance is a difficult trait to test in the greenhouse. For most plant diseases, the correlation between greenhouse and field performance is low. Yet, none of the selected lines have been tested under field conditions in disease nurseries.

In contrast to the confusion which surrounds disease resistance, cell selection has been used very effectively in selecting for amino acid overproduction, particularly for threonine and tryptophan. The only drawback to this approach is its limitation to free amino acids. Although Hibberd and Green (1982) increased free threonine 75–100 times, total threonine content was increased only 33–59%. Seed proteins represent a significant source of amino acids. Molecular approaches which can alter the amino acid profile of seed storage proteins may ultimately produce more dramatic results in improving nutritional quality.

Finally, Huitema *et al.* (1989) developed chrysanthemums which can flower at reduced temperatures by screening for plant regenerations at 6 °C. Mutants which could flower at 12 °C were selected. The results are surprising, since no direct selection for low temperature flowering was induced. The overall complexity of low temperature tolerance in crop plants, from seed germination, vegetative development, pollen production and fertilization to seed pod development and maturity, would suggest that a direct cell selection approach would not be useful.

In conclusion, cell selection has been effective in the development of certain agronomically useful mutations. The efficacy of the approach has varied depending on the selection system and mechanism of resistance. Numerous reports of mutant recovery have not been supported by rigorous testing of the selected material under field conditions.

## 3.11 REFERENCES

Arcioni, S., M. Pezzotti and F. Damiani (1987). *In vitro* selection of alfalfa plants resistant to *Fusarium oxysporum* f. sp. *medicaginis*. *Theor. Appl. Genet.*, **74**, 700–705.

Behnke, M. (1979). Selection of potato callus for resistance to culture filtrates of *Phytophthora infestans* and regeneration of resistant plants. *Theor. Appl. Genet.*, **55**, 69–71.

Behnke, M. (1980). General resistance to late blight of *Solanum tuberosum* plants regenerated from callus resistant to culture filtrates of *Phytophthora infestans*. *Theor. Appl. Genet.*, **56**, 151–152.

Ben-Hayyim, G. and Y. Goffer (1989). Plantlet regeneration from a NaCl-selected salt-tolerant callus culture of Shamouti orange (*Citrus sinensis* L. Osbeck). *Plant Cell Reprod.*, **8**, 680–683.

Bennetzen, J. L. and T. L. Adams (1984). Selection and characterization of cadmium-resistant suspension cultures of the wild tomato *Lycopersicon peruvianum*. *Plant Cell Reprod.*, **3**, 258–261.

Bourgin, J. P. (1986). Isolation and characterization of mutant cell lines and plants: auxotrophs and other conditional lethal mutants. In *Cell Culture and Somatic Cell Genetics of Plants*, ed. I. E. Vasil, vol. 3, pp. 475–498. Academic Press, New York.

Brettell, R. I. S. and E. Thomas (1980). Reversion of Texas male-sterile cytoplasm maize in culture to give fertile, T-toxin resistant plants. *Theor. Appl. Genet.*, **58**, 55–58.

Carlson, P. S. (1973). Methionine sulfoximine-resistant mutants of tobacco. *Science (Washington, D.C.)*, **180**, 1366–1368.

Chaleff, R. S. (1980). Further characterization of picloram tolerant mutants of *Nictiana tabacum*. *Theor. Appl. Genet.*, **58**, 91–95.

Chaleff, R. S. (1981). *Genetics of Higher Plants: Applications of Cell Culture*. Cambridge University Press, New York.

Chaleff, R. S. (1983a). Considerations of developmental biology for the plant cell geneticist. In *Genetic Engineering of Plants*, ed. T. Kosuge, C. P. Meredith and A. Hollaneder, pp. 257–270. Plenum Press, New York.

Chaleff, R. S. (1983b). Isolation of agronomically useful mutants from plant cell cultures. *Science (Washington, D.C.)*, **219**, 676–682.

Chaleff, R. S. (1986). Isolation and characterization of mutant cell lines and plants: herbicide-resistant mutants. In *Cell Culture and Somatic Cell Genetics of Plants*, ed. I. E. Vasil, vol. 3, pp. 499–512. Academic Press, New York.

Chaleff, R. S. and M. F. Parsons (1978). Direct selection *in vitro* for herbicide-resistant mutants of *Nicotiana tabacum*. *Proc. Natl. Acad. Sci. USA*, **75**, 5104–5107.

Chaleff, R. S. and T. B. Ray (1984). Herbicide-resistant mutants from tobacco cell cultures. *Science (Washington, D.C.)*, **223**, 1148–1151.

Chen, T. H. H. and L. V. Gusta (1986). Isolation and characterization of mutant cell lines and plants: cold tolerance. In *Cell Culture and Somatic Cell Genetics of Plants*, ed. I. E. Vasil, vol. 3, pp. 527–535. Academic Press, New York.

Connor, A. J. and C. P. Meredith (1985). Large scale selection of aluminium-resistant mutants from plant cell culture: expression and inheritance in seedlings. *Theor. Appl. Genet.*, **71**, 159–165.

Creason, G. L. and R. S. Chaleff (1988). A second mutant enhances resistance of a tobacco mutant of sulfonylurea herbicides. *Theor. Appl. Genet.*, **76**, 177–182.

Csépiö, A., P. Medgyesy, É. Hideg, S. Demeter, L. Márton and P. Maliga (1985). Triazine-resistant *Nicotiana* mutants from photomixotrophic cell cultures. *Mol. Gen. Genet.*, **200**, 508–510.

Dix, P. H. (1986). Cell line selection. In *Plant Cell Culture Technology*, ed. M. M. Yeoman, pp. 143–201. Blackwell, Boston.

Duncan, D. R. and J. M. Widholm (1986). Cell selection for crop improvement. In *Plant Breeding Reviews*, ed. J. Janick, vol. 4, pp. 153–173. AVI, Westport.

Evans, D. A. and W. R. Sharp (1983). Single gene mutations in tomato plants regenerated from tissue culture. *Science (Washington, D.C.)*, **221**, 949–951.

Evans, D. A., W. R. Sharp and H. P. Medina-Filho (1984). Somaclonal and gametoclonal variation. *Am. J. Bot.*, **71**, 759–774.

Evans, D. E., N. E. Rothnie, J. P. Sang, M. V. Palmer, D. L. Mulcahy, M. B. Singh and R. B. Knox (1988). Correlations between gametophytic (pollen) and sporophytic (seed) generations for polyunsaturated fatty acids in oilseed rape *Brassica napus* L. *Theor. Appl. Genet.*, **76**, 411–419.

Flick, C. E. (1983). Isolation of mutants from cell culture. In *Handbook of Plant Cell Culture*, ed. D. A. Evans, W. R. Sharp, P. V. Ammirato and Y. Yamada, vol. 1, pp. 393–441. Macmillan, New York.

Flick, C. E. and D. A. Evans (1984). Mutant selection. In *Plant–Microbe Interactions: Perspectives*, ed. T. Kosuge and E. W. Nester, vol. 1, pp. 168–172. Macmillan, New York.

Freytag, A. H., J. A. Wrather and A. W. Erichsen (1990). Salt tolerant sugarbeet progeny from tissue cultures challenged with multiple salts. *Plant Cell Reprod.*, **8**, 647–650.

Gengenbach, B. G., C. E. Green and C. M. Donovan (1977). Inheritance of selected pathotoxin resistance in maize plants regenerated from cell culture. *Proc. Natl. Acad. Sci. USA*, **74**, 5113–5117.

Hartman, C. L., T. J. McCoy and T. R. Knous (1984). Selection of alfalfa (*Medicago sativa*) cell lines and regeneration of plants resistant to the toxin(s) produced by *Fusarium oxysporum* f. sp. *Medicaginis*. *Plant Sci. Lett.*, **34**, 183–194.

Hasegawa, P. M., R. A. Bressan and A. K. Handa (1980). Growth characteristics of NaCl-selected and nonselected cells of *Nicotiana tabacum* L. *Plant Cell Physiol.*, **21**, 1347–1355.

Hibberd, K. A. and C. E. Green (1982). Inheritance and expression of lysine plus threonine resistance selected in maize tissue culture. *Proc. Natl. Acad. Sci. USA*, **79**, 559–563.

Hibberd, K. A., T. Walter, C. E. Green and B. G. Gengenbach (1980). Selection and characterization of a feedback-insensitive tissue culture of maize. *Planta*, **148**, 183–187.

Hibberd, K. A., P. C. Anderson and M. Barker (1986). Tryptophan overproducer mutants of cereal crops. *US Pat.* 4 581 847.

Hickok, L. G. and O. J. Schwarz (1986). An *in vitro* whole plant selection system: paraquat tolerant mutants in the fern *Ceratopteris*. *Theor. Appl. Genet.*, **72**, 302–306.

Hoffmann, F., E. Thomas and G. Wenzel (1982). Anther culture as a breeding tool in rape. Progeny analyses of androgenetic lines and induced mutants from haploid cultures. *Theor. Appl. Genet.*, **61**, 225–232.

Hughes, K. (1983). Selection for herbicide resistance. In *Handbook of Plant Cell Culture*, ed. D. A. Evans, W. R. Sharp, P. V. Ammirato and Y. Yamada, vol. 1, pp. 442–460. Macmillan, New York.

Huitema, J. B. M., W. Preil, G. C. Gussenhove and M. Schneidereit (1989). Methods for the selection of low temperature tolerant mutants of *Chrysanthemum morifolium* Ramat. by using irradiated cell suspension cultures. Selection of regenerants *in vivo* under suboptimal temperature conditions. *Plant Breed.*, **102**, 140–147.

Jackson, P. J., E. J. Roth, P. R. McClure and C. M. Naranjo (1984). Selection, isolation and characterization of cadmium-resistant *Datura innoxia* suspension cultures. *Plant Physiol.*, **75**, 914–918.

Jacobs, M., I. Negrutiu, R. Dirks and D. Cammaerts (1987). Selection programmes for isolation and analysis of mutants in plant cell cultures. In *Plant Biology*, ed. C. E. Green, D. A. Somers, W. P. Hackett and D. D. Biesboer, vol. 3, pp. 243–264. Liss, New York.

Kononowicz, A. K., K. Floryanowicz-Czekalska, J. Clithero, A. Meyers, P. M. Hasegawa and R. A Bressan. (1990a). Chromosome number and DNA content of tobacco cells adapted to NaCl. *Plant Cell Reprod.*, **8**, 672–675.

Kononowicz, A. K., P. M. Hasegawa and R. A. Bressan (1990b). Chromosome number and nuclear DNA content of plants regenerated from salt adapted plant cells. *Plant Cell Reprod.*, **8**, 676–679.

Larkin, P. J. and W. R. Scowcroft (1981). Somaclonal variation — a novel source of variability from cell cultures for plant improvement. *Theor. Appl. Genet.*, **60**, 197–214.

Ling, D. H., P. Vidhyaseharan, E. S. Borromeo, F. J. Zapata and T. W. Mew (1985). *In vitro* screening of rice germplasm for resistance to brown spot disease using phytotoxin. *Theor. Appl. Genet.*, **71**, 133–135

MacDonald, M. V. and D. S Ingram (1986). Toward the selection *in vitro* for resistance to *Alternaria brassicicola* (Schw.) Wilts., in *Brassica napus* ssp. *oleifera* (Metzg.) Sinsk., winter oilseed rape. *New Phytol.*, **104**, 621–629.

Matern, U., G. Strobel and J. Shepard (1978). Reaction to phytotoxins in a potato population derived from mesophyll protoplasts. *Proc. Natl. Acad. Sci. USA*, **75**, 4935–4939.

McCoy, T. J. (1987). Characterization of alfalfa (*Medicago sativa* L.) plants regenerated from selected NaCl tolerant lines. *Plant Cell Reprod.*, **6**, 417–422.

McHughen, A. and M. Swartz (1984). A tissue culture derived salt tolerant line of flax (*Linum usitatissimum*). *J. Plant Physiol.*, **117**, 109–117.

Miller, O. K. and K. Hughes (1980). Selection of paraquat-resistant variants of tobacco from cell cultures. *In Vitro*, **16**, 1085–1091.

Mulinix, C. A. and A. F. Iezzoni (1988). Microgametophytic selection in two alfalfa (*Medicago sativa* L.) clones. *Theor. Appl. Genet.*, **75**, 917–922.

Nabors, M. W., S. E. Gibbs, C. S. Bernstein and M. E. Meis (1980). NaCl tolerant tobacco plants from cultured cells. *Z. Pflanzenphysiol.*, **97**, 13–17.

Nabors, M. W., C. S. Kroskey and D. M. McHugh (1982). Green spots are predictors of high callus growth rates and shoot formation in normal and in salt stressed tissue cultures of oat (*Avena sativa* L.). *Z. Pflanzenphysiol.*, **105**, 341–349.

Negrutiu, I., A. Cattoir-Reynearts, I. Vergruggen and M. Jacobs (1984). Lysine overproducer mutants with altered dihydrodipicolinate synthase from protoplast culture of *Nicotiana sylvestris* (Spegazzini and Comes). *Theor. Appl. Genet.*, **68**, 11–20.

Newsholme, D. M., M. V. MacDonald and D. S. Ingram (1989). Studies of selection *in vitro* for novel resistance to phytotoxic products of *Leptosphaeria maculans* (Desm.) Ces. & De Not. in secondary embryogenic lines of *Brassica napus* ssp. *oleifera* (Metzg.) Sinsk., winter oilseed rape. *New Phytol.*, **113**, 117–126.

Ottaviano, E., M. Sari-Gorla and D. L. Mulcahy (1980). Pollen tube growth rates in *Zea mays*: implications for genetic improvement of crops. *Science (Washington, D.C.)*, **210**, 437–438.

Polsoni, L., L. S. Knott and W. D. Beversdorf (1988). Large-scale microspore culture technique for mutation–selection studies in *Brassica napus*. *Can. J. Bot.*, **66**, 1681–1685.

Rahman, M. M. and K. Kaul (1989). Differentiation of sodium chloride tolerant cell lines of tomato (*Lycopersicon esculentum* Mill.) cv. Jet Star. *J. Plant Physiol.*, **133**, 710–712.

Rain, D. N. and P. S. Carlson (1978). Herbicide-tolerant tobacco mutants selected *in situ* and recovered *via* regeneration from cell culture. *Genet. Res.*, **32**, 85–89.

Rains, D. W., S. S. Croughan and T. P. Croughan (1986). Isolation and characterization of mutant cell lines and plants: salt tolerance. In *Cell Culture and Somatic Cell Genetics of Plants*, ed. I. E. Vasil, vol. 3, pp. 537–547. Academic Press, New York.

Reisch, B., S. H. Duke and E. T. Bingham (1981). Selection and characterization of ethionine-resistant alfalfa (*Medicago sativa* L.) cell lines. *Theor. Appl. Genet.*, **59**, 89–94.

Rines, H. W. and H. H. Luke (1985). Selection and regeneration of toxin-insensitive plants from tissue cultures of oats (*Avena sativa*) susceptible to *Helminthosporium victoriae*. *Theor. Appl. Genet.*, **71**, 16–21.

Sacristán, M. D. (1982). Resistance responses to *Phoma lingam* of plants regenerated from selected cell and embryogenic cultures of haploid *Brassica napus*. *Theor. Appl. Genet.*, **61**, 193–200.

Sacristán, M. D. (1986). Isolation and characterization of mutant cell lines and plants: disease resistance. In *Cell Culture and Somatic Cell Genetics of Plants*, ed. I. E. Vasil, vol. 3, pp. 513–525. Academic Press, New York.

Sacher, R. F., D. L. Mulcahy and R. C. Staples (1983). Developmental selection during self pollination of *Lycopersicon* × *Solanum* $F_1$ for salt tolerance of $F_2$. In *Pollen: Biology and Implications for Plant Breeding*, ed. D. L. Mulcahy and E. Ottaviano, pp. 329–342. Elsevier, New York.

Searcy, K. B. and D. L. Mulcahy (1985). Pollen selection and the gametophytic expression of metal tolerance in *Silene dioica* (Caryophyllaceae) and *Mimulus guttatus* (Scrophulariaceae). *Am. J. Bot.*, **72**, 1700–1706.

Shaner, D., T. Malefyt and P. Anderson (1985). Herbicide-resistant maize through cell culture selection. In *BCPC Monograph 32: Biotechnology and its Application to Agriculture*, ed. L. G. Copping and P. Rodgers, pp. 45–50. British Crop Protection Council.

Singer, S. R. and C. N. McDaniel (1984). Selection of amitrole-tolerant tobacco calli and the expression of this tolerance in regenerated plants and progeny. *Theor. Appl. Genet.*, **67**, 427–432.

Sjödin, C. and K. Glimelius (1989). Differences in response to the toxin sirodesmin PL produced by *Phoma lingam* (Tode ex Fr.) Desm. on protoplasts, cell aggregates and intact plants of resistant and susceptible *Brassica* accessions. *Theor. Appl. Genet.*, **77**, 76–80.

Smith, G. (1985). Sporophytic screening and gametophytic verification of phytotoxin tolerance in sugar beet (*Beta vulgaris* L.). In *Biotechnology and Ecology of Pollen*, ed. D. L. Mulcahy, G. B. Mulcahy and E. Ottaviano, pp. 83–88. Springer, New York.

Swanson, E. B., M. P. Coumans, G. L. Brown, J. D. Patel and W. D. Beversdorf (1988). The characterization of herbicide tolerant plants in *Brassica nopus* L. after *in vitro* selection of microspores and protoplasts. *Plant Cell Reprod.*, **7**, 83–87.

Swanson, E. B., M. J. Herrgesell, M. Arnoldo, D. W. Sippell and R. S. C. Wong (1989). Microspore mutagenesis and selection: Canola plants with field tolerance to the imidazolinones. *Theor. Appl. Genet.*, **78**, 525–530.

Tal, M. (1983). In *Handbook of Plant Cell Culture*, ed. D. A. Evans, W. R. Sharp, P. V. Ammirato and Y. Yamada, vol. 1, pp. 461–488. Macmillan, New York.

Thanutong, P., I. Furusawa and M. Yamamoto (1983). Resistant tobacco plants from protoplast-derived calluses selected for their resistance to *Pseudomonas* and *Alternaria* toxins. *Theor. Appl. Genet.*, **66**, 209–215.

Thomas, B. R. and D. Pratt (1982). Isolation of paraquat-tolerant mutants from tomato cell cultures. *Theor. Appl. Genet.*, **63**, 169–176.

Widholm, J. M. (1972). Cultured *Nicotiana tabacum* cells with an altered anthranilate synthetase which is less sensitive to feedback inhibition. *Biochim. Biophys. Acta*, **261**, 52–58.

Windsor, J. A., L. E. Davis and A. G. Stephenson (1987). The relationship between pollen load and fruit maturation and the effect of pollen load on offspring vigor in *Cucurbita pepo*. *Am. Nat.*, **129**, 643–656.

Yamada, Y. and F. Sato (1983). In *Handbook of Plant Cell Culture*, ed. D. A. Evans, W. R. Sharp, P. V. Ammirato and Y. Yamada, vol. 1, pp. 489–500. Macmillan, New York.

Ye, J. M., K. N. Kao, B. L. Harvey and B. G. Rossnagel (1987). Screening salt-tolerant barley genotypes *via* $F_1$ anther culture in salt stress media. *Theor. Appl. Genet.*, **74**, 426–429.

Zamir, D., S. D. Tanksley and R. A. Jones (1982). Haploid selection for low temperature tolerance of tomato pollen. *Genetics*, **101**, 129–137.

# 4

# Bioreactors for the Mass Cultivation of Plant Cells

ALAN H. SCRAGG
*Bristol Polytechnic, UK*



## 4.1 INTRODUCTION

The biotechnological exploitation of plant cell cultures for the production of commercially interesting compounds has been proposed for some time. Routier and Nickell (1956) first discussed the use of plant cell cultures for the commercial production of compounds normally extracted from plants. Despite advances in organic chemistry, the plant kingdom still remains a source of a wide range of phytochemicals, including pharmaceuticals, insecticides, flavours, fragrances and colours (Balandrin and Klocke, 1988). Farnsworth and Morris (1976) estimated that some 25% of prescriptions in the USA contained active ingredients obtained from plants. Plant cell cultures have been shown to be capable of producing a wide range of phytochemicals, in particular pharmaceuticals (Vickery and Vickery, 1981), and have therefore been considered as an alternative to the agricultural production of such compounds (Barz and Ellis, 1981; Berlin, 1984, 1988; Dougall, 1979; Nickell, 1980; Staba, 1977). In addition to the production of secondary metabolites, plant cell cultures have been used for biotransformations (Alfermann *et al.*, 1985; Veliky and Jones, 1981) and for the production of specific enzymes (Esquivel *et al.*, 1988). To date the Mitsui Petrochemical Industries production of shikonin, a dye and pharmaceutical, from cultures of *Lithospermum erythrorhizon* is the only published commercial process (Curtin, 1983) but recently processes for the production of ginseng biomass (Furuya *et al.*, 1984) and berberine have reached industrial production scale (Fujita, 1987). Any industrial process will require the efficient growth of plant cells in large volumes, whether they are required for secondary product accumulation, biotransformations or enzyme production. Also growth in bioreactors allows a more precise control of parameters such as oxygen and carbon dioxide, pH and agitation than is achieved in flasks. The mass cultivation of plant cells can best be regarded as the cultivation of cell suspensions in volumes larger than those possible in shake flasks, and will involve the cultivation in bioreactors of 2 L volume and above (Scragg and Fowler, 1985).

The large scale cultivation of plant cells started in 1959 with the work of Tulecke and Nickell (1959), who cultivated a number of cell lines in simple aspirated glass carboys. The research was directed towards the use of plant cell culture as a supply of food during space travel. Other researchers used roller bottles (Martin, 1980), a technology borrowed from animal cell cultures, for the growth of plant cells, but in all cases the bioreactors were simple in design.

The first commercial bioreactor used for plant cells was the Microferm New Brunswick of 7.5 and 15 L used by Byrne and Koch (1962) to grow *Daucus carota* cells. The 1970s saw considerable interest in the cultivation of tobacco cells for the production of a high nicotine, low tar product. Much of the research was carried out in Japan (Kato *et al.*, 1972, 1975), where large stirred tank bioreactors were used at low stirrer speeds to cultivate tobacco cells in both batch and continuous systems (Table 1). During this period only one report of the large scale growth of cells other than tobacco can be found. Hahlbrock *et al.* (1974) reported the growth of *Glycine max* and *Petroselinium* in a 300 L stirred tank bioreactor.

In 1972 Mandels (1972) had recognized that plant cells were more sensitive to shear stress than microorganisms and Dalton (1978) suggested that the problems encountered in growing plant cells in stirred tank bioreactors may be related to their shear sensitivity. Although designed as early as 1955 by LeFrancois (LeFrancois *et al.*, 1955) it was the 1970s that saw the development of the airlift bioreactor for the growth of microorganisms, in particular the very large scale ICI single cell process. One advantage of the airlift bioreactor was its low shear stress characteristics, which led to its use for both animal and plant cell cultures. In 1977 both Zenk (Zenk *et al.*, 1977) and Wagner and Vogelmann (1977) used airlift bioreactors to cultivate *Catharanthus roseus* and *Morinda citrifolia* respectively. Wagner and Vogelmann (1977) also compared growth and anthraquinone production by *M. citrifolia* grown in a range of bioreactor designs and concluded that the airlift design gave the best results (Figure 1). Since then a number of cultures have been grown in airlift bioreactors

**Table 1** Bioreactor Types Used to Cultivate Plant Cells

| *Bioreactor type* | *Volume* (L) | *Cell line* | *Year* |
|---|---|---|---|
| Stirred tank | 30 | *N. tabacum* | 1972 |
| | 130, 600 | *N. tabacum* | 1972 |
| | 10, 30, 300 | *Glycine max* | 1974 |
| | 15, 65 | *N. tabacum* | 1975 |
| | 15, 30, 230 | *N. tabacum* | 1977 |
| | 15, 500 | *N. tabacum* | 1977 |
| | 10 | *Morinda citrifolia* | 1977 |
| | 7.5 | *Catharanthus roseus* | 1981 |
| | 20 000 | *N. tabacum* | 1982 |
| | 500, 750 | *Lithospermum erythrorhizon* | 1983 |
| | 30 | *Panax ginseng* | 1984 |
| | 32 | *Colum blumei* | 1985 |
| | 14 | *Catharanthus roseus* | 1986 |
| | 70, 750, 800 | *Catharanthus roseus*<br>*N. tabacum*<br>*Solanum demissum* | 1986 |
| | 2000, 20 000 | *Panax ginseng* | 1986 |
| | 5000 | *Catharanthus roseus* | 1986 |
| | 1000 | *Lithospermum erythrorhizon* | 1986 |
| Airlift | 10 | *Morinda citrifolia*<br>*Catharanthus roseus* | 1977 |
| | 20 | *Catharanthus roseus* | 1977 |
| | 30 | *Catharanthus roseus* | 1977 |
| | 10, 100 | *Catharanthus roseus* | 1981 |
| | 20 | *Trispterygium* | 1981 |
| | 200 | *Digitalis lanata* | 1983 |
| | 5, 10 | *Catharanthus roseus* | 1984 |
| | 20 | *Berberis wilsonae* | 1985 |
| | 80 | *Catharanthus roseus* | 1987 |
| | 80 | *Helianthus annuus* | 1989 |
| Rotating drum | 2.5–1000 | *Lithospermum erythrorhizon* | 1983 |
| Tayler–Couette bioreactor | 2.5 | *Beta vulgaris* | 1987 |
| Membrane, stirred bioreactor | 21 | *Thalictrum rogosum* | 1988 |

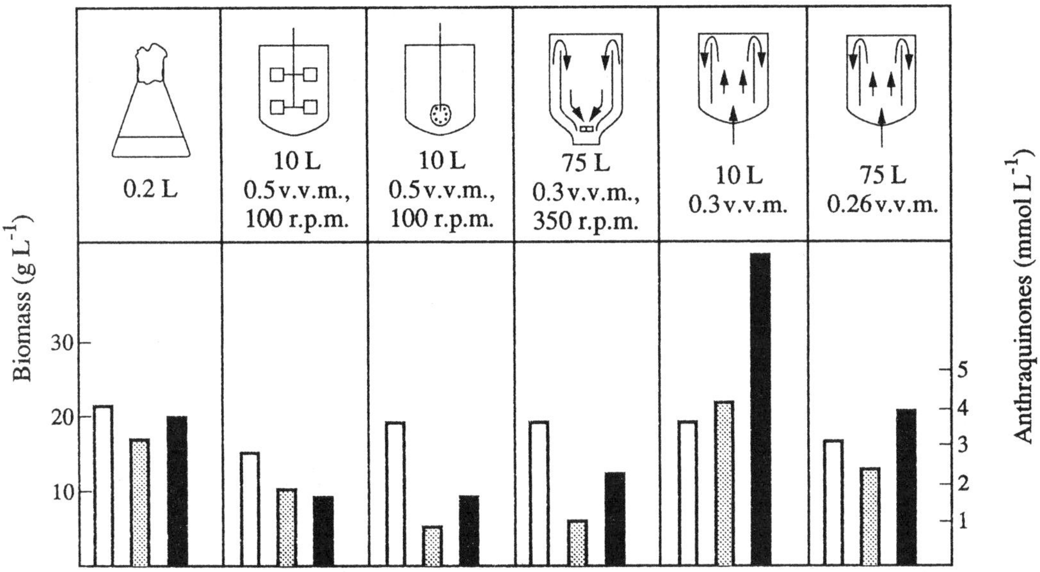


**Figure 1** The effect of bioreactor design on the production of anthraquinones by *Morinda citrifolia*. Biomass $g\,L^{-1}$ (□); anthraquinone $mmol\,L^{-1}$ (▩); productivity $mmol\,h^{-1}$ (■); (reproduced from Wagner and Vogelmann, 1977, by permission of Springer-Verlag)

of up to 200 L in volume (Alfermann *et al.*, 1985; Breuling *et al.*, 1985; Scragg *et al.*, 1987a; Smart and Fowler, 1984). This did not mean that the stirred tank bioreactor was discarded; in fact a number of cultures have been successfully grown in such vessels (Table 1) (Berlin, 1986; Drapeau *et al.*, 1987; Fujita and Tabata, 1987; Furuya *et al.*, 1984; Pareilleux and Vinas, 1983; Payne *et al.*, 1989; Schiel and Berlin, 1987; Ulbrich *et al.*, 1985). The Mitsui shikonin process uses two stirred tanks of 500 to 750 L volume and Berlin (1986) has reported the use of a 5000 L stirred tank for the growth of *C. roseus.*

In this chapter the characteristics of plant cell suspensions and how these affect their growth in bioreactors are covered.

## 4.2 CHARACTERISTICS OF PLANT CELL SUSPENSIONS

For workers not acquainted with plant cell suspensions, they would appear at first to be very similar to microbial cultures. Although in some aspects plant cell suspensions may be treated as microbial in nature, they do have particular characteristics which can affect their growth in bioreactors. Some of these characteristics are listed in Table 2 along with some of the consequences for bioreactor cultivation.

**Table 2** The Characteristics of Plant Cell and Microbial Cultures

| *Characterization* | *Microbial cells* | *Plant cell suspension* | *Consequences for mass cultivation* |
|---|---|---|---|
| Size | 2–10 μm | 10–200 μm | Rapid sedimentation; shear sensitivity |
| Individual cells | Often | Not often; generally form aggregates up to 2 mm in diameter | Rapid sedimentation; large sampling ports required; formation of microenvironments |
| Growth rate | Rapid; doubling times of 1–2 h | Slow; doubling times of 2–5 d | Long culture runs; problems of maintaining sterility; low productivity |
| Inoculation density | Small | 5–20% | Large inoculation vessels required; reduces scale-up |
| Shear stress sensitivity | Insensitivity | Sensitive/tolerant | Slow stirred speeds |
| Aeration requirements | High | Low | Low oxygen demand; $O_2$ supply less critical; low $k_La$ vessels can be used |

### 4.2.1 Size of Plant Cells

Plant cells are large compared with most microorganisms, being in general 40–200 μm long and 10–40 μm in width. The cells in culture are bounded by a rigid cellulose-based cell wall. The size and shape of individual cells in a culture often varies considerably and this can be altered by cultural conditions (Scragg *et al.*, 1988; Tanaka, 1987).

Early in the growth cycle the cells are often small and rounded, containing a dense cytoplasm with a small vacuole. As the growth cycle proceeds, the cell enlarges to a point where it stops increasing in dry weight, but expands rapidly with the development of a large vacuole which can take up over 90% of the cell volume. The process of rapid cell expansion can be seen during the growth cycle if both wet and dry weight of the cells are estimated (Figure 2). Plant cells in suspension, once they have divided and produced a cross wall, may separate. However, in most cases plant cells in suspension are found in groups or aggregates which can number from tens to thousands, and be of 2 mm or more in diameter. A typical size distribution of a plant cell suspension can be seen in Figure 3. Whether the formation of aggregates is due to nonseparation upon cell division or true aggregation after division is not known. The aggregate structure can be loose in nature (Scragg, 1990a) and may be held together by extracellular polysaccharides, which plant cells are known to produce in culture (Hale *et al.*, 1987). Cultural conditions such as aeration and medium constituents can affect the size and frequency of aggregates in culture. Attempts have been made to reduce aggregation by the addition of enzymes (pectinases) (Henshaw *et al.*, 1966) or by the reduction of calcium in the medium (calcium is required to form the middle lamella between cells as calcium pectate), but with little success. Mechanical methods such as higher shaker speeds or higher impeller speeds in stirred tank bioreactors often reduce the average aggregate size. Cultures that have been established for a considerable time (2–5 years) often show a reduction in aggregate size distribution over that time. The alteration in aggregate size distribution does not appear to greatly affect the growth rate of cultures. Individual viable cells are not often found in plant cell suspensions.

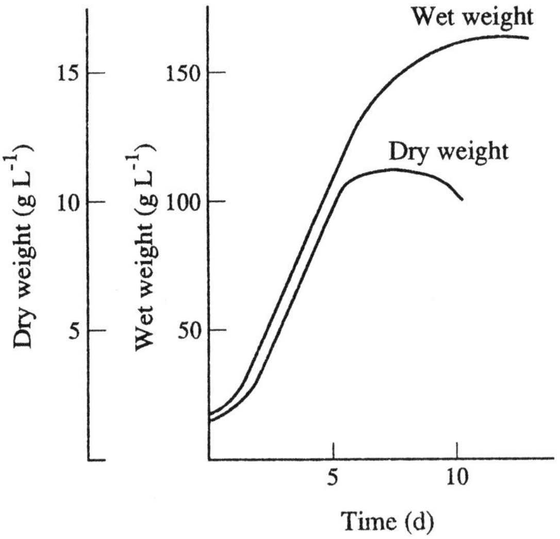


**Figure 2** A typical growth pattern of a plant cell suspension

The large size of individual cells and the aggregates makes for a very rapid settling rate (Figure 4). In bioreactors this means that if aeration or mixing is stopped the cells settle out rapidly and will become anoxic quickly. This will also mean that plant cell aggregates will settle out in areas of poor mixing, and around obstructions. The aggregate sizes will also require large port sizes (at least 8 mm) for sampling, exit and inoculation ports, and the sample size will need to be large (50 mL) to avoid unrepresentative sampling. As the bioreactor will be run for 2–3 weeks, even with samples taken every 2 d considerable volume of culture will be removed, and therefore bioreactors of below 5 L in volume are difficult to operate.

It has been suggested that the aggregate structure found in plant cell suspensions can also influence secondary product accumulation (Schuler, 1981). Aggregates, in particular the larger ones, will produce a gradient of nutrients across the particle in the manner of fungal pellets. Depletion of oxygen or

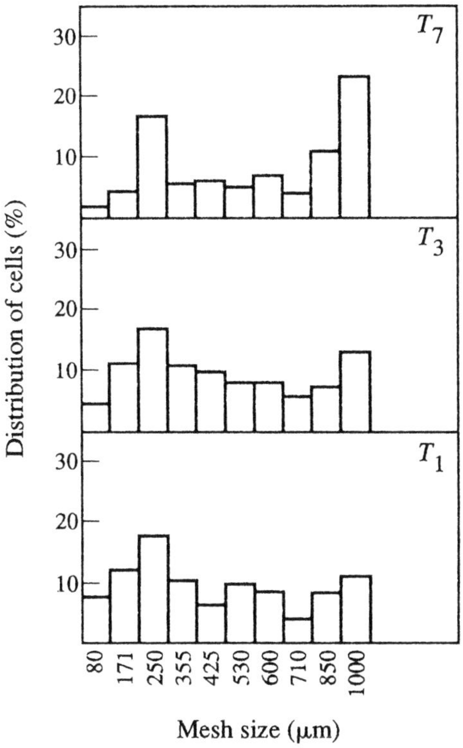


**Figure 3** Size distribution of the aggregates of a *Helianthus annuus* culture grown in a 7 L airlift bioreactor, after 1, 3 and 7 d culture (Scragg, unpublished data)

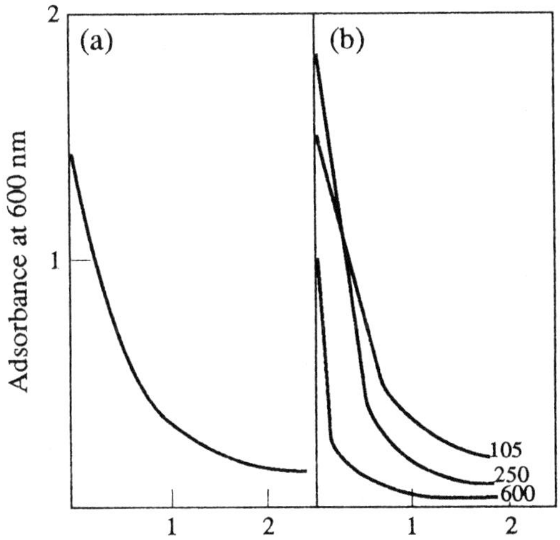


**Figure 4** The settling rates of a *Catharanthus roseus* culture and aggregates separated from the cultures (reproduced from Scragg *et al.*, 1987b, by permission of Elsevier, London)

glucose may induce or encourage secondary product accumulation. There is some evidence of higher serpentine levels being found in aggregates of a particular size range, but this appears very dependent on the cell line.

### 4.2.2 Growth Rates

Plant cells in suspension have a very slow growth rate when compared with most microorganisms, with doubling times of 2–6 d compared with 2–6 h for microorganisms. The low growth rate appears not to be due to the low activities of the enzymes present in plant cells, but is rather a reflection of the cell's growth cycle, where the cytoplasm can represent less than 10% of the cell's volume. The most rapidly growing culture that has been reported is tobacco (Noguchi *et al.*, 1982) with a doubling time of 18 h (Table 3), but a typical doubling time would be 2–3 d. The growth rate of a suspension will often increase considerably the longer the culture is kept in culture (Figure 5). In some cases the growth rate is reduced when the cultures are grown in bioreactors, whereas in other cases little effect on growth rate has been found (Scragg *et al.*, 1987a). The practical consequences of this slow growth rate is that the bioreactor runs will be of 2–3 weeks rather than days. This requires considerable attention to the maintenance of sterility, and other problems such as water loss and sampling volume associated with long bioreactor runs can also occur. In our own experiences the larger bioreactors require pressure testing prior to each run in order to eliminate faulty seals and find leaks. The vessel is then filled with sterile medium 1–3 d before inoculation, and samples taken of the inoculation and the vessel for sterility checks. Given these extensive precautions, the number of cultures lost to microbial contamination can be reduced to an acceptable level (less than 5%). The long bioreactor runs also mean that the number of runs that an individual bioreactor can accommodate in a year is limited to perhaps one per month. Thus the number of experiments is reduced and to cover a range of conditions a number of bioreactors are required. On a production scale, the development of an inoculum will take a considerable time and the overall productivity per bioreactor will be reduced. As many products of interest are accumulated after growth has ceased, the possibility of reducing the bioreactor run time is limited.

**Table 3** Example of the Growth Rates of Plant Cell Suspensions in Bioreactors

| *Cell line* | *Bioreactors* | *Specific growth rate* $\mu(d^{-1})$ | *Doubling time* $t_d(d)$ |
|---|---|---|---|
| *Nicotiana tabacum* | 600 L STR | 0.42 | 1.7 |
| *Nicotiana tabacum* | 15, 30, 130 L STR | 0.96 | 0.72 |
| *Nicotiana tabacum* | 15, 500 L STR | 1.10 | 0.63 |
| *Morinda citrifolia* | 10 L airlift | 0.48 | 1.4 |
| *Catharanthus roseus* | 30 L airlift | 0.32 | 2.2 |
| *Helianthus annuus* | 80 L airlift | 0.30 | 2.3 |
| *Picrasma quassioides* | 7 L airlift | 0.12 | 5.9 |
| *Quassia amara* | 7 L airlift | 0.13 | 5.2 |

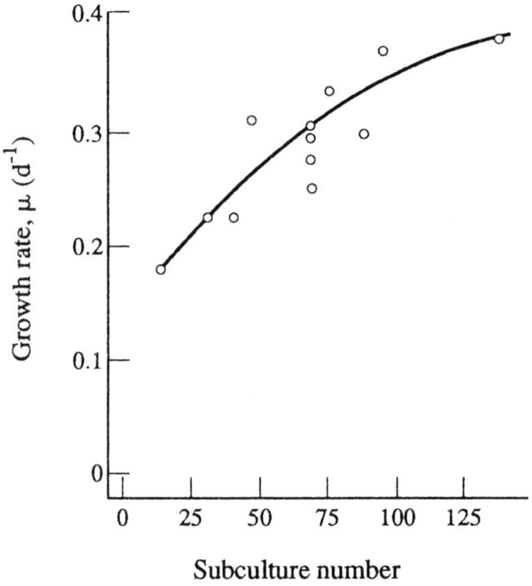


**Figure 5** The change in growth rate with subculture number for *Catharanthus roseus* (redrawn from Morris *et al.*, 1989, with permission of Kluwer, Dordrecht)

### 4.2.3 Inoculation Density

Unlike most microorganisms, plant cells have a critical inoculation density below which growth will not occur. An example is shown in Figure 6. Plant cells appear to require either cell-to-cell contact or substances produced by the cells and exported into the medium before they will grow. These conditions can be provided by the addition of a large inoculum, generally 10% (v/v), or in some cases addition of conditioned medium (medium in which cells have already grown). This means that a bioreactor inoculum will need to be large and on scale-up the steps will have to be small (10% inoculum) and will thus take longer.

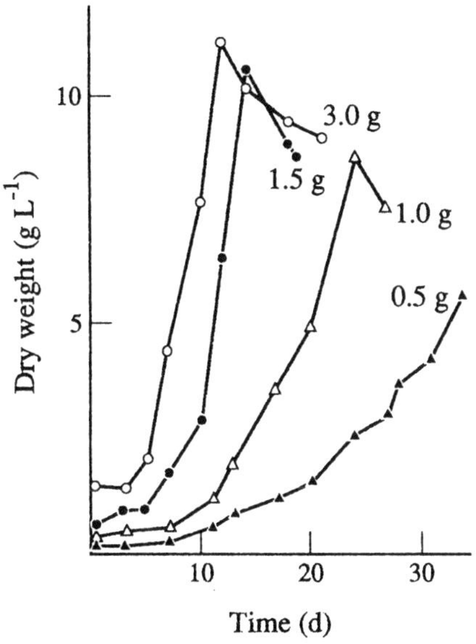


**Figure 6** The effect of inoculation density on the growth of *Picrasma quassiodides* culture (unpublished data of R. D. Steward). The weight of inoculum in wet weight is given for each curve

### 4.2.4 Foam Production

The culture of plant cells in bioreactors, in particular airlift bioreactors, is often accompanied by the formation of a crust or 'meringue' at the top of the vessel (Fowler, 1982). The meringue can cover the whole surface of the culture and can stop the circulation in airlift bioreactors. The structure of the meringue is shown in Figure 7 and appears to be a polysaccharide matrix in which cells are trapped. The trapping of a considerable number of cells can cause problems when taking representative samples. The meringue appears to form on foam which develops as the culture begins to grow. The foam is probably due to the proteins present in the culture medium, as plant cells do export considerable amounts of protein. The initial foam is probably stabilized and built-up by the extracellular polysaccharides that plant cell cultures are known to produce (Hale *et al.*, 1987).

Antifoams have been used to control meringue formation with varying results. Wang and Staba (1963) used a silicone-based antifoam at concentrations up to 1000 p.p.m. with spearmint cultures with no adverse effects. In contrast, silicone antifoam at 100 p.p.m. was found to be toxic with spinach cultures (Dalton, 1978). The effect of two antifoams, a silicone-based antifoam and polypropylene glycol on the growth of *C. roseus* and its culture in an airlift has been determined (Bond *et al.*, 1987). It was concluded that neither antifoam inhibited growth but the polypropylene glycol proved to be more efficient at suppressing meringue formation. As the diameter of the bioreactor increases, the problem of meringue decreases.

### 4.2.5 Aeration

The low volumetric metabolic rate of plant cells gives the suspension a low oxygen requirement. The requirement is in the region of 1–10 mmol L$^{-1}$ h$^{-1}$ compared with values of 5–200 mmol L$^{-1}$ h$^{-1}$ for microorganisms (Table 4).

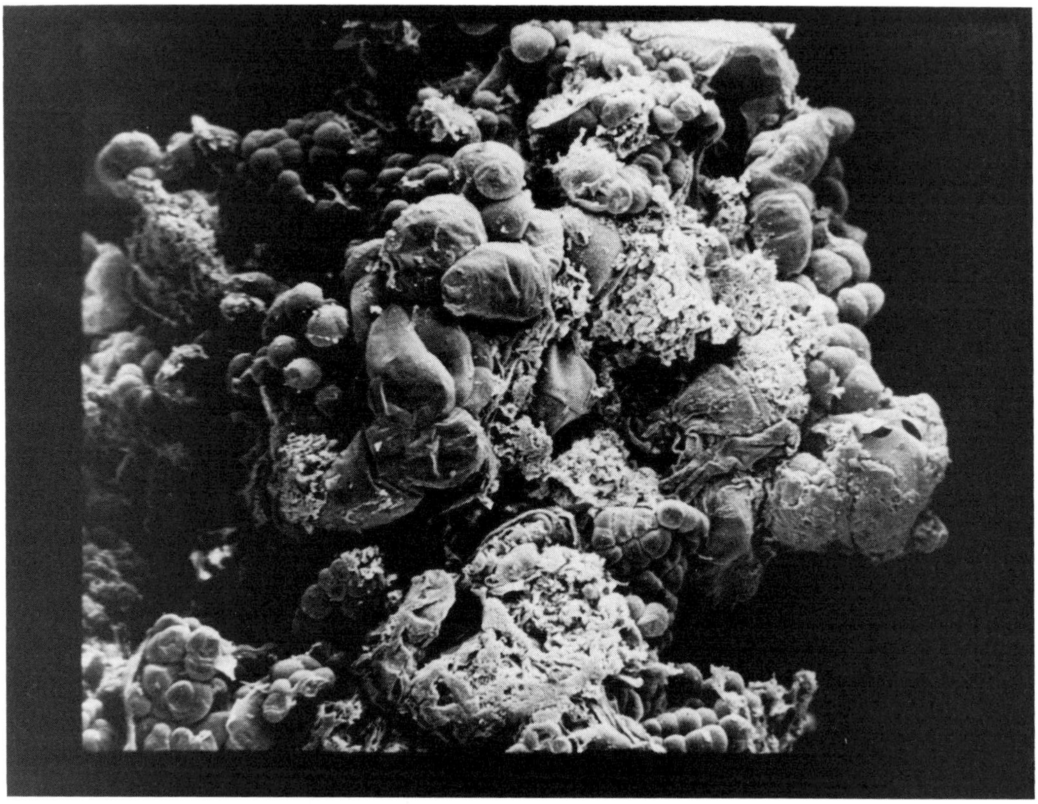

**Figure 7** A scanning electron micrograph of a *Catharanthus roseus* meringue

**Table 4** Examples of Respiration Rates of Plant Cell Cultures

| *Cell line* | *Uptake rate* mmol $O_2$ (g dry wt)$^{-1}$ h$^{-1}$ | *Cell line* | *Uptake rate* mmol $O_2$ (g dry wt)$^{-1}$ h$^{-1}$ |
|---|---|---|---|
| *Acer pseudoplatanus* | 0.2 | *Helianthus annuus* | 0.26–0.65 |
| *Nicotiana tabacum* | 0.03 | *Saccharum* | 1.1–3.6 |
| *Catharanthus roseus* | 0.3–0.7 | | |

Oxygen is the least soluble nutrient for aerobic cultures and as such it is often the growth-limiting substrate. Much of the working and design of bioreactors, aeration and agitation, is towards maximizing the oxygen supply. Thus, with a low oxygen requirement its supply may not be a critical feature in the growth of plant cells in bioreactors. The measure of the rate of supply of oxygen is the $k_La$ (volumetric oxygen transfer coefficient). Kato *et al.* (1975) working with tobacco suspensions in a modified stirred tank bioreactor, showed an effect of aeration and $k_La$ on growth. Subsequently it was shown that $k_La$ values could affect the growth of *C. roseus* cultures (Pareilleux and Vinas, 1983). In a recent study (Figure 8) an optimum was found for growth and alkaloid accumulation by *C. roseus* (Boysan *et al.*, 1988; Leckie *et al.*, 1991). It has been shown that growth and serpentine accumulation by *C. roseus* was not affected by initial dissolved oxygen values down to 20% (Scragg *et al.*, 1987a), although the value dropped to zero at mid log phase.

High aeration rates have been shown to reduce growth in bioreactors (Smart and Fowler, 1984; Pareilleux and Vinas, 1983; Ducos *et al.*, 1988). Hegarty *et al.* (1986) have shown that high aeration rates of 0.57 and 0.186 v.v.m. reduced the growth of *C. roseus* (Table 5) by stripping off carbon dioxide. Addition of carbon dioxide to the aeration gas restored the growth to normal. The level of carbon dioxide added to the aeration gas was in the order of 1%; above 5% carbon dioxide, growth inhibition was found. The implications of these observations are that, in gas-driven bioreactors, care has to be taken to ensure that high gassing rates used for mixing do not strip off carbon dioxide or other volatiles and reduce growth.

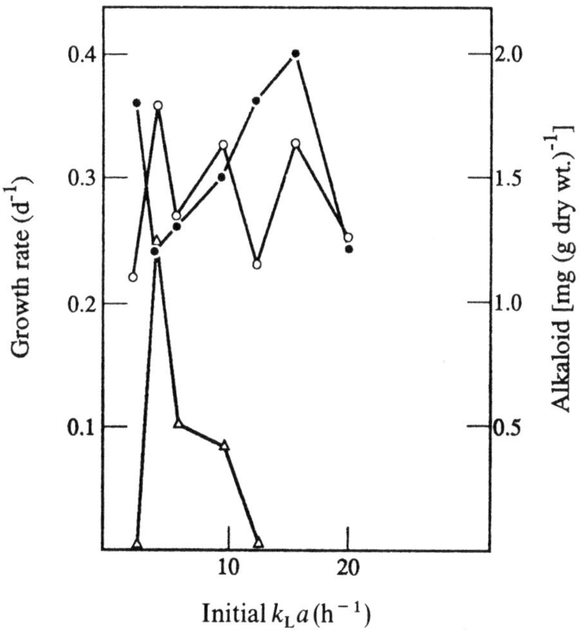


**Figure 8** The effect of initial $k_La$ on the growth and alkaloid accumulation by *Catharanthus roseus* IDI cultures (Leckie *et al.*, 1991). (○) dry weight $g\,L^{-1}$; (●) serpentine mg $(g\,dry\,wt.)^{-1}$; (△) ajmalicine mg $(g\,dry\,wt.)^{-1}$

**Table 5** The Effect of Aeration Rates on the Growth and Doubling Time of a *Catharanthus roseus* Culture[a]

| *Aeration rates* ($L\,min^{-1}$) | *Mean doubling times* $t_d$ (h) | *Growth rate* ($d^{-1}$) |
|---|---|---|
| 2 | $39 \pm 2$ | 0.42 |
| 3 | $46 \pm 2$ | 0.36 |
| 4 | $58 \pm 17$ | 0.29 |
| 6 | $72 \pm 12$ | 0.23 |

[a]Reproduced from Hegarty *et al.* (1986) with permission of the publishers.

### 4.2.6 Plant Cell Culture Viscosity

In order to increase the productivity of slow-growing plant cell suspensions, a high cell density is required. The theoretical limits have been suggested by Tanaka (1987) at $60\,g\,L^{-1}$ (at 10% of the cell volume being dry weight, solid cells would equal $100\,g\,L^{-1}$). In shake flasks values of $35\text{–}40\,g\,L^{-1}$ have been achieved with batch cultures of *C. roseus* (Scragg *et al.*, 1987a), and fed-batch of *Coleus* (Ulbrich *et al.*, 1985). At these high cell densities it has been suggested that the culture may be sufficiently viscous to impair mixing (Tanaka, 1981, 1982).

The first study on the rheology of plant cell suspensions used tobacco cells (Kato *et al.*, 1978) and concluded that the cells produced a high viscosity, whereas the culture filtrate changed little throughout the growth period. The viscosity was measured using a Brookfield-type rotational viscometer. Wagner and Vogelmann (1977), again using a rotational viscometer, showed that the rheological properties of *C. roseus* and *Morinda citrifolia* cultures were complicated, showing shear thinning and thixotropic behaviour. Tanaka (1982) used a similar method to Kato *et al.* (1978) to measure the rheological properties of cultures of *Cudronia tricuspidate*, *Vinca rosea* and *Agrostemma githago* and found these to be dependent on the size, specific gravity and cell concentration. The cultures were also found to be non-Newtonian and pseudoplastic. In a second study Vogelmann (1981), using a roughened horseshoe impeller, showed that *C. roseus* cultures showed yield stress, Bingham viscosity and thixotropic behaviour, although a second culture was Newtonian rather than thixotropic.

Owing to the aggregated nature of plant cell suspensions it has been shown that the normal form of rotational viscometer was unsuitable for the measurement of the visocosity of plant cell suspensions (Scragg *et al.*, 1986). However, using a rotational viscometer the medium was shown, at least with *C. roseus*, to be Newtonian and of low viscosity. This was perhaps unexpected as many cultures are suspected of producing extracellular polysaccharides (Hale *et al.*, 1987). In the case of *C. roseus* the polysaccharide probably remained attached to the cells and therefore is not found free in the medium. The rheometer measuring system was modified to accommodate either an anchor or turbine (Rushton) impeller. Although these impellers do not produce laminar flow required for the accurate measurement of viscosity, they eliminate many of the problems associated with rheometers such as settling, particle disruption and separation. This problem was first encountered by Roels *et al.* (1974) when investigating the rheology of mycellial pellets. The turbine impeller perhaps is closer to the conditions found in a bioreactor than a rotational rheometer. The disadvantage of the impellers is their narrow range of measurement due to turbulence formation. However, with such impellers, suspensions of *C. roseus*, *Helianthus annuus* and *Acer pseudoplanatus* at 10–40 g $L^{-1}$ (dry weight) concentrations were shown to be Bingham with viscosities in the range 3–25 mPa (Figure 9). Therefore, probably due to the particulate nature of plant cell suspensions, the viscosity of high density cultures would appear to be less of a problem than was at first thought.

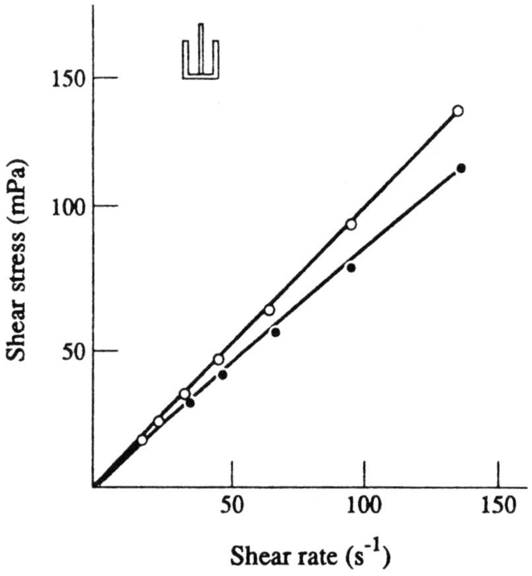


**Figure 9** The shear characteristics of a *Catharanthus roseus* culture using an anchor impeller. (○) medium; (●) medium plus cells (100 g $L^{-1}$ wet weight) (Reproduced from Scragg *et al.*, 1988 with permission of Butterworth, Stoneham)

### 4.2.7 Shear Sensitivity

Although in early cultures stirred tank bioreactors were used to cultivate plant sell suspensions, it was suggested that the difficulties associated with the cultivation of plant cell suspensions was due to their sensitivity to shear stress (Dalton, 1978; Mandels, 1972). Plant cells, because of their large size, rigid cell wall and extensive vacuole, have been regarded as sensitive to shear stress for a considerable time. The shear stress generated by the impeller in a stirred tank bioreactor was thought to be responsible for the difficulties of growing plant cell suspensions in such vessels (Smart, 1984). The low shear stress characteristics of the airlift bioreactor and its development on a large commercial scale encouraged its use for plant cell suspensions. The range of vessels used for the cultivation is shown in Table 1. Zenk *et al.* (1977) and Wagner and Vogelmann (1977) were the first to the use the airlift bioreactor for the cultivation of *C. roseus* and *M. citrifolia*. The design has proved successful for the cultivation of plant cells and other shear sensitive cells such as animal cell cultures. Despite this work, few data were available concerning the nature and extent of the shear stress sensitivity of plant cells.

Recently, research has been carried out on the response of plant cell cultures to a high shear environment. Using a Couette-type shearing device, the effect of various shear rates on *Nicotiana tabacum* cultures has been determined (Hooker *et al.*, 1989). At a shear rate of $1193\,s^{-1}$ it was shown that viability, as measured by reduction of tetrazolium salts, was lost after 12 h treatment. Many of the problems in this area derived from the difficulty in accurately measuring shear stress and viability. The viability of plant cells is generally determined by a measure of the integrity of the cell membrane by using dyes such as Evans blue or fluorescein diacetate (Widholm, 1972). However, the integrity of the cell membrane does not necessarily mean that the cell is capable of growth and division. Therefore, one of the best functional measures of culture viability is its ability to grow and divide, and this function has been tested against exposure to high levels of shear stress (Scragg *et al.*, 1988). Plant cells have been exposed to an impeller speed of 1000 r.p.m., an average shear rate of $167\,s^{-1}$ and a maximum shear rate of $1500\,s^{-1}$ for up to 5 h in a 3 L stirred tank bioreactor. Samples were removed at intervals and inoculated into fresh medium, and growth followed. The result using a culture of *C. roseus* is shown in Figure 10, where it is clear that the

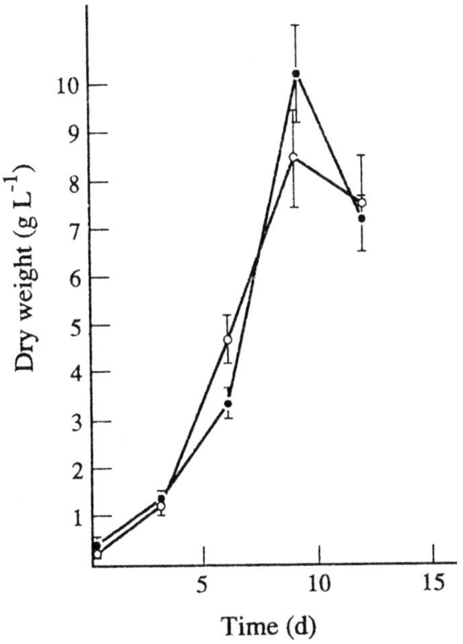


**Figure 10** The growth of *Catharanthus roseus* before and after 5 h exposure to 1000 r.p.m. in a 3 L bioreactor (○) before, (●) after (reproduced from Scragg *et al.*, 1988, with permission of Butterworth, Stoneham)

**Table 6** The Effects of Shear Stress on the Viability of Plant Cell Suspensions[a]

| *Cell line* | *Loss of weight (%) after 5 h at 1000 r.p.m.* *Dry wt.* | *Wet wt.* | |
|---|---|---|---|
| *Datura stramonium* | 40 | 70 | Sensitive |
| *Helianthus annus* | 6 | 44 | Tolerant |
| *Vitis* spp. | 28 | 10 | Tolerant |
| *Solanum* spp. | 6 | 20 | Tolerant |
| *Nicotiana tabacum* | 91 | 96 | Sensitive |
| *Picrasma quassioides* | | | |
| 1st year | — | — | Sensitive |
| 2nd year | 20 | 47 | Tolerant |
| *Catharanthus roseus* | | | |
| 0 line | 0 | 4 | Tolerant |
| *Catharanthus roseus* C87N | 30 | 40 | Tolerant |
| *Catharanthus roseus* C87P | | | |
| 1st year | 35 | 47 | Sensitive |
| 2nd year | 23 | 27 | Tolerant |
| *Catharanthus roseus* IDI | 26 | 46 | Tolerant |

[a]Reproduced from Scragg (1990b), with permission of the publishers.

culture was not affected by up to 5 h exposure. A range of plant cell suspensions have been tested in a similar manner (Table 6), and it is clear that most plant cell suspensions are more stress tolerant than was at first thought. Another interesting feature of the shear stress tolerance was its dynamic nature. Cultures tested early in their history were found to be shear sensitive but when tested later become shear tolerant. The development of this tolerance appears to be related to improvements in growth and growth rates of the cultures, but not to the aggregated nature of the culture. This has allowed the consideration of the stirred tank bioreactor as an alternative to the airlift bioreactor. Many of the cultures tested for shear tolerance have subsequently been grown successfully in stirred tank bioreactors.

In addition to the effect of exposure to high shear stress for short periods, *C. roseus* cultures have been exposed to high shear stress for extended periods. The results of growth of *C. roseus* cultures at 500 and 1000 r.p.m. in a 3 L stirred tank bioreactor are shown in Figure 11. At the highest speed, growth was reduced somewhat and cell expansion reduced considerably, but nevertheless growth occurred. It is hoped to extend this study to other cultures, but it does suggest that the normal stirred tank can be used for plant cell cultures at quite high operating speeds. The higher speeds may be required to mix high density cultures. However, cultivation of *N. tabacum* cell cultures in 3 L bioreactors at impeller speeds of up to 200 r.p.m. showed that with a normal flat bladed impeller, growth was stopped at the higher speed (Hooker *et al.*, 1990).

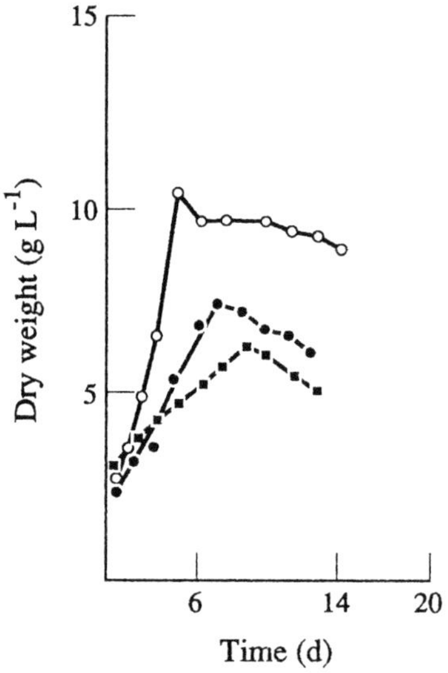


**Figure 11** The growth of *Catharanthus roseus* IDI at 500 and 1000 r.p.m. in a 3 L bioreactor (○) shake flask; (●) 500 r.p.m.; (■) 1000 r.p.m. (reproduced from Scragg, 1990b, with permission of Clarendon Press, Oxford)

### 4.2.8 Mixing

In a normal stirred tank bioreactor the impeller has a dual function; to break up the sparged air bubbles in order to increase oxygen transfer and to mix the culture. In plant cell suspensions mixing of the culture is probably more important because of the low oxygen requirements of such cultures. Mixing can be affected by the culture viscosity and one has to bear in mind the shear sensitivity of the cultures. Under conditions of high cell density, higher impeller speeds or higher aeration rates are required for good mixing and oxygen supply. With the possible sensitivity of plant cells to shear stress, these high impeller speeds and aeration rates may result in cell damage. Neither stirred tank bioreactors nor airlift bioreactors give completely satisfactory mixing at high cell densities.

To reduce these problems a number of impeller designs have been investigated. Furuya *et al*, (1984) used three impellers, a disc Rushton turbine, an anchor and angled disc turbine, and concluded that the angled disc turbine gave the best growth ratio and dry weight increase from *Panax ginseng* cultures, in a 30 L bioreactor. A spiral impeller was found to be most effective for the culture of high levels of *Coleus* (Ulbrich *et al.*, 1985) when compared with turbine and anchor impellers (Figure 12). Inclined blade impeller systems have also been used at high speeds (300 r.p.m.) in stirred tank bioreactors (Leckie, 1989). 30° and 60° inclined impellers have been used with *C. roseus*

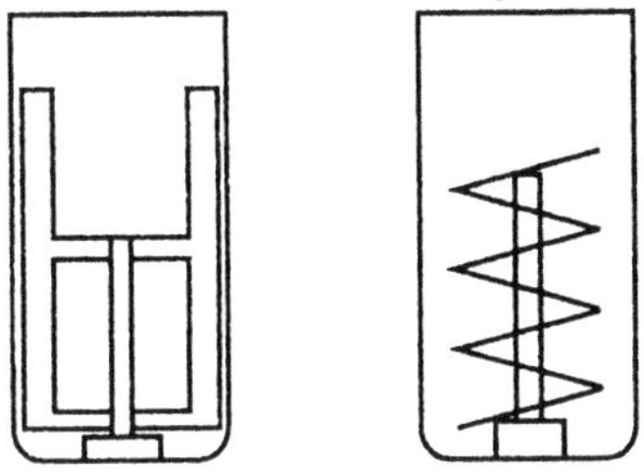


**Figure 12** Impeller designs used in a stirred tank bioreactor to cultivate *Coleus* cell suspensions (Ulbrich *et al.*, 1986)

cultures and the 60° impeller in particular gave less shear, and thus better growth. The inclined impeller also has an advantage in that it has a high pumping ability, which should be useful with high density cultures.

Other impeller designs have been used for plant suspension cultures. Cultures of *Glycine max* and *Pinus elliottii* have been grown in 1.25 and 2.5 L working volume bioreactors with flat bladed, marine and cell-lift impellers run at 30–80 r.p.m. (Treat *et al.*, 1989). The cell-lift impeller shown in Figure 13(a) yielded a similar biomass to the flat bladed impeller, but improved viability and increased aggregate size. In another study using *N. tabacum* suspension cultures and a 5 L bioreactor, a large flat bladed and a sail impeller were tried (Hooker *et al.*, 1990). The large flat bladed impeller was a considerable improvement over the normal flat bladed impeller in terms of growth and biomass yield (Figure 13b).

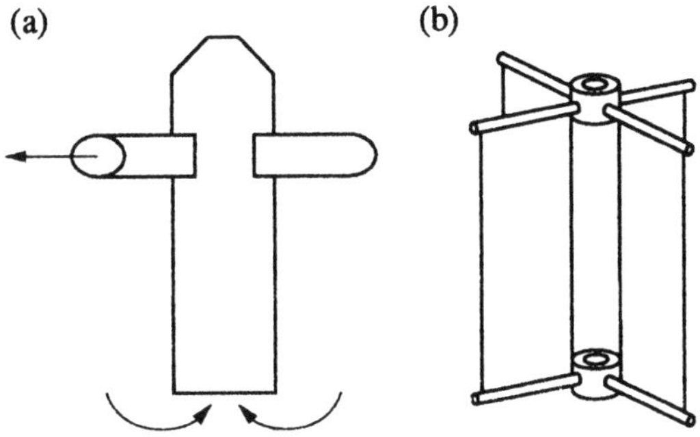


**Figure 13** Alternative impeller designs: (a) the cell-lift impeller used to cultivate *Glycine max* and *Pinus elliotti* (Treat *et al.*, 1989); (b) a large flat bladed impeller used with *N. tabacum* cultures (Hooker *et al.*, 1990)

## 4.3 BIOREACTOR CONFIGURATION

There are several basic bioreactor configurations that could be used with aerobic suspension cultures (Figure 14) and choosing an optimal bioreactor configuration for a process depends on a number of factors. Some of the more important factors are oxygen transfer, mixing and the magnitude of the acceptable shear rates. In many cases a successful bioreactor design involves a balance of potentially conflicting factors. In the case of plant cell suspensions the bioreactor has to provide good mixing at low shear stress levels with a moderate oxygen requirement.

With the considerations described above, a wide range of bioreactor configurations have been used to cultivate plant cell suspensions (Table 1). The initial reports on the mass cultivation of plant cells described the use of simple sparged carboys or roller bottles, technology from animal cell cultures. The simple bioreactors were soon replaced by stirred tank bioreactors run at low impeller speeds. The low shear characteristics of the airlift bioreactor encouraged its adoption in 1977 for the cultivation of plant cell suspension. Airlift bioreactors up to 80 L in volume have been successfully used for the cultivation of a wide range of cell lines (Breuling *et al.*, 1985; Scragg *et al.*, 1987a; Scragg, 1990b) and remain one of the first choices in bioreactor design. However, the use of stirred tank bioreactors has not ceased and recently Berlin (1986) and Berlin *et al.* (1985) have

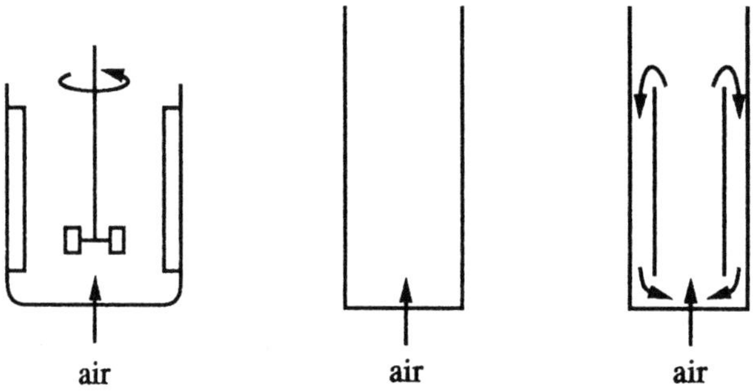


**Figure 14** Basic bioreactor designs

reported the growth of *N. tabacum*, *C. roseus* and *Solarum demissum* in bioreactors of 70, 750 and 8000 L. These bioreactors were fitted with paddle impellers, which were operated at below 100 r.p.m. Schiel and Berlin (1989) have used a sequential series of bioreactors of 25, 70, 300 and 5000 L (Figure 15) to cultivate *C. roseus* cultures for the production of ajmalicine. Here again the impeller was a turbine and was operated at 60–80 r.p.m. Fujita and Tabata (1987) have used a 1000 L stirred tank to grow cultures of *Lithospermum erythrorhizon*.

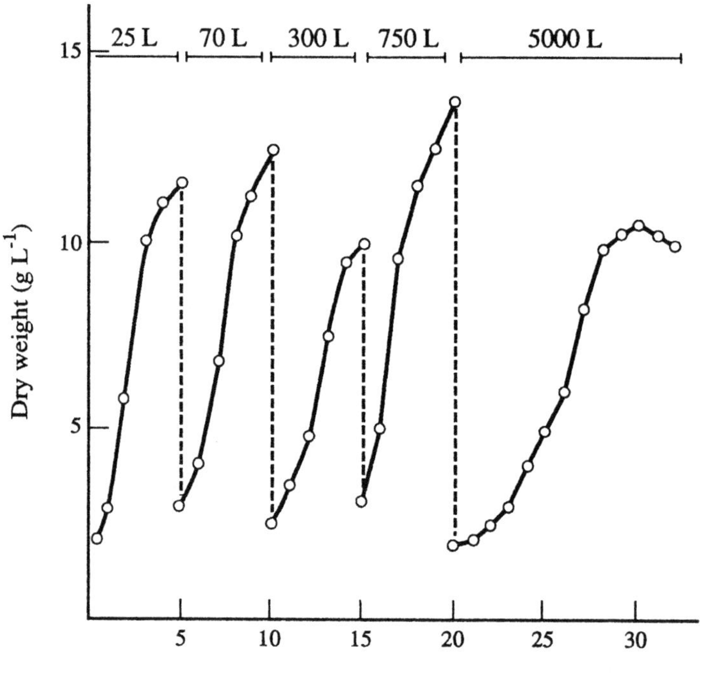


**Figure 15** The scale-up of *Catharanthus roseus* culture (reproduced from Schiel and Berlin, 1989, with permission of Kluwer, Dordrecht)

Alternative designs to the airlift and stirred tank bioreactor have been used to cultivate plant cells and attempt to achieve good mixing under low shear conditions. Tanaka *et al.* (1983) reported the use of a rotating drum bioreactor (Figure 16), which has been used to cultivate *C. roseus* and *L. erythrorhizon* up to volumes of 1000 L. The design has proved to be superior to stirred tank bioreactors for the production of shikonin from *L. erythrorhizon* (Tanaka, 1987).

A bioreactor based on Taylor–Couette flow, where mixing was achieved by the vortices between two concentric rotating cylinders has been tested (Janes *et al.*, 1987). Aeration is provided by the rotating inner cylinder, which is also a gas-permeable membrane, and this provides a bubble free bioreactor. The bioreactor has successfully been used to cultivate *Beta vulgaris* cells.

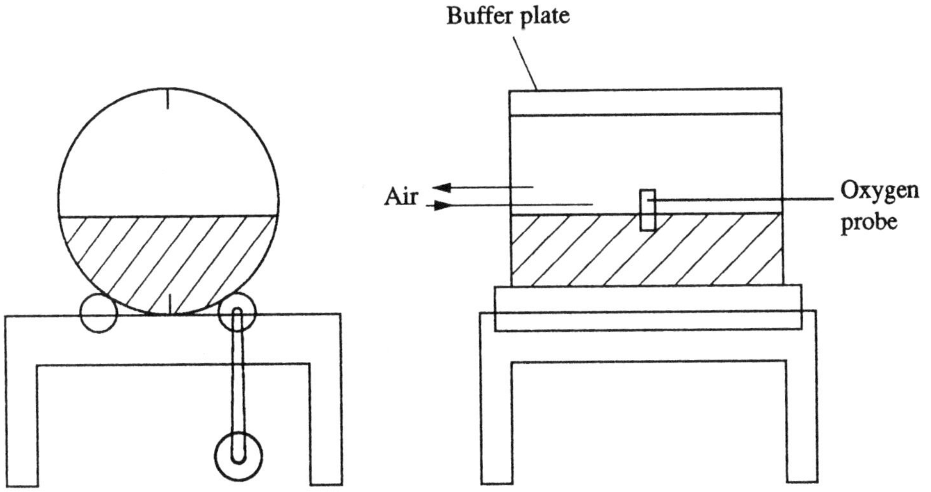


**Figure 16** A schematic diagram of the rotating drum fermenter (reproduced from Tanaka *et al.*, 1983, with permission of Wiley, New York)

Another bioreactor designed to provide bubble free aeration has been reported by Piehl *et al.* (1988). The supply of oxygen is provided by a rotating coil of membranes constructed from hydrophobic porous polypropylene. The membranes are permeable to oxygen, which diffuses into the culture as air is passed through the coil. The rotation of the coil helps with mixing and gas diffusion and 1 and 21 L bioreactors have been used to cultivate *Thalictrum rugosum*, which yielded a biomass level of 50 g $L^{-1}$ dry weight and maintained a dissolved oxygen level of 30% throughout the culture period.

## 4.4 BIOREACTOR OPERATION

A commercial process requires high productivities (g $L^{-1} d^{-1}$), high product yield and high product concentrations to be economic. Productivities determine the size of the bioreactor used and product accumulation determines the manner in which the bioreactor is operated. Many of the

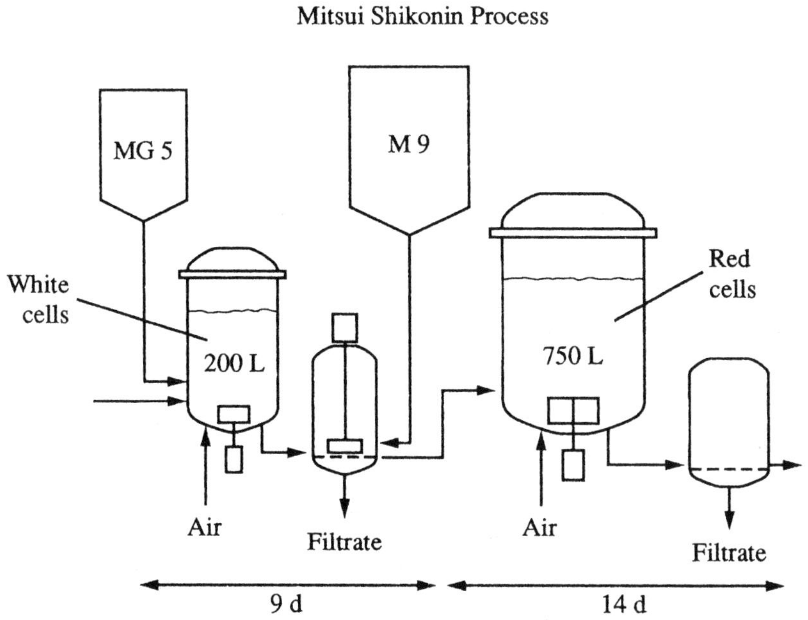


**Figure 17** The Mitsui process for the production of shikonin from cultures of *Lithospermum erythrorhizon* (Curtin, 1983)

**Table 7** Variation in Process Formats for Plant Cell Cultures

| | | |
|---|---|---|
| Single-stage | Batch | Product either growth or nongrowth related |
| Two-stage | Batch (nongrowth related product) | Batch (change in conditions to accumulate product) |
| | Draw-fill | Batch |
| | Continuous | Batch |

products of commercial interest produced by plant cell cultures are secondary products and are often produced after growth has ceased. Under conditions where growth and product accumulation are separate it can be advantageous to optimize growth and production separately. Often product accumulation is stimulated by using a separate medium (Zenk *et al.*, 1987) as employed in the recent shikonin process (Curtin, 1983; Figure 17). Whether using a single or two-stage process, the bioreactor can be operated as a batch, fed-batch, draw-fill and continuous system (Table 7).

Batch cultivation is characterized by changing physiological and environmental conditions and is the most widely used system. If a high biomass level is required, as a high initial carbon level may inhibit growth or if the production of the product is maximal at a specific growth rate, then fed-batch operation, where the growth-limiting nutrient is added over an extended period, can be used. The draw-fill technique operates by removing up to 90% of the culture and replacing this with fresh medium. This technique removes the need for inoculating the vessel and hence shortens the time normally used to clean out the bioreactor, resterilize it and inoculate. Continuous culture operations allow the cells to grow at a constant rate and gives a continuous supply of cells. Because of the slow growth and aggregated nature of plant cell cultures, continuous culture has proved difficult. There have been a number of reports on continuous culture on a laboratory scale but only one of 20 000 L on an industrial scale.

In conclusion it is clear that the large scale cultivation of plant cell suspensions is now possible for many cell lines. No one bioreactor design or process format is optimal for all cell lines; the design will have to be chosen with regard to the characteristics of the particular cell line.

## 4.5 REFERENCES

Alfermann, A. W., H Spieler and E. Reinhard (1985). Biotransformation of cardiac glycosides by *Digitalis* cell cultures in airlift reactors. In *Primary and Secondary Metabolism of Plant Cell Cultures*, ed. B. Deus-Neumann, W. Barz and E. Reinhard, pp. 316–322. Springer-Verlag, Berlin.

Balandrin, M. F. and J. A. Klocke (1988). Medicinal, aromatic and industrial materials from plants. In *Biotechnology in Agriculture and Forestry*, ed. Y. P. S. Bajaj, vol. 4, pp. 3–36. Springer-Verlag, Berlin.

Barz, W. and B. E. Ellis (1981). Plant cell cultures and their biotechnological potential. *Ber. Dtsch. Bot. Ges.*, **94**, 1–22.

Berlin, J. (1984). Plant cell cultures — a future source of natural products. *Endeavour*, **8**, 5–8.

Berlin, J. (1986). Large-scale fermentation of transformed and nontransformed plant cell cultures for the production of useful compounds. *Symbiosis*, **2**, 55–65.

Berlin, J. (1988). Formation of secondary metabolites in cultured plant cells and its impact on pharmacy. In *Biotechnology in Agriculture and Forestry*, ed. Y. P. S. Bajaj, vol. 4, pp. 37–59. Springer-Verlag, Berlin.

Berlin, J., V. Wray, E. Forsche, H.-G. Reng, H. Schüler, R. Luckinger and H.-P. Mühlbach (1985). Production of potato spindle tuber viroid (PSTV) by large-scale fermentation of PVTS-infected potato cell suspension cultures. *J. Exp. Bot.*, **36**, 1985–1995.

Bond, P. A., P. Hegarty and A. H. Scragg (1987). The use of antifoams in the mass cultivation of plant cells. In *Proceedings of the 4th European Congress of Biotechnology*, ed. O. M. Neijssel, R. R. Van-der-Meet and K. Ch. A. M. Leyben, pp. 440–443. Elsevier, Amsterdam.

Boysan, F., K. R. Cliffe, F. Leckie and A. H. Scragg (1988). The growth of *Catharanthus roseus* in stirred tank bioreactors. In *Fluid Dynamics*, ed. R. King, pp. 245–258. Elsevier, London.

Breuling, M., A. W. Alfermann and E. Reinhard (1985). Cultivation of cell cultures of *Berberis wilsonae* in 20 L airlift bioreactors. *Plant Cell Reports*, **4**, 200–223.

Byrne, A. F. and M. B. Koch (1962). Food production by submerged cultures of plant tissue cells. *Science (Washington, D.C.)*, **135**, 215–226.

Curtin, M. E. (1983). Harvesting profitable products from plant tissue culture. *Biotechnology*, **1**, 649–657.

Dalton, C. C. (1978). The culture of plant cells in fermenters. *Heliosynthase of Aquaculture seminar de Martiques CNRS*, pp. 1–11.

Dougall, D. K. (1979). Production of biologicals by plant cell cultures. In *Cell Substances*, ed. J. C. Petricciani, H. E. Hopps and P. J. Chapple, pp. 135–145. Plenum, New York.

Drapeau, D., H. W. Blanch and C. R. Wilke (1987). Ajmalicine, serpentine and catharamthine accumulation in *Catharanthus roseus* bioreactor cultures. *Planta Med.*, **52**, 373–376.

Ducos, J. P., G. Feron and A. Pareilleux (1988). Effect of aeration rate and influence of $P_{CO_2}$ in large-scale cultures of *Catharanthus roseus* cells. *Appl. Microbiol. Biotechnol.*, **25**, 101–105.

Esquivel, M. G., M. M. R. Fonseca, J. M. Novais, J. M. P. Cabral and M. S. S. Pais (1988). Continuous coagulation of milk using immobilized cells of *Cynara cardunculus*. In *Plant Cell Biotechnology*, ed. M. S. S. Pais, F. Muvituna and J. M. Novais, pp. 379–387. Springer-Verlag, Berlin.

Farnsworth, N. R. and R. W. Morris (1976). Higher plants — the sleeping giant of drug development. *Am. J. Pharm.*, **148**, 46–52.

Fowler, M. W. (1982). The large-scale cultivation of plant cells. *Prog. Ind. Microbiol.*, **17**, 209–220.

Fujita, Y. and M. Tabata (1987). Secondary metabolites from plant cells — pharmaceutical applications and progress in commercial production. In *Plant Tissue and Cell Culture*, ed. C. E. Green, D. A. Somers, W. P. Hackett and D. D. Biesboer, pp. 169–185. Liss, New York.

Furuya, T., T. Yoshikawa, Y. Orihara and H. Oda (1984). Studies on the culture conditions for *Panax ginseng* in jar fermenters. *J. Nat. Prod.*, **47**, 70–75.

Hahlbrock, K., J. Ebel and A. Oaks (1974). Determination of specific growth stages of plant cell cultures by monitoring conductivity changes in the medium. *Planta*, **118**, 75–84.

Hale, A. D., C. J. Pollock and S. J. Dalton (1987). Polysaccharide production in liquid cell suspension cultures of *Phleum praterse* L. *Plant Cell Rep.*, **6**, 435–438.

Hegarty, P. K., N. J. Smart, A. H. Scragg and M. W. Fowler (1986). The aeration of *Catharanthus roseus* L. G. Don suspension cultures in airlift bioreactors: the inhibition effect at high aeration rates on culture growth. *J. Exp. Bot.*, **37**, 1911–1920.

Henshaw, G. G., K. K. Jha, A. R. Mehta, D. J. Shakeshaft and H. E. Street (1966). Studies on the growth in culture of plant cells. *J. Exp. Bot.*, **17**, 362–377.

Hooker, B. S., J. M. Lee and G. An (1989). Response of plant tissue culture to a high shear environment. *Enzyme Microb. Technol.*, **11**, 484–490.

Hooker, B. S., J. M. Lee and G. An (1990). Cultivation of plant cells in a stirred vessel: effect of impeller design. *Biotechnol. Bioeng.*, **35**, 296–304.

Janes, D. A., N. H. Thomas and J. A. Callow (1987). Demonstration of a bubble-free annular-vortex membrane bioreactor for batch culture of red beet cells. *Biotechnol. Technol.*, **1**, 257–262.

Kato, K., Y. Shiozawa, A. Yamada, K. Nishida and M. Noguchi (1972). A jar fermenter culture of *Nicotiana tabacum* L. cell suspensions. *Agric. Biol. Chem.*, **36**, 899–902.

Kato, A., Y. Shimizu and S. Nagai (1975). Effect of initial $k_La$ on the growth of tobacco cells in batch culture. *J. Ferment. Technol.*, **53**, 744–751.

Kato, K., S. Kowazoe and Y. Soh (1978). Viscosity of the broth of tobacco cells in suspension culture. *J. Ferment. Technol.*, **56**, 224–228.

Kurz, W. G. W., K. B. Chatson, F. Constabel, J. P. Kutney, L. S. L. Choi, P. Kolodziejczyk, S. K. Sleigh, K. L. Stuart and B. R. Worth (1981). Alkaloid production in *Catharanthus roseus* cell culture VIII. *Planta Med.*, **42**, 22–31.

Leckie, F. (1989). Ph.D. Thesis, University of Sheffield.

Leckie, F., A. H. Scragg and K. C. Cliffe (1991). An investigation into the role of initial $k_La$ on the growth and alkaloid accumulation by cultures of *Catharanthus roseus*. *Biotechnol. Bioeng.*, **37**, 364–370.

LeFrancois, L., C. G. Mariller and J. V. Mejane (1955). Effectionements aux procedes de cultures pongiques et de fermentations industrielles. *Fr. Pat.* 1 102 200.

Mandels, M. (1972). The culture of plant cells. *Adv. Biochem. Eng.*, **2**, 201–215.

Martin, S. M. (1980). Mass culture systems for plant cell suspensions. In *Plant Tissue Culture as a Source of Biochemicals*, ed. E. J. Staba, pp. 149–166. CRC Press, Boca Raton, FL.

Morris, P., K. Rudge, R. Cresswell and M. W. Fowler (1989). Regulation of product synthesis in cell cultures of *Catharanthus roseus* V. Long-term maintenance of cells on a production medium. *Plant Cell, Tissue and Organ Culture*, **17**, 79–90.

Nickell, L. G. (1980). Products. In *Plant Tissue Culture as a Source of Biochemicals*, ed. E. J. Staba, pp. 256–269. CRC Press, Boca Raton, FL.

Noguchi, M., T. Matsumoto, Y. Hirata, K. Yamamoto, A. Katsuyama, A. Kato, S. Azechi and K. Katoh (1982). Improvements of growth rates of plant cell cultures. In *Plant Tissue Culture and its Biotechnological Applications*, ed. W. Barz, E. Reinhard and M. H. Zeuk, pp. 85–94. Springer-Verlag, Berlin.

Pareilleux, A. and R. Vinas (1983). Influence of the aeration rate on the growth yield in suspension cultures of *Catharanthus roseus* L. G. Don. *J. Ferment. Technol.*, **61**, 429–433.

Payne, G. F., M. L. Shuler and P. Brodelius (1989). Large scale plant cell culture. In *Large Scale Cell Culture Technology*, ed. B. K. Lydersen, pp. 193–229. Hanser, New York.

Piehl, G.-W., J. Berlin, G. Mollenschott and J. Lehmann (1988). Growth and alkaloid production of a cell suspension culture of *Thalictrum rugosum* in shake flasks and membrane-stirrer reactors with bubble-free aeration. *Appl. Microbiol. Biotechnol.*, **29**, 456–461.

Roels, J. A., J. Van Den Berg and R. M. Voncken (1974). The rheology of mycelial broth. *Biotechnol. Bioeng.*, **16**, 181–208.

Routier, J. B. and L. G. Nickell (1956). Cultivation of Plant Tissue. *US Pat.* 2 747 334.

Schiel, O. and J. Berlin (1987). Large scale fermentation and alkaloid production of cell suspension cultures of *Catharanthus roseus*. *Plant Cell Tissue and Organ Culture*, **8**, 153–161.

Schuler, M. L. (1981). Production of secondary metabolites from plant tissue culture. *Ann. N.Y. Acad. Sci.*, **369**, 65–69.

Scragg, A. H. (1990a). Large scale cultivation of *Helianthus annuus* cell suspensions. *Enzyme Microb. Technol.*, **12**, 82–85.

Scragg, A. H. (1990b). Fermentation systems for plant cells. In *Secondary Products from Plant Tissue Culture*, ed. B. V. Charlwood and M. J. C. Rhodes, pp. 243–263. Clarendon Press, Oxford.

Scragg, A. H. and M. W. Fowler (1985). The mass culture of plant cells. In *Cell Culture and Somatic Cell Genetics of Plant*, ed. I. Vasil, vol. 2, pp. 103–118. Academic Press, London.

Scragg, A. H., E. J. Allan, P. A. Bond and N. J. Smart (1986). Rheological properties of plant cell suspension cultures. In *Secondary Metabolism in Plant Cell Cultures*, ed. P. Morris, A. H. Scragg, A. Stafford and M. W. Fowler, pp. 178–194. Cambridge University Press, London.

Scragg, A. H., P. Morris, E. J. Allan, P. Bond and M. W. Fowler (1987a). Effect of scale-up on serpentine formation by *Catharanthus roseus* suspension cultures. *Enzyme Microb. Technol.*, **9**, 619–624.

Scragg, A. H., P. Bond, F. Leckie, R. C., Cresswell and M. W. Fowler (1987b). Growth and product formation by plant cell suspensions cultivated in bioreactors. In *Bioreactors and Biotransformations*, ed. G. W. Moody and P. B. Baker, pp. 12–25. Elsevier, London.

Scragg, A. H., E. J. Allan and F. Leckie (1988). Effect of shear on the viability of plant cell suspensions. *Enzyme Microb. Technol.*, **12**, 362–367.

Smart, N. J. (1984). Plant cell technology as a route to natural products. *Lab. Pract.*, Jan., 11–35.

Smart, N. J. and M. W. Fowler (1984). Effect of aeration on large-scale cultures of plant cells. *Biotechnol. Lett.*, **3**, 171–176.

Staba, E. J. (1977). Tissue culture and pharmacy. In *Plant Cell Tissue and Organ Culture*, ed. J. Reinart and Y. S. Bajaj, pp. 694–704. Springer-Verlag, Berlin.

Tanaka, H. (1981). Technological problems in cultivation of plant cells at high density. *Biotechnol. Bioeng.*, **23**, 1203–1218.

Tanaka, H. (1982). Some properties of pseudocells of plant cells. *Biotechnol. Bioeng.*, **14**, 2591–2596.

Tanaka, H. (1987). Large-scale cultivation of plant cells at high density: a review. *Process Biochem*, Aug., 106–113.

Tanaka. H., F. Nishijima, M. Suwa and T. Iwamoto (1983). Rotating drum fermenter for plant cell suspension cultures. *Biotechnol. Bioeng.*, **25**, 2359–2365

Treat, W. J., C. R. Engler and E. J. Soltes (1989). Culture of photomixotropic soybean and pine in a modified fermentor using a novel impeller. *Biotechnol. Bioeng.*, **34**, 1191–1202.

Tulecke, W. and L. G. Nickell (1959). Production of large amounts of plant tissue by submerged culture. *Science* (*Washington, D.C*), **130**, 863–864.

Ulbrich, B., W. Weisner and H. Arens (1985). Large scale production of rosemarinic acid from plant cultures of *Coleus blumei*. In *Secondary Metabolism of Plant Cell Culture*, ed. B. Deus-Neumann, W. Barz and E. Reinhard, pp. 293–303. Springer-Verlag, Berlin.

Veliky, I. A. and A. Jones (1981). Bioconversion of gitoxiganin by immobilized plant cells in a column bioreactor. *Biotechnol. Lett.*, **3**, 551–554.

Vickery, M. L. and B. Vickery (1981). *Secondary Plant Metabolism*, Macmillan Press, London.

Vogelmann, H. (1981). Aspects on scale-up and mass cultivation of plant tissue culture. In *Advances in Biotechnology*, ed. M. Moo-Young, pp. 117–121. Pergamon Press, Oxford.

Wagner, F. and H. Vogelmann (1977). Cultivation of plant tissue cultures in bioreactors and formation of secondary products. In *Plant Tissue and its Biotechnological Applications*, ed. W. Barz, E. Reinhard and M. H. Zeuk, pp. 245–252. Springer-Verlag, Berlin.

Wang, C. J. and E. J. Staba (1963). Peppermint and spearmint tissue culture II. Dual-carboy culture of spearmint tissues. *J. Pharm. Sci.*, **52**, 1058–1062.

Widholm, J. M. (1972). The use of fluorescein diacetate and phenosafranine for determining the viability of cultured plant cells. *Stain Technol.*, **47**, 189–194.

Zenk, M. H, E. El-Shagi, H. Arens, J. Stockigt, E. W. Weiler and B. Deus (1977). Formation of the indole alkaloids serpentine and gimaliane in cell suspension cultures of *Catharanthus roseus*. In *Plant Tissue Culture and its Biotechnical Applications*, ed. W. Barz, E. Reinhard and M. H. Zeuk, pp. 27–43. Springer-Verlag, Berlin.

# 5

# Immobilized Plant Cells

PAUL D. WILLIAMS and FERDA MAVITUNA
*University of Manchester Institute of Science and Technology, UK*



## 5.1 INTRODUCTION

Immobilization of biocatalysts offers many potential advantages and is now a well-established technique, with the history of enzyme immobilization going back over 25 years and including many industrial applications. The immobilization of microorganisms is less well developed in terms of large scale applications, but is widely used in the laboratory. With this background it was inevitable that immobilization techniques should be applied to plant cell cultures and much work has been carried out to establish methods for plant cell immobilization and suitable bioreactors for use with the immobilized cultures.

Although many of the methods used for immobilization are common to both microorganisms and plant cell cultures, it is worth noting that there are major practical differences between culturing suspensions of plant cells and microorganisms. Individual plant cells are generally larger than microbial cells and in suspension culture are usually found as aggregates, which may be up to several millimetres in diameter. They also show greater susceptibility to damage from mechanical stresses such as sustained shear, and from environmental stresses such as changes in temperature and oxygen concentration. Any method for the immobilization of plant cells must therefore take into account these characteristics.

In general the advantages of immobilizing biocatalyst may be summarized (Brodelius, 1984) as follows: (i) the biocatalyst is easily recovered and can be used over an extended period of time; (ii) the desired product is easily separated from the catalyst; (iii) the continuous operation of a process is readily achieved; and (iv) the immobilized catalyst often shows increased stability. These benefits are equally applicable to microbial systems but, more specifically, immobilization offers a solution to some of the physiological requirements and some of the process engineering problems particularly associated with plant cell cultures.

Physiological requirements for plant cell cultures for the production of secondary metabolites appear to include cell to cell contact to allow transfer of materials from one cell to another. Such contact may help induce cytodifferentiation, which is related to secondary metabolism (Yeoman *et al.*, 1982). There is also evidence that plant secondary metabolites are produced at higher concentrations in slow-growing cultures (Kurz and Constabel, 1985). In an immobilized system, growth and production phases can be decoupled and controlled by chemical and physical stress conditions. This allows cells to be retained in the bioreactor for extended periods, with alternating rejuvenation/growth and secondary metabolite production cycles. The slow growth of plant cell cultures gives a long lead time before beginning bioreactor operation and it is therefore of advantage to extend the productivity of a bioreactor as far as possible.

Process engineering problems may develop from the tendency of plant cells to aggregate, which can lead to blockages in pipes and openings and to the culture rapidly sedimenting, if it is not continually agitated. However, the shear sensitivity of the cultures means that mechanical agitation may be detrimental to cells and that cultures cannot be transported using conventional pumps, without significant loss of viability. Again, immobilization may be a solution to these problems and may offer a microenvironment protected from sustained shear.

The main disadvantage of immobilization is that it is only of use with cell lines which excrete the product of interest into the culture medium. Attempts to induce the release of products which are normally retained within the cell, by such techniques as permeabilization, have generally decreased cell viability to an undesirable extent, although Brodelius and Nilsson (1983) have produced encouraging results with reversible permeabilization of *Catharanthus roseus* using dimethyl sulfoxide (DMSO). Systems for inducing product release may eventually further increase the applicability of immobilization, but even without such methods the range of secondary metabolites produced by immobilized plant cell cultures is extensive.

Brodelius (1985a) suggests that the potential role of immobilized plant cells for the large scale production of secondary metabolites cannot be fully evaluated, until such biological problems as low productivities and genetic instability have been addressed. However, although improvements in these areas are of key importance to the commercialization of plant tissue culture products, immobilization offers sufficient advantages to suggest that it will have important applications in future developments in this technology.

## 5.2 SYSTEMS FOR IMMOBILIZED CULTURES

### 5.2.1 Immobilization Techniques

Any immobilization method selected for plant cells should be harmless to the cells, easy to carry out under aseptic conditions, capable of operating for long periods and, particularly for large scale applications, low in cost (Mavituna *et al.*, 1987). In practice this has meant the use of some sort of entrapment immobilization in almost all cases.

Table 1 gives a number of examples of the systems of immobilization which have been used with plant cells, together with the associated plant species and their products. This list is not exhaustive, but aims to indicate the variety of cultures which have been immobilized and the popularity of the various methods of immobilization.

Entrapment methods which have been used with plant cell cultures can be categorized after Novais (1988) into: (i) gel entrapment by ionic network formation; (ii) gel entrapment by precipitation; (iii) gel entrapment by polymerization; and (iv) entrapment in preformed structures. The most widely used form of immobilization with plant cell cultures is entrapment by ionic network formation, especially in the form of alginate beads. Alginate is a polysaccharide which forms a

**Table 1** Some Examples of Immobilized Plant Cell Systems

| *Plant species* | *Substrate/ precursor* | *Product* | *Immobilization method* | *Ref.* |
|---|---|---|---|---|
| **Bioconversions** | | | | |
| *Catharanthus roseus* | Cathenamine | Ajmalicine | Agarose | Felix *et al.*, 1981 |
| *Digitalis lanata* | Digitoxin | Digoxin | Alginate | Brodelius *et al.*, 1981 |
| *Daucus carota* | Digitoxigenin | Periplogenin | Alginate | Jones and Veliky, 1981 |
| *Daucus carota* | Gitoxigenin | 5-Hydroxy-gitoxigenin | Alginate | Veliky and Jones, 1981 |
| *Mentha spicata* | (−)-Menthone | (+)-Neomenthol | Polyacrylamide | Galun *et al.*, 1983 |
| *Mucuna pruriens* | L-Tyrosine | L-DOPA | Alginate | Wichers *et al.*, 1983 |
| *Papaver somniferum* | Codeinone | Codeine | Alginate | Furuya *et al.*, 1984<br>Furusaki *et al.*, 1988 |
| *Papaver somniferum* | Codeinone | Codeine | Polyurethane foam | Corchete and Yeoman, 1989 |
| *Nicotiana tabacum* | Acetoacetic esters | 3-Hydroxy-butanoates | Alginate | Naoshima *et al.*, 1989 |
| *Nicotiana tabacum* | Keto esters | Hydroxy esters | Alginate | Naoshima and Akakabe, 1989 |
| **Synthesis from precursors** | | | | |
| *Catharanthus roseus* | Tryptamine, Secologanin | Ajmalicine | Alginate, agarose, Agar, carrageenan | Brodelius *et al.*, 1979<br>Brodelius and Nilsson, 1980 |
| *Capsicum frutescens* | Isocapric acid, Vanillylamine, Valine, ferulic acid | Capsaicin | Polyurethane foam | Brodelius and Nilsson, 1983<br>Lindsey *et al.*, 1983<br>Lindsey and Yeoman, 1984<br>Mavituna *et al.*, 1987 |
| *Datura innoxia* | Ornithine | Scopolamine | Alginate | Lindsey, 1982 |
| *Nicotiana tabacum* | Phenylalanine | Caffeoyl-putriscine | Alginate | Berlin *et al.*, 1989 |
| *Coffea arabica* | Theobromine | Caffeine | Membrane | Lang *et al.*, 1990 |
| ***De novo* synthesis** | | | | |
| *Morinda citrifolia* | | Anthraquinones | Alginate | Brodelius *et al.*, 1980 |
| *Catharanthus roseus* | | Ajmalicine, Serpentine | Alginate, agarose, Agar, carrageenan | Brodelius and Nilsson, 1980<br>Brodelius *et al.*, 1981<br>Lambe and Rosevear, 1983 |
| *Catharanthus roseus* | | Strictosidine lactam, Ajmalicine, Epivindolinine, Tabasonine, Catharanthine | Polyester fibres | Rho *et al.*, 1990 |
| *Solanum aviculare* | | Steroid glycosides | Polyphenylene oxide | Jirku *et al.*, 1981<br>Macek *et al.*, 1981 |
| *Glycine max* | | Phenolics | Hollow fibres | Shuler, 1981 |
| *Catharanthus roseus* | | Enzymes | Agarose, Polyurethane | Felix and Mosbach, 1982 |
| *Vicia faba* | | Ethane | Alginate | Schnabl *et al.*, 1983 |
| *Nicotiana tabacum* | | Epiandrosterone | Alginate | Tsuchiya, 1983 |
| *Lavandula vera* | | Pigments | Polyurethane gel | Tanaka *et al.*, 1984<br>Nakajima *et al.*, 1986 |
| *Capsicum frutescens* | | Capsaicin | Polyurethane foam | Lindsey and Yeoman, 1984<br>Wilkinson *et al.*, 1988 |
| *Apium graveolens* | | Phthalides | Alginate | Watts and Collin, 1985 |
| *Salvia miltiorrhiza* | | Cryptotan-shinone | Alginate | Miyasaka *et al.*, 1986 |
| *Coffea arabica* | | Methylxanthine | Alginate | Haldimann and Brodelius, 1987 |
| *Thalictrum minus* | | Berberine | Alginate | Kobayashi *et al.*, 1987 |
| *Dioscorea deltoidea* | | Diosgenin | Polyurethane foam | Ishida, 1988 |
| *Ginkgo biloba* | | Ginkgolides | Polyester fibre | Carrier *et al.*, 1990 |

reasonably stable gel in the presence of multivalent cations, with calcium being commonly used. Beads of alginate-containing cells are formed by dripping a cell/sodium alginate solution into a calcium chloride solution. This method of immobilization has the advantage of being easily reversible, by the addition of a calcium chelating agent, such as EDTA, which disrupts the gel by solubilizing the bound $Ca^{2+}$. However, the associated disadvantage is that media containing other calcium chelating agents, such as phosphates, may give gel disruption over long run times. $\kappa$-Carrageenan gels formed in a similar manner, using either calcium or potassium, can also be used for the entrapment of plant cells.

Preparations of agar and agarose (a purified agar) can be used to trap plant cells by precipitation. The polysaccharides form gels when a heated aqueous solution is cooled, and various commercial preparations are available with different gelling temperatures. The gel can then be mechanically broken down into particles or, alternatively, the gel can be dispersed into particles whilst still liquid by mixing in a hydrophobic phase. When particles of the desired size are obtained the whole mixture is cooled to give solidification (Nilsson *et al.*, 1983).

Gel entrapment by polymerization is most commonly carried out using polyacrylamide. However, the toxicity of the initiator and cross-linking agents used in the polymerization has in some cases caused a loss of cell viability (Cabral *et al.*, 1984). Galun and coworkers (1983) overcame this problem by suspending plant cells in an aqueous solution of prepolymerized linear polyacrylamide, partially substituted with acrylhydrazide groups. Controlled amounts of a dialdehyde were then used to cross link the gel, which was then mechanically disintegrated into particles. Another method of avoiding toxicity effects has been to mix the cells in a viscous polysaccharide solution (alginate or xanthan gum) to provide protection before the cells are entrapped in polyacrylamide (Rosevear and Lambe, 1982).

Entrapment in preformed structures involves some form of open network through which nutrient medium may pass, but which entraps plant cells or cell aggregates. Hollow fibre reactors have been used by Shuler (1981) to immobilize plant cells, with the cells being held between the fibres whilst the medium is rapidly recirculated through the fibres. Preformed reticulated polyurethane foam can also be used as a support matrix for immobilized plant cells, with the cells being entrapped within the network either by physical restriction or by attachment to the foam matrix. Such reticulated foam matrices have been used both in the form of particles (Ishida, 1988; Lindsey *et al.*, 1983) and as sheets (Williams *et al.*, 1987). The efficiency of entrapment in foam networks has been shown to be dependent on the matching of foam pore size with plant cell aggregate size (Park and Mavituna, 1987), and mechanical methods have been developed to artificially alter aggregate size without affecting long term culture viability (Williams *et al.*, 1988).

Entrapment is by far the most generally used type of immobilization, but other systems have been tried. Cabral and coworkers (1984) used gelatinous, hydrous metal oxides ($Ti^{IV}$ and $Zr^{IV}$) to form partial covalent bonds between cells. Plant cells have also been covalently linked to polyphenylene oxide beads (Jirku *et al.*, 1981).

### 5.2.2 Bioreactor Configurations

The main body of experimental work with immobilized plant cells has been carried out under simple batch conditions in shake flasks and relatively little work has been carried out on the design of bioreactors specifically for use with immobilized plant cells. In some ways, reactor design for immobilized cells should be simpler than for plant cell suspension cultures, as the problems associated with stress tolerance are minimized. However, problems of microbial contamination remain and, if immobilized reactors are to be run for extended time periods, can even be exacerbated. Also, in cases where immobilization takes place *in situ*, the reactor design has to accommodate both freely suspended and immobilized cells (Mavituna *et al.*, 1987). As yet no product from immobilized plant cells has been produced on a large scale and the design of bioreactors suitable for scale-up is still under investigation.

Packed bed reactors have been a favoured system for use with plant cells immobilized in alginate beads. Brodelius and coworkers (1979) used a packed bed of alginate beads to give a continuous production of ajmalicine from *Catharanthus roseus* and later Brodelius and Nilsson (1983) used a similar arrangement to show the effects of DMSO permeabilization of *Catharanthus* cells. Packed beds have also been used with cells immobilized with polyphenylene oxide (Jirku *et al.*, 1981) and agarose (Brodelius and Nilsson, 1983).

Fluidized beds have provided an alternative reactor configuration for use with alginate entrapped cells (Hamilton *et al.*, 1984). The large scale aseptic production of gel beads presents problems for the scale-up of such reactors, which have not been completely overcome, although Hulst and coworkers (1985) and Brodelius (1984) have devised improvements on the laboratory technique which allow for increased numbers of beads to be produced. The system described by Hulst uses a vibrating nozzle, which breaks up a jet of alginate into uniform droplets by mechanical vibrations. In the Brodelius system alginate is forced, under low pressure, through a set of six nozzles to form the droplets.

Another type of immobilized plant cell reactor, the membrane reactor, has the cells physically separated from the liquid medium by a membrane which allows the free passage of substrates, nutrients and cell products. Hollow fibre membrane reactors were originally developed for animal cell culture, but have been successfully used with plant cells. Prenosil and Pedersen (1983) have described their use for the culture of cells of *Glycine max, Daucus carota* and *Petunia hybrida.* A concentrated suspension of cells is introduced into the shell side of the reactor and medium, aerated in a separate reservoir, is circulated through the fibres. The fibre membrane provides a barrier against bacterial and other contaminants and the plant cells grow in the extra fibre spaces, without disrupting the integrity of the fibres. Other membrane reactor systems have been used by Shuler and coworkers (1986) for the production of phenolics from *Nicotiana tabacum* and for the production of caffeine from *Coffea arabica* (Lang *et al.*, 1990). Such reactors have several advantages, including good control of fluid dynamics and flow distribution (Shuler *et al.*, 1984), and the inherent advantage of using a membrane, of improved protection against contamination. However, membrane reactors are also expensive, liable to fouling, have problems with gas transfer and are also difficult to inoculate (Shuler *et al.*, 1986).

A flat bed reactor design was employed by Lindsey and Yeoman (1983), where cells were seated on a substratum of polypropylene foam matting, for the production of capsaicin from *Capsicum frutescens.* This work was taken further in the development of a system of immobilization using reticulate polyurethane foam (Lindsey *et al.*, 1983) and in reactor design for use with this method of immobilization. Initially these bioreactors were in the form of circulating beds of particulate foam (Mavituna *et al.*, 1987), but this was later adapted to a design using sheets of foam (Park, 1986). In this design the sheets were suspended as vertical baffles in a stirred reactor, so that the plant cells were incorporated into the network by the stirring to give solid sheets of cells. Sheets of stainless steel mesh have also been used in this type of bioreactor (Park, 1986). The main advantage of bioreactors using polyurethane foam or stainless steel mesh is the simplicity of the immobilization step, which is carried out *in situ* in the bioreactor vessel, decreasing the risk of contamination and reducing the problems of scale-up (Mavituna *et al.*, 1987).

## 5.3 CHARACTERISTICS OF IMMOBILIZED PLANT CELLS

### 5.3.1 Viability of Immobilized Plant Cells

It is important to know whether a particular method of immobilization adversely affects the biosynthetic ability of the plant cells and if not, how long this ability can be maintained. Except for the very simple biotransformations, the formation of the required compounds depends on sustaining the viability of the cells after immobilization. However, the definition of 'viability' can be subjective, and in this context all that is required may be that the cells are capable of producing a specific metabolite. The inability to biosynthesize the compound of interest does not necessarily imply that the cells are not viable; usually there are various other reasons, often unknown. Therefore, especially in the cases of failure to obtain the product, the viability of the cells after immobilization and during employment should be checked. The methods used to test the viability include staining, measurement of growth, cell division, respiration, substrate uptake, plasmolysis and NMR spectra.

#### *5.3.1.1 Staining*

The most commonly used staining techniques are the uptake and hydrolysis of fluorescein diacetate (FDA) (Widholm, 1972) or phenosafranin, exclusion of Evans blue, or the triphenyl tetrazolium chloride (TTC) method (Towill and Mazur, 1975). However, care is needed in the quantification of viability based on FDA and Evans blue staining results. If the cells are in aggregates, it will be the outer shell which can be observed for staining. It is difficult to view the interior of the aggregate without inflicting some damage on the cells. Therefore, for the immobilized or aggregated cells the TTC method, which is based on the production of red formazan, may give more reliable results. However, Law and Mavituna (1990) have suggested that the FDA and TTC methods may yield varying results depending on the physiology of the cultures and the medium composition.

Lindsey (1982) used FDA to test cell viability after the immobilization of various cultures in alginate, agar and reticulate polyurethane foam matrix and compared the results with the viability of suspension and flatbed cultures. The best results were obtained with polyurethane foam immobilization at 70–80% viability and worst were with agar entrapment at 50–60% viability. Using FDA, Wilkinson *et al.* (1990) found that a suspension of *Capsium frutescens* cells in batch culture were almost completely dead eight weeks after inoculation, whereas the cells immobilized in polyurethane foam matrix remained over 90% viable in batch culture for over 12 weeks, without any change in medium.

### *5.3.1.2 Respiration and substrate uptake*

Oxygen consumption and carbon dioxide evolution rates, as well as uptake rates of other nutrients from the medium, such as sucrose, phosphate, nitrate and ammonium, by the immobilized cells, have been compared with the rates obtained from freely suspended cells as a measure of viability after immobilization. Such comparisons, however, should be based on specific rates or the same amount of plant cells in immobilized and suspension cultures. The comparison is further complicated by the fact that the observed rates in immobilized cell systems can be lower than those in suspension cultures, not only because of a possible lowering of viability but also as a result of diffusional limitations in the immobilization matrix (Brodelius, 1988; Rho *et al.*, 1990; Robins *et al.*, 1986; Wilkinson *et al.*, 1990). Brodelius (1985b) has shown that the relative respiration of *C. roseus* immobilized in alginate increases after dissolving the polymer.

Brodelius and Nilsson (1980) have observed respiratory activity and growth of *C. roseus* immobilized in alginate, agar, agarose, and carrageenan but not in gelatin, polyacrylamide, alginate–gelatin or agarose–gelatin. Nakajima *et al.* (1985) used respiration measurements for the cell viability of *Lavandula vera* entrapped in agar, carrageenan and alginate, which produced blue pigments.

The uptake of phosphate, nitrate, ammonium and sucrose by *Capsicum frutescens* immobilized in a polyurethane foam matrix has been studied by Lindsey and Yeoman (1984). After a lag phase of four days all four nutrients were removed from the medium at approximately linear rates. In the case of alginate-entrapped cells, phosphate is sequestered by calcium ions present in the gel, as evidenced by the removal of about 17% of the total phosphate over a nine day period by cell-free alginate particles (Lindsey, 1982).

### *5.3.1.3 Growth and division*

Growth and cell division are the most important criteria indicating cell viability, even if they may not be desirable in an immobilized cell system. It is very difficult to measure the number of cells in an immobilized or aggregated state. As a measure of cell division, Brodelius (1985a, 1985b) suggests cutting thin slices of gel-entrapped cells and, after fixation, staining the chromosomes with carbol–fuchsin. Subsequently the red-stained nuclei are counted under the microscope and, out of at least 500 nuclei, those in mitosis are determined and expressed as a percentage of the total.

Parr *et al.* (1984) have developed a nondestructive method to determine the volume of space occupied by intact cells in an immobilization matrix. Their method relies on determining the differential dilutions of two molecules, one excluded by the plasmalemma, such as $^{14}C$-labelled, nonmetabolizable saccharide, and the other molecule able to diffuse freely throughout the cell, such as $^{3}H$ water. The method differentiates between the live and dead cell volume due to the presence or otherwise of intact plasmalemma. The 'viable volume' is that part expressed as a percentage of a defined space which excludes mannitol but is accessible to water. The method gave a close correlation with the dry weight during the growth and expansion phase of *Beta vulgaris* immobilized in a reticulate foam matrix, and much more accurately indicated necrosis in later phases.

Straightforward fresh and dry weight determinations are widely used, assuming the dry weight of the matrix stays constant during incubation and drying.

### *5.3.1.4 NMR spectra*

Phosphorus nuclear magnetic resonance (NMR) is a noninvasive technique for the study of intracellular metabolism. The levels of the major phosphorylated metabolites, such as ATP, and the uptake and storage of inorganic phosphate in different intracellular compartments can be monitored. Furthermore, the cytoplasmic and vacuolar pH values can be directly obtained from the spectra (Brodelius and Vogel, 1984). Brodelius (1985a) gives the $^{31}P$ NMR spectra of free, agarose-entrapped and alginate-entrapped *C. roseus* cells, which are very similar.

#### *5.3.1.5 Plasmolysis*

The integrity of the cell membrane can be checked by exposing cells to osmotic stress. Glycerol and sorbitol are generally used as plasmolyzing agents, and are followed by staining with a dye such as phenosaffranin which penetrates the damaged membrane. This method also suffers from the common visual observation problems of the staining methods applied to aggregated or immobilized cells.

#### *5.3.1.6 Scanning electron microscopy*

Scanning electron microscope (SEM) studies of immobilized cells can indicate the intactness of the cell, the distribution of cells in the immobilization matrix and visual differences between the various methods of immobilization. Robins *et al.* (1986) have recorded various SEM micrographs of immobilized cells of *Beta vulgaris* and *Humulus lupulus* in polyurethane foam, *Cinchona pubescens* in nylon matrix and *Daucus carota* in alginate.

### 5.3.2 Growth of Immobilized Cells

The main reason for immobilizing plant cells is to separate growth from the production phase and to prolong the use of cells in a stationary or very slow growing state so that the production of secondary metabolities is encouraged (Lindsey and Yeoman, 1983, 1985; Yeoman *et al.*, 1980, 1982). Therefore, after allowing some initial growth to fill the available space in the gel and in the pores, or on the surface of the support structures, growth of the immobilized cells is not desirable. This is generally achieved by switching to a medium which lacks the compounds necessary for growth. However, growth of immobilized cells can be encouraged in order to test the viability, to compare them with suspension cultures or to rejuvenate and repopulate the immobilization matrix if necessary.

Growth of immobilized cells may be affected due to diffusional problems, mechanical stress, or substrate and product interactions with the immobilization matrix due to charged groups. Excessive growth can disrupt the gel beads and cells can leak into the medium and grow as a suspension culture. Robins *et al.* (1986) found that with *D. carota* in alginate, the loss of cells from gel beads became a serious problem when the cells had grown to occupy only 30–40% of the bead volume. At higher levels disintegration occurred. Cell leakage can also be observed with other methods of immobilization.

In a comparison of the various methods of immobilization, Nilsson (1983) reported growth of *C. roseus* cells in alginate, carrageenan, agar and agarose, but not in gelatin or polyacrylamide. Cell growth patterns in agar and agarose were similar to those of suspension cultures. Mavituna and Park (1985) found that the maximum specific growth rate of *Capsicum frutescens* in suspension culture using dry weight measurements was $0.245\,d^{-1}$, and when immobilized in polyurethane foam matrix, it was reduced to $0.144\,d^{-1}$. Rho *et al.* (1990) found that the doubling time for *C. roseus* increased slightly from 2.0–2.3 d to 2.6–4.6 d when immobilized on the surface of a nonwoven polyester short fibre cloth. Brodelius *et al.* (1979), Majerus and Pareilleux (1986) and Payne *et al.* (1988) have all observed slower growth with immobilized cells of *C. roseus* than cells in suspension. According to Hamilton *et al.* (1984) the growth of alginate-immobilized cells of *D. carota* and *Petunia hybrida* was similar to free cells. Therefore, the growth characteristics of immobilized plant cells when compared to the free cells seem to depend on the method of immobilization, the plant species and physicochemical environment, especially the medium composition.

### 5.3.3 Biosynthetic Capacity

The biosynthetic capacity of immobilized plant cells, just like that of freely suspended cells, can be divided into three categories: biotransformations, *de novo* synthesis and synthesis from precursors.

#### *5.3.3.1 Bioconversions*

Biotransformation reactions are usually stereospecific and involve oxidation, reduction, hydroxylation, methylation, demethylation, acetylation, isomerization, glycosilation, esterification, epoxidation or saponification (Furuya, 1978; Kurz and Constabel, 1985). The chemical compounds which can undergo biotransformation are varied and include aromatics, steroids, alkaloids, coumarins and terpenoids (Alfermann *et al.*, 1980; Morris *et al.*, 1985). These compounds do not necessarily have to be natural intermediates in the plant metabolism. Biotransformations by plant cell cultures are attractive only if the transformation cannot be performed by chemical synthesis or microorganisms. In most biotransformation studies, in order to provide the nutrients for the growth of the cells prior to biotransformation or for the intermittent rejuvenation a number of basal media have been used, including those of Gamborg (1970), Linsmaier and Skoog (1965), Murashige and Skoog (1962) and Nitsch and Nitsch (1969).

The earliest study of biotransformations using immobilized plant cells was the 12$\beta$-hydroxylation of digitoxin to digoxin by *Digitalis lanata* entrapped in alginate (Brodelius *et al.*, 1979). It was found that the efficiency of biotransformation was not different from that of freely suspended cells and the hydroxylation continued for 33 d. Since then, the 12$\beta$-hydroxylation of digitoxin derivatives to the corresponding digoxin derivatives have become the most extensively studied biotransformation by immobilized plant cells. In a study, *D. lanata* entrapped in alginate converted most of the added digitoxin to purpurea glycoside A and only a small amount to digoxin and deacetyllanatoside C. $\beta$-Methyldigitoxin was hydroxylated to $\beta$-methyldigoxin (Alfermann *et al.*, 1983). The hydroxylating activity of the immobilized cells was about 50% of the suspended cells, although the activity of the immobilized cells was retained for 61 d, by which time the gel had deteriorated. However, the hydroxylation of $\beta$-methyldigitoxin could be carried out for up to 180 d with alginate-entrapped *D. lanata* by Moritz *et al.* (1982).

*Daucus carota* cells immobilized in alginate are capable of hydroxylating digitoxigenin to periplogenin even though neither of these compounds are ordinarily found in carrot (Jones and Veliky, 1981). *Mucuna pruriens* cells entrapped in alginate were used by Wichers *et al.* (1983) for the biotransformation of L-tyrosine to L-DOPA. Additional $Ca^{2+}$ in the medium to stabilize the beads severely inhibited the synthesis of L-DOPA. Galun *et al.* (1983) studied the biotransformation of (−)-menthone to (+)-neomenthol and of (+)-pulegone to (+)-isomenthone using *Mentha* cells immobilized in polyacrylamide hydrazide. The rate of biotransformation was similar in both freely suspended and immobilized cultures. Gamma irradiation of *Mentha* cells resulted in nondividing cells which could be used for continuous biotransformation of monoterpenes without growth (Galun *et al.*, 1985). Furuya *et al.* (1984) demonstrated that *Papaver somniferum* cells immobilized in alginate could transform (−)-codeinone to (−)-codeine much more efficiently (70.4% conversion over three days) than the freely suspended cells (60.8% conversion).

Since the early 1980s, several other biotransformations have been studied using immobilized plant cells, as summarized in Table 1. One of the current interests is the bioconversion of xenobiotics (pesticides) by plant cell cultures since the higher plants have a pronounced ability to metabolize and degrade xenobiotics (Barz *et al.*, 1990; Harms and Kottutz, 1990).

In most cases with immobilized plant cells, the results suggest that the biotransformations proceed in a more or less identical manner to those of freely suspended cells. Although the conversion rates may sometimes be slower, the increased stability of the enzyme system may compensate for this. It is important to establish whether the cells will need rejuvenation for prolonged operation, and if so how often and with which medium. It is also necessary to determine the optimum concentration and the rate of addition of substrate, the possible toxicities of the substrates and products, and the permeability of the cells to these compounds.

#### 5.3.3.2 *Synthesis from precursors*

In a number of experimental systems, it was shown that the enzymes of secondary metabolism did not work at their maximum rates because the concentrations of precursors, cosubstrates and other necessary intracellular compounds were too low (Luckner, 1990). Therefore, an increase in the intracellular concentrations of precursors and other rate-limiting compounds by addition to the medium may result in increased rates of secondary metabolite synthesis. However, the choice of the precursors requires at least some understanding of the biochemical pathways involved. As a result, there are only a few examples of production of secondary metabolites from precursors, especially using immobilized cells. Some of the experiments with precursor feeding were not successful, probably due to the strict compartmentalization and channelling of these compounds in the producer cell.

The production of various indole alkaloids from the precursors tryptamine and secologanin by immobilized *C. roseus* has been studied extensively (Brodelius *et al.*, 1979; Brodelius and Nilsson,

1980, 1983). Cells immobilized in agar, agarose and carrageenan produced ajmalicine isomers at about the same rate as the freely suspended cells and cells entrapped in alginate had increased synthesis at up to 169% of that observed for freely suspended cells. In these studies, the addition of precursors increased the product yield 12 times after five days of incubation in comparison to the *de novo* synthesis after two weeks of incubation. *Datura innoxia* immobilized in alginate increased the accumulation of scopolamine when fed with the precursor ornithine (Lindsey, 1982).

The yield of capsaicin was increased by supplying phenylalanine, vanillyamine, valine, ferulic acid and isocapric acid to *Capsicum frutescens* cells immobilized in polyurethane foam matrices (Mavituna *et al.*, 1987).

With or without the precursors, the addition of inhibitors or saturators (as cosubstrates) of metabolic pathways competing for the rate-limiting compounds may also increase the product yields. The addition of sinapic acid to *C. frutescens* cells immobilized in a polyurethane foam matrix increased the yield of capsaicin (Park, 1986; Wilkinson, 1987).

### *5.3.3.3* **De novo** ***synthesis***

Plant cell cultures produce a wide range of compounds (Phillipson, 1990; Neumann *et al.*, 1985; Staba, 1980). In addition to the products of primary and secondary metabolism, it has been reported that 85 novel compounds have been isolated from about 30 different plant cell cultures (Ruyter and Stockigt, 1989). There is considerable interest in the production of such compounds, especially the secondary metabolites by immobilized plant cells. Table 1 gives a summary of some of the model studies.

Cells of *C. roseus* have been immobilized for the production of indole alkaloids in gels (Brodelius *et al.*, 1981; Brodelius and Nilsson, 1980; Lambe and Rosevear, 1983; Majerus and Pareilleux, 1986; Rosevear and Lambe, 1982), in membrane (Kargi, 1988; Payne *et al.*, 1988), in cotton fibre (Payne *et al.*, 1988), in fibreglass mats (Facchini and DiCosmo, 1990) and in a cloth of nonwoven polyester short fibres (Archambault *et al.*, 1990; Rho *et al.*, 1990). Only alginate entrapment has been reported to increase the production of indole alkaloids in immobilized cultures in comparison to the suspension cultures. In experiments with polyacrylamide-entrapped *C. roseus*, Rosevear and Lambe (1982) observed that the synthesis of ajmalicine and serpentine was slow at the beginning, but it increased after 40 d and was maintained for another 110 d. The total yield of products after 150 d was much higher than the maximum yield obtained with suspended cells.

Inhibition of secondary metabolic activity was observed for cultures immobilized in membrane and cotton fibres. Immobilizing *C. roseus* in fibreglass mats resulted in a decreased specific accumulation of tryptamine, catharanthine and ajmalcine relative to the suspension cultures. No alkaloids were detected in the medium indicating that cell leakage did not occur. Alginate-entrapped cells of *Morinda citrifolia* produced about 10 times as much anthraquinones after 21 d of incubation as did freely suspended cells in a medium lacking hormones in order to limit growth (Brodelius *et al.*, 1980). Cells of *Glycine max* and *Daucus carota* entrapped in hollow fibre reactors could synthesize phenolics continuously for 30 d (Shuler, 1981; Jose *et al.*, 1983). *G. max* cells immobilized in a membrane reactor with pore size of 125 $\mu$m could be operated for 110 d for the synthesis of phenolics, resulting in the productivity being about four times higher, on cell dry weight basis, in comparison to the suspension cultures (Shuler and Hallsby, 1983). *Solanum aviculare* cells covalently coupled to polyphenylene oxide beads were able to synthesize steroid glycosides during 11 d of incubation (Jirku *et al.*, 1981).

Lindsey *et al.* (1983) found that the capsaicin production was increased by two orders of magnitude after the immobilization of *Capsicum frutescens* cells in the pores of polyurethane foam matrix. When produced, capsaicin, the pungent chilli flavour, is excreted into the medium. In a circulating bed reactor developed for plant cells immobilized in polyurethane foam matrices (Mavituna *et al.*, 1987), it was discovered that reducing the dissolved oxygen concentration to between 0–20% of saturation level resulted in the appearance of capsaicin in the medium. After production, if the dissolved oxygen concentration of the medium was returned to between 60–100% saturation, capsaicin disappeared from the medium (Wilkinson *et al.*, 1988).

Carrier *et al.* (1990) studied the production of ginkgolides by *Ginkgo biloba* cells immobilized on the surface of a nonwoven short fibre polyester material. Only very small amounts of ginkgolides were detected in the cells extracted after harvesting since the product is intracellular.

Comparison of experimental data from different research groups on the effect of immobilization for *de novo* synthesis is difficult because of the diversity of the systems used and the style with which the data are reported.

## 5.4 MASS TRANSFER

### 5.4.1 Background

In an immobilized cell system, substrates will have to be supplied from the bulk liquid to the cells in the innermost parts of the immobilized cell matrix in order to prevent starvation. Likewise, the products, assuming that they are excreted, will have to be transported from the vicinity of the producing cells to the bulk liquid from which they can be recovered without affecting the intactness of the immobilized cell matrix. There are several resistances along the path of mass transfer between the bulk liquid and the cells inside the immobilized cell matrix which can slow down the rate of mass transfer, leading to the development of concentration gradients. If the mixing in the bulk liquid is good, it can be assumed that the concentrations of the substrates and the products will be uniform everywhere in the bulk liquid. The transferring compounds will have to go through a stagnant liquid film surrounding the immobilized cell matrix *via* molecular diffusion. The thickness of this film will be determined by the hydrodynamic conditions in the bulk liquid and for a well-mixed bulk liquid this film can be assumed to be thin enough not to cause a mass transfer problem. The transfer of compounds through the immobilized cell matrix is usually assumed to be by molecular diffusion in order to simplify the theoretical analysis of the phenomenon. However, the microstructure of the immobilization matrix may bring other types of transfer into action. In gel entrapment, individual cells or cell aggregates are surrounded with the gel material which itself poses a resistance to mass transfer. Since most gels are swollen with aqueous liquid, the mass transfer through them can again be assumed to be by molecular diffusion, albeit perhaps at a slower rate than that in the liquid medium. In other immobilization methods, such as entrapment in a reticulated foam matrix, in membranes, or surface immobilization in fibre mats, the immobilized cell matrix is composed of cell aggregates with capillaries filled with liquid. In some cases, mucus layers were observed around the cell aggregates which would act as a resistance to mass transfer. In the aggregates, the cytoplasm of most of the individual cells will be connected to each other through the plasmadesmata. All this leads to a more complicated mechanism for mass transfer, including active transport and capillary action.

Although immobilization naturally decreases the rates of transport of substrates and products through the immobilized cell matrix, the effect of this on plant cell metabolism is not immediately obvious. The plant cells metabolize much more slowly compared with the average microbial cultures; therefore, they may cope with a slower rate of mass transfer in the immobilized state. In order to check whether mass transfer is the rate-limiting step in the activity of immobilized plant cells, a regime analysis must be carried out (Kossen and Oosterhuis, 1985). An example of such a study can be found in the regime analysis of freely suspended cultures of *C. roseus* for a scale-up exercise by ten Hoopen *et al.* (1988), in which the characteristic time concept was used to compare the rates of different phenomena such as growth, oxygen uptake and mass transfer.

The point value of the rate of mass transfer through the immobilized cell matrix with molecular diffusion is directly proportional to the effective diffusion coefficient and the local concentration gradient. However, the extent of mass transfer limitation and its effect on the biological reaction rates in the immobilized cell matrix will depend on the effective diffusion coefficients of the transferring solutes in the immobilization matrix, the concentration of cells, the biological rates and kinetics and the shape and dimensions of the immobilized plant cell matrices. It has been suggested that the mass transfer limitation in immobilized plant cell systems can be advantageous if lack of nutrients can serve as a stress factor resulting in secondary metabolite synthesis.

### 5.4.2 Effect on Physiology

Individual plant cells are normally large compared to microbial cells, ranging from 30 to 100 $\mu$m in diameter. Even in freely suspended cultures, they have a tendency to exist in aggregates of two to 200 cells. In large aggregates and in certain immobilization systems which allow continous cell to cell contact over a considerable distance, concentration profiles develop as a result of mass transfer limitations. Therefore, cells in the centre of such aggregates experience a different environment than those on the outer regions. Cells in the centre are typically enlarged, have less starch and show partial differentiation (Street *et al.*, 1965; Withers, 1976). Such cells are called the 'feeder' cells since their products are thought to be stimulatory to peripheral cells. When hormonal concentrations are favourable, embryoids will arise from single cells which are on the surface of the aggregates (McWilliam *et al.*, 1974; Street and Withers, 1974). Cell to cell contact and a heterogeneous physicochemical environment in aggregates may be leading to redifferentiation which in turn seems to be necessary for the production of secondary metabolites (Sahai and Shuler, 1984; Shuler, 1981; Takayama *et al.*, 1977; Yeoman *et al.*, 1980; Yeoman *et al.*, 1982).

### 5.4.3 Effective Diffusion Coefficients

In the investigation of mass transfer limitations in immobilized cell systems it is important to know the effective diffusion coefficients of transferring compounds in the immobilization matrix. Using radioactively labelled glucose, a set of experiments were performed by Mavituna *et al.* (1987) to determine the time dependent glucose concentration profiles in callus and reticulated foam entrapped cells of *C. frutescens*. Applying one-dimensional molecular diffusion theory, the effective diffusion coefficients for glucose in callus and in immobilized cell matrix were found to lie in the ranges of $0.28 \times 10^{-10}$ to $2.8 \times 10^{-10}$ $m^2 s^{-1}$ and $13.9 \times 10^{-10}$ to $139 \times 10^{-10}$ $m^2 s^{-1}$, respectively at 25 °C (Wilkinson *et al.*, 1990). The effective diffusion coefficient of glucose was measured to be $6.83 \times 10^{-10} m^2 s^{-1}$ at 30 °C in 2% Ca alginate gel beads (Tanaka *et al.*, 1984), and $4.8 \times 10^{-10} m^2 s^{-1}$ at 30 °C in 3% carrageenan (Nguyen and Luong, 1986). The diffusion coefficient of glucose in pure water at 30 °C is $6.8 \times 10^{-10}$ $m^2 s^{-1}$. The effective diffusion coefficient is dependent not only on the type of gel or other medium of immobilization, but also on the composition of the gel and the cell loading (Hannoun and Stephanopoulos, 1986; Klein and Manecke, 1982). Rhodes *et al.* (1985) suggested that the formation of a thin mucilaginous film by the secretion of polysaccharides by the cells immobilized in polyurethane foam would reduce the diffusion coefficients further.

### 5.4.4 Oxygen Uptake Rates

Most of the medium components are soluble in aqueous medium so that large enough concentrations can be maintained in the bulk liquid, if necessary, to create the requisite driving force to attain a sufficient rate of mass transfer into the immobilized cell matrix. The problem, however, emerges with the materials of low solubility in aqueous medium, such as oxygen. In such cases the low mass transfer may not meet the rate of biological demand. This demand, which is often expressed in terms of a Michaelis–Menten type of kinetics, is charaterized by the maximum specific uptake rate, the saturation constant and the critical substrate concentration (Fonseca *et al.*, 1988). The typical oxygen uptakes rates (Payne *et al.* 1987) for plant cells are considerably less than those experienced with microbial cells. In some cases, high oxygen concentration is reported to be toxic to the metabolic activity of the cell (Pareilleux and Chaubet, 1981; Turner and Quartely, 1956).

Hulst *et al.* (1985) studied the influence of the support material, cell loading and the bead diameter on the respiration rate of *Daucus carota* cells immobilized in alginate. Assuming that the dissolved oxygen concentration at the bead surface was about 95% of saturation, they calculated that for good diffusion characteristics the bead diameter may be as high as 1.4 and 3.8 mm for cell densities of 50 and 5% (w/w), respectively.

*Thalictrum minus* cells are reported to consume more oxygen when producing berberine, which is excreted to the medium (Tabata, 1988). Alginate-entrapped cells were found to turn black owing to the insufficient supply of oxygen and they failed to produce berberine. Therefore, a special bioreactor was devised in which the alginate beads were piled up in a column, through which the medium was circulated, and were exposed to air periodically (Kobayashi *et al.*, 1987).

Robins *et al.* (1986) have investigated the oxygen uptake rates of *D. carota*, *B. vulgaris* and *Cinchona pubescens* immobilized in alginate and reticulated foam particles and concluded that there was oxygen transfer limitation. Using an oxygen microelectrode, they measured the oxygen gradient within the foam particles. The results showed that in a fully loaded particle, the oxygen concentration diminished rapidly to a value of less than 5% of saturation within the outermost 1 mm of the 8 × 8 × 8 mm polyurethane foam cube. However, only after 10–12 weeks of culturing without medium change did necrotic regions appear at the centre of the foams. They suggested that cells at the centre were possibly supplied with their metabolic requirements *via* symplastic connections between cells grown *in situ*, a situation that would not exist in gel beads.

Mavituna *et al.* (1987) have found that the apparent maximum specific oxygen uptake rate for *Capsicum frutescens* cells immobilized in polyurethane foam particles was $1.0 \times 10^{-3}$ g oxygen g dry weight$^{-1}$ h$^{-1}$ compared with $2.0 \times 10^{-3}$ g oxygen g dry weight$^{-1}$ h$^{-1}$ obtained with freely suspended cells. They have also discovered that a reduction in the dissolved oxygen concentration of the medium results in the production of capsaicin by *C. frutescens* entrapped in polyurethane foam particles (Wilkinson *et al.*, 1988).

Since the relationship between metabolism and dissolved oxygen concentration is complex, a conclusion cannot be reached about the effect of reduced availability of oxygen in immobilized plant cell systems on secondary metabolite production and growth. Further examples include beneficial effects of high oxygen concentration on berberine and jatrorhizine production by *Berberis wilsonae* (Breuling *et al.*, 1985), on alkaloid production by *C. roseus* (Scragg *et al.*, 1987) and on anthraquinone production by *Morinda citrifolia* (Wagner and Vogelmann, 1977). However, decreased dissolved oxygen concentration resulted in increased phenolics production in *Cynara cardunculus* (Lima-Costa *et al.*, 1988) and in thiophenes production by *Tagetes* species (Ketel *et al.*, 1987).

### 5.4.5 Light

Although most plant cell cultures are not fully competent of photosynthesis and are rarely exposed to photosynthetically significant light intensities, the metabolism of cultures can be affected by periodic exposure to light, and the quality and intensity of the light are significant (Seibert and Kadkade, 1980). In immobilized cell cultures only the outer cell layers in the immobilized matrix may receive some light. This may be advantageous in the cases where some precursors are formed in light and some in dark conditions, such as the *Catharanthus* alkaloids. The supply of light to the interior of the immobilized cell matrix may be possible by the use of optical fibres.

### 5.4.6 Carbon Dioxide

Scragg *et al.* (1987) found in one of their cell lines of *C. roseus* that the addition of 4% $CO_2$ to the sparging gas was necessary to retain alkaloid productivity. If other cultures also benefit from increased $CO_2$ concentrations, then immobilization may be advantageous in that the $CO_2$ produced by the cells will be more concentrated in the immobilization matrix than the bulk liquid, where it is stripped out by aeration.

## 5.5 PRODUCT RELEASE

In order to fully utilize the advantages of immobilized cell systems the products must be excreted to the bulk liquid. This allows the cells to be exploited over long periods with continuous or intermittent recovery of the products from the bulk liquid. When a product is intracelluar, the only warrant for immobilization would be a very significant increase in production when compared with suspension cultures. Unfortunately, many products appear to be intracellular, stored in the vacuoles, although they are transportable in the whole plant. The transport of metabolites into vacuolar space or into the extracellular medium is still very little understood (Brodelius, 1990; Matile, 1990).

### 5.5.1 Spontaneous Release

Many secondary metabolites are spontaneously released by plant cells grown in culture. Parr (1988) lists some examples of such products, which include various alkaloids, such as indoles, pyridines, quinolines, benzylisoquinoline, quinolizidines, anthraquinones, capsaicin, opines, phenolics and terpenoids.

In certain cases immobilization appears to induce a spontaneous release of the products that are normally stored within the cells in suspension (Brodelius *et al.*, 1981). For instance indole alkaloids were released into the medium from entrapped cells of *C. roseus* (Rosevear and Lambe, 1982). There is always the possibility that the immobilization method damages the intactness of the cell membrane, making it permeable. If possible, a suitable cell line which is capable of spontaneous product release must be sought and then used in immobilized cell systems.

An understanding of the transport across the plasma membrane separating the cell interior from the external medium and the tonoplast surrounding the vacuole should help in creating the suitable physicochemical environment which would induce spontaneous release of the products of interest. Both active and passive transport have been suggested as the underlying mechanisms. Some evidence has been presented for the active transport of secondary metabolites over membranes involving carrier proteins, for example, for indole, isoquinoline, quinolizidine and pyrrolizidine alkaloids into vaculoes isolated from cell cultures (Brodelius, 1990). It has been proposed that neutral alkaloid molecules can freely diffuse across membranes but protonated alkaloid cations cannot. In such cases, extracellular alkaloid pH can have an effect on the transport (Brodelius, 1990).

### 5.5.2 Permeabilization

Various attempts have been made to increase the permeability of plant cell membranes, and these have been reviewed by Brodelius (1990), Felix (1982) and Parr (1988). Chemical permeabilization involves various surface active chemicals such as DMSO, phenethyl alcohol, chloroform, triton X-100 and hexadecyltrimethylammonium bromide. All these compounds induce product release at the expense of cell viability. Electroporation has been investigated with cell cultures of *C. rubrum* and *T. rugosum*; product release was achieved but the cell viability was decreased (Brodelius *et al.*, 1988). Other permeabilization methods include ultrasonication and ionophoretic release, in which the cells are subjected to a low constant current in a specially designed device. The results of permeabilization experiments are generally discouraging, because it appears to be very difficult to relase vacuolar compounds into the medium by permeabilizing the plasma membrane and the tonoplast without killing the cells at the same time.

## 5.6 CONCLUDING REMARKS

The ability of plant cells to synthesize a vast range of compounds far exceeds the biosynthetic diversity of other kingdoms. Despite the very few industrial applications so far, by increased understanding of plant cell biochemistry, physiology and molecular biology, some strategies may be developed (Yamada and Hashimoto, 1990) leading to further processes based on plant cell cultures. The significant advantages offered by immobilized cell systems suggest that they will have an important role to play in these developments.

## 5.7 REFERENCES

Alfermann, A., W. Bergmann, C. Figur, U. Helmbold, D. Schwantag, I. Schuller and E. Reinhard (1983). In *Plant Biotechnology*, ed. S. Mantell and H. Smith, Society for Experimental Biology Seminar Series, vol. XVIII, pp. 67–74, Cambridge University Press, Cambridge.

Alfermann, A., I. Schuller and E. Reinhard (1980). *Plant Med.*, **40**, 218.

Archambault, J., B. Volesky and W. Kurz (1990). Development of bioreactors for the culture of surface-immobilized plant cells, *Biotechnol. Bioeng.*, **35**, 702–711.

Barz, W., M. Jordan and G. Metschulat (1990). Bioconversion of xenobiotics (pesticides) in plant cell cultures. In *Progress in Plant Cellular and Molecular Biology*, ed. H. Nijkamp, L. van der Plas and J. van Aartrijk, pp. 631–639, Kluwer Academic, Dordrecht.

Berlin, J., B. Martin, J. Nowak, L. Witte, V. Wray and D. Strack (1989). *Z. Naturforsch, C: Biosci.*, **44C**, 249–254.

Breuling, M., A. Alfermann and E. Reinhard (1985). Cultivation of cell cultures of *Berberis wilsonae* in 20 L airlift bioreactors. *Plant Cell Rep.*, **4**, 220.

Brodelius, P. (1984). Immobilization of cultured plant cells and protoplasts. In *Cell Culture and Somatic Cell Genetics of Plants*, ed. I. K. Vasil, vol. 1, pp. 535–546. Academic Press, New York.

Brodelius, P. (1985a). Immobilized plant cells. In *Applications of Isolated Enzymes and Immobilized Cells to Biotechnology*, ed. A. Laskin, pp. 109–148, Addison-Wesley, Massachusetts.

Brodelius, P. (1985b). Immobilized plant cells: preparation and biosynthetic capacity. In *Immobilized Cells and Enzymes: A Practical Approach*, ed. J. Woodward, chap. 8, IRL Press, Oxford.

Brodelius, P. (1988). Immobilized plant cells as a source of biochemicals. In *Bioreactor Immobilized Enzymes and Cells: Fundamentals and Applications*, ed. M. Moo-Young, pp. 167–196, Elsevier Applied Science, London.

Brodelius, P. (1990). Transport and accumulation of secondary metabolites. In *Progress in Plant Cellular and Molecular Biology*, ed. H. Nijkamp, L. van der Plas and J. van Aartrijk, pp. 567–576. Kluwer Academic, Dordrecht.

Brodelius, P., B. Deus, K. Mosbach and M. Zenk (1979). Immobilized plant cells for the production and transformation of natural products. *FEBS Lett.*, **103**, 93–97.

Brodelius, P., B. Deus, K. Mosbach and M. Zenk (1980). The potential use of immobilized plant cells for the production and transformation of natural products. *Enzyme Eng.*, **5**, 373–381.

Brodelius, P., C. Funk and R. Shillito (1988). *Plant Cell Rep.*, **7**, 186–188.

Brodelius, P. and K. Nilsson (1980). Entrapment of cells in different matrices, a comparative study. *FEBS Letts*, **122**, 312–316.

Brodelius, P. and K. Nilsson (1983). Permeabilization of immobilized plant cells, resulting in release of intracellularly stored products with preserved cell viability. *J. Appl. Microbiol. Technol.*, **17**, 275–280.

Brodelius, P. and H. Vogel (1984). *Ann. N. Y. Acad. Sci.*, **434**, 496–500.

Brodelius, P., M. Zenk, B. Deus and K. Mosbach (1981). Catalysts for the production and transformation of natural products having their origin in higher plants. *Eur. Pat. Appl.*, 22 434.

Cabral, J., P. Fevereiro, J. Novais and M. Pais (1984). Comparison of immobilization methods for plant cells and protoplasts. *Ann. N. Y. Acad. Sci.*, **434**, 501–503.

Carrier, D., P. Coulombe, M. Mancini, R. Neufeld, M. Weber and J. Archambauld (1990). Immobilized *Ginkgo biloba* cell cultures. In *Progress in Plant Cellular and Molecular Biology*, ed. H. Nijkamp, L. van der Plas and J. van Aartrijk, pp. 614–618, Kluwer Academic, Dordrecht.

Corchete, P. and M. Yeoman (1989). Biotransformation of (−)-codeinone to (−)-codeine by *Papaver somniferum* cells immobilized in reticulate polyurethane foam. *Plant Cell Rep.*, **8**, 128–131.

Facchini, P. and F. DiCosmo (1990). Immobilization of cultured *C. roseus* cells using a fibreglass substratum. *Appl. Microbiol. Biotechnol.*, **33**, 36–42.

Felix, H. (1982). Permeabilized cells. *Anal. Biochem.*, **120**, 211.

Felix, H., P. Brodelius and K. Mosbach (1981). Enzyme activities of the primary and secondary metabolism of simultaneously permeabilized and immobilized plant cells. *Anal. Biochem.*, **116**, 462–470.

Felix, H. and K. Mosbach (1982). Enhanced stability of enzymes in permeabilized and immobilized cells. *Biotechnol. Lett.*, **4**, 181–186.

Fonseca, M., F. Mavituna and P. Brodelius (1988). Engineering aspects of plant cell culture. In *Plant Cell Biotechnology*, ed. M. Pais, F. Mavituna and J. Novais, pp. 389–401, Springer, Berlin.

Furusaki, S., T. Nozawa, T. Isohara and T. Furuya (1988). Influence of substrate transport on the activity of immobilized *Papaver somniferum* cells. *Appl. Microbiol. Biotechnol.*, **29**, 437–441.

Furuya, T. (1978). Biotransformation by plant cell cultures. In *Frontiers of Plant Tissue Culture 1978*, ed. T. Thorpe, pp. 191–200. International Association of Plant Tissue Culture, Calgary, Canada.

Furuya, T., T. Yoshikawa and M. Taira (1984). *Phytochemistry*, **23**, 999–1001.

Galun, E., D. Aviv, A. Dantes and A. Freeman (1983). Biotransformation by plant cells immobilized in cross-linked polyacrylamide hydrazide; monoterpene reduction by entrapped *Mentha* cells. *Planta Med.*, **49**, 9–13.

Galun, E., D. Aviv, A. Dantes and A. Freeman (1985). Biotransformation by division-arrested and immobilized plant cells: bioconversion of monoterpenes by gamma-irradiated suspended and entrapped cells of *Mentha* and *Nicotiana*. *Planta Med.*, **45**, 511–514.

Gamborg, O. L. (1970). *Plant Physiol.*, **45**, 372.

Haldimann, D. and P. Brodelius (1987). Redirecting cellular metabolism by immobilization of cultured plant cells: a model study with *Coffea arabica*. *Phytochemistry*, **26**, 5 and 1431–1434.

Hamilton, R., H. Pedersen and C. Chin (1984). Immobilized plant cells for the production of biochemicals. *Biotechnol. Bioeng. Symp.*, **14**, 383–396.

Hannoun, B. and G. Stephanopoulos (1986). Diffusion coefficients of glucose and ethanol in cell-free and cell-occupied calcium alginate membranes. *Biotechnol. Bioeng.*, **28**, 829–835.

Harms, H. and E. Kottutz (1990). Bioconversion of xenobiotics in different plant systems: cell suspension cultures, root cultures and intact plants. In *Progress in Plant Cellular and Molecular Biology*, ed. H. Nijkamp, L. van der Plas and J. van Aartrijk. pp. 650–655. Kluwer Academic, Dordrecht.

Hulst, A., J. Tramper, K. Riet and J. Westerbeek (1985). A new technique for the production of immobilized biocatalyst in large quantities. *Biotechnol. Bioeng.*, **27**, 870–876.

Ishida, B. (1988). Improved diosgenin production in *Dioscorea deltoidea* cell cultures by immobilized in polyurethane foam. *Plant Cell Rep.*, **7**, 270–273.

Jirku, V., T. Macek, T. Vanek, V. Krumphanzl and K. Kubanek (1981). Continuous production of steroid glycoalkaloids by immobilized plant cells. *Biotechnol. Lett.*, **3**, 447–450.

Jones, A. and I. Veliky (1981). Effect of medium constituents on the viability of immobilized plant cells. *Can. J. Bot.*, **59**, 2095–2101.

Jose, W., H. Pedersen and C. Chin (1983). *Ann. N. Y. Acad. Sci.*, **413**, 409–412.

Kargi, F. (1988). Alkaloid formation by *Catharanthus roseus* cells in a packed column biofilm reactor. *Biotechnol. Lett.*, **10**, 181–186.

Ketel, D., A. Hulst, H. Gruppen, H. Breteler and J. Tramper (1987). Effects of immobilization and environmental stress on growth and production of nonpolar metabolites of *Tagetes minuta* cells. *Enzyme Microbiol. Technol.*, **9**, 303–307.

Klein, J. and G. Manecke (1982). New developments in the preparation and characterization of polymer-bound biocatalysts. *Enzyme Eng.*, **6**, 181–189.

Kobayashi, Y., H. Fukui and M. Tabata (1987). An immobilized cell culture system for berberine production by *Thalictrum minus* cells. *Plant Cell Rep.*, **6**, 185–186.

Kossen, N. and N. Oosterhuis (1985). Modelling and scaling-up of bioreactors. In *Biotechnology*, ed. H. Rehm and G. Reed, vol. 2, ed. H. Brauer, pp. 571–605, VCH, Weinheim.

Kurz, W. and F. Constabel (1985). Aspects affecting biosynthesis and biotransformation of secondary metabolites in plant cell cultures. *CRC Crit. Rev. Biotechnol.*, **2**, 105–118.

Lambe, C. and A. Rosevear (1983). *Proc. Biotech. 83, London, May, 1983*, pp. 565–576, Online, Northwood.

Lang, J., K. Yoon and J. Prenosil (1990). The effects of precursor feeding on alkaloid production in *Coffea arabica* plant cell cultures in a novel membrane reactor. In *Proceedings of the European Congress on Biotechnology, 5th*, ed. C. Christiansen, L. Munck and J. Villadsen, pp. 132–135, Mundsgaard, Copenhagen.

Law, D. and F. Mavituna (1990). Contradictions between two establised viability tests. In *Abstracts of the International Congress on Plant Tissue and Cell Culture, VIIth*, Amsterdam.

Lima-Costa, E., J. Novais, M. Pais and J. Cabral (1988). Effect of aeration on *Cynara cardunculus* plant cell cultures. In *Plant Cell Biotechnology*, ed. M. Pais, F. Mavituna and J. Novias, pp. 343–351, Springer, Berlin.

Lindsey, K. (1982). Studies on the growth and metabolism of plant cells cultured on fixed bed reactors. Ph.D. Thesis, Edinburgh University, UK.

Lindsey, K. and M. Yeoman (1983). Novel experimental systems for studying the production of secondary metabolites by plant tissue cultures. In *Plant Biotechnology*, ed. S. Mantell and H. Smith, pp. 39–66. Cambridge University Press.

Lindsey, K. and M. Yeoman (1984). The viability and biosynthetic activity of cells of *Capsicum frutescens* Mill. cv. annuum immobilized in reticulate polyurethane. *J. Exp. Bot.*, **35**, 1684–1696.

Lindsey, K. and M. Yeoman (1985). Immobilized plant cells. In *Plant Cell Culture Technology*, ed. M. Yeoman, pp. 229–267. Blackwell, Oxford.
Lindsey, K., M. Yeoman, G. Black and F. Mavituna (1983). A novel method for the immobilization and culture of plant cells. *FEBS Lett.*, **155**, 143–149.
Linsmaier, E. M. and F. Skoog (1965). Organic growth factor requirements of tobacco tissue cultures. *Physiol. Plant*, **18**, 100–127.
Luckner, M. (1990). *Secondary Metabolism in Microorganisms, Plant and Animals*, Fischer and Springer, Jena and Berlin.
Macek, T., V. Jirku, T. Vanek and I. Benes (1981). Immobilization of plant cells on polyphenylene oxide. In *1st Int. Conf. Chem. Biotechnol. Biol. Act. Nat. Prod.*, ed. B. Atanasova, **3**, 158–161. Bulgaria Academy of Sciences, Sofia.
Majerus, F. and A. Pareilleux (1986). Alkaloid accumulation in Ca alginate entrapped cells of *Catharanthus roseus*: using a limiting growth medium. *Plant Cell Rep.*, **5**, 302–305.
Matile, P. (1990). The toxic compartment of plant cells. In *Progress in Plant Cellular and Molecular Biology*, ed. H. Nijkamp, L. van der Plas and J. van Aartrijk, pp. 557–566. Kluwer Academic, Dordrecht.
Mavituna, F. and J. Park (1985). Growth of immobilized plant cells in reticulate polyurethane foam matrices. *Biotechnol. Lett.*, **7**, 637–640.
Mavituna F., J. Park and D. Gardner (1987a). Determination of effective diffusion coefficient of glucose in callus tissue. *Chem. Eng. J. (Lausanne)*, **34**, B1–B5.
Mavituna F., J. Park, P. Williams and A. Wilkinson (1987). Characteristics of immobilized plant cell reactors. In *Plant and Animal Cells: Process Possibilities.*, ed. C. Webb and F. Mavituna, pp. 92–115. Horwood, Chichester.
McWilliam, A., S. Smith and H. Street (1974). *Ann. Bot. (London)*, **38**, 243–250.
Miyasaka, H., M. Nasu, T. Yamamoto, Y. Endo and K. Yoneda (1986). Production of cryptotanshinone and ferruginol by immobilized cultured cells of *Salvia miltiorrhiza. Phytochemistry*, **25**, 7 and 1621–1624.
Moritz, S., I. Schuller, C. Figur, A. Alfermann and E. Reinhard (1982). In *Proc. Int. Congr. Plant Tissue Cell Cult. 5th, 1982,* Fujiwara, pp. 401–402, Japanese Association of Plant Tissue Cultures, Tokyo.
Morris, P., A. Scragg, N. Smart and A. Stafford (1985). Secondary product formation by cell suspension cultures. In *Plant Cell Culture: A Practical Approach*, ed. R. Dixon, pp. 127–167, IRL Press, Oxford.
Murashige, T. and F. Skoog (1962). A revised medium for rapid growth and bio-assays with tobacco cultures. *Physiol. Plant*, **15**, 473–497.
Nakajima, H., K. Sonomoto, H. Morikawa, F. Sato, K. Ichimura, Y. Yamada and A. Tanaka (1985). Entrapment of *Lavandula vera* cells with synthetic resin prepolymers and its application to pigment production. *Appl. Microbiol. Biotechnol.*, **24**, 266–270.
Nakajima, H., K. Sonomoto, N. Usui, F. Sato, Y. Yamada, A. Tanaka and S. Fukui (1985). Entrapment of *Lavandula vera* cells and production of pigments by entrapped cells. *J. Biotechnol.*, **2**, 107–117.
Naoshima, Y. and Y. Akakabe (1989). Biotransformation of some keto esters through the consecutive reuse of immobilized *Nicotiana tabacum* cells. *J. Org. Chem.*, **54**, 4237–4239.
Naoshima, Y., Y. Akakabe and F. Watanabe (1989). Biotransformation of acetoacetic esters with immobilized cells of *Nicotiana tabacum. Agric. Biol. Chem.*, **53 (2)**, 545–547.
Neumann, K., W. Barz and E. Reinhard (1985). *Primary and Secondary Metabolism of Plant Cell Cultures.* Springer, Berlin.
Nguyen, A. and J. Luong (1986). Diffusion in κ-carrageenan gel beads. *Biotechnol. Bioeng.*, **28**, 1261–1267.
Nilsson, K. (1983). Immobilized animal and plant cells: preparation and potential biotechnological applications. Ph.D. Thesis, University of Lund, Sweden.
Nilsson, K., S. Birnbaum, S. Flygare, L. Linse, U. Schroder, U. Jeppsson, P. Larsson, K. Mosbach and P. Brodelius (1983). A general method for the immobilization of cells with preserved viability. *Eur. J. Appl. Microbiol. Biotechnol.*, **17**, 319–326.
Nitsch, J. P. and C. P. Nitsch (1969). Haploid plants from pollen grains. *Science (Washington, D.C.)*, **163**, 85–87.
Novais, J. (1988). Methods of immobilization of plant cells. In *Plant Cell Biotechnology*, ed. M. Pais, F. Mavituna and J. Novais, pp. 353–363. NATO ASI Series, Springer, New York.
Pareilleux, A. and N. Chaubet (1981). *Eur. J. Appl. Microboil. Biotechnol.*, **11**, 222–225.
Park, J. (1986). Plant tissue culture and plant cell immobilization. Ph.D. Thesis, University of Manchester Institute of Science and Technology, UK.
Park, J. and F. Mavituna (1987). Factors affecting the immobilization of plant cells in biomass support particles. In *Plant and Animal Cells: Process Possibilities.*, ed. C. Webb and F. Mavituna, pp. 295–303. Horwood, Chichester.
Parr, A. (1988). Secondary products from plant cell culture. *Biotechnology in Agriculture*, pp. 1–34. Liss, New York.
Parr, A., J. Smith, R. Robins and M. Rhodes (1984). Apparent free space and cell volume estimation: a nondestructive method for assessing the growth and membrane integrity/viability of immobilized plant cells. *Plant Cell Rep.*, **3**, 161–164.
Payne, G., M. Shuler and P. Brodelius (1987). Large scale plant cell culture. In *Large Scale Cell Culture Technology*, ed. B. Lydersen, pp. 194–229, Hanser, Munich.
Payne, G., N. Payne and M. Schuler (1988). Bioreactor considerations for secondary metabolite production from plant cell tissue cultures: indole alkaloids from *Catharanthus roseus. Biotechnol. Bioeng.*, **31**, 905–912.
Phillipson, J. (1990). Possibilities of finding new products from plant cell cultures. In *Progress in Plant Cellular and Molecular Biology*, ed. H. Nijkamp, L. van der Plas and J. van Aartrijk, pp. 592–600. Kluwer Academic, Dordrecht.
Prenosil, J. and H. Pedersen (1983). Immobilized plant cell reactors. *Enzyme Microb. Technol.*, **5**, 323–331.
Rho, D., C. Bedard and J. Archambault (1990). Physiological aspects of surface-immobilized *Catharanthus roseus* cells. *Appl. Microbiol. Biotechnol.*, **33**, 59–65.
Rhodes, M., R. Robins, R. Turner and J. Smith (1985). Mucilaginous film production by plant cells immobilized in a polyurethane or nylon matrix. *Can. J. Bot.*, **63**, 2357–2363.
Robins, R., A. Parr, S. Richards and M. Rhodes (1986). Studies of environmental features of immobilized plant cells. In *Secondary Metabolism in Plant Cell Cultures*, ed. P. Morris, A. Scragg, A. Stafford and M. Fowler, pp. 162–207. Cambridge University Press, Cambridge.
Rosevear, A. and C. Lambe (1982). *Eur. Pat. Appl.*, 82 301 571.4.
Ruyter, C. and J. Stockigt (1989). *GIT Fachz. Lab.*, **33**, 283–293.
Sahai, O. and M. Shuler (1984). *Biotechnol. Bioeng.*, **26**, 27–36.
Schnabl, H., C. Elbert and R. Youngman (1983). Release of ethane from immobilized plant cell protoplasts in response to chemical treatment. *Physiol. Plant.*, **59**, 46–49.

Schnabl, H. and R. Youngman (1985). Immobilization of plant cell protoplasts inhibits enzymic lipid peroxidation. *Plant Sci.*, **40**, 65–69.

Scragg. A., P. Morris, E. Allan, P. Bond, P. Hegarty, N. Smart and M. Fowler (1987). The effect of scale-up on plant cell culture performance. In *Process Possibilities for Plant and Animal Cell Culture*, ed. C. Webb and F. Mavituna, pp. 78–91, Horwood, Chichester.

Seibert, M. and P. Kadkade (1980). Environmental factors: A, Light. In *Plant Tissue Culture as a Source of Biochemicals*, ed. E. Staba, p. 123, CRC Press, Boca Raton.

Shuler, M. (1981). Production of secondary metabolites from plant tissue culture — problems and prospects. *Ann. N. Y. Acad. Sci.*, **369**, 65–79.

Shuler, M. and Hallsby (1983). Paper 76b, AIChE, 1983, Summer National Meeting, Denver, USA.

Shuler, M., G. Payne and G. Hallsby (1984). Prospects and problems in the large scale production of metabolites from plant cell tissue cultures. *J. Am. Oil Chem. Soc.*, **61**, 1724–1728.

Shuler, M., G. Hallsby, G. Payne and T. Cho (1986). Bioreactors for immobilized plant cell cultures. *Ann. N. Y. Acad. Sci.*, **469**, 270–278.

Staba, E. (1980). *Plant Tissue Culture as a Source of Biochemicals*. CRC Press, Boca Raton.

Street, H., G. Henshaw and M. Buiatti (1965). *Chem. Ind. N. Z.*, **1**, 27–33.

Street, H. and L. Withers (1974). In *Tissue Culture and Plant Science*, ed. H. Street, pp. 71–100. Academic Press, London.

Tabata, M. (1988). Secretion of secondary products by plant cell cultures. In *Proc. Int. Biotechnol. Symp., 8th, 1988*, ed. G. Durand, L. Bobichon and J. Florent, pp. 167–178, Société Francais de Microbiologie, Paris.

Takayama, S. *et al.* (1977). *Physiol. Plant.*, **41**, 313–320.

Tanaka, A., K. Sonomoto and S. Fukui (1984). Various applications of living cells immobilized by prepolymer method. *Ann. N. Y. Acad. Sci.*, **434**, 479–489.

ten Hoopen, H., W. van Gulik, J. Meijer and K. Luyben (1988). Scale-up of plant cell cultures. In *Proc. Int. Biotechnol. Symp., 8th, 1988*, ed. G. Durand, L. Bobichon and J. Florent, pp. 179–191, Scoiété Francais de Microbiologie, Paris.

Towill, E. and P. Mazur (1975). Studies on the reduction of 2,3,5-triphenyltetrazolium chloride as a viability assay for a plant tissue cultures. *Can. J. Bot.*, **53**, 1097–1102.

Tsuchiya, T. (1983). Steroid modification by immobilized plant cells. *Jpn. Pat. Appl.*, 83/34 870.

Turner, E. and J. Quartley (1956). *J. Exp. Bot.*, **7**, 362–371.

Veliky, I. and A. Jones (1981). Bioconversion of gitoxigenin by immobilized plant cells in a column bioreactor. *Biotechnol. Lett.*, **3(10)**, 551–554.

Wagner, F. and H. Vogelmann (1977). In *Plant Tissue Culture and its Biotechnological Applications*, ed. W. Bartz, E. Reinhard and M. Zenk, pp. 245–252, Springer, New York.

Watts, M. and H. Collin (1985). Growth and nutrient uptake by immobilized tissue cells of celery (*Apium graveolens*). *Plant Sci.*, **42**, 67–72.

Wichers, H., T. Malingre and H. Huizing (1983). The effect of some environmental factors on the production of L-DOPA by alginate entrapped cells of *Mucuna pruriens*. *Planta*, **158**, 482–486.

Widholm, J. (1972). The use of fluorescein diacetate and phenosafranine for determining viability of cultured plant cells. *Stain Technol.*, **47**, 189–194.

Wilkinson, A. (1987). Bioreactor design for chemicals from immobilized plant cells. Ph.D. Thesis. University of Manchester Institute of Science and Technology, UK.

Wilkinson, A., J. Park, P. Williams and F. Mavituna (1990). Immobilization of plant cells and bioreactor design. In *Proc. APBioChEC '90*, April 1990, Kyungju, Korea.

Wilkinson, A., P. Williams and F. Mavituna (1988). The effect of oxygen stress on secondary metabolite production by immobilized plant cells in bioreactors. In *Plant Cell Biotechnology*, ed. M. Pais, F. Mavituna and J. Novais, pp. 373–377, Springer, Berlin.

Williams, P., A. Wilkinson and F. Mavituna (1987). Comparison of the activity of immobilized and freely suspended cells of *Capsicum frutescens*. In *Proceedings of 4th European Congress on Biotechnology*, ed. O. Neijssel, R. van der Meer and K. Luyben, vol. 2, pp. 410–411. Elsevier, Amsterdam.

Williams, P., A. Wilkinson, J. Lewis, G. Black and F. Mavituna (1988). A method for the rapid production of fine plant cell suspension cultures. *Plant Cell Rep.*, **7**, 459–462.

Withers, L. (1976). *J. Exp. Bot.*, **27**, 1073–1084.

Yamada, Y. and T. Hashimoto (1990). Possibilities for improving yields of secondary metabolites in plant cell cultures. In *Progress in Plant Cellular and Molecular Biology*, ed. H. Nijkamp, L. van der Plas and J. van Aartrijk, pp. 547–556. Kluwer Academic, Dordrecht.

Yeoman, M. M., K. Lindsey, M. B. Miedzybrodzka and W. R. McLauchlan (1982). Accumulation of secondary products as a facet of differentiation in plant cell and tissue cultures. In *Differentiation in Vitro*, ed. M. M. Yeoman and D. E. S. Truman, British Society for Cell Biology, Symposium 4, pp. 65–82. Cambridge University Press, Cambridge.

Yeoman, M. M., M. B. Miedzybrodzka, K. Lindsey and W. R. McLauchlan (1980). The synthetic potential of cultured plant cells. In *Plant Cell Cultures: Results and Perspectives*, ed. O. Cifferi and B. Parisi, pp. 327–343, Elsevier, Amsterdam.

# 6

# Plant Cell Culture, Process Systems and Product Synthesis

MICHAEL W. FOWLER and ANGELA M. STAFFORD
*University of Sheffield, UK*



## 6.1 INTRODUCTION

The commercial production of natural products from the some form of cell or tissue culture has long been a goal of plant biotechnology; indeed, the first reference to such a potential was made over 30 years ago. While since that time a great deal of effort has been applied, only a handful of processes have come to commercial fruition, principally in Japan, and even these operate at the limits of economic viability.

In spite of the apparent lack of progress at the commercial level it is generally agreed that the potential for the synthesis of high value natural products from plant cell cultures is tremendous (Stafford and Fowler, 1991). Of all groups of living organisms the plant kingdom probably possesses the most wide-ranging biochemistry in terms of types of molecular structures synthesized, coupled with a highly versatile enzymology. The range of opportunities presented by this biochemical Aladdin's cave is equally wide and covers all sectors of the chemical, pharmaceutical and food industries. Recently there have also been indications of potentials in diagnostics and related areas, principally through the use of cell culture derived plant enzymes (Stafford and Fowler, 1991).

Plant cell cultures could potentially contribute in at least four ways to the commercial synthesis of natural products: (i) as a new production route to established products, for instance morphine, quinine and digoxin; (ii) as a route of synthesis to a novel product from plants difficult to grow, and as a means of maintaining germplasm from rare plants of commercial significance. (This is particularly important in relation to recent developments in ethnopharmacology and the screening of rain forests for novel products: Farnsworth, 1990); (iii) as a source of novel chemicals in their own right; and (iv) of increasing importance as sources of enzymes and biotransformation systems either on their own or in combination with chemical syntheses.

Certain chapters in this volume have focused on specific aspects of plant cell and tissue culture for product synthesis. It is the purpose of this chapter to place these into context and to link them with other key aspects such as product synthesis, yield improvement, process format, biotransformation potentials and so on. Before doing this, however, it is important to place plant cell culture itself in context and briefly review the reasons behind considering this technology as an alternative production system for plant cell derived natural products. These are: (i) independence from various environmental factors, such as climate, pests, geographical and seasonal constraints; (ii) defined production systems, with production as and when, and at the level required, giving a close control over market supply; (iii) more consistent product quality and yield; and (iv) freedom from political interference. There may also be additional advantages in downstream recovery and efficiency of product purification, a situation true of enzyme products.

While great progress has been made in recent years in plant cell and tissue culture the approaches used are still highly empirical. However, the application of sophisticated techniques of molecular biology and the directed manipulation of particular gene sequences, and in turn proteins, will do much to change this situation. In future it is possible to envisage the enhancement of particular enzyme systems where these impose a rate limitation on a particular pathway, as well as providing wholly new or alternative pathways with possibilities of enhanced or altered patterns of product synthesis. Such approaches will greatly improve the chances of developing viable commercial processes.

## 6.2 PLANT CELLS AND TISSUE CULTURE

The culture of plant cells and tissues has become largely routine, and, while difficulties with some species are unquestionably encountered, it can be said that within reasonable limits it is possible to place any plant species in culture. This applies not only to higher plants but also to lower plants, such as mosses and liverworts, where there are increasing numbers of reports of successful culture initiation. A number of excellent texts exist on the practicalities of cell and tissue culture and the reader is referred to these for futher details (Dixon, 1985; Pollard and Walker, 1990). The literature abounds with a great array of different nutrient formulations, further emphasizing the largely empirical nature of the subject. Often these formulations have been developed with very different purposes in mind, such as product synthesis or shoot formation. Some of the more esoteric formulations along with those more routinely used, are summarized by Stepan-Sarkissian (1990a).

While a great deal of attention has been paid in recent years to the use of cell cultures in commercial process systems, the potentials of other culture systems must not be forgotten, including hairy roots, shoot cultures and immobilized cells and tissues. Immobilized systems have been covered in Chapter 5 of this volume by Williams and Mavituna, while the others will be covered in later sections of this chapter. From a methodological standpoint the underlying principles of operating these systems, however, are not vastly different from those for cell cultures.

## 6.3 CELL CULTURES FOR BIOACTIVE COMPOUNDS

### 6.3.1 The Range of Substances Synthesized by Cell Cultures

Plant cells grown in culture produce a wide range of molecular species (see Table 1). However, it is often far from the case that they produce either qualitatively or quantitatively the same substances found in the parent plant from which they were established, and it is perhaps naive to think that this should be the case. The plant cell situated inside the whole plant body has a very different physiological and biochemical environment to the plant cell supported on an agar bed or floating in a liquid. Within the parent plant, chemical gradients exist which are all part of the complex morphogenetic and temporal relationships of the developing plant body. Such a situation is obviously not found in a cell suspension culture, but to varying degrees may be found in callus, tissue or organ cultures. Support for the view that synthesis of many of the complex 'secondary' substances produced by plants is in some way integrated with differentiation and development comes from the observation that often a degree of tissue differentiation or organ development has to be observed before synthesis of secondary metabolites becomes obvious (Misawa, 1985). One of the great challenges facing plant molecular biologists is to unravel and find methods of uncoupling the complex mechanisms linking differentiation with secondary metabolite synthesis, and so enable commercial production of natural products in simple cell culture. It is the strongly held view of the authors that commercial viability in this area lies in achieving the most simple operating conditions possible which essentially lie in suspended cell cultures, preferably growing in continuous or cell recycle systems.

**Table 1** Classes of Secondary Metabolites Produced by Plant Cell Cultures and Some Examples

| *Chemical class* | *Representative product* | *Species* |
|---|---|---|
| ALKALOID | | |
| Indole alkaloid | Catharanthine | *Catharanthus roseus* |
| Isoquinoline alkaloid | Protoberberines (*e.g.* jatrorrhizine) | *Berberis* spp. |
| Quinolizidine alkaloid | Sparteine | *Lupinus polyphyllus* |
| Pyridine alkaloid | Nicotine | *Nicotiana tabacum* |
| Tropane alkaloid | Hyoscyamine | *Hyoscyamus niger* |
| Steroid alkaloid | Solanidine | *Solanum tuberosum* |
| Purine alkaloid | Caffeine | *Coffea arabica* |
| TERPENOID | | |
| Monoterpenoid | Anethole | *Pimpinella anisum* |
| Sesquiterpenoid | Paniculide | *Andrographis paniculata* |
| Steroid | Diosgenin | *Dioscorea deltoidea* |
| PHENYLPROPANOIDS | | |
| Flavonoids | Anthocyanins | *Vitis vinifera* |
| Anthraquinones | Various | *Galium mollugo* |
| Naphthaquinones | Shikonin | *Lithospermum erythrorhizon* |

Of greater long term interest and opportunity are those substances which appear novel to cell cultures and which have not previously been reported from the parent plant from which the cell culture was established. It is in this area also that plant cell culture has much to offer alongside ethnopharmacology (Battersby, 1990). Quite a wide range of novel substances have been reported from plant cell cultures (Table 2) covering a variety of chemical structures. Many of these substances have some degree of biological activity. It has been estimated that something like 1500 plus new chemical structures are reported from plants each year, of which a significant number have some form of biological activity (Farnsworth, 1990). The possibility of finding novel, commercially interesting structures is therefore high.

That cell cultures produce novel substances is perhaps not surprising. Plant cells possess a large amount of genetic material, including DNA encoding information, which, while redundant to the current environment of the intact plant, has not been discarded. Placing plant cells in culture imposes

**Table 2** Examples of Novel Substances Reported in Plant Cell Cultures (Berlin, 1983; Kreis and Reinhard, 1989; Ruyter and Stockigt, 1989)

| *Chemical class* | *Compound* | *Plant species* |
|---|---|---|
| Alkaloid | Armorine | *Stephania cepharantha* |
| Alkaloid | Norcepharadione | *Stephania cepharantha* |
| Alkaloid | Rauglucin | *Rauwolfia serpentina* |
| Alkaloid | Epchrosin | *Ochrosia elliptica* |
| Terpenoid | Dihomovalerate | *Valeriana wallichii* |
| Terpenoid | Paniculide A | *Andrographis paniculata* |
| Terpenoid | Honokiol | *Thuja occidentalis* |
| Terpenoid | Tarennoside | *Gardenia jasminoides* |
| Anthraquinone | Lucidin | *Morinda citrifolia* |
| Phenylpropanoid | Rutacultin | *Ruta graveolens* |
| Phenylpropanoid | Podoverin | *Podophyllum versipella* |

an environment and stress situation very different from that found in the intact plant. It is hardly surprising that the cells respond to this by displaying very different metabolic behaviour, which in turn results in marked differences in natural product synthesis compared to the parent plant. Little is understood of the mechanisms which may need to be manipulated to relax or modify the controls which 'normally' prevent expression of this DNA. A veritable treasure trove of novel molecular species representing many commercial opportunities awaits the group that can achieve this.

### 6.3.2 Yield Enhancement

Having identified a cell line which synthesizes a desired substance, the next key question, and one which has proved of great difficulty, relates to increasing the yield of the product. This may be seen in two ways: amount of product formed as a proportion of the total dry matter synthesized by the culture and the yield per litre of culture medium. Both are important criteria of performance by the culture but the latter has a particular importance in determining the economic viability of the final process.

Many different approaches have been used in attempts to increase product yield. They include chemical, physical and genetic strategies.

### 6.3.3 Nutrient Formulation

Most of the constituents of plant cell culture media have been subjected to manipulation in attempts to enhance culture productivity, and evidence suggests that, depending upon the products in question, many of these components are important determinants of their accumulation. A number of good reviews covering this area have been published (Mantell and Smith, 1983; Misawa, 1985; Stafford *et al.*, 1986).

Hormone composition is often a critical factor in secondary product accumulation. The type and concentration of auxin and cytokinin, usually in combination, have also been shown to affect cell division, cell enlargement and root or shoot differentiation. In certain cases the biochemical 'differentiation' which allows secondary product accumulation may exclude cell division, and *vice versa*, giving rise to the concept of 'production' media. In most of these, hormone composition is an important factor. For instance, indole alkaloid accumulation in *Catharanthus roseus* was generally depressed by the addition of 2,4-D to media (Zenk *et al.*, 1977b). The substitution of 2,4-D by another auxin, IAA, allowed high levels of indole alkaloid accumulation in nondividing cultures, while the inclusion of NAA allowed simultaneous alkaloid accumulation and cell division (Morris, 1986).

The effectiveness of a particular production medium will depend upon a combination of various factors. Zenk *et al.* (1977a) found that the Murashige and Skoog medium gave the best yield of serpentine out of a wide range of other 'basal media'; but as each basal medium is a mixture of 10–20 different minerals salts and vitamins, the most important factors with respect to secondary product formation are usually difficult to ascertain. In cell suspensions, the nitrogen source, sugar type and concentration, and phosphate level have all been shown to influence the accumulation of plant secondary metabolites belonging to many chemical classes. The overall mineral salt concentration may be an important factor: Inagaki and coworkers (1991) found that half strength B5 salts were optimum for both biomass and verbascoside production in cell cultures of *Leucosceptrum japonicum*, over a range of one twentieth to twice the normal salt concentration.

Of the carbon sources, sucrose is the most popular for plant tissue cultures, although growth will be supported by other disaccharides such as maltose, lactose, cellobiose, melebiose and trehalose (Herouart *et al.*,1991) as well as monosaccharides such as glucose and fructose. Catharanthine production in cell suspension cultures of *Catharanthus roseus* was enhanced by incorporating lactose as the carbohydrate source (Smith *et al.*, 1987). The level of sugar has been shown to affect the productivity of secondary metabolite accumulating cultures; thus, sucrose concentrations of 2.5% and 7.5% in *Coleus blumei* media brought about rosmarinic yields of 0.8 g $L^{-1}$ and 3.3 g $L^{-1}$, respectively (Misawa, 1985). For indole alkaloid accumulation in cell cultures of *Catharanthus roseus*, 8% surose was found to be optimal in the concentration range tested of 4–12% (Knobloch and Berlin, 1980). Yields of benzophenanthridine alkaloids from suspension cultures of *Eschscholtzia californica* were increased 10-fold to around 150 mg $L^{-1}$ by increasing the medium sucrose concentration to 8% (Berlin *et al.*, 1983). As with many secondary products, the profile of alkaloid accumulation exhibited a downturn in stationary phase; the products disappeared within a period of four days, a good example of the fact that the levels of secondary products accumulated in cell cultures should be viewed as a function of product biosynthesis and degradation.

Nitrogen concentration has been found to affect the level of proteinaceous or amino acid products in cell suspension cultures. Thus, Misawa (1985) found that the yields of proteinase inhibitors and antiplant virus substances in *Scopolia japonica* and *Phytolacca americana*, respectively, were significantly affected by the levels of nitrogen compounds incorporated in the medium. Likewise, the level of L-glutamine accumulated in cells of *Symphytum officinale* could be increased by elevating the nitrogen concentration of the medium.

While higher levels of phosphate have often been found to enhance cell growth, they may have a negative effect upon secondary product formation. Sasse *et al.* (1982) observed that phosphate decreased the level of alkaloids in cultured cells of tobacco, *Catharanthus roseus* and *Peganum harmala*. Following transfer into phosphate-containing media, the activities of enzymes channelling primary metabolites into secondary biosynthetic pathways, such as phenylalanine ammonia lyase, were also reduced. Phosphate levels also affected anthocyanin accumulation in *Daucus carota* and *Vitis vinifera* cell cultures, and anthraquinone accumulation in suspensions of *Galium mollugo* (for a review see Stafford *et al.*, 1986).

An exhaustive investigation of medium composition was carried out by research workers developing the first commercial plant cell culture process for Mitsui Petrochemical Industries in Japan. Their objective was to enhance the level of shikonin pigments in cultures of *Lithospermum erythrorhizon*, by medium manipulation and strain selection. Having selected a very high pigment-producing callus strain (M18) by repeated analytical screening, the strain was transferred to a liquid medium with the result that no shikonins were formed. It was a found that this loss of product was due to the absence of an agar component from the liquid medium, an acidic polysaccharide agaropectin (Fukui *et al.*, 1983). This shows that no possibility can be ignored when designing and developing good production media.

### 6.3.4 Physical Conditions

Environmental conditions such as light, temperature, medium pH and aeration have been examined for their effect upon secondary product accumulation in many types of cultures. In particular, the importance of aeration and agitation, especially at large scale, have been dealt with in numerous reviews (Martin, 1980; Scragg and Fowler, 1985). Kreis and Reinhard (1989) described the influence of dissolved oxygen levels on protoberberine production in selected cultures of *Berberis wilsoniae*. Optimum dissolved oxygen levels of 50% allowed an alkaloid yield of around $3\,g\,L^{-1}$ culture after 20 days growth in an airlift bioreactor. Higher aeration rates produced a dramatic decrease in alkaloid productivity. It is evident that airlift and stirred tank bioreactors can allow similar secondary product levels in cultured plant cells; however, in stirred tank vessels the characteristics of the stirrer may be critical (Kreis and Reinhard, 1989).

Gassing regimes are likely to be as critical for the growth and productivity of differentiated cultures as for cell suspension cultures. Greidziak and coworkers (1990) found that the $O_2$ tensions, $CO_2$ concentration and irradiation had significant effects upon the growth and cardenolide formation in somatic embryo cultures of *Digitalis lanata*, grown in gaslift fermentors.

### 6.3.5 Elicitation

Upon microbial or fungal attack, many plants respond by producing phytoalexins at the site of infection. The phenomenon has been widely explored in cell cultures as well as in whole plants, and though true phytoalexins are mainly flavonoid compounds, secondary metabolites belonging to other chemical classes can be induced using a similar strategy.

Elicitors such as fungal homogenates and even inorganic salts such as copper sulphate have been shown to induce a range of plant secondary products. For example, sanguinarine is a benzophenanthridine alkaloid produced by cultures of a number of plant species including *Papaver somniferum*, the opium poppy. This compound is of interest because it has demonstrated antibiotic activity against the bacteria which cause plaque formation on teeth, and it has been included as the active ingredient in a number of dental care products (Constabel, 1990). Treatment of opium poppy cell suspensions with a homogenate of *Botrytis* mycelium resulted in a remarkable accumulation of sanguinarine of up to 3% of the plant cell dry weight. One of the significant features of successful elicitation treatments is that they do not require transfer of cells to a new production medium, and do not need a lengthy culture period following induction; elicitation usually brings about a very rapid response. In the case described above, the accumulation of sanguinarine occurred within three days after induction with fungal homogenate compared with the several weeks often required in production media.

The discovery of an effective but undefined elicitor such as fungal homogenate leads logically to the definition of the active ingredient within that crude mixture. Constabel (1990) reported that chitin simulated the effect of *Botrytis* homogenate and that elicitation of sanguinarine production in poppy cell cultures resulted (Tyler *et al.*, 1988). An advantage of chitin was its relatively nondestructive effect upon plant cell suspensions; the use of crude microbial and fungal homogenates often leads to a percentage of cell death accompanying the increased levels of secondary product accumulation.

Further examples of elicitors and their effects are provided in Table 3.

**Table 3** Elicitation of Secondary Product Accumulation in Plant Cell Cultures

| *Species* | *Elicitor* | *Product* | *Induced yield* | *Ref.* |
|---|---|---|---|---|
| *Papaver somniferum* | *Botrytis* spp. | Sanguinarine | 0.01–2.9% | Eilert *et al.* (1985) |
| *Catharanthus roseus* | *Pythium vexans* | Catharanthine | $0–17\,\mu g\,L^{-1}$ | Nef *et al.* (1991) |
| | | Ajmalicine | $17–90\,\mu g\,L^{-1}$ | |
| *Ruta graveolens* | Various fungal preparations | Acridone alkaloids | Not given | Baumert *et al.* (1990) |
| *Cinchona ledgeriana* | *Phytophthora cinnamonii* | Anthraquinones | $15–110\,mg\,g^{-1}$ dry wt | Wijnsma *et al.* (1985) |
| *Bideus pilosa* | *Pythium aphanidermatum* | Aromatic polyalkynes | $0–3.2\,\mu g\,g^{-1}$ dry wt | DiCosmo *et al.* (1982) |
| *Glycine max* | *Phytophthora megasperma* | Glyceollin | $0–200\,nM\,g^{-1}$ fresh weight | Ebel *et al.* (1976) |

It should be said that elicitation does not always work. There is a definite plant cell species *versus* elicitor type interaction which means that the ultimate selection of elicitor for a particular plant cell culture system depends upon an empirical approach. However, for certain classes of plant products, particularly those which from whole plant work can be described as phytoalexins, the approach holds promise.

### 6.3.6 Induction

A wide range of unrelated substances have been found to somehow 'induce' secondary metabolism in cultured plant cells. This heterogeneous collection includes agaropectin (mentioned above), a component of agar which was found to be necessary for shikonin production in *Lithospermum* cell lines, as well as miscellaneous polysaccharides, inorganic compounds such as vanadyl sulphate, colchicine and herbicides, and physical conditions such as osmotic stress.

In an interesting study following on the discovery of the effect of agaropectin, Fukui and coworkers (1990) found that an endogenous polysaccharide induced shikonin biosynthesis in *Lithospermum* cell cultures. The polysaccharides were isolated from shikonin-producing cells cultured on a production medium, and were shown to induce shikonin biosynthesis in a growth medium. Shikonin biosynthesis was also induced by a pectinase treatment, and it was suggested that the polysaccharides may be released from cell walls of *Lithospermum* cells under certain conditions such as on the M9 production medium used, which was deficient in ammonium. The active polysaccharides were observed to vary widely in molecular weight and sugar composition; they consisted mainly of galacturonic acid, galactose, arabinose and glucose, with minor constituents rhamnose, xylose and mannose.

Osmotic stress imposed by increasing the medium mannitol concentration was found to stimulate indole alkaloid production in *Catharanthus roseus* cell cultures (Rudge and Morris, 1986). Mannitol was also found to stimulate product accumulation in grape (*Vitis vinifera*) cultures (Bao Do and Cormier, 1991). A combination of 88 mM sucrose with 165 mM mannitol brought about an increase in anthocyanin concentration in the culture, with the number of pigmented cells increasing from 48% to 84% of the total cell population. The osmotic potential of the culture medium mainly affected the yield of peonidin 3-glucoside, the major anthocyanin constituent. As peonidin 3-glucoside can be derived from cyanidin 3-glucoside by methylation, it was hypothesized that high osmotic potential may stimulate anthocyanin methylation as well as increasing anthocyanin level *per se*.

On a rather different note, a sustained induction of secondary metabolite level using colchicine was achieved with cell suspensions of *Valeriana wallichii*. A 70-fold increase in valepotriate accumulation was effected following treatment with 0.05% colchicine, and the increase in product level was sustained for over a year (Becker and Chavadej, 1985). The basis of this increase was never fully investigated, but in view of the fact that following colchicine treatment, the initially higher ploidy level was not maintained, it was concluded that the valepotriate increase may have been due to gene amplification or to cell selection. This treatment had the added effect of producing a culture exhibiting a new spectrum of products: six new diene valepotriates were detected which could not be detected in the parent plant (Becker *et al.*, 1984).

### 6.3.7 Precursor Feeding

In 1965, Chan and Staba were the first to investigate the effects of precursor feeding on alkaloid formation. Since the first 'production media' were developed by Zenk in 1977, precursor feeding has been an obvious and popular approach to increasing secondary metabolite production in plant cell cultures. The concept, extrapolated from years of work on microbial systems, is based upon the idea that any compound which is an intermediate in or at the beginning of a secondary biosynthetic route stands a good chance of increasing the yield of the final product. Approaches vary from immediate precursor feeding, in which case the process may be described more accurately as a one-step biotransformation, to primary metabolite feeding.

Zenk *et al.* (1977b) was one of the first to analyze the effects of primary metabolite feeding, when he included tryptophan as a primary precursor of indole alkaloids in some of his production media. In one cell line the strategy worked well, with a three-fold increase in serpentine level stimulated by tryptophan at 500 mg $L^{-1}$ and in another cell line tryptophan at any concentration depressed the level of serpentine; tryptophan at 1000 mg $L^{-1}$ decreased serpentine from 60 mg $L^{-1}$ in controls to 20 mg $L^{-1}$. This cell line difference in response to precursor feeding was a significant finding; it reflects the genetic and so-called 'spigenetic' differences which can arise in culture, and the resulting disparity in the regulation of secondary metabolism in cultured cells.

Since this early report, many workers contemplating trying to enhance secondary product levels in plant cell cultures have explored the possibilities of precursor feeding. As serious consideration must be given to the economics of any potentially commercial plant cell culture process, precursor feeding is most likely to be adopted by those faced with the task of inducing high levels of secondary product formation in intrinsically low- producing cultures. An informative review describing examples of such products is the one by Berlin (1986), in which he discusses products which are spontaneously produced by plant cell cultures compared with those only detected in trace amounts.

Monoterpenoid and sesquiterpenoid essential oils are only ever accumulated by cell suspension cultures in very low yields unless the cultures are fed with precursors. For a comprehensive review of essential oil production in plant cell and tissue cultures see Mulder-Krieger *et al.*, (1988). From precursor-feeding studies it can be shown that cell cultures derived from essential oil producing species are able to transform exogenous substrates to a range of monoterpene products, *via* numerous types of transformation including esterification, glycosidation, reduction, hydrolysis, isomerization and so on (Suga and Hirata, 1990). Common problems associated with the use of monoterpene substrates are poor solubility in aqueous media combined with phytotoxicity, and attempts to overcome these obstacles have resulted in the investigation of two-phase culture systems. For instance, Cormier and Ambid (1987) found that the addition of Miglyol 821, a $C_8/C_{10}$ triglyceride, to cell cultures of Muscat grape reduced the toxicity of exogenous geraniol and allowed a five-fold increase in substrate load. Nascent products could be recovered from the lipophilic phase and the persistence of products in the culture was much improved. However, the further turnover of monoterpene biotransformation products by esterification, glycosidation or $\beta$-oxidation is a problem which must still be resolved.

Podophyllotoxin is a cytotoxic lignan isolated from the rhizomes of *Podophyllum* species. The toxin provides the substrate for semisynthetic drugs, such as etoposide and teniposide, and it has attracted interest as a target for plant cell or tissue cultures because of the relatively poor availability of the whole plants. Cell suspensions of *Podophyllum hexandrum* have been established which accumulate up to 0.1% podophyllotoxin (Van Uden *et al.*, 1989). Out of a range of seven intermediates of the phenylpropenoid pathway only coniferin, a glucosylated intermediate, was found to enhance podophylloxin accumulation by a factor of between 4.5 and 12 (Van Uden *et al.*, 1990). In a subsequent study it was found that by introducing coniferyl alcohol (the aglycone of coniferin) as a complex with $\beta$-cyclodextrin, this otherwise ineffectual precursor produced a significant increase in podophyllotoxin compared with controls (Woerdenbag *et al.*, 1990) .The concept of using cyclodextrins as a means of introducing relatively insoluble precursors to cultures deserves further attention; these cyclic oligosaccharides are able to form inclusion complexes with apolar ligands, bringing about alterations in the physicochemical properties of ligands, such as water solubility. In the particular example described above, the water solubility of coniferyl alcohol increased from 0.15 mM to about 3.4 mM in the presence of $\beta$-cyclodextrin.

The biotransformation of $\beta$-methyldigitoxin to $\beta$-methyldigoxin using *Digitalis lanata* cell suspensions has been a reality for well over a decade (Reinhard, 1974). Yields approaching 95% efficiency in large scale (*e.g.* 20 L) bioreactors have been obtained, and the process has been evaluated in batch, semicontinuous, freely suspended and immobilized cells. Some further examples of successful precursor feeding are provided in Table 4.

**Table 4** Examples of Precursor Feeding Resulting in Increased Secondary Product Yields in Plant Cell Cultures

| *Species* | *Feed precursor* | *Secondary product* | *Yield increase* | *Ref.* |
|---|---|---|---|---|
| *Tripterygium wilfordii* | Farnesol (100 $\mu g\,mL^{-1}$) | Tripdiolide | ×4 | Misawa (1985) |
| *Catharanthus roseus* | Secaloganin (250 $mg\,mL^{-1}$) | Serpentine | ×25 | Stafford and Smith (1986) |
| *Dioscorea deltoidea* | Cholesterol (100 $mg\,L^{-1}$) | Diosgenin | ×2 | Chowdhury and Chaturvedi (1979) |
| *Coleus blumei* | Phenylalanine (500 $mg\,L^{-1}$) | Rosmarinic acid | ×2 | Zenk *et al.* (1977a) |
| *Peganum harmala* | Tryptamine (5 mM) | Serotonin | ×100 | Sasse *et al.* (1987) |

### 6.3.8 Approaches Relating to Levels of Differentiation

Because of the frequent low yields of undifferentiated plant cell suspension cultures, in recent years the possibility of obtaining differentiated plant culture systems as the basis for commercial processes has attracted considerable attention. Shoot, root, and embryo cultures of many species have been developed and while generally the scope for large scale culture is not as great as for easily manipulated plant cell suspensions, some of the results emerging from these differentiated systems are potentially promising.

The basis for the interest in differentiated systems is the knowledge, culled from numerous observations, that when low- or non-producing undifferentiated plant callus or suspension cultures are allowed to differentiate, the newly formed shoots, roots or embryos usually regain their ability to accumulate secondary products. It has long been realized that low productivity in cell cultures rarely reflects a genetic abnormality but rather is a function of the biochemical status. Misawa (1985) discusses the phenomenon of redifferentiation and secondary metabolism quoting *Papaver somniferum*, *Digitalis* and *Pyrethrum* cultures as examples.

Monoterpenoids and sesquiterpenoids are used widely by the food and cosmetics industries and to a lesser extent by the pharmaceutical industry, and some of these 'essential oils' have a high value; for example, Bulgarian rose oil commands a price of \$11 000 $kg^{-1}$. These compounds are volatile and often highly susceptible to oxidation, and are therefore unlikely to accumulate to significant levels in cell suspension cultures. However, it has been shown that shoot cultures hold some promise as sources of essential oils. Charlwood and Moustou (1988) have demonstrated that essential oil producing shoot-proliferating callus of *Pelargonium* can be cultivated in submerged cultures of up to 1 L volume, and more recently Spencer *et al.* (1990) have investigated shooty 'teratomas' of *Mentha* species as a source of essential oils, in culture volumes of up 2 L. The cultured shoots of both species differentiated glands capable of accumulating essential oils, and in case of the *Mentha* cultures the chemical profile was similar to that of the intact plants; however, in both cases the overall yields were low compared with those of the whole plant.

Much more attention has been given to the potential of root cultures, notably 'hairy root' cultures, as production systems. Many commercially interesting secondary metabolites are synthesized in the root and then accumulated elsewhere in the plant, as in the case of the nicotine (pyridine) and tropane alkaloids of members of the *Solanaceae* family. Solanaceous plants have attracted considerable interest in the context of 'hairy root' cultures, not only because root cultures of these species accumulate high levels of their typical alkaloids, but also because the family is very susceptible to infection by the hairy root bacterium *Agrobacterium rhizogenes*, and produces rapidly growing transformed roots with relative ease. Many other dicotyledonous plant families are also susceptible to infection and the literature covering secondary product accumulation by hairy root cultures is now vast.

Table 5 provides examples of hairy root cultures which have been found to accumulate products of industrial interest.

**Table 5** Hairy Root Cultures Producing Metabolites with Applications in Industry (From Flores *et al.*, 1987; Hamill *et al.*, 1987; Hitton and Rhodes, 1990; Yonemitsu *et al.*, 1990)

| *Species* | *Metabolite* | *Application* |
|---|---|---|
| *Datura stramonium* | Hyoscyamine | Anticholinergic |
| *Duboisia myoporoides* | Scopolamine (hyoscine) | Treatment of motion sickness |
| *Atropa belladonna* | Atropine | Pupil dilation in ophthalmic practise |
| *Lithospermum erythrorhizon* | Shikonin | Antiinflammatory pigment |
| *Beta vulgaris* | Betacyanin, betaxanthin | Pigment |
| *Panax ginseng* | Ginseng saponins | Tonic |
| *Catharanthus roseus* | Catharanthine | Substrate for synthesis of antileukaemic alkaloids (e.g. vinblastine) |
| *Cinchona ledgeriana* | Quinoline alkaloids | Antimalarial agent |
| *Lobelia inflata* | Lobeline | Respiratory stimulant |

Somatic embryos derived from embryogenic cell suspension cultures are potentially an additional source of useful secondary products. A medium change may be required to trigger embryogenesis in the undifferentiated culture. Somatic embryos of *Digitalis lanata* capable of cardiac glycoside biosynthesis were grown in 5 L bioreactors and yields of up to 1.1 $\mu$mol cardenolides $g^{-1}$ dry weight were obtained (Greidziak *et al.*, 1990). Many important products have been shown to accumulate in developing embryos, including cocoa lipids in *Theobroma cacao*, $\gamma$-linolenic acid in seeds of borage and evening primrose, and alkaloids in *Papaver somniferum* somatic embryos. However, in many systems the maturation rate of the embryos may be slow and in the case of cocoa lipids and morphine alkaloids the onset of product accumulation *in vitro* occurs at a relatively late developmental stage. The wider application of somatic embryo production systems has as yet to be realized.

### 6.3.9 Genetic Approaches

The approaches to product enchancement given above relate to the manipulation of culture conditions, with the exception of 'hairy root' systems which of course require a genetic transformation step for their induction. The common feature with all these approaches is that they are empirical. Also fundamentally empirical is the vast amount of screening and selection work which has been carried out over the last 20 years. Many successes have been obtained by random screening, as in the case of the shikonin production system. The highest-yielding *Lithospermum* cell lines were obtained *via* visual and analytical selection, and formed the basis of the Mitsui commercial process. Numerous reviews and articles have dealt with this general subject area (Fujita and Tabata, 1987; Widholm, 1980; Zenk *et al.*, 1977b), and this aspect will not be covered here.

Given the relatively slow growth of plant cell and tissue cultures, microbial systems expressing recombinant plant proteins are an obvious first choice when contemplating the production of important plant-derived proteins for industrial use. This approach was used by Unilever to clone thaumatin, a proteinaceous sweetener naturally obtained from the plant *Thaumatococcus daniellii* (Edens *et al.*, 1982). The cloning of plant proteins into *E. coli* is not guaranteed success, however, due to the tendency of the host organism to process recombinant proteins into dense inclusion bodies. The subsequent protein recovery steps required to arrive at a functional product can be lengthy and inefficient. Nowadays yeast hosts provide a very attractive alternative, combining the advantages of high growth rates and secretory systems for recombinant protein products.

For the vast majority of valuable plant products, which are nonproteinaceous, the cloning option is unrealistic. Recombinant DNA technology has not yet reached the stage at which large numbers of foreign genes can be transferred to and coordinately expressed in any host, and our knowledge of plant secondary metabolism and the regulation of secondary biosynthetic routes is still very poor. However, a few secondary product pathways have received in depth attention in recent years, with the result that the concept of genetically modifying the rate at which secondary products can be produced by plants and their cell cultures is fast becoming a reality.

A number of genes encoding enzymes with a key regulatory role in secondary biosynthetic pathways have now been cloned and characterized. These include the genes coding for chalcone synthase and phenylalanine ammonia lyase involved in the regulation of phenolic biosynthesis (Kreuzaler *et al.*, 1983), tryptophan decarboxylase (De Luca *et al.*, 1989) and strictosidine synthase (Kutchan *et al.*, 1988) of the indole alkaloid pathway and the hyoscyamine 6$\beta$-hydroxylase enzyme (Hashimoto *et al.*, 1990). Hashimoto and coworkers have progressed from purifying the hyoscyamine 6$\beta$-hydroxylase enzyme from cultured roots of *Hyoscyamus niger*, through to isolating the cDNA encoding the enzyme and inserting the cDNA into tobacco callus. They were able to show that the transgenic tobacco calli expressed the cDNA; the transformed tissues exhibited 6$\beta$-hydroxylase activity and produced a 38 kDa protein which cross reacted with 6$\beta$-hydroxylase specific antibody on Western blots. Following the further introduction of the 6$\beta$-hydroxylase gene into *H. niger* plants which naturally express the enzyme, Northern hybridization showed that the 6$\beta$-hydroxylase mRNA was widespread in roots but absent in cultured cell suspensions, stems and leaves.

This type of strategy not only provides insight into tissue and cell specific enzyme expression in plants, but also provides us with the opportunity to increase secondary metabolite production in transgenic plants and their cultures.

Constabel (1990) describes a similar investigation, this time an attempt to overcome the inability of *Catharanthus roseus* cell suspensions to produce the important dimeric indole alkaloid, vinblastine. While elicitation of cell cultures with *Pythium* homogenates, as described earlier in this chapter, resulted in the induction of catharanthine, the other substrate, vindoline, required for the condensation step to anhydrovinblastine could only be detected in differentiated plants. A key enzyme involved in the formation of vindoline, acetyl coenzyme A:deacetylvindoline *O*-acetyl transferase, has been purified and characterized. The ultimate objective is to reintroduce this gene into cultured cells of *C. roseus* and monitor its expression.

Hamill *et al.* (1990), working with cell cultures of *Solanaceae* species, attempted to modify the expression of the pyridine alkaloid pathway culminating in nicotine biosynthesis. They tackled the cloning of two genes, encoding putrescine methyl transferase and ornithine decarboxylase, and are aiming to reinsert these into tobacco cell and tissue cultures under the control of a high expression promoter (such as the CaMV 35S promoter).

Alternative strategies dependent on the techniques of genetic manipulation are becoming more widely applicable to the task of secondary product yield enchancement in plants and plant cell cultures. For example, the availability of primary metabolites such as amino acids to secondary metabolism can in theory be increased by modifying regulatory enzyme characteristics. Site-directed or random mutagenesis followed by screening could allow the detection and engineering of mutant enzymes that play a key role in amino acid production, and which are no longer sensitive to feedback inhibition by amino acids.

The development of a new *Petunia* flower colour by transforming a mutant plant with a maize gene encoding one of the anthocyanin pathway enzymes (dihydroflavonol 4-reductase; Van der Krol *et al.*, 1988) illustrates the potential of a further strategy. Introducing an enzyme with different substrate specificites into, for example, steroid-producing plants or cell cultures could give rise to new biosynthetic branches and high yields of modified products.

Finally, antisense mRNA technology, having been used successfully to inhibit flower pigment formation in *Petunia* (Meyer *et al.*, 1987), could be applied to the suppression of metabolic side branches which compete with important secondary biosynthetic pathways.

## 6.4 MASS CULTURE SYSTEMS

Scragg has earlier (Chapter 4) dealt in some depth with the biochemical engineering aspects of growing plant cells on a large scale. Suffice at this point therefore to highlight some of the key points insofar as they relate to process development and operation.

### 6.4.1 Bioreactor Conformation

Plant cells have been grown in a surprisingly wide variety of bioreactor types and conformations (Spier and Fowler, 1983). In the early days there was a general view that conventional stirred tank reactors (CSTR) were unsuitable for the growth of plant cells because of the high levels of shear produced. Classical studies by Wagner and coworkers (Wagner and Vogelman, 1977) with cells of *Morinda citrifolia* provided strong support for this view. In consequence, much of the early work on large scale growth was carried out with air driven reactors, typically of draught tube format (Figure 1) (Scragg and Fowler, 1985), which develop comparatively low levels of shear. In recent years, however, there has been a return to CSTRs and numerous examples of large scale growth in this type of reactor now exist. Indeed, the process formats used in those processes which have been commercialized have involved the use of CSTRs. What has brought about this change? No verified explanation exists. However, it is possible that during cell selection for other criteria, such as improved productivity, unwitting selection has also been made for cells which exhibit higher and higher levels of shear tolerance (Fowler, 1987). That is not to say that all plant cell lines will grow under high shear conditions; there are still good examples of shear intolerant plant cell lines (Leckie *et al.*, 1990).

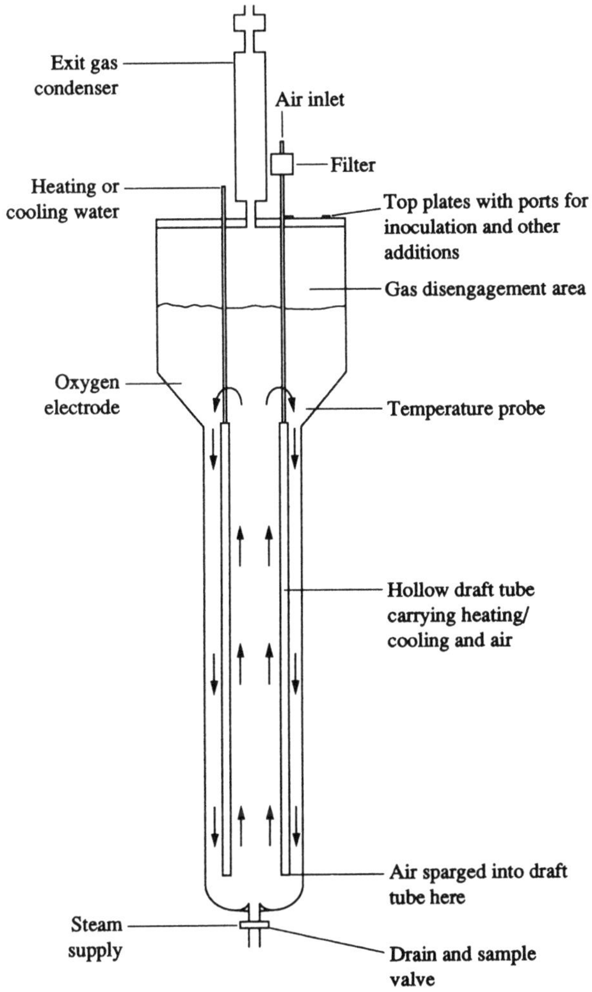


**Figure 1** Diagram of draught tube airlift fermenter used to grow plant cells (as per Scragg and Fowler, 1985)

A particularly interesting approach to large scale plant cell culture was the use by Ulbrich *et al.* (1985) of an Archimedian screw in 100 L tower reactors to grow *Coleus* cells for rosmarinic acid production. Characterized by low shear conditions and with good mixing, the system produced very high biomass and product levels, in fact some of the highest ever published for plant cell cultures. Effective gassing was maintained by multiple entry points throughout the reactor.

Another approach developed by Tanaka (1983) and others has been to use rotating drum or thin film bioreactors, where the cell culture is effectively confined to a layer a few millimetres thick around the reactor. Such systems have the advantage of a high surface area, which aids gas transfer, and good mixing of biomass and nutrient. Unfortunately volume productivities do not compare with more conventional reactor designs.

Given the quite large amount of effort which has gone into investigating different reactor conformations for plant cell growth it is perhaps salutary that all the processes which have achieved commercialization, or are near it, utilize conventional stirred tank reactors, albeit on occasion with modified paddle systems and slower stirring speeds to reduce shear. This is perhaps not surprising in that there is a wealth of knowledge available with stirred tank reactors, which are also often already freely available (Charles, 1985). Equally, if the increased performance of a different, and perhaps novel, reactor design is only marginal there is little commercial incentive to change from a CSTR. Innovation is not necessarily related to commercial potential.

### 6.4.2 Scale

Plant cells have been grown in bioreactors of quite considerable volume. Westphal (1990) has recently reported growing *Rauwolfia* cells in stirred tank reactors of up to 75 $m^3$. There are a number of examples of growth at smaller volumes and some are given in Table 6.

**Table 6** Large Scale Processes Based upon Plant Cell Cultures (Hashimoto and Azechi, 1988; Kreis and Reinhard, 1989; Westphal, 1990)

| *Scale* (L) | *Configuration* | *Species* | *Product* |
|---|---|---|---|
| 85 | Airlift | *Catharanthus roseus* | Indole alkaloids |
| 200 | Module spiral stirrer (Archimedian screw) | *Coleus blumei* | Rosmarinic acid |
| 300 | Airlift | *Digitalis lanata* | Cardiac glycosides |
| 1000 | Stirred tank, rotary drum | *Lithospermum erythrorhizon* | Shikonin |
| 2000 | Stirred tank | *Panax ginseng* | Saponins |
| 20 000 | Stirred tank | *Nicotiana tabacum* | Nicotine, biomass |
| 75 000 | Stirred tank | *Rauwolfia serpentina* | Indole alkaloids |

### 6.4.3 Process Format

Two general process formats are in widespread use for plant cell culture products, of single stage and two stage. The former is typically applicable to the production of biomass, primary metabolites or in some cases enzymes, while two stage, albeit in a variety of forms, is typically utilized where the synthesis of a desired product is uncoupled from growth and division, but more associated with tissue or organ development (see earlier). A wide range of potential process formats exist to cover this situation, a number of which have been explored for plant cell culture (see Tables 7 and 8). With a two stage process, the first part in general is geared to producing large amounts of biomass as quickly as possible, and it is only in the second part that conditions are so arranged that natural product synthesis, not biomass production, becomes the major focus. A good example of such a two stage approach is to be sęen in the process developed by Mitsui & Co. for the production of the red dye shikonin from *Lithospermum erythrorhizon*. An outline diagram of the process is shown in Figure 2. The first or seed stage is carried out in a 200 L stirred tank reactor and is directed towards rapid growth and biomass production. This first stage takes of the order of 8–9 d and is then followed by a second stage carried out in a 750 L stirred tank reactor which takes up to 14 d. This second stage is characterized by very low levels of cell division and growth but an active synthesis of shikonin.

**Table 7** Potential Process Formats for Plant Cell Culture

| *Single stage* | *Two stage* | |
|---|---|---|
| | *First stage* | *Second stage* |
| Batch | Batch | Batch |
| Fed batch/semicontinuous | Continuous | Batch |
| Continuous | Continuous/batch | Immobilized |
| | Fed batch | Fed batch |

**Table 8** Process Formats for Plant Cell Culture Systems which are either Operational or at Pilot Plant Stage

| *Product* | *Process format* | *Single or first stage* | *Second stage* | *Ref.* |
|---|---|---|---|---|
| Shikonin | Two stage | Batch culture<br>Airlift reactor | Batch culture<br>Airlift reactor | Curtin (1983) |
| Rosmarinic acid | Two stage | Fed batch culture<br>Spiral reactor | Batch culture<br>Spiral reactor | Ulbrich and Arens (1985) |
| Digoxin | Two stage | (i) Continuous culture<br>Airlift reactor<br>(ii) Batch culture<br>Airlift reactor | Batch culture<br>Airlift reactor<br>Immobilized cell culture | Reinhard and Alfermann (1980)<br>Alfermann *et al.* (1983) |
| Tobacco | Two stage | Batch culture<br>Stirred tank reactor | Batch culture<br>Stirred tank reactor | Noguchi *et al.* (1977) |
| Geraniol | Two stage | Batch culture<br>Stirred tank reactor | Batch culture<br>Stirred tank reactor | Mitsui Petrochemical Industries (unpublished) |
| Ginseng biomass | Single stage | Batch culture<br>Stirred tank reactor | — | Furuya (1982) |

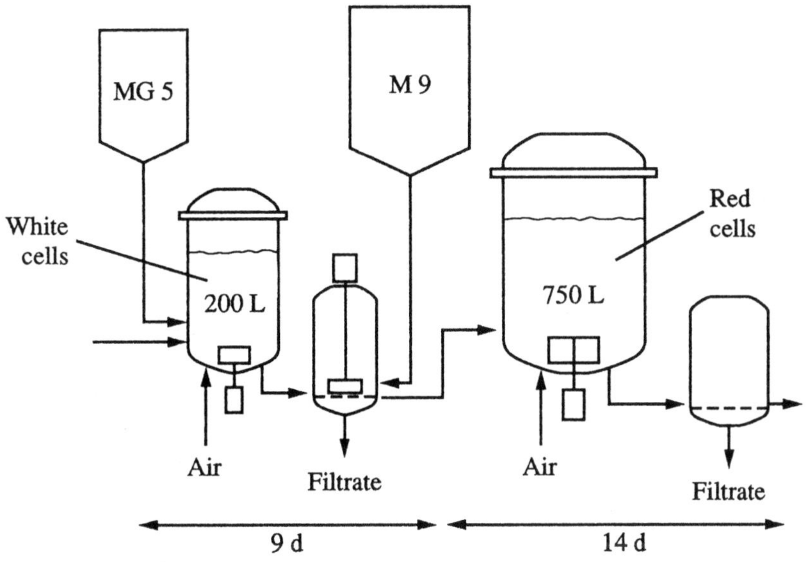


**Figure 2** Outline of the process developed by Mitsui & Co. for shikonin production from cells of *Lithospermum erythrorhizon* (after Curtin, 1983)

## 6.5 ALTERNATIVE PROCESS SYSTEMS

The discussion above has focused principally on process systems and reactors built around large scale growth of cell cultures. The apparent coupling of secondary metabolite synthesis either with zero growth or differentiation/development has led to the exploration of a wide range of other process approaches and reactor formats. Two areas are of particular interest and promise for the future: immobilized cell systems and hairy roots.

### 6.5.1 Immobilized Cells and Tissues

In many respects, immobilized cell technology represents an almost ideal process system. The process is able to operate continuously, tight control is readily achieved, downtime of reactors is reduced to a minimum and long lag phases for cell growth are avoided. Product recovery can be readily achieved from spent or recycled medium and growth conditions coupled to recycling of medium can be closely controlled to maximum economic efficiency.

Much effort has been put into the development of immobilized plant cell systems and this has been covered in detail in Chapter 5 of this volume by Williams and Mavituna. In spite of all the effort expended it is, however, still the case that no immobilized plant cell process has yet been brought to commercial fruition. This should not, however, detract from the potentials of the approach, which still remain. There are three principal areas where progress is required towards commercial application: (i) release of product into the medium; (ii) increased productivity; and (iii) control of genetic instability.

Product release into the medium is proving a difficult problem. While a range of chemicals have been used in attempts to increase the permeability of cell membranes and walls, most result in a sufficient loss in cell viability to render the culture uneconomic as a long term production system. Of all the materials tested for increased permeabilization without major effect on cell viability, dimethyl sulfoxide possibly shows the greatest promise (Brodelius and Nilsson, 1983). This apparent lack of success with permeabilization is to a degree frustrating in that immobilized plant cells produce a wide range of useful secondary metabolites. Methods to enable release of these into the medium could open up immobilized cell technology to a whole range of product opportunities.

A wide variety of bioreactor conformations have been tested for immobilized cell technology in combination with an equally wide range of support systems. These have all been extensively covered by Williams and Mavituna in Chapter 5.

A potentially very important application for immobilized cell systems lies in biotransformation processes. Possibly the best-characterized example of this is the 12-hydroxylation of $\beta$-methyl digitoxin to $\beta$-methyl digoxin by cells of the foxglove *Digitalis lanata*. Extensive studies by Reinhard and coworkers (see, for example, Alfermann *et al.*, 1983) have shown that by using a two stage approach (see later) it is possible to achieve very high transformation rates using in fact both suspended and immobilized cells. The approach used has been to grow cells of *D. lanata* in batch or continuous culture on a medium formulated for rapid growth and which produces cells with little transforming ability. Transfer of the cells to a second medium in which they could either be suspended or immobilized then induced a 12-hydroxylation capability which could be harnessed for $\beta$-methyl digoxin production. An important aspect of this particular process is that almost 90% conversion could be achieved and the majority of the product is released into the external medium.

Many of the advantages that accrue to immobilized cell technology for direct product synthesis also apply to biotransformations, and it is in this mode that immobilized cell technology possibly has its greatest future.

### 6.5.2 Hairy Roots

Reference has already been made to hairy roots as a means of developing high levels of productivity of individual substances. While on a small scale hairy root systems present some interesting potentials, process scale-up raises a whole range of very different problems. Because of the nature of the tissue system, traditional technologies of stirred tank reactors in which nutrients and biomass are mixed together with paddles are not applicable, and other approaches have had to be explored.

Wilson and coworkers, for instance (Wilson *et al.*, 1990), have reported on 'droplet' and submerged reactors for hairy root growth. The principle used in the droplet reactor was essentially to supply nutrients through a droplet or misting dispersion system from the top of the reactor (see Figure 3), which would then flow down over the root mass. Such an approach was tested with bioreactors of 2, 10 and 500 L. In addition, a totally submerged mode is also possible, although few details of this have so far been released. Given the very different nature of these bioreactors to conventional systems, which latter are directed towards rapid mixing and high gas transfer levels, particular attention was paid to monitoring nutrient input levels and gas availability.

As sampling a hairy root culture for biomass production is also much more complicated than, for example, sampling a cell suspension culture, a different way of measuring biomass levels had to be adopted. This was achieved by mounting the reactor on load cells and taking weight measurements of the reactor drained of medium at varying intervals of time. As can be seen in Figure 3, the nutrient feed line is part of a recycle system, the reservoir having sufficient capacity to hold the total medium as the rig is drained for biomass weight measurement. Gas transfer is aided by the high surface area of the small number of droplets released into the bioreactor from the head plate. Given the need for tight control over gassing regimes indicated by other work (Hegarty *et al.*, 1986), this is one aspect of the operation which may need further attention in the future.

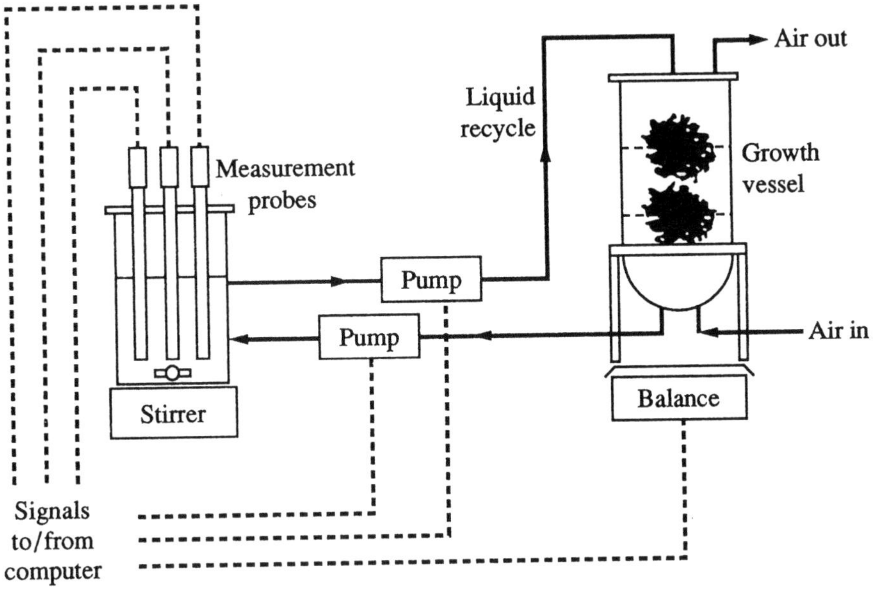


**Figure 3** Droplet reactor as developed by Wilson *et al.* (1990) for transformed root cultures

Little performance data are available for hairy roots grown on anything like a large scale at present. What data exist do not indicate anything out of the ordinary, and indeed biomass productivity levels appear low in comparison with data for cell cultures. Data of Wilson *et al.* (1990) show maximum biomass levels of approximately 10 g $L^{-1}$ dry weight after some 20–25 d of culture, as compared with levels of greater than 25–30 g $L^{-1}$ dry weight produced over periods of 10–14 d which are regularly achieved with cell cultures. At this point one should add the caveat that this comparison only relates to biomass productivity and that the situation may be different in relation to secondary metabolite synthesis (Berlin *et al.*, 1990). Hairy root cultures of many species have yielded products of industrial interest (see Table 5), and some of these perform very favourably compared with plant cell suspension cultures.

A recurring theme through all discussions on large scale cell culture technology is the need for bioreactor designs which couple high growth rates with high biomass levels, good gas control and mixing. While CSTR systems broadly support these needs there are many indications that there is much to be learnt about their use for plant cell growth on a large scale and that 'ideal' conditions for their use have yet to be identified. Equally, the need to utilize tissue or organ cultures for the synthesis of some products adds a different dimension of complexity. While CSTRs in some form will probably be the system of choice for the majority of future process systems, at least for biomass and enzyme production, this is obviously an area much in need of further work and development.

## 6.6 PRODUCT RECOVERY

Little attention has so far been paid in the literature to product recovery from plant cell/tissue process systems. In traditional microbial fermentations this is recognized as an important cost component of the overall system and indeed can represent upwards of 50% or more overall process cost (Hacking, 1986).

As with other large scale biological production processes, plant cell processes suffer from the usual problem of a small amount of product contained in a large amount of biomass or water. Plant cell biomass is on the whole more convenient to deal with than whole plant biomass and does not need the same rigorous treatment to break up tissues and woody, waxy structures. Recovery of products from biomass would presumably follow classical lines of steam distillation or organic extraction.

For reasons partly outlined above, product release into the cell broth is probably the most desirable situation for the recovery and initial purification of a number of products. Unfortunately, as mentioned above, comparatively few desirable plant products are released into the medium. Where solubilization techniques have been attempted they have been relatively unsuccessful as regards long term cell viability, thereby reducing the potentials, for the moment, of immobilized cell technology.

The initial phase of recovery from the medium will almost certainly involve water removal and concentration. In the case of large liquid volume this can be expensive and also time consuming, raising concerns over product stability. Once concentration has been achieved a number of approaches exist, but principally absorption and partition/extraction are the most conventional.

An alternative approach to product recovery, and most suited to small volume applications, is to carry out continuous or discontinuous recovery while the culture is actively growing. Some work has been published on these strategies and one or two examples will serve to illustrate the potentials.

In the case of absorption, ion exchange resins (Amberlite) have been utilized to good effect to recover indole alkaloids released into the medium from cell cultures of *Catharanthus roseus* and *Cinchona ledgeriana* (Brodelius, 1990). Effective recovery was recorded in both cases and in the case of *C. ledgeriana* the productivity of anthraquinones was increased from about 1 mg $L^{-1}$ $d^{-1}$ to 20 mg $L^{-1}$ $d^{-1}$, presumably due to relief of end product inhibition within the cell mass. Activated charcoal and reversed phase silica have also been used for recovery by absorption with varying effect.

Another approach has been to use a solvent second phase, of particular value where substances of low water solubility are the product targets.

## 6.7 PRODUCTIVITY AND ECONOMICS

'Technical feasibility does not necessarily encompass economic viability' (Stafford and Fowler, 1991). At the end of the day it is the market place and demand which will govern the success or otherwise of attempts to develop commercial processes around plant cell culture systems.

Questions of economic viability have long been raised about plant cell culture and a number of authors have attempted detailed analysis. Perhaps the most comprehensive study was carried out in 1980 by Goldstein *et al.* Unfortunately no detailed information was available at that time for large scale plant cell processes and so in some respects the calculations were somewhat theoretical. They did, however, serve to focus on key areas of the overall process where improvements would have to be made if overall economic viability was to be achieved. A number of the key points raised by Goldstein and colleagues are as valid today as they were 10 years ago.

Firstly, there is a need for a high cell growth rate. Of the overall process time from setting up the initial seed inoculum to harvesting the product the cell lines may only be 'productive' for some 15% of that time (Fowler *et al.*, 1990). For the other 85% of the time, however, substantial process costs are still incurred and these have to be 'paid for' by the 15% of productive time. The faster the growth rate and the more the balance can be moved from the nonproductive obviously the better. Secondly, high biomass levels and high volumetric productivities are essential. Biomass levels of $>40$ g $L^{-1}$ are typically necessary to place a plant cell culture process in an economic frame, together with volume productivities in excess of 1 g $L^{-1}$ and preferably much higher. It is probably already obvious from these few simple comments that there is no ready answer. Each system has to be judged on its merits.

Costs of around $1500–3000 per kg of refined product are thought to be achievable for plant cell cultures. In the area of traditional products this still leaves some quite interesting targets and in the case of novel products provides a cost floor against which other methods of synthesis have to be measured.

There is one consideration with plant cell products which is often overlooked in cost and strategic analysis. Many plant products are extremely potent, for instance certain perfumes. A small amount therefore goes a long way. While the unit price might then be extremely high the overall market volume may be extremely low. A case in point is the perfume jasmine, which commands a very high price indeed but for which the market volume is only measured in kilogrammes (Curtin, 1983).

Until a number of plant cell processes are operational in a fully commercial mode and certain data are available, attempting process costings and analyses is largely an academic exercise, although it does provide some insights and guidelines.

## 6.8 FUTURE POTENTIALS AND NOVEL PRODUCTS

The technology of plant cell growth has progressed rapidly over the last decade to the point where a number of process systems have achieved or are approaching industrial operation. Technical feasibility has therefore been demonstrated up to quite large scales of operation. Unfortunately, questions still remain over economic viability, with a particular need for enhanced productivities and simplified process technology. In the longer term it is likely that novel products will emerge rather than the technology be used for traditional products.

Two areas of rapid growth and great potential relate to the use of plant cell derived enzymes in complex biotransformations or the separation of chiral synthons. Plant cell technology has much to contribute in combination with chemical synthesis and microbial transformation in these areas. The technical literature already contains many references to potentials in this area (for reviews see Stepan-Sarkissian, 1990a; Suga, 1990), which when coupled to techniques of cell and enzyme immobilization present a wide range of opportunities for the future.

A radical and somewhat divergent area for the exploitation of plant cell culture technology lies in the use of plant cell cultures for the synthesis of human and animal proteins and peptides. Much effort has already been directed at the cloning of animal genes, for example that coding for human serum albumin, into crop plants for production in the field (Van Montagu, 1991). More recently, Sijmons *et al.* (1990) have shown a similar potential with cell cultures. Production of high value animal proteins in this way could overcome many of the disadvantages seen with animal cell culture, genetically engineered microbial production, or through genetically engineered crops. The system would present no risks to health or raise such concerns as seen with microbial production, in the sense that the cell culture is nonpathogenic; concerns about release into the field or environment would not be raised as the whole process is contained; the product may be excreted into the cell culture medium, aiding product recovery; and, importantly for a number of animal proteins, plants have the ability to glycosylate proteins, a feature not typically seen with bacteria, and only in a qualitatively limited fashion in yeasts. Table 9 provides some examples of animal protein expression in transgenic plant tissues.

**Table 9** Recombinant Proteins Expressed in Plants (Hiatt *et al.*, 1989; Sijmons *et al.*, 1990; Vandekerckhove *et al.*, 1989)

| *Recombinant product* | *Host species* | *Tissue* | *Yield* |
|---|---|---|---|
| Leu-enkephalin | *Arabidopsis thaliana* | Seed | 200 nmol/g seed |
| Leu-eukephalin | *Brassica napus* | Seed | 50 nmol/g seed |
| IgG | *Nicotiana tabacum* | Leaf | 13 μg/mg protein |
| Human serum albumin | *Solanum tuberosum* | Leaf | 20 μg/mg protein |
| Human serum albumin | *Nicotiana tabacum* | Cell suspension | 0.25 μg/mg medium protein |

Having gone through a low period of interest in the late 1980s there is evidence of renewed interest and activity in plant cell technology. A variety of process systems have been well tested, processes are in commercial operation, a more directed approach to process development is beginning to emerge particularly through the use of molecular biology techniques to improve culture performance, and a wide range of opportunities are beginning to emerge in plant enzymology. While it is difficult to predict the shape of a plant cell biotechnology industry of the future, there is now perhaps little question that there will be one, and most probably an innovative and exciting one.

## 6.9 REFERENCES


Alfermann, A. W., W. Bergmann, C. Figur, V. Helmbold, D. Schwantag, I. Schulter and E. Reinhard (1983). Biotransformation of β-methyl digitoxin to β-methyl digoxin by cell cultures of *Digitalis lanata*. In *Plant Biotechnology*, ed. S. H. Mantell and H. Smith, pp. 67–74. Cambridge University Press, Cambridge.

Bao Do, C. and F. Cormier (1991). Accumulation of peonidin 3-glucoside enhanced by osmotic stress in grape (*Vitis vinifera* L.) cell suspension. *Plant Cell, Tissue Organ Cult.*, **24**, 49–54.

Battersby, A. R. (1990). Introduction. In *Bioactive Compounds from Plants*, CIBA Foundation Symposium 154, ed. D. J. Chadwick and J. Marsh, p. 1. CIBA Foundation, London.

Baumert, A., W. Maier and D. Groger (1990). Characterization of acridone specific enzymes and elicitor stimulation of acridone formation in cell cultures of *Ruta graveolens*. *Planta Med.*, **56**, 494.

Becker, H., S. Chavadej, P. W. Thies and E. Finner (1984). The structure of new valepotriates from tissue cultures of *Valeriana wallichii*. *Planta Med.*, **50**, 245–248.

Becker, H. and S. Chavadej (1985), Valepotriate production of normal and colchicine-treated cell suspension cultures of *Valeriana wallichii*. *J. Nat. Prod.*, **48**, 17–21.

Berlin, J. (1983). Naturstoffe aus pflanzlichen Zellculturen. *Chemie in Unserer Zeit*, **17**, 77–84.

Berlin, J., E. Forche, V. Wray, J. Hammer and W. Hosel (1983). Formation of benzophenanthridine alkaloids by suspension cultures of *Eschscholtzia californica*. *Z. Naturforsch.*, **38**, 346–352.

Berlin, J. (1986). Secondary products from cell cultures. In *Biotechnology*, ed. H.-J. Rehm and G. Reed, vol. 4, pp. 629–658. VCH Verlagsgesellschaft, Weinheim.

Berlin, J., C. Mollenschott, N. Greidziak, S. Erdogan and I. Kuzovkina (1990). Secondary product formation in suspension and hairy root cultures. In *Progress in Plant Cellular and Molecular Biology*, ed. H. J. J. Nijkamp, L. H. H. Van der Plas and J. Van Aartrijk, pp. 763–768. Kluwer Academic, Dordrecht.

Brodelius, P. and K. Nilsson (1983). *Eur. J. Appl. Microbiol. Biotechnol.*, **17**, 275–280.

Brodelius, P. (1990). Transport and accumulation of secondary metabolites. In *Progress in Plant Cellular and Molecular Biology*, ed. H. J. J. Nijkamp, L. H. H. Van der Plas and J. Van Aartrijk, pp. 567–568. Kluwer Academic, Dordrecht.
Charlwood, B. V. and C. Moustou (1988). Essential oil accumulation in shoot proliferation cultures of *Pelargonium* spp. In *Manipulating Secondary Metabolism in Culture*, ed. R. J. Robins and M. J. C. Rhodes, pp. 187–194. Cambridge University Press, Cambridge.
Chowdhury, A. R. and H. C. Chaturvedi (1979). Cholesterol and biosynthesis of diosgenin by tuber callus of *Dioscorea deltoidea*. *Curr. Sci.*, **49**, 237–238.
Constabel, F. (1990). Medicinal plant biotechnology. *Planta Med.*, **56**, 421–425.
Cormier, F. and C. Ambid (1987). Extractive bioconversion of geraniol by a *Vitis vinifera* cell suspension employing a two-phase system. *Plant Cell Rep.*, **6**, 427–430.
Curtin, M. E. (1983). Harvesting profitable products from plant tissue culture. *Biotechnology*, **1**, 649–657.
De Luca, V., C. Marineau and N. Brisson (1989). Molecular cloning and analysis of cDNA encoding a plant tryptophan decarboxylase: comparison with animal dopa decarboxylases. *Proc. Natl. Acad. Sci. USA*, **86**, 2582–2586.
DiCosmo, F., R. Norton and G. H. N. Towers (1982). Fungal culture–filtrate elicits aromatic polyalkynes in plant tissue culture. *Naturwissenschaften*, **69**, 550–551.
Dixon, R. A. (1985). *Plant Cell Culture, a Practical Approach*. IRL Press, Oxford, Washington DC.
Ebel, J., A. R. Ayers and P. Albersheim (1976). Host–pathogen interactions XII. Response of suspension-cultured soybean cells to the elicitor isolated from *Phytophthora megasperma* var. *Sojai*, fungal pathogen of soybeans. *Plant Physiol.*, **57**, 775–779.
Edens, L., L. Heslinga, R. Klok, A. M. Ledeboer, J. Maat, M. Toonen, C. Visser and C. T. Verrips (1982). Cloning of thaumatin gene in *E. coli* Antonie var Leeuwenhoek. *J. Microbiol. Serol.*, **48**, 303–304.
Eilert, U., W. G. W. Kurz and F. Constabel (1985). Stimulation of sanguinarine accumulation in *Papaver somniferum* cell cultures by fungal elicitors. *J. Plant Physiol.*, **119**, 65–76.
Farnsworth, N. R. (1990). The role of ethnopharmacology in drug development. In *Bioactive Compounds from Plants*, CIBA Foundation Symposium 154, ed. D. J. Chadwick and J. Marsh, pp. 2–21. CIBA Foundation, London.
Flores, H. E., M. W. Hoy and J. J. Pickard (1987). Secondary metabolites from root cultures. *Tibtech*, **5**, 64–69.
Fowler, M. W. (1987). Process systems and approaches for large scale plant cell culture. In *Plant Tissue and Cell Culture*, ed. C. E. Green, D. A. Somers, W. P. Hackett and D. D. Biesboer, pp. 459–471. Liss, New York.
Fujita, Y. and M. Tabata (1987). Secondary metabolites from plant cells — pharmaceutical applications and progress in commercial production. In *Plant Tissue and Cell Culture*, ed. C. E. Green, D. A. Somers, W. P. Hackett and D. D. Biesboer, pp. 169–185. Liss, New York.
Fukui, H., N. Yoshikawa and M. Tabata (1983). Induction of shikonin formation by agar in *Lithospermum erythrorhizon* cell suspension cultures. *Phytochemistry*, **22**, 2451–2453.
Fukui, H., M. Tani and M. Tabata (1990). Induction of shikonin biosynthesis by endogenous polysaccharides in *Lithospermum erythrorhizon* cell suspension cultures. *Plant Cell Rep.*, **9**, 73–76.
Furuya, T. (1982). In *Plant Tissue Cult., Proc. Int. Congr. Plant Tissue Cell Cult., 5th, 1982*, p. 301. Maruzen Co. Ltd., Tokyo.
Greidziak, N., B. Diettrich and M. Luckner (1990). Batch cultures of somatic embryos of *Digitalis lanata* in gaslight fermenters. Development and cardenolide accumulation. *Planta Med.*, **56**, 175–178.
Hacking, A. J. (1986). *Economic Aspects of Biotechnology*, pp. 306. Cambridge University Press, Cambridge.
Hamill, J. D., A. J. Parr, M. J. C. Rhodes, R. J. Robins and N. J. Walton (1987). New routes to plant secondary products. *Biotechnology*, **5**, 800–804.
Hashimoto, T. and S. Azechi (1988). Bioreactors for the large scale culture of plant cells. In *Biotechnology in Agriculture and Forestry*, ed. Y. P. S. Bajaj, vol. 4, pp. 104–122. Springer-Verlag, Berlin.
Hashimoto, T., J. Matsuda, S. Okabe, Y. Amano, D. J. Yun, A. Hayoshi and T. Yamada (1990). Molecular cloning, and tissue and cell specific expression of hyoscyamine 6$\beta$-hydroxylase. In *Progress in Plant Cellular and Molecular Biology*, ed. H. J. J. Nijkamp, L. H. W. Van der Plas and J. Van Aartrijk, pp. 775–782. Kluwer Academic, Dordrecht.
Hegarty, P. K., N. J. Smart. A. H. Scragg and M. W. Fowler (1986). The aeration of *Catharanthus roseus* L. G. Don suspension cultures in airlift bioreactors: the inhibition effect at high aeration rates on culture growth. *J. Exp. Bot.*, **37**, 1911–1920.
Herouart, D., R. S., Sangwan and B. S. Sangwan-Norreel (1991). Selection and characteristics of a lactose-adapted *Datura innoxia* cell line grown in suspension culture. *Plant Cell, Tissue Organ Cult.*, **24**, 97–104.
Hiatt, A., R. Cafferkey and K. Bowdish (1989). Production of antibodies in transgenic plants. *Nature* (*London*), **342**, 76–78.
Hilton, M. G. and M. J. C. Rhodes (1990). Growth and hyoscyamine production of 'hairy root' cultures of *Datura stramonium* in a modified stirred tank reactor. *Appl. Microbiol. Biotechnol.*, **33**, 132–138.
Inagaki, N., N. Nishimura, M. Okada and H. Mitsuhashi (1991). Verbascoside production in plant cell cultures. *Plant Cell Rep.*, **9**, 484–487.
Knobloch, K. H. and J. Berlin (1980). Influence of medium composition on the formation of secondary compounds in cell suspension cultures of *Catharanthus roseus* L. G. Don. *Z. Naturforsch.*, **35c**, 551–556.
Kreis, W. and E. Reinhard (1989). The production of secondary metabolites by plant cells cultivated in bioreactors. *Planta Med.*, **55**, 409–416.
Kreuzaler, F., H. Ragg, E. Fautz, D. N. Kuhn and K. Hahlbrock (1983). UV induction of chalcone synthase mRNA in cell suspension cultures of *Petroselinum hortense*. *Proc. Natl. Acad. Sci. USA*. **80**, 2591–2593.
Kutchan, T. M., N. Hampp, F. Lottspeich, K. Beyreuther and M. H. Zenk (1988). The cDNA clone for strictosidine synthase from *Rauwolfia serpentina*. DNA sequence determination and expression in *E. coli*. *FEBS Lett.*, **237**, 40–44.
Leckie, F., A. H. Scragg and K. C. Cliffe (1990). The effect of continuous high shear stress on plant cell suspension cultures. In *Progress in Plant Cellular and Molecular Biology*, ed. H. J. J. Nijkamp, L. H. W. Van der Plas and J. Van Aartrijk, pp. 689–693. Kluwer Academic, Dordrecht.
Mantell, S. H. and H. Smith (1983). Cultural factors that influence secondary metabolite accumulations in plant cell and tissue cultures. In *Plant Biotechnology*, ed. S. H. Mantell and H. Smith, pp. 75–108. Cambridge University Press, Cambridge.
Martin, S. M. (1980). Mass culture systems for plant cell suspension. In *Plant Tissue as a Source of Biochemicals*, ed. E. J. Staba, pp. 149–166. CRC Press, Boca Raton, Florida.
Meyer, P., I. Heidmann, G. Forkmann and H. Saedler (1987). A new petunia flower colour generated by transformation of a mutant with a maize gene. *Nature*, (*London*),. **330**, 677–678.
Misawa, M. (1985). Production of useful plant metabolites. In *Advances in Biochemical Engineering/Biotechnology*, ed. A. Fiechter. vol. 31, pp. 59–88. Springer-Verlag, Berlin.

Morris, P. (1986). Regulation of product synthesis in cell cultures of *Catharanthus roseus* II: comparison of production media. *Planta Med.*, **52**, 121–126.

Mulder-Krieger, Rh., R. Verpoorte, A. B. Svendsen and J. J. C. Scheffer (1988). Production of essential oils and flavours in plant cell and tissue cultures. A review. *Plant Cell, Tissue Organ Cult.*, **13**, 85–154.

Nef, C., B. Rio and H. Chrestin (1991). Induction of catharanthine synthesis and stimulation of major indole alkaloid production by *Catharanthus roseus* cells under nongrowth-altering treatment with *Pythium vexans* extracts. *Plant Cell Rep.*, **10**, 26–29.

Noguchi, M., T. Matsumoto, Y. Hirata, K. Yasamoto, A. Katsuyama, A. Kato, A. Azechi and K. Kato (1977). Improvement of growth rates of plant cell cultures. In *Plant Tissue Culture and its Biotechnological Application*, ed. W. Barz., E. Reinhard and M. H. Zenk, pp. 85–94. Springer-Verlag, Berlin.

Pollard, J. W. and J. M. Walker (1990). *Methods in Molecular Biology*, vol. 6, p. 597. Humana Press, Clifton, New Jersey.

Reinhard, E. (1974). Biotransformation by plant tissue cultures. In *Tissue Culture and Plant Science*, H. E. Street, pp. 433–459. Academic Press, London.

Reinhard, E. and W. Alfermann (1980). Biotransformation by plant cell cultures. *Adv. Biochem. Eng.*, **16**, 49–84.

Rudge, K. and P. Morris (1986). The effect of osmotic stress on growth and alkaloid accumulation in *Catharanthus roseus*. In *Secondary Metabolism in Plant Cell Cultures*, ed. P. Morris, A. H. Scragg, A. Stafford and M. W. Fowler, pp. 75–81. Cambridge University Press, Cambridge.

Ruyter, C. M. and J. Stockigt (1989). Novel natural products from plant cell and tissue culture — an update. *GIT Fachz. Lab.*, **4**, 283–293.

Sasse, F., K.-H. Knobloch and J. Berlin (1982). Induction of secondary metabolism in cell suspension cultures of *Catharanthus roseus, Nicotiana tabacum* and *Peganum harnala*. In *Tissue Cult. Plant Sci., Proc. Int. Congr. Plant Tissue Cell Cult., 5th, 1982*, ed. A. Fujiwara, pp. 343–344. Maruzen Co. Ltd., Tokyo.

Sasse, F., L. Witte and J. Berlin (1987). Biotransformation of tryptamine to serotonin by cell suspension cultures of *Peganum harnala*. Planta Med., **53**, 354–359.

Scragg, A. H. and M. W. Fowler (1985). The mass culture of plant cells. In *Cell Culture and Somatic Cell Genetics of Plants*, vol. 2, pp. 103–128. Academic Press, New York.

Sijmons, P. C., B. M. M. Bekker, T. C. V. Schrammeijer, Th. C. Verwoerd, P. J. M. Van Den Elzen and A. Hoekema (1990). Production of correctly processed human serum albumin in transgenic plants. *Biotechnology*, **8**, 217–221.

Smith, J. I., A. A. Quesnel, N. J. Smart, M. Misawa and W. G. W. Kurz (1987). The development of a single stage growth and indole alkaloid production medium for *Catharanthus roseus* L. G. Don suspension cultures. *Enzyme Microb. Technol.*, **9**, 466–469.

Spencer, A., J. D. Hamill, J. Reynolds and M. J. C. Rhodes (1990). Production of terpenes by transformed differentiated shoot cultures of *Mentha piperita citrata* and *M. piperita vulgaris*. In *Progress in Plant Cellular and Molecular Biology*, ed. H. J. J. Nijkamp, L. H. W. Van der Plas and J. Van Aartrijk, pp. 619–624. Kluwer Academic, Dordrecht.

Spier, R. E. and M. W. Fowler (1983). Animal and plant cell culture. In *Comprehensive Biotechnology*, ed. A. T. Bull and H. Dalton, vol. 1, pp. 301–330. Pergamon Press, Oxford.

Stafford, A. and L. Smith (1986). Effects of modification of the primary precursor level by selection and feeding on indole alkaloid accumulation in *Catharanthus roseus*. In *Secondary Metabolism in Plant Cell Cultures*, ed. P. Morris, A. H. Scragg, A. Stafford and M. W. Fowler, pp. 250–256. Cambridge University Press, Cambridge.

Stafford, A., P. Morris and M. W. Fowler (1986). Plant cell biotechnology; a perspective. *Enzyme Microb. Technol.*, **8**, 577–587.

Stafford, A. and M. W. Fowler (1991). Plant cell culture and product opportunities. *Agro-Industry Hi-Tech.*, **2**, 19–23.

Stepan-Sarkissian, G. (1990a). Biotransformation by plant cell cultures. In *Plant Cell and Tissue Culture*, ed. A. Stafford and G. Warren, pp. 163–204. Open University Press, Milton Keynes.

Stepan-Sarkissian, G. (1990b). Selection of media for tissue and cell culture. In *Methods in Molecular Biology*, vol. 6, pp. 1–12. Humana Press, Clifton, New Jersey.

Suga, T. and T. Hirata (1990). Biotransformation of exogenous substrates by plant cell cultures. *Phytochemistry*, **29**, 2393–2406.

Tanaka, H. (1983). Rotating drum fermenter for plant cell suspension cultures. *Biotechnol. Bioeng.*, **25**, 2359–2370.

Tyler, R. T., U. Eilert, C. O. Rijnders, A. A. Roewer and W. G. W. Kurz (1980). Semicontinuous production of sanguinarine and dihydrosanguinarine by *Papaver somniferum* L. cell suspension cultures treated with fungal homogenate. *Plant Cell Rep.*, **7**, 410–413.

Ulbrich, B., W. Wiesner and H. Arens (1984). Large scale production of rosmarinic acid from plant cell cultures of *Coleus blumei*. In *Primary and Secondary Metabolism of Plant Cell Cultures*, ed. K. H. Weumann, W. Barz and E. Reinhard, pp. 293–303. Springer-Verlag, Berlin.

Vandekerckhove, J., J. Van Damme, M. Van Lijsebettens, J. Botterman, M. De Block, M. Vandeviele, A. De Clercq, J. Leemans, M. Van Montagu and E. Krebbers (1989). Enkephalins produced in transgenic plants using modified 2S seed storage proteins. *Biotechnology*, **7**, 929–932.

Van der Krol, A. R., P. E. Lenting, J. Veenstra, I. M. Van der Meer, A. G. M. Gerats, J. N. M. Mol and A. R. Stuitje (1988). An antisense chalcone synthase gene in transgenic plants inhibits flower pigmentation. *Nature* (*London*), **333**, 866–869.

Van Montagu, M. (1990). 'New plants': future in agriculture. *Agro-Industry Hi-Tech.*, **1**, 8–15.

Van Uden, W., N. Pras, J. F. Visser and Th. M. Malingre (1989). Detection and identification of podophyllotoxin produced by cell cultures derived from *Podophyllum hexandrum* Royle. *Plant Cell Rep.*, **8**, 165–168.

Van Uden, W., N. Pras and Th. M. Malingre (1990). On the improvement of the podophyllotoxin production by phenylpropanoid precursor feeding to cell cultures of *Podophyllum hexaandrum* Royle. *Plant Cell, Tissue Organ Cult.*, **23**, 217–224.

Wagner, F. and H. Vogelmann (1977). Cultivation of plant tissue cultures in bioreactors and formation of secondary metabolites. In *Plant Tissue and its Biotechnological Application*, ed. W. Barz, E. Reinhard and M. H. Zenk, pp. 245–252. Springer-Verlag, Berlin.

Westphal, K. (1990). Large scale production of new biologically active compounds in plant cell cultures. In *Progress in Plant Cellular and Molecular Biology*, ed. H. J. J. Nijkamp, L. H. W. Van der Plas and J. Van Aartrijk, pp. 601–608. Kluwer Academic, Dordrecht.

Widholm, J. M. (1980). Selection of plant cell lines which accumulate certain compounds. In *Plant Tissue Culture as a Source of Biochemicals*, ed. J. Staba, pp. 99–113. CRC Press, Boca Raton, Florida.

Wijnsma, R., J. T. K. A. Go, I. N. van Weerden, P. A. A. Harkes, R. Verpoorte and A. Baerheim Svendsen (1985). Anthraquinones as phytoalexins in cell and tissue cultures of *Cinchona* spec. *Plant Cell Rep.*, **4**, 241–244.

Wilson, P. D. G., M. G. Hilton, P. T. H. Mechan, C. R. Waspe and M. J. C. Rhodes (1990). The cultivation of transformed roots from laboratory to pilot plant. In *Progress in Plant Cellular and Molecular Biology*, ed. H. J. J. Nijkamp, L. H. H. Van der Plas and J. Van Aartrijk, pp. 700–705. Kluwer Academic, Dordrecht.

Woerdenbag, H. G., H. J. Van Uden, H. W. Frijlink, C. F. Lerk, N. Pras and Th. M. Malingre (1990). Increased podophyllotoxin production in *Podophyllum hexandrum* cell suspension cultures after feeding coniferyl alcohol as a β-cyclodextrin complex. *Plant Cell Rep.*, **9**, 97–100.

Yonemitsu, H., K. Shimomura, M. Satake, S. Mochida, M. Tanaka, T. Endo and A. Kaji (1990). Lobeline production by hairy root culture of *Lobelia inflata* L. *Plant Cell Rep.*, **9**, 307–310.

Zenk, M. H., H. El-Shagi and B. Ulbrich (1977a). Production of rosmarinic acid by cell suspension cultures of *Coleus blumei*. *Naturwissenschaften*, **64**, 585–586.

Zenk, M. H., H. El-Shagi, H. Arens, J. Stockigt, E. W. Weiler and B. Deus (1977b). Formation of indole alkaloids serpentine and ajmalicine in cell suspension cultures of *Catharanthus roseus*. In *Plant Tissue Culture and its Biotechnological Application*, ed. W. Barz, E. Reinhard and M. H. Zenk, pp. 17–43. Springer-Verlag, Berlin.

# 7

# Plant Regeneration from Cultured Protoplasts of Higher Plants


SERGIO J. OCHATT

*INRA, Angers, France*

and

JOHN B. POWER

*University of Nottingham, UK*

## 7.1 TOTIPOTENCY, MORPHOGENESIS AND THE CENTRAL ROLE OF PLANT REGENERATION FROM TISSUE CULTURES

It has been known for 30 years now that whole plants can be recovered from cells cultured *in vitro* (Reinert, 1959; Steward *et al.*, 1958), and there has been a steady advancement in the theoretical and practical study of the regeneration process, particularly in the context of its exploitation for technological applications. Precisely because of this driving force, the developmental biology inherent to the morphogenic process *in vitro* has received only scant attention and therefore a significant proportion of the literature on plant cell culture concentrates on empirical approaches to the production of regenerated plants (Ranch and Pace, 1988).

Early this century, Haberlandt (1902) put forward the notion that individual nucleated plant cells have the genetic capacity to be converted to whole plants either directly or through an intervening callus stage. This phenomenon has been termed 'totipotence', whilst the term 'regeneration' has been widely used in the context of tissue culture to denote the recovery of a whole organism from cells, tissues, organs, meristems or zygotic embryos, cultivated *in vitro*.

The recovery of plants from *in vitro* cultures may be most readily achieved and characterized using tissues already at a level of organization and which retains developmental competence. Thus, explanted material is almost certainly predetermined towards plant organization. Totipotency is perhaps most appropriately demonstrated by the regeneration of complete plants from truly single cells: protoplasts.

In theory, therefore, all cells are totipotent (Steward *et al.*, 1958) but, in practice there is always going to be substantial variation in its expression. This expression of totipotency, in the differentiated state, is referred to as 'morphogenetic competence'. It is the inability in some species/genotypes to express totipotency (through organogenesis or somatic embryogenesis) which often confounds and limits the capacity for plant regeneration. Such variability in part derives from the physiological state of the target cells and is probably either due to a differential receptivity to the prevailing cultural conditions or by an inability on the part of the researcher to define precisely the appropriate requirements. With a progressive reduction in tissue organization, an alteration in the differentiation status of the cells may also be expected as cells, *in vitro*, are not exposed to the stabilizing and inductive influences from neighbouring cells/tissues of the *in vivo* environment. Consequently, the differentiated cell state which confers morphogenetic competence may not be easily retained/regained in a relatively unorganized state. In this respect, though, the theory of totipotence predicts that the competent state is never irreversibly lost, and also that the conditions leading to the induction and maintenance of such a differentiated state may be supplied artificially, presumably by complementing the current metabolic activitites of the cell (Ranch and Pace, 1988).

Against this background, the three main components with the largest influence on competence for plant regeneration, *in vitro*, are genotype (the genetic base), the ontogenetic state of the explant source and the cultural environment (medium composition and growth conditions). All three have typically been used to characterize and establish the baseline morphogenic responses for a given species. In this chapter these parameters will be discussed for tissue cultures where isolated protoplasts are the primary cell source for plant regeneration.

## 7.2 PROTOPLASTS AS A SOURCE FOR REGENERABLE TISSUE CULTURES

The term protoplasts refers to all components of a plant cell, excluding the cell wall, but they do not simply represent 'wall-less cells' as was originally envisaged. Even when protoplasts were first isolated mechanically, variable yields and a lack of reproducibility between successive protoplast isolations led to a rapid recognition that the use of cell wall degrading enzymes, was a preferable method for the release of large numbers of uniform protoplasts from most plant tissues.

Ironically, it is precisely such a treatment of cells with crude enzyme preparations that leads to a deleterious effect on protoplast viability, due mainly to the many undefined, contaminating compounds inevitably found in such preparations. In addition, the high osmotic pressure conditions required for early stage cultures and the accumulation of metabolic products during culture also adversely affect cell viability. Thus, although protoplasts can now be isolated from virtually any plant species and any type of tissue source, the ability to isolate protoplasts, capable of sustained mitotic division with subsequent callus proliferation and plant regeneration, is still limited to a relatively small (but nonetheless increasing) number of species.

The use of protoplasts has now become a baseline for the successful genetic improvement of many species (Pental and Cocking, 1985). Protoplasts also offer the possibility for the indirect transformation of plants (Gunn and Day, 1986), although this approach has been somewhat eclipsed since it was shown recently that transformed plants could also be obtained following coculturing of explants with disarmed vectors of agrobacteria (Horsch *et al.*, 1988). This has, in turn, opened up the possibility of transforming any plant species, capable of regeneration from complex explants. There is, however, one field where protoplast technology remains unparalelled by most other biotechnological approaches to plant breeding; protoplast fusion and somatic hybridization technologies provide the opportunity for the bypassing of reproductive isolation barriers and, therefore, facilitating gene-flow between species. To achieve such goals, the development of efficient protoplast to plant systems stands as a prerequisite.

### 7.2.1 Protoplast Isolation

Cocking (1960) was the first to use enzymatic methods for the release of plant protoplasts by applying an extract of hydrolytic enzymes to tomato root tips. Many enzyme combinations have since been used to isolate plant protoplasts, and most hydrolytic enzymes are now readily available commercially. Cellulases and/or hemicellulases are crucial for the release of protoplasts, in conjunction with pectinases (mixed procedure), the latter required primarily for cell separation in tissues in order that the cellulase can gain access and ensure efficient cell wall degradation (Table 1). In addition, former studies on the isolation of protoplasts were based on the use of a sequential rather than a mixed enzymatic treatment. For this sequential procedure, tissues are macerated with pectinases first, to release cells, and then the cells are incubated in cellulases to digest the cell wall and release the protoplasts (Fowke and Gamborg, 1980).

**Table 1** Manufacturers and Sources of Commercial Enzymes most Frequently used for Protoplast Isolation

| *Enzyme* | *Organism* | *Commercial source* |
|---|---|---|
| *Cellulases* | | |
| Driselase | Basidiomycetes | Kyowa Hakko Kogyo Co., Japan |
| Onozuka RS, Onozuka | *Trichoderma viride* | Yakult Honsha Co. R-10 Ltd., Japan |
| Cellulysin | *T. viride* | Calbiochem, USA |
| Cellulase | *Aspergillus niger* | Sigma Chemical Co. |
| Meicelase CESB | *T. viride* | Meiji Seika Kaisha Ltd., Japan |
| *Hemicellulases* | | |
| Rhozyme HP-150 | *Aspergillus niger* | Rohm and Hass Co., USA |
| Hemicellulase | *A. niger* | Sigma Chemical Co. |
| *Pectinases* | | |
| Macerozyme R-10 | *Rhizopus spp.* | Yakult Honsha Co. Ltd., Japan |
| Macerase | *Rhizopus spp.* | Calbiochem, USA |
| Pectolyase Y-23 | *Aspergillus japonicus* | Seishim Pharmaceutical Co. Ltd., Japan |
| Pectinol AC | *A. niger* | Corning Glass, USA |
| Pectinase | *A. niger* | Sigma Chemical Co. |
| PATE | Pectic acid-acetyl transferase | Hoescht, Germany |

### 7.2.2 The Importance of Source Tissues for Protoplast Isolation

While leaf tissues have been used most frequently for protoplast isolation, many other plant tissues have also been employed as protoplast sources, in experiments with dicotyledonous species. In this respect, growth conditions (*ex vitro versus in vitro*) for the donor tissue were shown to play a key role in the efficiency of isolation and subsequent cultural responses when these were taken from intact plants (Evans and Bravo, 1983; Gamborg *et al.*, 1981). It is not surprising, therefore, that an ever increasing number of workers have elected, particularly for recalcitrant genotypes, to use *in vitro* shoot cultures in order not only to generate a continuous supply of leaf material for large-scale protoplast isolations but also to provide uniform and repeatable growth conditions of the protoplast source tissues. It is interesting to note that, both for woody and herbaceous plant genotypes, young tissues have consistently proved to be the best sources of protoplasts. In this context, axenic shoot cultures are also valuable in providing juvenile phase tissues of woody species (McCown and Russell, 1987; Ochatt, 1990; Revilla *et al.*, 1987).

For protoplast isolation, from undifferentiated callus and cell suspension cultures, the physiological status of the donor material is also important, with optimum responses being gained when cells are converted to protoplasts during their exponential phase of growth (Ochatt *et al.*, 1989a; Power *et al.*, 1989).

On the other hand, for monocotyledonous species, sustained divisions have never been demonstrated in mesophyll protoplasts and, to date, the only regenerable protoplast systems reported amongst these species used embryogenic suspension cultures (most frequently initiated from seeds or seed parts) as the protoplast source (Vasil, 1987, 1988).

### 7.2.3 Treatments of Donor Tissues Prior to Enzymatic Digestion

Preincubation treatments of source tissues play an important role in protoplast isolation and culture, by enhancing viability and coupled, for recalcitrant species in particular, with an enhancement of cultural responses for such protoplasts. In this respect, a plasmolysis period, typically of 1 h, prior to incubation in enzyme, in a solution containing the same osmoticum (*e.g.* inorganic salts plus a sugar or a sugar alcohol) as to be used later, during enzymatic digestion, has improved the efficiency of protoplast isolation from leaves, callus and cell suspension cultures, and for a large number of species (Gamborg *et al.*, 1981; Ochatt *et al.*, 1989a).

In addition, and in order to improve the action of enzymes on tissues, large callus portions are chopped into small (125 $mm^3$) pieces or sliced into thin (1.0 mm wide) slices, prior to enzyme digestion. For leaf and petal tissues, they can be either digested intact, or they (including root tips) may be chopped into thin strips (usually 1–2 mm wide), or have their lower epidermis removed by peeling using fine forceps, prior to enzyme incubation. Alternatively leaves are bruised with a soft brush or with the cutting edge of a scalpel (Fowke and Constabel, 1985; Power *et al.*, 1989). Carborundum (an abrasive) and cutinases have also been used to remove the leaf epidermis (Power *et al.*, 1989). In addition, the lower epidermis can be easily removed if turgor is reduced in leaf cells and therefore many authors incubate donor plants in the dark, for 12–24 h, prior to protoplast isolation (Evans and Bravo, 1983).

Specific plant growth conditions were necessary for succesful plant regeneration from mesophyll protoplast cultures of potato, *Solanum tuberosum* (Shepard and Totten, 1977). Plants raised from potato tubers, were maintained in high light intensity (15 000 lux), with a 12 h light period, until 4–10 days prior to the isolation of protoplasts. Then the plants were transferred to a growth room with a 6 h light photoperiod of a lower light intensity (7000 lux). Throughout this period, the plants were maintained at 24 °C and 70–75% relative humidity, and they were fertilized daily with a dilute fertilizer solution (1.0 g $L^{-1}$ of an N:P:K, 20:20:20 fertilizer). According to the authors (Shepard and Totten, 1977), variation from this procedure would inevitably result in a reduced survival of the potato protoplasts.

Preincubation of tissues, *in vitro*, has been used in combination with growth of donor plants under optimum conditions to increase the release and viability of protoplasts of alfalfa (Kao and Michayluk, 1980). Protoplast yield was enhanced markedly if leaflets were taken and peeled, after the dark period, from young plants maintained at 21 °C, with 12 h illumination (12 000 lux) and fertilized weekly. Coupling this with dark preincubation of the peeled leaflets, for 36–48 h, in a modified cell culture medium enriched with sugars, antioxidants and growth regulators resulted in a significantly higher plating efficiency for the cultured alfalfa protoplasts (Kao and Michayluk, 1980).

In *Asparagus*, cells are released by a hand homogenizer (they are ground in 0.7 M mannitol) prior to treatment with enzymes (Bui-Dang-Ha and Mackenzie, 1973).

As for other protoplast sources, Duhoux (1980) isolated protoplasts from pollen grains of *Cupressus arizonica* by using high levels (to 12.5%) of wall-degrading enzymes for up to 20 h. Prior to enzyme treatment, pollen grains were hydrated in sterile tap water until the exine layer swelled and burst. Upon transfer of the pollen to the enzyme mixture, the cellulosic intine was digested but complete removal of the intine layer was difficult because of its resynthesis during the 20 h isolation period (Duhoux, 1980). Protoplasts were also isolated from pollen tetrads, and then used for experiments of gametosomatic hybridization by fusing them with diploid protoplasts of tobacco (Pirrie and Power, 1986) and *Petunia* (Lee and Power, 1988). For both these reports, pollen tetrads were incubated in a single-enzyme solution (2.0% Driselase in 9% w/v mannitol solution containing the CPW inorganic salts of Power *et al.*, 1989) and for a short time (20 min–1 h). Redenbaugh *et al.* (1980), conversely, reported that pollen tetrad protoplasts could be produced after 30 h digestion. Other tissues, such as aleurone and root tips also required much higher concentrations of enzymes and/or longer incubation times for the successful release of protoplasts (Evans and Bravo, 1983), whilst Cocking (1985) successfully isolated protoplasts from root hairs of numerous species after digestion periods that ranged between 3 and 20 min.

### 7.2.4 Conditions for Tissue Incubation and Protoplast Release

The precise methods used to ascertain optimum conditions for protoplast release are mostly empirical (Evans and Bravo, 1983; Gamborg *et al.*, 1981; Power *et al.*, 1989). In this context, and irrespective of source tissue, most isolation procedures are carried out at a temperature of about 25 °C, in the dark or under a constant, low light intensity illumination (100–500 lux) and for periods ranging between 4 and 20 h. Superimposed on this, a shaking of incubation mixtures (30–60 cycles $min^{-1}$) has also proven to be beneficial for the release of protoplasts, particularly from cell suspensions or finely chopped callus cultures (Ochatt, 1990; Ochatt *et al.*, 1989a).

Enzyme solutions are also frequently supplemented with buffers (to give pH values 5.5–6.0) such as MES (2-[*N*-morpholino]ethanesulfonic acid); since these enhance the viability of protoplasts by minimizing the shift to more acidic pH values in the enzyme solutions that may occur naturally during tissue digestion (Gamborg, 1976). In addition, those cells that are damaged or lysed during isolation will release hydrolytic enzymes capable of damaging the remaining cells/protoplasts. This problem can be countered by adding compounds, that negate the effects of contaminating proteins, present in the enzyme preparations and lysis products, such as potassium dextran sulfate (usually added at a concentration of 0.5%, w/v). The addition also, to the enzyme mixture, of antioxidants such as PVP (polyvinylpyrrolidone) has proved to have beneficial effects on the resulting viability of protoplasts particularly for recalcitrant genotypes. In this respect, and specifically for deciduous tree species, the addition of 1.0% (w/v) PVP-10 (average MW 10 000) appeared to be an essential requisite for the isolation of large numbers of viable protoplasts, especially those from leaf mesophyll tissues (Ochatt *et al.*, 1989a; Revilla *et al.*, 1987).

Protoplasts released directly into standard cell culture medium will burst. Hence the pressure that is mechanically supported by the plant cell wall must be replaced with an appropriate osmotic pressure. The osmotic pressure between cell interior and exterior must be balanced or transfer of cells to a plasmolyzing solution will induce stress on the plant cell. It has been observed that when osmotic pressure is too high, metabolism and growth are negatively affected. This can be monitored by a reduced uptake of amino acids across the plasma membrane (Ruesink, 1978) and a reduced cell wall regeneration (Pearce and Cocking, 1973). The synthesis of new cell wall material is influenced by both the type and concentration of osmoticum. Hence, protoplasts should be just slightly plasmolysed during isolation.

### 7.2.5 Protoplast Purification

Following enzyme incubation, a mixture of protoplasts, cells and cellular debris is usually obtained and needs to be purified in order to obtain debris-free protoplasts. For this purpose, a number of approaches have been developed and with varying degrees of success (Gamborg *et al.*, 1981; Power *et al.*, 1989).

The most frequently used purification technique consists of filtration and centrifugation. Protoplasts, in enzyme solution, are passed through a metal or nylon sieve (30–100 μm pore size depending on size of protoplasts), so as to retain undigested cells, cell clumps, and any vascular tissues. Thereafter, protoplasts and cell debris (vacuoles, plastids, storage crystals) are recovered from the filtrate and then centrifuged at a speed duration sufficient to sediment most protoplasts whilst leaving cell fragments in suspension. Centrifugation is typically carried out at 100 g (5–10 min) whereupon the supernatant is discarded and the protoplasts are subsequently resuspended in enzyme-free salts and osmoticum washing medium (Power *et al.*, 1989). The inclusion of salts in the washing solution serves to compensate for any ionic loss that might have occurred in the course of isolation (Cocking, 1972).

Centrifugation and washing may be repeated several times until a purified protoplast preparation is obtained. Protoplasts are finally suspended at a known plating density (see Section 7.3.2.2) and cultured in an appropriate medium. This filtration-centrifugation technique has the advantage, as compared to other purification techniques, of using the same osmotic-based solution throughout the entire isolation/purification procedure, thereby avoiding any possible, osmotic shock, and has been successfully used for the isolation of protoplasts from cell suspension and callus cultures of several species (Evans and Bravo, 1983).

On the other hand, for more fragile protoplast preparations (*e.g.* mesophyll protoplasts) this method results in excessive cell breakage during filtration and therefore should not be used. For such systems, floatation procedures have been developed to purify protoplasts (Ochatt *et al.*, 1989a; Revilla *et al.*, 1987).

As protoplasts have a relatively low density, when compared to organelles or cellular fragments, many types of density gradients or dense solutions (*e.g.* of sucrose) have been used to sediment debris whilst facilitating the recovery of protoplasts as discrete layers. A solution of a sugar (generally 20–23% sucrose) or sugar alcohol (most frequently sorbitol) may be combined with the enzyme protoplast mixture and, by using an appropriate, empirically determined, speed of centrifugation, protoplasts are recovered at the top of the centrifuge tube. Since floating protoplasts usually accumulate as a very thin layer, some workers have used Babcock bottles (Ochatt and Caso, 1986; Shepard and Totten, 1977). These have a long, thin neck, which ensures that a thicker more easily handled layer of purified protoplasts is formed (Gamborg *et al.*, 1981). The concentration of sucrose used varied between 0.3 and 0.6 M, and with centrifugation speeds of up to 350*g* (such as for protoplasts of many *Solanum* and *Nicotiana* species; Gamborg *et al.*, 1981; Power *et al.*, 1989), but reduced to 100*g* or less for fragile protoplasts such as those of tomato (Evans and Bravo, 1983), *Prunus* and *Pyrus* (Ochatt, 1990). Centrifugation for 3–10 min is normal. The sucrose floatation method has sometimes been combined with a filtration step prior to centrifugation and, although it results in less protoplast breakage than the filtration-centrifugation procedure, it may, in some cases, damage protoplasts due to osmotic shock changes. The most critical variables for protoplast purification, using this method, are sucrose concentration and centrifugation speed, as both can affect the stability and, by implication, also the viability of more fragile protoplasts.

Alternative gradients have been employed to aid in protoplast isolation and purification; two step gradients of polysucrose Ficoll (with 6% on top and 9% below), dissolved in MS medium with 7% sorbitol, have been used to purify carrot cell suspension protoplasts following centrifugation for 5 min, at 150*g* (Gosch *et al.*, 1975). Ficoll gradients have also been used for purification of tobacco and other protoplast systems (Larkin, 1976; Watts *et al.*, 1974).

In addition, discountinous gradients have also been used to purify protoplasts of several species, with intact protoplasts being recovered as a layer between two interphases (Ochatt *et al.*, 1989a; Power *et al.*, 1989; Revilla *et al.*, 1987). In general, these methods involve the layering of the protoplasts in enzyme-free isolation solution, on top of the same solution but supplemented with a high molecular weight subtance, such as Percoll or Ficoll. As these high molecular weight substances are osmotically inert, their use may be preferable to using sucrose for floatation, since the same osmotic strength is maintained throughout the isolation procedure thereby enhancing viability. A drawback to these approaches, for those species with delicate protoplasts, is that such substances must be washed away, by repeated resuspension and centrifugation, following purification. They are therefore, likely to be damaged during this process (Ochatt *et al.*, 1989a).

An alternative strategy, serial sieving of the protoplast-enzyme suspension through meshes of decreasing pore size (*e.g.* 100 μm followed by sieves of 80, 64, 53, 45, 30 and 20 μm pore size, depending on the average diameter of protoplasts in the preparation), can be used for systems with fragile protoplasts that do not float readily, such as those isolated from leaf mesophyll tissues of several deciduous woody species (Revilla *et al.*, 1987). This method, however, normally determines a large loss of isolated protoplasts during sieving and results in nonfully purified protoplast preparations.

### 7.2.6 Determination of Protoplast Purity, Viability, Yield and Size

Immediately after isolation and purification, protoplasts are resuspended either in enzyme-free isolation solution or in an appropriate culture medium and the presence/absence of cell walls, as well as protoplast size, yield and viability are determined.

#### *7.2.6.1 Assessment of the presence/absence of cell walls*

For the assessment of cell wall in protoplast preparations several strategies are followed. Sometimes, circumstancial evidence alone is used, such as truly spherical form of protoplasts and/or the absence of birefringence when they are observed under the microscope (Evans and Bravo, 1983; Fowke and Constabel, 1985). More accurate, however, is the use of fluorescing molecules. Calcoflour White at 0.1% (w/v) (Galbraith, 1981; Nagata and Takebe, 1970), in the protoplast suspension medium, has been used most frequently, coupled with examination under UV light (Power *et al.*, 1989). Also, the fluorescent brightening agent, Tinapol, was used for fluorescence assessments for the presence of cellulose (Cocking, 1985).

For cultural assessments, only protoplast preparations that are completely free of debris and undigested cells should be used, since low frequency regeneration of plants could be from cells or spheroplasts.

#### *7.2.6.2 Determination of protoplast viability*

In order as to determine viability, albeit inaccurately, an active cyclosis (Pelcher *et al.*, 1974), an even distribution of organelles (or, for leaf protoplasts, the red autofluorescence of chlorophyll, under UV light), the variation of size with osmotic changes and the photosynthetic activity (Kanai and Edwards, 1973) have all been employed. Staining provides more consistency in results and in this respect, phenosafranine (specific for dead protoplasts) (Evans and Bravo, 1983), the exclusion of Evans blue dye (Kanai and Edwards, 1973) and fluorescein diacetate (FDA) (Widholm, 1972) have been used to assess viability of freshly isolated protoplasts. FDA has been adopted by a majority of workers. Aliquots (usually 100 $\mu$L) of the protoplast suspension, in washing or culture medium, are mixed with an equal volume of appropriate culture medium with FDA (0.1 mL of a 5.0 mg mL$^{-1}$ stock in acetone per 10 mL of medium) and are observed microscopically under UV light. Molecules of FDA, a nonfluorescing compound, pass freely across the plasma membrane but are cleaved, by the action of esterases (Larkin, 1976; Widholm, 1972). As a result, the released fluorescein is retained only within a protoplast with an intact membrane so that metabolically active (*i.e.* viable) protoplasts fluoresce yellow-green (within 5 min) by excitement of the accumulated fluorescein with the UV light.

For all protoplast systems, at least 500 protoplasts should be counted for viabililty assessments. For this, different samples are normally observed and the results expressed as a percentage. For 'good' systems, viability of freshly isolated protoplasts should never be below 50–60%. In this context, several pre-incubation treatments of donor tissues have been shown to improve viability of freshly isolated protoplasts (see Section 7.2.3). The percentage viability normally drops during initial culture of protoplasts, markedly so for woody species protoplasts (Ochatt, 1990).

#### *7.2.6.3 Assessment of protoplast yield*

Using a haemocytometer, the number of protoplasts in an aliquot sample taken from a known volume of a suspension of purified protoplasts in culture medium are counted. For all subsequent culture and/or fusion manipulations the density of the protoplast population is expressed in terms of the number of protoplasts mL$^{-1}$ of suspension, and this is brought to the density required for culture/fusion by dilutions with appropriate culture medium. The yield, as a measurement of isolation efficiency, is expressed as the number of protoplasts produced per g fresh weight of source tissue digested (Fowke and Constabel, 1985; Gamborg *et al.*, 1981; Ochatt *et al.*, 1989a; Power *et al.*, 1989).

#### *7.2.6.4 Determination of protoplast size*

The characterization of protoplasts, in terms of size, is a prerequisite for their successful manipulation in the context of fusion and/or the subsequent fluorescence activated cell sorting of heterokaryons (Power *et al.*, 1989). There is also a clear relationship between protoplast size and their ability to withstand electric pulses in the context of electroporation either for foreign gene uptake (Shillito *et al.*, 1985) or for the enhancement of protoplast cultural responses (see Section 7.5). In addition, and specifically for protoplasts of rosaceous top-fruit trees species of the genera *Prunus*, *Pyrus* and *Malus*, a relationship was also found between protoplast diameter and the requirement for use of a semisolid or liquid medium for sustained growth during the initial culture stages. Protoplasts smaller than 30 $\mu$m in diameter, (such as those of *Prunus* in general) grew best in agarose-solidified medium whilst protoplasts beyond this size (*e.g. Pyrus, Malus*) exhibited sustained growth in liquid medium, provided dishes were constantly shaken (Ochatt, 1990).

Size is determined by measuring at least 200 protoplasts per sample, using a micrometric scale attached to the microscope eyepiece. For dicotyledonous species, leaf mesophyll protoplasts tend to be smaller in size than protoplasts isolated from callus or cell suspension cultures of the same species, with a typical size of 10–40 μm in diameter and more than 30 μm in diameter, respectively. For monocotyledonous species, protoplasts are normally smaller than 30 μm, irrespective of the protoplast source tissue.

## 7.3 PROTOPLAST CULTURE

### 7.3.1 Medium Composition

#### *7.3.1.1 Basal media used for protoplast culture*

Usually, the composition of those media used for protoplast culture is a modification of a routinely used standard cell culture medium. The nutritional requirements of cultured tissues and cells are very similar to those of the protoplast counterpart. Cell culture media, most commonly used as a basis for protoplast media, are MS medium (Murashige and Skoog, 1962), B5 medium (Gamborg *et al.*, 1968), KM medium (Kao and Michayluk, 1975), and many (minor but critical) modifications thereof.

In terms of the modification of individual media components, key alterations have been proposed in the concentration of ammonium, the addition of microelements and organic components of the medium in order to achieve successful cultural responses for some protoplast systems (Evans and Bravo, 1983; Gamborg *et al.*, 1981).

Ammonium ions, in particular, have been found to be detrimental to protoplast survival, and many media have been devised that have either a reduced ammonium concentration (*e.g.* for paper mulberry, Oka and Ohyama, 1985), or that lack ammonium (*e.g.* for some genotypes of potato, Upadhya, 1975; tomato, Zapata *et al.*, 1981; tobacco, Caboche, 1980; for all *Pyrus* genotypes thus far examined, Ochatt and Caso, 1986; Ochatt and Power, 1988a, 1988c; and for many other woody plant protoplast systems, (see Section 7.5.5). Conversely, the concentration of calcium ions, which may be important in terms of membrane stability, has sometimes been increased 2–4-fold over the concentration normally used for cell cultures (Eriksson, 1977).

#### *7.3.1.2 Carbon sources and choice of osmoticum*

Since protoplasts are nonphotosynthetic, a carbon source is needed for sustained growth (Evans and Bravo, 1983; Fowke and Constabel, 1985; Power *et al.*, 1989). Glucose may be the preferred carbon source for several protoplast systems (Gamborg, 1977), since this sugar is rapidly depleted in the culture medium by cultured protoplasts, a depletion which is directly proportional to the reduced requirements for a high osmotic solution when the cell wall is resynthesized. However, an alternative or supplementary carbon source, such as sucrose, may be preferred or even necessary for some species (Michayluk and Kao, 1975). Similarly, a nonmetabolizable osmoticum, such as mannitol or sorbitol, may be necessary in some cases (Von Arnold and Eriksson, 1977). Indeed, in recent years mannitol has been found to be the appropriate osmoticum for a large number of plant genotypes and for both herbaceous and woody plant species (Ochatt *et al.*, 1989a; Power *et al.*, 1989). There are, in addition, several species, including also woody and herbaceous individuals, where a mixture of several different carbon sources were found to provide optimum results (Gamborg *et al.*, 1981; Ochatt and Caso, 1986).

Osmotic pressure is manipulated by adding various sugars or sugar alcohols to the isolation (*cf.* Section 7.2.4) and culture solutions used for protoplasts. Mannitol, sorbitol, glucose and sucrose have all been frequently used. Mannitol and sorbitol, separately or in combination, have been used most often (Evans and Bravo, 1983), although mixtures including various osmotic agents simultaneously (mannitol, sorbitol, xylitol, inositol, sucrose) have been used for successful culture of leaf mesophyll protoplasts of both potato (Gamborg *et al.*, 1981) and wild pear (Ochatt and Caso, 1986). Glucose has been used successfully as an alternative to all of these for isolation of protoplasts from cultured tobacco cells (Evans and Bravo, 1983; Kao and Michayluk, 1974).

Solutions of mineral salts, such as KCl and $CaCl_2$ have also been used as osmotic agents (Horine and Ruesink, 1972) even though evidence that they are preferable to mannitol or sorbitol has not been given.

The effective osmotic concentration will depend on the endogenous cell osmotic pressures at the time of isolation and these are markedly influenced by environmental growth conditions (Shepard and Totten, 1977; Wallin *et al.*, 1977) and age (Kao and Michayluk, 1980) of donor tissues.

Concentrations of 0.23–0.90 M mannitol have been successfully used during culture of protoplasts of a large number of species (Evans and Bravo, 1983; Fowke and Constabel, 1985; Ochatt *et al.*, 1989a; Power *et al.*, 1989).

### *7.3.1.3 The role of organic compounds*

As far as the organic constituents of the medium are concerned, numerous organic components have been added to protoplast culture media. In most cases, and particularly for many herbaceous dicotyledonous species, vitamin (*i.e.* thiamine·HCl, pyridoxine·HCl and nicotinic acid) requirements seem to be the same for both cells and protoplasts (Evans and Bravo, 1983).

Against this background, Kao and Michayluk (1975) have suggested the addition of several vitamins, organic acids, sugars, sugar alcohols, and undefined nutrients such as casamino acids and coconut water for the culture of protoplasts at very low population densities. This medium, as reported by Kao and Michayluk (1975) and modifications thereof, is used successfully on a broad range of species, including cereals (Thompson *et al.*, 1986), legumes (Gilmour *et al.*, 1987), ornamentals (Power *et al.*, 1989) and fruit trees (Patat-Ochatt *et al.*, 1988). In many cases, though, many of these factors are not in fact essential for culture (division/plating efficiency) of protoplasts and, subsequently, many workers have eliminated some of these organics from their final culture medium, since no tangible benefits could be attributed to their use (*e.g.* Uchimiya and Murashige, 1976, for casein hydrolysate; also *cf.* Fowke and Constabel, 1985).

Differences in protoplast requirements for organic components can be dramatic, even for different varieties within a single species and irrespective of the protoplast source tissue (Ochatt, 1990, and references therein). For example rootstocks and scion genotypes of top-fruit trees of the genera *Malus* (Patat-Ochatt *et al.*, 1988) and *Pyrus* (Ochatt *et al.*, 1989a) differed in this respect, with rootstock protoplasts (apple and pear) requiring organically rich media as compared to the scion counterpart.

### *7.3.1.4 Growth regulators*

The form and concentration of growth regulators are the one media components that have been varied the most and are most clearly genotype/source specific (Evans and Bravo, 1983; Izhar and Power, 1977; Pental and Cocking, 1985). This is not surprising, since subtle changes in growth regulators and their application have been previously shown to have dramatic effects on the responses of cultured cells and/or tissues.

Both auxins and cytokinins are usually required for protoplast division. The auxin, 2,4-dichlorophenoxyacetic acid (2,4-D), is the growth regulator most commonly used in protoplast culture media; however, for some species, other auxins [such as 1-naphthaleneacetic acid (NAA) or 4-indol-3-ylbutyric acid (IBA)] are preferred. In addition to auxins, zeatin or 6-benzylaminopurine (BAP) are the two cytokinins most frequently reported to support division and sustained growth of cultured protoplasts and for numerous species. There are a few reports where only one type of growth regulator was required for cell division in regenerating protoplasts, *e.g.* only auxins were required by tobacco protoplasts (Uchimiya and Murashige, 1976), whilst mesophyll protoplasts, of some sour cherry (*Prunus cerasus* L.) genotypes, would enter division in a medium supplemented with zeatin alone (Ochatt and Power, 1988b). On the other hand, for most protoplast systems that require both auxin and cytokinin, an auxin/cytokinin ratio favouring the auxin is generally conducive to sustained division. Conversely, for some of the plant systems generally regarded as recalcitrant, in terms of protoplast culture and the recovery of plants from them, such as woody species in general, cytokinin-rich media were required to support the sustained growth of cultured protoplasts to the microcallus stage, *e.g.* as for several cherry (*Prunus* species) genotypes (Ochatt *et al.*, 1987; Ochatt and Power, 1988b; Ochatt *et al.*, 1989a).

## 7.3.2 Culture Conditions

### *7.3.2.1 Choice of culture conditions*

The composition of the culture medium and the optimum conditions for culture are obviously species- and sometimes variety/line-dependent (Izhar and Power, 1977; Pental and Cocking, 1985; Power *et al.*, 1989). The same is also true for light/dark conditions employed during culture. Most protoplast systems do not require light for the initiation of division and, some, are indeed light sensitive. The temperature range used normally is 25–30 °C, with the majority of species being kept at 25 °C (Evans and Bravo, 1983; Power *et al.*, 1989). However, with protoplasts of *Atropa belladonna* (Gosch *et al.*, 1975) and *Lycopersicon peruvianum* (Zapata *et al.*, 1977), temperature was critical for division (see Section 7.6.1). When light is required, dishes can be provided with a continuous fluorescent illumination of up to 1000 lux (Evans and Bravo, 1983; Ochatt *et al.*, 1989a; Power *et al.*, 1989). There are few examples where a photoperiodic light regime has proved necessary such as wild pear, *Pyrus communis* var. *pyraster* L. (Ochatt and Caso, 1986).

#### *7.3.2.2* ***Culture methods***

Protoplasts can be successfully cultured in a variety of ways which can depend on the nature of the experiments to be undertaken (Evans and Bravo, 1983; Gamborg *et al.*, 1981; Power *et al.*, 1989).

Of the various culture methods adopted the following are the most frequently used. (i) Plating in agar medium: a known volume of protoplasts, suspended at twice their required final culture density, is mixed with an equal volume of molten (40 °C), agar-solidified (1.2%, w/v) medium and, following setting of the agar, dishes are sealed (with Parafilm or Nescofilm) and maintained, inverted, in a culture room. (ii) Hanging/sitting drop culture: small droplets (20–40 $\mu$L) of protoplasts, in liquid medium, are dispensed with a pipette on the lid (or base) of a Petri dish, with a small volume of medium usually added (to counter dehydration) for hanging drop culture. Sealed dishes are maintained in a moist chamber (Potrykus *et al.*, 1976). (iii) Liquid on agar culture: a thin layer of agar (Difco-Bacto, pure) medium is dispensed and allowed to set, with an equal volume of protoplast suspension, at twice their optimum density, layered on the surface of the solidified medium. Overall protoplast plating density is thereby reduced by half. (iv) Agarose media culture: an autoclaved (4 × strength, low melting point) agarose suspension (2.4–4.0%, w/v) in distilled water is diluted (1:1) with the appropriate double-strength protoplast medium. This diluted agarose suspension is then mixed (1:1) with protoplasts in their single-strength medium but at twice their final plating density and then dispensed, as a thin layer, into Petri dishes (Shillito *et al.*, 1983). (v) Agarose bead culture: protoplasts, in molten (40 °C) agarose medium, are dispensed (4.0 mL) into small (3.5–5.0 cm diameter) Petri dishes and allowed to solidify. The agarose layer, thus formed, is then cut into 4 equal sized blocks and these are then transferred to larger dishes (9.0 cm diameter) containing liquid culture medium and of the same composition (Shillito *et al.*, 1983). (vi) Agarose drop culture: protoplasts, in agarose medium, are dispensed as droplets (50–150 $\mu$L each) onto the bottom of Petri dishes and, after solidification, are submerged in the same liquid medium. Alternatively, for protoplast systems with a high sensitivity to osmotic changes (*e.g.* as for *Prunus cerasus* L.; Ochatt and Power, 1988b), or when gradual reduction of the osmotic pressure of the medium is difficult to achieve (*cf.* Section 7.3.3.3) droplets can be directly submerged in osmoticum-free liquid medium from day zero of culture.

For agar culture, protoplasts are either allowed to regenerate cell walls in liquid culture before they are mixed with 0.6% agar (Coutts and Wood, 1977) or they are placed in agar medium from day zero (Cella and Galun, 1980). One method successfully applied to tobacco protoplasts is to mix protoplasts with an equal volume of medium prepared in agar maintained at 45 °C (Nagata and Takebe, 1971). Small aliquots of the protoplast-agar mixture are then poured into plates. Using this method, protoplasts remain in a fixed position, avoiding protoplast clumping and hence permitting the monitoring of separate clones.

The agar culture technique has been coupled with the concept of feeder protoplasts, in order to increase the plating efficiency of recalcitrant species. For this, a preparation of freshly isolated protoplasts is X-irradiated to block cell division, then mixed with a low density of normal, viable protoplasts, and the mixture of viable and inviable protoplasts is then poured with 0.6% agar in Petri dishes (Raveh *et al.*, 1973). Alternatively, a feeder layer can be used, whereby feeder cells, X-irradiated and hence incapable of division, are mixed with 0.6% agar and plated. Subsequently, viable protoplasts are mixed also with 0.6% agar and layered on top of the X-irradiated feeder cells (Cella and Galun, 1980).

Coculturing protoplasts of a more recalcitrant species with a reliable, fast-growing protoplast preparation is another technique that has been reported to improve the plating efficiency. The fast-growing protoplasts presumably provide the other species with growth factors and undefined diffusible chemicals which aid in regeneration of a cell wall and cell division (Evans, 1979). Similarly, albino cells (protoplasts) have been used as nurse cultures for single-transferred protoplasts, since colonies derived from the latter can be visually distinguished from albino colonies by their colour (Menczel *et al.*, 1978).

#### 7.3.2.3 *Initial plating density for protoplast culture*

In terms of the initial plating density requirement, protoplasts of most systems are cultured within the density range of $5 \times 10^2$ to $1 \times 10^6$ protoplasts $mL^{-1}$ medium (Evans and Bravo, 1983; Gamborg *et al.*, 1981). For the, so-called, recalcitrant species, where protoplast to protoplast relationships seem to play a key role in the cultural responses, a high initial plating density (typically $> 1 \times 10^5$ protoplasts $mL^{-1}$ medium) is required as for both cereal (Vasil, 1987) and woody plant species protoplasts (Ochatt, 1990).

Recently, interest has been generated in the culture of individual, isolated protoplasts, as this approach would be particularly useful in the context of genetic manipulation as based on cell to cell fusion or selection from single cell (protoplast) clones following mutagenesis. In this context, Kao and Michayluk (1975) undertook extensive assessments on the nutritional requirements of cultured cells and protoplasts and at nine different initial plating densities (*i.e.* between 2 and 5600 cells $mL^{-1}$), using *Vicia hajastana* cells. This technique was successfully extended for the culture of single tobacco protoplasts/heterokaryons following protoplast fusion (Gleba, 1978; Kao, 1977). Raveh and Galun (1975) used irradiated feeder cells as an aid to culture protoplasts of tobacco at a low density, and by using a feeder layer, they were able to successfully culture 5–50 protoplasts $mL^{-1}$ of medium. This method permits an evaluation of single, protoplast-derived clones. In addition, a microculture system was developed (Koop and Schweiger, 1985), that permitted the successful culture of individual protoplasts of *Nicotiana tabacum* cv Xanthi leading, ultimately, to the recovery of plants. This involved the selection of individual protoplasts from a suspension (in culture medium) under microscopical observation and their subsequent transfer into microdroplets was (20–80 nL) of fully-defined medium. Each microdroplet was contained within a separate drop (1 $\mu$L) of mineral oil, and 50 such microdroplets were placed on a coverslip. For culture, coverslips were kept in a moist chamber (to avoid evaporation and hence changes in the medium concentration) (Koop and Schweiger, 1985). More recently, this same strategy was successfully applied for plant regeneration from hypocotyl protoplasts of the more difficult to regenerate species, *Brassica napus* (Spangenberg *et al.*, 1986). The potential of this methodology for a broader range of species, in the context of culture of individual heterokaryons and recovery of somatic hybrid plants, plus a summarized description of the experimental setup and protocol to be followed has been reported recently (Schweiger *et al.*, 1987).

### 7.3.3 Initial Growth Responses of Cultured Protoplasts and the Importance of Genotype

#### 7.3.3.1 *Cell wall regeneration*

It is now accepted that wall regeneration in protoplasts is a prerequisite for cell division; but nuclear division, though, has been observed in the absence of cell wall regeneration for tobacco protoplasts (Meyer and Herth, 1978).

The process of wall regeneration has been followed, using electron microscopy, with cultured tobacco protoplasts (Fowke and Gamborg, 1980; Nagata and Yamaki, 1973). These studies suggest that the cell membranes, in recently isolated protoplasts, contain protruding microtubules around which the newly synthesized cellulose microfibrils are arranged. This random cellulose matrix eventually becomes thicker and the fibrils become oriented parallel to the plasmalemma, resulting in a continuous cell wall. The regularity of cell wall regeneration as well as the duration of the lag period prior to the onset of cellulose synthesis and wall formation depend partly on the plant species and also on the degree of differentiation of the donor cells used for protoplast isolation. In this respect, mesophyll, callus and cell suspension protoplasts of most Solanaceous genotypes (*e.g.* of the genera *Nicotiana*, *Petunia*, *Datura*, *Solanum*) and many *Brassica* all form cell walls very quickly (*e.g.* within 24–40 h) (Evans and Bravo, 1983; Gamborg *et al.*, 1981; Power *et al.*, 1989). On the other hand, cereal protoplasts (Vasil, 1987; 1988) and mesophyll protoplasts of legumes (Davey and Power, 1988) may require up to four days for cell wall regeneration, whilst an even longer time (typically seven or more days) is needed for most of the woody plant protoplast systems (McCown and Russell, 1987; Ochatt *et al.*, 1989a; Vardi and Galun, 1988).

Protoplast budding (incomplete cell wall resynthesis) has been observed frequently in protoplasts and for a number of species, and is the result of the action of osmotic pressure on the weakened areas in the newly synthesized cell wall (Fowke and Gamborg, 1980). Protoplast budding occurs when a suboptimal culture medium is used, thereby showing that composition of the medium is also important (Evans and Bravo, 1983). In this respect, high concentrations of sucrose and sorbitol inhibit cell wall regeneration for potato protoplasts (Shepard and Totten, 1977), whilst evidence of a specific requirement for growth regulators was provided, in both herbaceous (Izhar and Power, 1977; Uchimiya and Murashige, 1976) and woody species (Ochatt *et al.*, 1989a).

#### *7.3.3.2 Division of protoplast-derived cells*

Reproducible division has been observed in protoplasts of far fewer species than those exhibiting wall formation. Many factors are important in the initiation and maintenance of cell division in cultured protoplasts, including genotype, culture medium (particularly growth regulators), environmental culture conditions (light and temperature) and physiological status (age) and fertilizer/sanitary treatments of the donor tissue used for protoplast isolation (Evans and Bravo, 1983; Gamborg *et al.*, 1981; Power *et al.*, 1989).

Protoplasts isolated from cell cultures containing fast-growing cells (*i.e.* already in a dedifferentiated state at the time of isolation), often undergo first cell divisions sooner (sometimes at half the time) than leaf mesophyll protoplasts (Ochatt *et al.*, 1989a; Power *et al.*, 1989). Protoplast systems of woody plant genotypes, for example, typically undergo a lag phase that is rarely shorter than seven days prior to the onset of division (Ochatt, 1990), and as long as of 25 days, for cotton protoplasts (Bhojwani *et al.*, 1977).

Incomplete cytokinesis can occur during first division, resulting in spontaneous fusion and production of multinucleate protoplasts (Fowke and Constabel, 1985). This response, particulary for woody plant protoplast systems, is frequently associated with the addition, to the protoplast culture medium, of supraoptimal concentrations of growth regulators (especially cytokinins; Ochatt *et al.*, 1989a). In most cases, such multinucleate protoplasts thus produced will not undergo continued growth (Evans and Bravo, 1983).

First cell division usually occurs within 2–7 days of culture of healthy protoplasts, with subsequent division proceeding more rapidly and resulting in multicellular clumps (Gamborg *et al.*, 1981). However, for nearly all of the so-called recalcitrant protoplast systems (such as for monocotyledonous and woody species), a lag phase is observed prior to the onset of first division.

Particular relevance should also be attached to the general amenability to *in vitro* tissue culture techniques of the genotype under investigation, as a clear direct relationship exists between this and their likely responsiveness at the protoplast culture level (Evans and Bravo, 1983; McCown and Russell, 1987).

Higher plant protoplasts, particulary those of top-fruit trees and woody species, exhibit a marked decline in viability during culture usually, irrespective of the culture medium employed (Ochatt, 1990). However, this can be turned to advantage in determining the choice of the most suitable medium and conditions for culture. For this, a representative sample of protoplasts are stored at 4 °C in the dark, and are suspended in the osmoticum alone. The baseline decline in viability (assessed every 48 h) is then monitored. This, in turn, will provide a reference point for the progressive optimization of media components and culture conditions, whilst simultaneously placing all further viability determinations, during culture, on a comparative basis (Ochatt *et al.*, 1989a).

#### *7.3.3.3 Cell colony formation and reduction of osmotic pressure of the protoplast culture medium*

Fresh culture medium is added, typically every 1–2 weeks, when protoplasts are growing rapidly. Visible microcallus formation usually occurs in the initial culture medium within 2–4 weeks from the onset of first mitotic divisions. This first division is more or less synchronous.

When the regenerated cells grow rapidly, the osmotic pressure can be slowly reduced when fresh culture medium is added. The procedure for the reduction of the osmotic pressure of the initial culture medium will vary dependant on the culture method initially adopted. Thus, for protoplasts in semisolid layers/blocks these are cut into smaller blocks (*i.e.* into quarters or by halves, depending on initial size) and transferred to fresh medium of a reduced osmotic pressure (the volume of which should be just enough to cover the blocks), whilst for protoplasts in agarose droplets, the surrounding liquid medium covering the droplets, is removed and replaced by an appropriate volume of fresh liquid medium with a reduced concentration of osmoticum. For protoplasts, in droplets, that are submerged in osmoticum-free medium, from day zero, the osmotic pressure within the droplets is gradually reduced due to diffusion from the surrounding osmoticum-free medium into the droplets.

Protoplasts are generally left undisturbed throughout the culture period, until microcalli have been proliferated. For liquid cultures, a reduction of osmotic pressure is achieved, firstly, by the addition, to each dish, of the appropriate volume of the osmoticum-free medium counterpart, and for subsequent dilutions (when cell colonies are present) by the removal of all the original medium and its replacement with the osmoticum-free medium.

For most protoplast systems, the first reduction in osmotic pressure is usually initiated at or immediately after first division, and the procedure repeated when protoplast-derived colonies comprise of 10, 20, 30 and 50 cells. For systems where protoplasts prove more sensitive to osmotic shocks, such as for mesophyll protoplasts in general (Evans and Bravo, 1983; Power *et al.*, 1989) and particularly for those of rosaceous woody species; Ochatt *et al.*, 1989a) and ornamental *Compositae* (Al-Atabee and Power, 1987, 1989), the osmotic pressure of the culture medium is reduced more slowly and the number of successive steps increased accordingly.

Most reports indicate that the use of a dilution medium with an osmotic pressure approximately 70–75% of that of the previous step will usually support sustained growth (Gamborg *et al.*, 1981; Power *et al.*, 1989). However, the optimum time and rate for the successive dilutions of osmotic pressure is quite variable depending on the genotype and must therefore be determined experimentally. A more rapid reduction in osmotic pressure is indicated by browning and/or growth arrest of dividing cells. The optimum extent/rate of dilution can be decided upon by the duration of the lag phase before the reinitiation of sustained growth following dilution (Ochatt, 1990).

#### *7.3.3.4 Assessment of cultural responses: A redefinition of protoplast plating efficiency*

The responses of cultured protoplasts are assessed and normally expressed as a percentage plating efficiency (dividing protoplasts) based on the initial number of plated protoplasts (Evans and Bravo, 1983; Gamborg *et al.*, 1981; Power *et al.*, 1989), but some confusion has arisen as to the exact definition of the term 'plating efficiency'. In this context, some authors have applied this term as defined above (Evans and Bravo, 1983), whilst others used it to express the proportion (%) of initially plated protoplasts that proliferated to the microcallus stage (Ochatt *et al.*, 1989a). Others (*cf.* Fowke and Constabel, 1985, for a review) have linked the assessments to various (specified) timings but usually the developmental stage at which such a plating efficiency assessment was performed was never stated. A general approach in this respect is, therefore, clearly needed in order to compare protoplast systems. In this respect, perhaps the best way to sort out this inconsistency would be to define the percent plating efficiency, for all protoplast systems, at three main stages, based not on the timing (which will obviously depend on the genotype and source) but allied to a distinct developmental stage post first mitotic division: (i) initial plating efficiency (IPE) is defined as the percentage of the originally plated protoplasts that had regenerated a cell wall and undergone at least one mitotic division; (ii) intermediate plating efficiency (MPE) is defined as the percentage of the originally plated protoplasts that had divided to give colonies of 10 cells; and (iii) final plating efficiency (FPE) is defined as the percentage of the originally plated protoplasts that had proliferated to (visible) microcalli (1–2 mm diameter).

Attention has to be paid to the MPE stage for media comparisons and evaluations since, for many protoplast systems (particularly for woody species), the 10 cell colony stage is the developmental threshold beyond which growth can be arrested in many dividing colonies when grown under suboptimal cultural conditions (Ochatt *et al.*, 1989a).

## 7.4 PROLIFERATION OF PROTOPLAST-DERIVED MICROCALLI

Protoplasts-derived microcalli of 100 or more cells (*i.e.* 1–2 mm diameter) are transferred (individually or in groups of up to 20–25) to semisolid medium (10.0 mL) either for direct plant regeneration assessments (for most herbaceous species; see Evans and Bravo, 1983; Power *et al.*, 1989), or for further proliferation as callus (for most woody species; Ochatt, 1990). This distinction is necessary, since protoplast microcalli of the tree species are most frequently prone to phenolic browning and are unable to exhibit differentiation if transferred to plant regeneration media at such an early developmental stage (Ochatt *et al.*, 1989a). This need to generate a well-established callus mass (*ca.* 1.0 mL), prior to attempts at plant regeneration, could be linked to the fact that, for woody species, plant regeneration from true callus is in fact rare (Bajaj, 1989). Moreover any reports, specifically for fruit trees describing plant regeneration from callus, are all based on the use of calli where portions of the initial explant are still present (James, 1987). In the absence of an explant portion, which is clearly the case for protoplast-derived callus, a longer time in culture seems to be required before heterogeneity and polarity, a prerequisite for organogenesis, can be established within the tissues. However, there is always the risk that long-term culture might in fact reduce organogenic competence (Evans and Bravo, 1983; Ochatt, 1990), therefore this further proliferation period must, consequently, be restricted to a minimum.

Protoplast-derived calli of most genotypes/sources are capable of sustained proliferation in either a MS-based (including N6 (Chu *et al.*, 1975) and WPM (Lloyd and McCowm, 1981) media) or B5-based (including Kao and Michayluk (1975) K8 and Km8 media) culture medium. Similarly, growth regulators are added as used for the proliferation of explant-derived calli of the respective genotype/source and a similar pattern is true in terms of the light and temperature requirements for culture.

## 7.5 PROTOPLAST ELECTROPORATION AND THE ENHANCEMENT OF CULTURAL RESPONSES

For a majority of regenerating protoplast systems resolved to date, it is desirable to improve their plating efficiency and specifically their efficiency with respect to plant regeneration. For the recalcitrant genotypes (cereals, woody species, some ornamentals), it is desirable to reduce the duration of the inevitable lag period of protoplasts prior to the onset of mitotic division. Recent studies, devised to evaluate the effects of electricity on cultural responses, of protoplasts of various tissue and species sources have shown that this is, indeed, possible. Isolated protoplasts have been subjected to an electroporation treatment whereby a high voltage, short duration DC electric pulse (as used for pulse-mediated pore formation, for transformation studies with herbaceous species; Shillito *et al.*, 1985) has proved successful (Rech *et al.*, 1987).

Protoplasts, suspended, at four times their final density required for culture, and in a buffer solution, are electrotreated. Treatments normally consist of three successive exponential DC pulses (250 V to 2000 V), at 10 s intervals, and for pulse durations of 10 to 50 $\mu$s, prior to protoplast culture. There is a marked influence of the electrical parameters on protoplast viability, related specifically to protoplast size. Smaller protoplasts ($<20\ \mu$m diameter) are typically less sensitive to electric pulses than larger protoplasts ($>40\ \mu$m diameter), irrespective of species (Rech *et al.*, 1987).

Voltages of 250 V to 1000 V and with pulse durations of 20–40 $\mu$s trigger an earlier onset of protoplast division and with a significant increase of protoplast plating efficiency for both herbaceous (*i.e. Glycine canescens, Solanum viarum*) and woody (*i.e. Solanum dulcamara, Pyrus communis, Prunus avium* × *pseudocerasus*) dicotyledonous systems (Rech *et al.*, 1987). Such responses have subsequently been shown to occur for the monocot, *Pennisetum squamulatum* (Gupta *et al.*, 1988). In all systems assessed, the recovery of a significantly larger number of microcalli from electroporated protoplasts as compared to untreated control protoplasts takes place.

Assessments of the electrical effects performed during the later cultural stages, with *Prunus avium* × *pseudocerasus, Pyrus communis* and *Solanum dulcamara* protoplasts, showed a clear carry-over effect of the electrotreatment of original protoplasts. This was manifested with respect to an increased callus proliferation ability and is coupled with a higher frequency of shoot bud regeneration and rooting ability in regenerated shoots of electrotreated protoplast origin (Chand *et al.*, 1988; Ochatt, 1990; Ochatt *et al.*, 1988a).

A separate series of experiments, where cell suspension protoplasts of *Solanum dulcamara* and *Prunus avium* × *pseudocerasus* (Colt cherry) were labelled with [methly$^3$H]thymidine, showed that its incorporation into the acid precipitable material (indicative of DNA synthesis in cultured protoplasts) was significantly higher for electroporated protoplasts as compared to untreated controls. Moreover, the addition, to the protoplast culture medium, of the inhibitor of cell wall synthesis, 2,6-dichlorobenzonitrile, showed that, for electroporated protoplasts, such an enchancement in the rate of DNA synthesis was not necessarily linked with cell wall synthesis and, hence, with the eventual timing of the first mitotic division (Rech *et al.*, 1988).

An analysis of the results from all these studies, involving electromanipulation of protoplasts prior to culture, have indicated that the most likely explanation for the observed responses lies in an electroporation-mediated enchancement of DNA synthesis which could in turn entail an earlier expression of genes controlling the early stages of differentiation. This may be coupled with permanent membrane modifications leading to a sustained capacity for a larger/more efficient uptake of the requisite media components. Transmission of this capacity to the tissues derived from electrotreated protoplasts could explain the increased biomass generation which is clearly sustained throughout culture and leads, ultimately, to an earlier onset of organogenesis with increased shoot and root formation ability (Ochatt, 1990). Further credibility for this explanation of long-term effects of electroporation, is provided by results, for Colt cherry, which showed that such a promotory carry over effect on division and plant regeneration persists over a large number of subculture passages (Ochatt *et al.*, 1988b).

An additional, and highly significant spin-off of the electroporation of protoplasts, lies in the context of plant breeding. Following electroporation of protoplasts there is a marked increase in the throughput of heterokaryons and such enhanced cultural responses, used as a selection strategy, have led to the successful production of somatic hybrid fruit tree rootstocks of *Prunus avium* × *pseudocerasus ( + ) Pyrus communis* var *pyraster* L., where parental protoplasts were electroporated prior to their chemically induced fusion (Ochatt *et al.*, 1989b). Such a strategy could be used, generally, as a basis for the selection of heterokaryon-derived tissues.

The general applicability of these electrical treatments, in enhancing growth and regeneration responses in protoplasts has yet to be assessed on a broad range of species, but is clearly of importance in any protoplast system where a long lag phase prior to first division occurs or where poor plating efficiency and/or limited shoot bud regeneration potential exists. The rooting of regenerated shoots can also be promoted by electroporation of protoplasts prior to their culture (Chand *et al.*, 1988; Ochatt *et al.*, 1988a). This is particulary significant for difficult to root species (*e.g.* trees), and much more so for those medicinal species where the synthesis of pharmaceutically important compounds occurs mainly (or only) in root tissues.

## 7.6 PLANT REGENERATION FROM PROTOPLAST-DERIVED CALLUS

Until quite recently, plant regeneration from isolated protoplasts was often regarded to be a phenomenon restricted to the family *Solanaceae* and, consequently, many species in the genera *Nicotiana*, *Petunia* and *Solanum* have been and still are used as model systems in terms of the development of protoplast technology for both basic and applied purposes (Evans and Bravo, 1983; Griesbach, 1988). In recent times, however, protoplast technology has gained considerable momentum and a large number of species from different families including both mono- and di-cotyledonous genotypes are now amenable to plant regeneration from cultured protoplasts.

At present, protoplasts isolated from callus, cell suspensions, mesophyll tissues, shoot tips, flower petals and somatic embryos have all been regenerated to complete plants for members of the *Solanaceae* (Evans and Bravo, 1983; Griesbach, 1988; Yarrow and Barsby, 1988) and several other families (Power *et al.* 1989; Yarrow and Barsby, 1988). Despite these efforts, legumes (Hammatt *et al.*, 1987), cereals (Vasil, 1988) and woody species in general (McCown and Russell, 1987; Ochatt, 1990) are still difficult groups of plants to work with and, for these groups of plants, successful plant regeneration from protoplasts is still restricted to a limited number of genotypes and tissue sources.

For most reports of plant regeneration from protoplasts, details of experiments comparing media compositions are not normally given and it is, therefore, impossible to identify generalized procedures for plant regeneration. A large proportion of authors have developed their own culture medium after having very limited or no success with traditional media formulations. This clearly suggests that composition of culture media may be more critical for cultured protoplasts than for the cultured explant counterpart. Conversely, in most reports, room temperature (25 °C) has been used during culture and regeneration, suggesting that temperature is not critical. However, optimum light intensity varies dramatically from dark to high (3000 lux) light intensity between different plant species.

As for *ex vitro* transfer of the protoplast-derived regenerated plants, the requirements are typically similar to those for successful soil transfer of *in vitro* produced plants of other origins (*cf.* Chapter 8, this volume, by Rice). Rosaceous fruit trees appeared to be an exception in this respect, with protoplast-derived plants generally behaving as weaker and more enfeebled than their micropropagated or explant-derived counterpart (Ochatt, 1990).

Against this background, successful results in terms of plant regeneration from cultured protoplasts of several families or groups of species will be considered separately hereafter.

### 7.6.1 Cultivated Vegetables

Table 2 provides a list of the most economically important species in this group where plant regeneration from cultured protoplasts has been achieved.

Most of the methodologies used for protoplast regeneration were different between the species and also at the donor tissue level. The reader should refer to the cited publications for specific details.

**Table 2** Plant Regeneration from Protoplasts of Cultivated Vegetables

| *Family and species* | *Common name* | *Protoplast Source*[a] | *Ref.* |
|---|---|---|---|
| CUCURBITACEAE | | | |
| *Cucumis sativus* | Cucumber | CT | Colijn-Hooymans *et al.* (1988); Jia *et al.* (1986) |
| CRUCIFERAE (SYN. BRASSICACEAE) | | | |
| *Brassica alboglabra* | Chinese kale | S | Pua (1987) |
| *B. carinata* | | CT | Chuong *et al.* (1987) |
| *B. juncea* | | L | Eapen *et al.* (1989) |
| *B. napus* | Rape | L | Kartha *et al.* (1974) |
| | | H | Spangenberg *et al.* (1986) |
| *B. nigra* | Black mustard | CS | Klimaszewska and Keller (1986) |
| *B. oleracea* | Cabbage | L | Bidney *et al.* (1983); Fu *et al.* (1985); Lillo and Shahin (1986); Nishio *et al.* (1987) |
| | | R | Xu *et al.* (1982) |
| | Cauliflower | H | Glimelius (1984) |
| | | CT | Vatsya and Bhaskaran (1982) |
| | Broccoli | L | Robertson and Earle (1986); Robertson *et al.* (1988) Lillo and Olsen (1989) |
| LEGUMINOSAE | | | |
| *Vigna aconitifolia* | Moth bean | L | Shekhawat and Galston (1983) |
| *Psophocarpus tetragonolobus* | Winged bean | CS | Wilson *et al.* (1985) |
| *Glycine max* | Soybean | CS | Wei and Xu (1988) |
| UMBELIFERAE | | | |
| *Daucus carota* | Carrot | ECS | Dudits (1984) |
| *Foeniculum vulgare* | Fennel | ECS | Miura and Tabata (1986) |
| SOLANACEAE | | | |
| *Solanum tuberosum* | Potato | L | Binding *et al.* (1978); Shepard and Totten (1977); see also Fowke and Gamborg (1985) and refs. therein |
| *S. melongena* | Eggplant | L | Bhatt and Fassuliotis (1981); Guri and Izhar (1984); Saxena *et al.* (1981) |
| | | CS | Gleddie *et al.* (1986) |
| *Lycopersicon esculentum* | Tomato | L | Morgan and Cocking (1985); Niedz *et al.* (1985); Shahin (1985); Tan *et al.* (1987); Zapata *et al.* (1981) |
| *Capsicum annuum* | Pepper | L | Diaz *et al.* (1988); Saxena *et al.* (1981) |
| CONVOLVULACEAE | | | |
| *Ipomoea batatas* | Sweet potato | PT, S | Sihachakr and Ducreux (1987) |
| ASTERACEAE | | | |
| *Lactuca sativa* | Lettuce | L | Engler and Grogan (1983) |
| | | L, CT, R | Berry *et al.* (1982) |
| HELIANTHACEAE | | | |
| *Helianthus annuus* | Sunflower | ST, L | Binding *et al.* (1981); Dupuis and Chagvardieff (1988) |
| LILIACEAE | | | |
| *Asparagus officinalis* | Asparagus | CD | Bui-Dang-Ha and Mackenzie (1973) |
| EUPHORBIACEAE | | | |
| *Manihot esculenta* | Cassava | L | Shahin and Shepard (1980) |
| ROSACEAE | | | |
| *Fragaria × ananassa* | Strawberry | L | Nyman and Wallin (1988) |

[a]Protoplast Source: CD = cladodes; CS = cell suspensions; CT = cotyledon; ECS = embryogenic cell suspension; H = hypocotyl; L = leaf mesophyll; PT = petiole; R = roots; S = stem; ST = shoot tip

For many of the Solanaceous species, protoplasts were cultured initially in the dark prior to transfer to a continuous diffuse (*e.g.* 400–500 lux) light regime for proliferation. For these systems also, protoplasts were cultured at room temperature (*i.e.* 25 °C), but for cell suspension protoplasts of the medicinal/ornamental species *Atropa belladonna* (Gosch *et al.*, 1975) and for mesophyll protoplasts of the wild tomato species *Lycopersicon peruvianum* (Muhlbach, 1980; Zapata *et al.*, 1977), elevated temperatures (28–29 °C) were required for successful culture leading to plant regeneration.

Among the various non-Solanaceous species where plant regeneration has been accomplished from protoplasts, several economically important crops are included, most of which are members of the *Cruciferae* (*i.e.* cauliflower, rapeseed, cabbage), while, for other families, the examples of success are more limited. Noteworthy in this respect, is the fact that the strawberry (*Fragaria × ananassa*

*Rosaceae*), a species traditionally regarded as recalcitrant, has recently been successfully regenerated from protoplasts (Nyman and Wallin, 1988).

### 7.6.2 Grain and Forage Legumes

The legumes in general are normally regarded as recalcitrant systems in terms of plant regeneration from protoplasts. In this context, this field was recently reviewed by Hammatt *et al.* (1987) and examples of successful regeneration are indeed still rare.

Protoplast-derived plants have been obtained from mesophyll protoplasts of *Vigna aconitifolia* (Shekhawat and Galston, 1983), *Medicago sativa* (Johnson *et al.*, 1981; Kao and Michayluk, 1980), and the wild relatives of alfalfa, *M. coerulea* and *M. glutinosa* (Power *et al.*, 1989). Similarly, plants were recovered from both mesophyll and cell suspension protoplasts of *Trifolium repens* and *T. pratense* (Davey and Power, 1988), and also for *T. rubens* and *T. hybridum* (Webb, 1988).

Plant regeneration was also realized for cotyledon, root and hypocotyl protoplasts of birdsfoot trefoil, *Lotus corniculatus* (Power *et al.*, 1989) and from cell suspension protoplasts of *Psophocarpus tetragonolobus* (Wilson *et al.*, 1985).

Arguably, the legume of most economic importance worldwide is the soybean, *Glycine max*. Reproducible plant regeneration from protoplasts of soybean, however, has only very recently been reported (Wei and Xu, 1988). By contrast, high frequencies of plant regeneration (0.2–1.2% of protoplasts regenerating into plants) have been achieved using cotyledonary protoplasts for *G. canescens* and *G. clandenstina* (Hammatt *et al.*, 1987).

### 7.6.3 Cereal and Forage Grasses

Although leaf mesophyll protoplasts have been found to be an amenable source for plant regeneration in dicotyledonous species, sustained divisions have never been demonstrated for mesophyll protoplasts of monocotyledonous genotypes, including the most important family the *Gramineae*. This is probably attributable to the rapid differentiation and subsequent senescence of leaf cells (Vasil, 1987) in the intact plant. Thus far, only rapidly growing and predominantly embryogenic suspension cultures have provided totipotent protoplast systems (Table 3).

On the other hand, a closer examination of the published information on the use of protoplast technology, as applied to the grasses and cereals, shows that changes in the nutrient components of the culture media and/or in the culture conditions are not crucial.

A problem that permeates this area is the fact that there is always a conflict operating in terms of developing cultures with the right characteristics needed for the ease of isolation of protoplasts, *i.e.* the

**Table 3** Plant Regeneration from Cultured Protoplasts of Cereal and Grasses

| *Botanic name* | *Common name* | *Ref.* |
|---|---|---|
| *Dactylis glomerata* | Orchard grass | Horn *et al.* (1988) |
| *Festuca arundinacea* | Tall fescue | Dalton (1988) |
| *Lolium multiflorum* | Italian ryegrass | Jones and Dale (1982); Dalton (1988) |
| *L. perenne* | Perennial ryegrass | Dalton (1988) |
| *Panicum maximum* | Guinea grass | Lu *et al.* (1981) |
| *P. miliaceum* | Proso millet | Heyser (1984) |
| *Oryza sativa* | Rice | Coulibaly and Demarly (1986); Fujimura *et al.* (1985); Kyozuka *et al.* (1987); Thompson *et al.* (1986); Yamada *et al.* (1986) |
| *Pennisetum americanum* | Pearl millet | Vasil and Vasil (1980) |
| *P. purpureum* | Napier grass | Vasil *et al.* (1983) |
| *Poa pratensis* | Kentucky bluegrass | Van der Valk and Zaal (1988) |
| *Polypogon fugax* | | Chen and Xia (1987) |
| *Saccharum officinalis* | | Srinivasan and Vasil (1986) |
| *Zea mays* | Maize | Kamo *et al.* (1987); Prioli and Sondhal (1989); Rhodes *et al.* (1988); Shillito *et al.* (1989); Vasil and Vasil (1987) |

more fast-growing a cell line is, the more quickly such a line will lose its morphogenetic potential. Thus, cultures must be routinely reinitiated at intervals, or parental stocks be stored at low temperatures (Vasil, 1988).

Tall fescue and Italian ryegrass protoplasts reliably and routinely yield plants which can be successfully established in soil (Dalton, 1988), as do protoplasts isolated from cell suspension cultures of orchardgrass (Horn *et al.*, 1988), sugarcane (Srinivasan and Vasil, 1986) and rice (Thompson *et al.*, 1986; see also Table 3). In rice, plant regeneration is enhanced by embedding protoplasts in agarose (Thompson *et al.*, 1986) and subjecting them to a heat shock treatment prior to culture (Thompson *et al.*, 1987).

By contrast, cereals such as wheat, barley, maize and triticale remain difficult, although protoplasts isolated from these, the most difficult of species, can now be cultured and induced to form callus. Important in this respect is the fact that protoplast-derived colonies of maize (Kamo *et al.*, 1987; Vasil and Vasil, 1987) have showed limited morphogenesis and more recently efficient regeneration of mature plants (Prioli and Sondhal, 1989; Shillito *et al.*, 1989), whilst those from barley will regenerate into albino plants only (Luhrs *et al.*, 1988, cited in Webb, 1988). The induction also of albino plantlets from protoplasts has also been reported in perennial ryegrass (Dalton, 1988) and *Panicum miliaceum* (Heyser, 1984). In this respect, the recovery of only albino plantlets from the protoplasts isolated from embryogenic cell suspensions is still a recurring and puzzling problem with many cereal species (Vasil, 1987, 1988; Table 3). The same is true for the production of mature, protoplast-derived plants which, to date, is limited to sugarcane (Srinivasan and Vasil, 1986), *Polypogon fugax* (Chen and Xia, 1987), rice (Coulibaly and Demarly, 1986; Fujimura *et al.*, 1985; Thompson *et al.*, 1986; Yamada *et al.*, 1986), maize (Vasil and Vasil, 1987), tall fescue (Dalton, 1988), guinea grass (Lu *et al.*, 1981), pearl millet (Vasil and Vasil, 1980), napier grass (Vasil *et al.*, 1983), orchardgrass (Horn *et al.*, 1988) and Italian ryegrass (Dalton, 1988).

### 7.6.4 Ornamentals

Protoplasts from over 50 species of ornamental plants have been regenerated into whole plants, including genera in both the monocotyledoneae and the dicotyledoneae, with approximately 60% of regenerable species members of either the genera *Petunia* or *Nicotiana*, within the *Solanaceae* (Griesbach, 1988). Successful plant regeneration, for non-Solanaceous ornamentals, is more restricted.

In this respect, families which include numerous ornamental species, such as the *Compositae*, still remain largely unresponsive, with relatively few examples of plant regeneration from cultured protoplasts. Thus for *Chrysanthemum*, arguably the most important ornamental species in economic terms, regeneration has proved extremely difficult, as reported by Otsuka *et al.* (1985) and Khalid *et al.* (1988). The latter authors succeeding only in terms of plant regeneration after triggering totipotence through electrofusion with protoplasts of an actively dividing but unrelated system (*Salpiglossis sinuata, Solanaceae*). Amongst the *Compositae*, the most successful results are those for *Gaillardia grandiflora* (Binding *et al.*, 1981) and for species in the genera *Dimorphotheca* and *Rudbeckia*, where complete plants were successfully regenerated and acclimated in soil (Al-Atabee and Power, 1987, 1990).

Other examples of successful production of protoplast-derived plants in non-Solanaceous genotypes include species of prime ornamental value such as *Pelargonium aridum*, *P.* × *hortorum* and *P. peltatum* (Yarrow *et al.*, 1987).

As far as monocots are concerned, *Hemerocallis* cv Autumn Blaze, the daylily, remains the only example where plants could be recovered from protoplasts (Fitter and Krikorian, 1981). Interestingly, in this case, the use of a rapidly growing, morphogenetically competent cell suspension culture as the protoplast source was the key to success, in common with all cereals and grasses (Section 7.5.3).

### 7.6.5 Woody Species (Trees and Shrubs)

The usefulness of protoplasts in the context of genetic improvement of plant species has been reviewed by Pental and Cocking (1985) and highlighted earlier in this chapter. In this respect, there is a particularly compelling case to develop protoplast technology as applied to woody species since their long life cycles and existing hybridization barriers makes conventional breeding approaches lengthy and, in general, inappropriate.

Trees are almost always highly heterozygous, outbreeding individuals and are normally propagated either *via* seeds, from open pollinations, or by asexual ( cuttings and/or micropropagation) means. Consequently, great benefits are to be derived from rapid breeding methods with some of the most promising and emergent approaches based on biochemical and physiological selection procedures applied to genetically manipulated tissue cultures (protoplasts) and the production of novel genotypes through somatic cell hybridization.

The last few years have seen the biotechnology of woody angiosperms and gymnosperms come to fruition, since major advances in this hitherto recalcitrant group of species have taken place. For example, not only has the production of somatic hybrid trees been achieved (Ochatt, 1990; Ochatt *et al.*, 1989a; Vardi and Galun, 1988, and references therein), but a foreign gene (*aroA* gene, encoding for tolerance to the herbicide glyphosate) of potential economic importance has been inserted and expressed in *Populus* (Fillatti *et al.*, 1988). This is against the background of a general improvement in ones ability to manipulate protoplasts of trees.

Several species in the genus *Citrus*, *Rutaceae*, provided the first examples of efficient protoplast to tree systems (*cf.* Vardi and Galun, 1988, and Section 7.6.5.2). Regeneration from cultured protoplasts is though, still limited, thus, when this subject area was reviewed by Sommer and Wetztein (1984), plant regeneration from protoplasts was restricted to a single example outside the *Citrus* species, namely the sunnhemp, *Crotalaria juncea*, a member of the *Fabaceae* (Rao *et al.*, 1982). By the end of 1986 (McCown and Russell, 1987), only five new examples of protoplast-based plant regeneration for woody plants, other than *Citrus*, were cited. At present, the list of non-*Citrus* woody plant genotypes regenerable from protoplasts has lengthened considerably and includes now over 20 species examples within the angiosperms and gymnosperms. For convenience, woody species will be considered in three broad groups, the gymnosperms (conifers), the tropical and subtropical species (including *Citrus*), and temperate angiosperms. In addition, a distinction has to be made, between forest trees, fruit trees and medicinal woody species (which are usually shrubs rather than trees).

### *7.6:5.1 Gymnosperms (conifers)*

Conifers comprise the most economically important group of forest species but there are few reports on successful isolation and culture of conifer protoplasts (see recent review by Kirby , 1988). Most workers observe only initial division with the formation of small cell colonies, in nearly all the systems examined to date.

The most encouraging results were those reported for Douglas fir (*Pseudotsuga menziesii*), by Kirby (1980), who recovered callus tissues from cotyledon protoplasts, for loblolly pine (*Pinus taeda*), with the recovery of somatic embryos from protoplasts isolated from embryonal suspensor masses (Gupta and Durzan, 1987), and the regeneration of somatic embryos from protoplasts isolated from an embryogenic suspension culture of white spruce, *Picea glauca* (Attree *et al.*, 1987). Although the germination ability of the embryoids, as produced in the last two reports, was never described these results do represent, nonetheless, a major step forward towards the establishment of plant regeneration from conifer protoplasts.

For all the experiments with gymnosperms, seedling tissues (cotyledons or cotyledon-derived tissue cultures) were used as the source of protoplasts.

### *7.6.5.2 Tropical and subtropical woody angiosperms*

This is the group of woody species where plant regeneration from cultured protoplasts was first accomplished, specifically for fruit tree species of the genus *Citrus* (Vardi, 1977). Members of this genus represent a significant exception amongst woody plant genotypes, in terms of their amenability to protoplast technology (isolation, culture, fusion) and were, for several years, the only examples of protoplast to tree systems available.

In 1975, Vardi *et al.* (1975) described the isolation and culture of orange (*Citrus sinensis* Osbeck. cv Shamouti) protoplasts with accompanied regeneration of somatic embryos. This, in turn, led to the first complete protocol for plant regeneration from nucellus callus protoplasts of *Citrus* (Vardi, 1977). Several other *Citrus* genotypes and those of closely allied genera now exhibit plant regeneration from protoplasts (Table 4). Regeneration, in all cases occurs through somatic embryogenesis with all protoplasts being isolated from seedling sources. Furthermore, in all but one example (Hidaka and Kajiura, 1988) protoplasts were isolated from nucellus-derived tissues. Procedures exist though, for the isolation of protoplasts from leaf tissues (Grosser and Chandler, 1987) and juice vesicles of *Citrus* fruits (Echeverria, 1987).

**Table 4** Protoplast to Tree Systems of Tropical and Subtropical Woody Species (Trees and Shrubs)

| *Botanic name* | *Variety* | *Common name* | *Protoplast source* | *Ref.* |
|---|---|---|---|---|
| FRUIT TREE SPECIES | | | | |
| *Citrus sinensis* | Shamouti | Orange | Nucellar callus | Vardi *et al.* (1975, 1982) |
| | Trovita | Orange | Nucellar callus | Kobayashi *et al.* (1983, 1985) |
| *C. aurantium* | Wild type | Sour orange | Nucellar callus | Vardi *et al.* (1982) |
| *C. paradisi* | Duncan | Grapefruit | Nucellar callus | Vardi *et al.* (1982) |
| *C. reticulata* | Dancy, Murcott, Ponkan | Mandarin | Nucellar callus | Vardi *et al.* (1982) |
| *C. limon* | Villafranca | Lemon | Nucellar callus | Vardi *et al.* (1982) |
| *C. jambhiri* | Wild type | Rough lemon | Nucellar callus | Vardi (personal communication, 1989) |
| *C. miti* | Wild type | Lime | Nucellar callus | Sim *et al.* (1988) |
| *Microcitrus sp.* | Wild type | | Nucellar callus | Vardi *et al.* (1986) |
| *Coffea canephora* | Wild type | Coffee | Somatic embryos | Schopke *et al.* (1987) |
| FOREST TREE SPECIES | | | | |
| *Crotalaria juncea* | Wild type | Sunnhemp | Cotyledon | Rao *et al.* (1982) |
| *Santalum album* | Wild type | Sandalwood | Embryogenic cell suspension | Rao and Ozias-Akins (1985) |
| *Liriodendron tulipifera* | Wild type | Yellow poplar | Seed embryo | Merkle and Sommer (1987) |
| *Pithecellobium dulce* | Wild type | | Seedling leaves | Saxena and Gill (1987) |
| MEDICINAL SPECIES | | | | |
| *Solanum dulcamara* | Wild type | Woody nightshade | Cell suspension from seedling leaves | Chand *et al.* (1988) |
| *Lycium barbarum* | Wild type | | Seedling leaves | Ratushnyak *et al.* (1989) |
| *Oxalis glaucifolia* | Wild type | | Callus from internode of mature plants | Ochatt *et al.* (1989c) |

The genetic uniformity of young protoplast-derived orange trees (*Citrus sinensis* Osbeck.) has been confirmed (Kobayashi, 1987), and, also, the somatic hybridization of two woody species reported, usually with protoplasts from embryogenic, nucellus-derived tissues of orange as one of the fusion partners (see review by Vardi and Galun, 1988, and references therein).

Protoplast research involving tropical fruit trees, other than *Citrus*, is very limited, with the successful production of protoplast-derived plants restricted to coffee, *Coffea canephora*, using protoplasts isolated from somatic embryos, these, in turn formed on callus derived from a nonseedling leaf explant. Protoplasts regenerated *via* somatic embryogenesis when cultured in a medium with a reduced ammonium level. Such embryoids were, in turn, germinated (albeit at a reduced rate) to whole plants (Schopke *et al.*, 1987).

Amongst nonfruit tropical tree species, reports on the successful establishment of protoplast regeneration are restricted to only four examples.

Rao *et al.* (1982) isolated protoplasts from cotyledons of *Crotalaria juncea* (sunnhemp), and these proliferated to callus when cultured in a low-ammonium medium, and in the light. Such calli gave somatic embryos in an auxin-free medium with plants recovered thereafter.

Another tree legume, *Pithecellobium dulce*, was regenerated, *via* organogenesis, from leaf protoplasts of *in vitro* germinated seeds which were cultured, again, in a low-ammonium ions medium (Saxena and Gill, 1987). Using seed embryos as a protoplast source with culture on a medium with reduced levels of ammonium, for initial protoplast culture, Merkle and Sommer (1987) succeeded in recovering whole plants, *via* embryogenesis, from the yellow poplar *Liriodendron tulipifera* (*Magnoliaceae*).

Rao and Ozias-Akins (1985) have regenerated sandalwood plants, *via* embryogenesis, from protoplasts isolated from embryogenic cell suspension cultures of nonseedling origin (*i.e.* initiated from callus induced on stem segments of a 20 year old tree). Culture of protoplasts of *Santalum album* in a medium with reduced ammonium levels was also a prerequisite for success in this system.

For the subtropical woody medicinal species (shrubs) there are only three successful examples of plant regeneration from protoplasts; the woody nightshade, *Solanum dulcamara* (Chand *et al.*, 1988), *Lycium barbarum* L., *Solanaceae* (Ratushnyak *et al.*, 1989) and *Oxalis glaucifolia* R. Knuth., *Oxalidaceae* (Ochatt *et al.*, 1989c), with seedling (for the *Solanaceae*) and nonseedling tissue (for *Oxalis*), being used as source. It is noteworthy that, although nonmesophyll protoplasts were employed, for both *S. dulcamara* and *O. glaucifolia*, regeneration of plants was *via* organogenesis (*cf.* Section 7.6.5.3.1).

### 7.6.5.3 Temperate woody angiosperms

It is for this group where the most advances had been made in the last four years in terms of developing and exploiting protoplast technology as a tool for breeding (Table 5). With one exception, the paper mulberry (Oka and Ohyama, 1985), nonseedling tissues have been used as the source for protoplasts in all of the systems within this group, as will be discussed below.

#### *(i) Fruit trees*

The paper mulberry, *Broussonetia kazinoki* (Moraceae), provided the first example of the successful recovery of protoplast-derived plants from mesophyll protoplasts of a woody species, with organogenesis as the pathway to plants (Oka and Ohyama, 1985). Protoplasts of this system were produced from *in vitro* germinated seedlings, and the use of a medium again with a reduced ammonium level was the key to success.

Ochatt and Caso (1986) achieved, for the first time, plant regeneration from protoplasts of a top-fruit tree, the wild pear (*Pyrus communis* var *pyraster* L.), using protoplasts of a nonseedling tissue. These authors compared leaves of field-grown plants with those of axenic shoot cultures (initiated from the same adult, 15 years old, trees) as protoplast sources. They found that leaves of *in vitro* rooted shoots released twice as many viable protoplasts as compared to field-grown sources. The use of an ammonium-free medium, coupled with the maintenance of protoplasts under a high light intensity (1.8 W/m$^2$) photoperiodic regime were essential. Following shoot regeneration, *via* organogenesis, rooting of these shoots was limited and inconsistent. Later experiments showed that protoclonal variation existed amongst the protoplast-derived wild pear shoots (Ochatt, 1987) and that an abnormal phenotype (with respect to callus proliferation capacity and leaf phenotype of regenerated shoots) was sometimes associated with changes in ploidy level. These changes were correlated with a lack of rooting ability in abnormal shoots.

Ochatt *et al.* (1987) reported plant regeneration from protoplasts of the commercially important temperate fruit tree genotype, the cherry rootstock Colt (*Prunus avium* × *pseudocerasus*). Shoots were produced, *via* organogenesis, from mesophyll protoplasts, but only roots (rhizogenesis) occurred for cell suspension protoplast-derived calli. Later on, shoots were though successfully produced from cell suspension protoplast-derived tissues (Ochatt *et al.*, 1988a). This was the first example of plant regeneration, *via* organogenesis, from nonmesophyll protoplasts for a woody species.

During 1988, several reports described on the induction of organogenesis from protoplasts of temperate fruit tree genotypes (Table 5). Whole plants were recovered from mesophyll protoplasts of several sour cherry clones, *Prunus cerasus* L. (Ochatt and Power, 1988b), Williams' Bon Chretien pear, *Pyrus communis* L. (Ochatt and Power, 1988c), Spartan and M.9 apple, *Malus* × *domestica* Borkh. (Patat-Ochatt *et al.*, 1988), and from leaf callus protoplasts of kiwifruit, *Actinidia chinensis* var

**Table 5** Protoplast to Tree Systems in Temperate Woody Species

| *Botanic name* | *Variety* | *Common name* | *Protoplast source* | *Ref.* |
|---|---|---|---|---|
| TEMPERATE FRUIT TREES | | | | |
| Seedling source | | | | |
| *Broussonetia kazinoki* | | Paper mulberry | Leaf | Oka and Ohyama (1985) |
| Nonseedling source | | | | |
| *Pyrus communis var pyraster L.* | Wild type | Wild pear | Leaf | Ochatt and Caso (1986); Ochatt (1987) |
| *P. communis* | Williams' | Common pear | Leaf | Ochatt and Power (1988c) |
| *Prunus avium* × *pseudocerasus* | Colt | Colt cherry | Leaf | Ochatt *et al.* (1987) |
| | | | Cell suspension | Ochatt *et al.* (1988a) |
| *P. cerasus* | CAB 4D, CAB 5H | Sour cherry | Leaf | Ochatt and Power (1988b) |
| *Malus* × *domestica* | Spartan, M.9 | Apple | Leaf | Patat-Ochatt *et al.* (1988) |
| | MM.106 | Apple | Leaf | Patat-Ochatt and Power (1990) |
| *Actinidia chinensis* | | Kiwifruit | Leaf | Tsai (1988) |
| TEMPERATE FOREST TREES | | | | |
| *Ulmus* × *Pioneer* | Wild type | Elm | Leaf callus | Sticklen *et al.* (1986) |
| *Populus alba* × *grandindentata* | Wild type Crandon | White aspen poplar | Leaf | Russell and McCown (1986, 1988) |
| *P. nigra* × *trichocarpa* | Wild type | Black poplar | Leaf | Russell and McCown (1988) |
| *P. tremula* | Erecta | European upright aspen | Leaf | Russell and McCown (1988) |

*chinensis* Planch. (Tsai, 1988). In addition, rhizogenesis was induced in tissues derived from embryo-callus protoplasts of Conference pear (Ochatt and Power, 1988a) which was subsequently extended to the limited regeneration of shoot buds on these protoplast-derived roots (Ochatt, 1990). For the apple rootstock MM.106, Patat-Ochatt *et al.* (1988) have regenerated nonrootable shoots first and, more recently, whole protoplast-derived trees (Patat-Ochatt and Power, 1990). For all the *Pyrus* protoplast systems assessed to date, the use of an ammonium-free medium was a key to success.

It is interesting to note that for sour cherry (Figure 1) and by implication other species, a novel approach to plant regeneration is *via* rhizogenesis as an intermediate step. For sour cherry it was found that protoplast-derived cultures of clones CAB 4D and CAB 5H, shoot buds were regenerated in decapitated protoplast-derived root segments (Ochatt and Power, 1988b). This same strategy was later extended, for the successful plant regeneration from protoplasts of other nominally recalcitrant genotypes, including ornamental species of the *Compositae* (Al-Atabee and Power, 1990).

Protoclonal variation in regenerants of Colt cherry has led to the recovery, for the first time, of salt/drought tolerant protoplast-derived plants (Ochatt and Power, 1989a, 1989b). This was done by

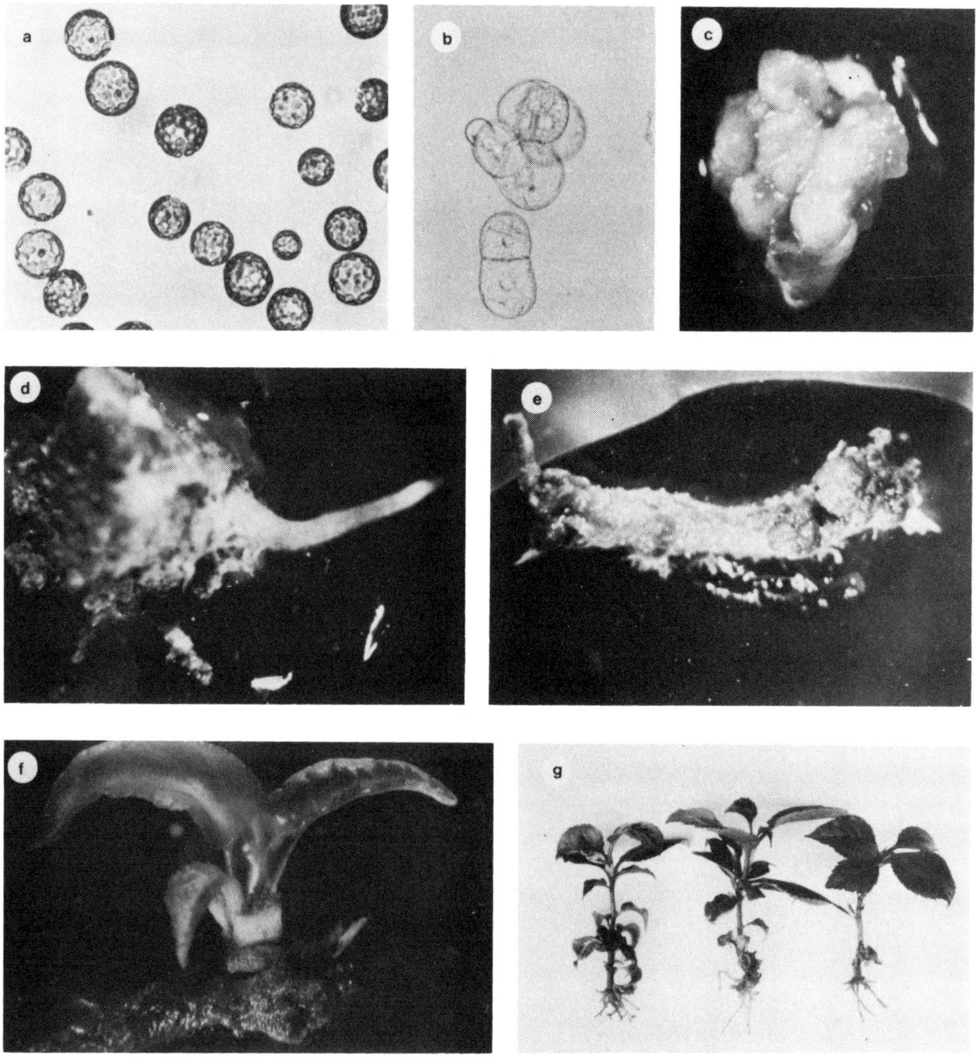


**Figure 1** Plant regeneration from mesophyll protoplasts of sour cherry clones (*Prunus cerasus* L.): (a) recently isolated protoplasts of clones CAB 4D; (b) first division and a 6-cell colony derived from protoplasts of clone CAB 5H (day 20 of culture); (c) a differentiated protoplast-derived callus of clone CAB 11E; (d) rhizogenesis in a protoplast-derived callus of clone CAB 4D (150 days of culture); (e) shoot bud induction in a protoplast-derived root of clone CAB 5H (165 days of culture); (f) shoot bud regeneration from a protoplast-derived root of clone CAB 4D (day 180 of culture); (g) protoplast-derived rooted plants of sour cherry (from left to right), clone CAB 4D, clone CAB 5H (regenerated *via* rhizogenesis), clone CAB 5H (regenerated *via* direct caulogenesis from callus)

using a direct recurrent selection strategy, whereby well-established cell lines, of protoplast origin, were subjected to recurrent cycles each involving a sequence of alternate, two successive three week subculture passages on a medium with stress-inducing agents (in this case NaCl, KCl, $Na_2SO_4$ or mannitol, to give a medium osmolality of up to 682 mOsm $kg^{-1}$), followed by two successive subculture passages on stress-free medium. At least three such cycles of recurrent selection were applied prior to any attempt at plant regeneration (from the putatively tolerant calli that survived this strategy) in a medium that also contained stress-including agents. Stability of the acquired stress tolerance was confirmed by an ability of protoplasts and explant-callus, taken from the first generation of regenerated tolerant trees, to proliferate under conditions of medium-induced stress (Ochatt and Power, 1989a). Protoplasts of the tolerant regenerant Colt cherry plants were subsequently used to show the existence of an inverse relationship between cell wall regeneration and tolerance to salinity, with protoplasts behaving as more tolerant than their respective cell counterparts (Ochatt and Power, 1989b). These results stressed the importance of regenerable protoplast systems for direct plant improvement, whilst they also represent a novel and useful tool to confirm the stability of new traits selected *in vitro*, particularly for species with lengthy life cycles (*i.e.* trees) or where only asexual means of propagation are available.

In addition, the temperate fruit tree species have provided the only example to date for woody species, excepting *Citrus*, of the production of somatic hybrid tree rootstocks (*Prunus avium* × *pseudocerasus* ( + ) *Pyrus communis* var *pyraster* L.) (Ochatt *et al.*, 1989b).

(*ii*) *Temperate forest trees*

In this group of woody species, a number of successful reports have emerged since 1986. Most of these have involved the culture of protoplasts, from nonseedling leaf tissues, in media with reduced ammonium ion concentrations.

In this context, the only report using nonmesophyll protoplasts is that of Sticklen *et al.* (1986), for *Ulmus* × Pioneer, *Ulmaceae*, who regenerated elm trees from leaf callus protoplasts (Table 5). All other examples are for poplar genotypes in the family *Salicaceae* (Table 5), with leaf mesophyll tissues as the protoplast source. Trees of *Populus alba* × *gradindentata* Crandon, and the white aspen/poplar, have been recovered (Russell and McCown, 1986), and later extended to black poplar, *P. nigra* × *trichocarpa*, and the European upright aspen, *P. tremula* Erecta (Russell and McCown, 1988).

For all *Populus* genotypes assessed, release of leaf mesophyll protoplasts was markedly enhanced by a rapid grinding of tissues, prior to enzyme incubation, followed by forceful washing of leaf debris after digestion (Russell and McCown, 1986, 1988).

In all these cases, shoot bud regeneration was *via* organogenesis. Protocols that were developed are highly genotype specific (McCown and Russell, 1987). There is, however, one feature in common for nearly all the forest and fruit temperate woody protoplast systems, namely the use, as protoplast source, of tissues that were either juvenile or reacted as juvenile-like plants. Indeed, most research involving mesophyll protoplasts of temperate woody species was based on the use of leaves from axenic shoot cultures as the protoplast source. In this respect, repeated subculture of axenic shoots, *in vitro*, has been shown to enhance the ratio of shoot multiplication and also to improve the rooting ability of the microcuttings thus produced, both features being associated with a juvenile physiological status (McCown and Russell, 1987; Ochatt *et al.*, 1989a).

A significant breakthrough in the protoplast culture conditions, which were successfully employed for culture of regenerable mesophyll protoplasts of several *Populus* hybrid genotypes (poplar and aspen), was the development of a floating disc culture technique (Russell and McCown, 1986, 1988). For this, protoplasts are grown in a monolayer together with a polyester screen disc at the surface of a liquid medium. Such a strategy was reported to aid the observation of developing protoplasts, the collection of data, the change of growth medium, and the handling of the protoplast-derived calli in general (Russell and McCown, 1986, 1988). As already mentioned, woody plant protoplasts are quite prone to phenolic oxidation and, dependant on size, to aggregation during the initial culture stages (Ochatt, 1990). In this context, and although it still has to be tested in a broad range of species, this novel cultural technique might help to negate both of those problems.

## 7.7 CONCLUDING REMARKS

After nearly three decades of basic protoplast research it is now clear that the potential for economic gain through protoplast manipulation will finally be fulfilled and the technology progressively accepted by plant breeders and seed companies.

Progress in the application of protoplast technology to the improvement of cultivated crops is still somewhat restricted by a lack of protoplast regeneration in some species despite notable advances and breakthroughs with respect to the cereals, tree fruit species and the woody ornamentals. Interestingly though, the small fruit crops (notably the grapevine; Krul, 1988), many vegetable crops and species of floral value and the monocotyledonous species in general are still recalcitrant.

Concomitant with the continuing breakthroughs in devising regeneration protocols is of course the development of new approaches for successful protoplast culture. The advent of, for example, agarose layers for increasing plating efficiency (Shillito *et al.*, 1983), the development of microchambers for culture of individual protoplasts/heterokaryons (Schweiger *et al.*, 1987), and the finding that electrical charges can dramatically stimulate protoplast division (Rech *et al.*, 1987) and plant regeneration from protoplasts (Ochatt *et al.*, 1989a), may finally resolve the remaining cultural problems of the 'recalcitrant species'. What is clear though is that there is still a paucity of knowledge as to the role that complex media play in triggering sustained division in protoplast-derived cells, and also of the relationships between factors and the genetic basis to plant regeneration.

The establishment of reproducible protoplast to plant systems is still largely achieved by empirical means and until this situation changes regeneration of plants from protoplasts will always be somewhat problematical.

## 7.8 REFERENCES


Al-Atabee, J. and J. B. Power (1987). Plant regeneration from protoplasts of *Dimorphotheca* and *Rudbeckia*. *Plant Cell Rep.*, **6**, 414–415.

Al-Atabee, J. and J. B. Power (1989). Protoplast isolation and plant regeneration in ornamental *Compositae*. *Acta Hortic*, **280**, 255–258.

Attree, S. M., F. Bekkaoui, D. I. Dustan and L. C. Fowke (1987). Regeneration of somatic embryos from protoplasts isolated from an embryogenic suspension culture of white spruce (*Picea glauca*). *Plant Cell Rep.*, **6**, 480–483.

Bajaj, Y. P. S. (ed.) (1986). *Biotechnology in Agriculture and Forestry 1, Trees I*. Springer-Verlag, Berlin.

Berry. S. F., D. Y. Lu, D. Pental and E. C. Cocking (1982). Regeneration of plants from protoplasts of *Lactuca sativa* L. *Z. Pflanzenphysiol.*, **108**, 31–38.

Bhatt, D. P. and G. Fassuliotis (1981). Plant regeneration from mesophyll protoplasts of eggplant. *Z. Pflanzenphysiol.*, **104**, 81–89.

Bhojwani, S. S., P. K. Evans and E. C. Cocking (1977). Protoplast technology in relation to crop plants: progress and problems. *Euphytica*, **26**, 343–360.

Bidney, D. L., J. F. Shepard and E. Kaleikau (1983). Regeneration of plants from mesophyll protoplasts of *Brassica oleracea*. *Protoplasma*, **117**, 89–92.

Binding, H., R. Nehls, O. Schieder, S. K. Sopory and G. Wenzel (1978). Regeneration of mesophyll protoplasts isolated from dihaploid clones of *Solanum tuberosum*. *Physiol. Plant.*, **43**, 52–54.

Binding, H., R. Nehls, R. Kock, J. Finger and G. Mordhorst (1981). Comparative studies on protoplast regeneration in herbaceous species of the dicotyledonaceae class. *Z. Pflanzenphysiol.*, **101**, 119–130.

Bui-Dang-Ha, K. and I. A. Mackenzie (1973). The division of protoplasts from *Asparagus officinalis* L. and their growth and differentiation. *Protoplasma*, **78**, 215–221.

Caboche, M. (1980). Nutritional requirements of protoplast-derived haploid tobacco cells grown at low densities in liquid medium. *Planta*, **149**, 7–18.

Cella, R. and E. Galun (1980). Utilization of irradiated carrot cell suspensions as feeder layer for cultured *Nicotiana* cells and protoplasts. *Plant Sci. Lett.*, **19**, 243–252.

Chand, P. K., S. J. Ochatt, E. L. Rech, J. B. Power and M. R. Davey (1988). Electroporation stimulates plant regeneration from protoplasts of the woody medicinal species *Solanum dulcamara* L. *J. Exp. Bot.*, **39**, 1267–1274.

Chen, D. and Z. Xia (1987). Mature plant regeneration from cultured protoplasts of *Polypogon fugax* Nees. ex Steud. *Sci. Sin.*, **30B**, 698–703.

Chu, C. C., C. C. Wang, C. S. Sun, C. Hsu, K. G. Yin, C. Y. Chu and F. Y. Bi (1975). Establishment of an efficient medium for another culture of rice through comparative experiments on the nitrogen sources. *Sci. Sin.*, **18**, 659–668.

Chuong, P. V., K. P. Pauls and W. V. Beversdorf (1987). Protoplast culture and plant regeneration from *Brassica carinata* Braun. *Plant Cell Rep.*, **6**, 67–69.

Cocking, E. C. (1960). A method for the isolation of plant protoplasts and vacuoles. *Nature*, **187**, 962–963.

Cocking, E. C. (1972). Plant cell protoplasts—isolation and development. *Annu. Rev. Plant Physiol.*, **23**, 29–50.

Cocking, E. C. (1985). Protoplasts from root hairs of crop plants. *Biotechnology*, **3**, 1104–1106.

Colijn-Hooymans, C. M., R. Bouwer and J. J. M. Dons (1988). Plant regeneration from cucumber protoplasts. In *Progress in Plant Protoplast Research*, ed. K. J. Puite, J. J. M. Dons, H. J. Huizing, A. J. Kool, M. Koornneef and F. A. Krens, pp. 91–94. Kluwer, Dordrecht.

Coulibaly, M. Y. and Y. Demarly (1986). Regeneration of plantlets from protoplasts of rice, *Oryza sativa* L. *Z. Pflanzenzuecht.*, **96**, 79–81.

Coutts, R. H. A. and K. R. Wood (1977). Improved isolation and culture methods for cucumber mesophyll protoplasts. *Plant Sci. Lett.*, **9**, 45–51.

Dalton, S. J. (1988). Plant regeneration from cell suspension protoplasts of *Festuca arundiancea* Schreb., *Lolium perenne* L. and *L. multiflorum* Lam. In *Progress in Plant Protoplast Research*, ed. K. J. Puite, J. J. M. Dons, H. J. Huizing, A. J. Kool, M. Koornneef and F. A. Krens, pp. 49–52. Kluwer, Dordrecht.

Davey, M. R. and J. B. Power (1988). Aspects of protoplasts culture and plant regeneration. In *Progress in Plant Protoplast Research*, ed. K. J. Puite, J. J. M. Dons, H. J. Huizing, A. J. Kool, M. Kornneef and F. A. Krens, pp. 15–26. Kluwer, Dordrecht.

Diaz, I., J. B. Power and J. R. Diaz-Ruiz (1988). Isolation, culture and regeneration of leaf protoplasts in the genus *Capsicum*. In *Progress in Plant Protoplast Research*, ed. K. J. Puite, J. J. M. Dons, H. J. Huizing, A. J. Kool, M. Kornneef and F. A. Krens, pp. 115–116. Kluwer, Dordrecht.

Dudits, D. (1984). Isolation and culture of protoplasts from carrot cell suspension cultures. In *Cell Culture and Somatic Cell Genetics*, ed. I. K. Vasil, vol. 1, pp. 391–397. Academic Press, New York.

Duhoux, E. (1980). Protoplast isolation of gymnosperm pollen. *Z. Pflanzenphysiol.*, **99**, 207–241.

Dupuis, J. M. and P. Chagvardieff (1988). Factors affecting protoplast plating efficiency and direct somatic embryogenesis in sunflower (*Helianthus annuus* L.). In *Progress in Plant Protoplast Research*, ed. K. J. Puite, J. J. M. Dons, H. J. Huizing, A. J. Kool, M. Kornneef and F. A. Krens, pp. 73–74. Kluwer, Dordrecht.

Eapen, S., V. Abraham, M. Gerdemann and O. Schieder (1989). Direct somatic embryogenesis, plant regeneration and evaluation of plants obtained from mesophyll protoplasts of *Brassica juncea*. *Ann. Bot.* (*London*), **63**, 369–372.

Echeverria, E. (1987). Preparation and characterization of protoplasts from *Citrus* juice vesicles. *J. Am. Soc. Hortic. Sci..*, **112**, 393–396.

Engler, D. E. and R. G. Grogan (1983). Variation in lettuce plants regenerated from protoplasts. *J. Hered.*, **75**, 426–430.

Eriksson, T. (1977), Technical advances in protoplast isolation and cultivation. In *Plant Tissue Culture and its Biotechnological Application*, ed. W. Barz, E. Reinhard and M. H. Zenk, pp. 313–322. Springer-Verlag, Berlin.

Evans, D. A. (1979). Chromosome stability in plants regenerated from mesophyll protoplasts of *Nicotiana* species. *Z. Pflanzenphysiol.*, **95**, 459–463.

Evans, D. A. and J. E. Bravo (1983). Protoplast isolation and culture. In *Handbook of Plant Cell Culture*, ed. D. A. Evans, W. R. Sharp, P. V. Ammirato and Y. Yamada, vol. 1, pp. 124–177. MacMillan, New York.

Fillatti, J. J., B. Haissig, B. McCown, L. Comai and D. Riemenschneider (1988). Development of glyphostate-tolerant *Populus* plants through expression of a mutant *aroA* gene from *Salmonella typhimurium*. In *Genetic Manipulation of Woody Plants, Basic Life Sciences*, ed. J. W. Hanover and D. E. Keathley, vol. 44, pp. 243–249. Plenum Press, New York.

Fitter, M. S. and A. D. Krikorian (1981). Recovery of totipotent cells and plantlet production from daylily protoplasts. *Ann. Bot. (London)*, **48**, 591–597.

Fowke, L. C. and O. L. Gamborg (1980). Applications of protoplasts to the study of plant cells. *Int. Rev. Cytol.*, **68**, 9–51.

Fowke, L. C. and F. Constabel (eds.) (1985). Plant Protoplasts. CRC Press, Boca Raton, Florida.

Fu, Y. Y., S. R. Jia and Y. Lin (1985). Plant regeneration from mesophyll protoplast culture of cabbage (*Brassica oleracea* var *capitata*). *Theor. Appl. Genet.*, **71**, 495–499.

Fujimura, T., M. Sakurai, H. Akagi, T. Negishi and A. Hirose (1985). Regeneration of rice plants from protoplasts. *Plant Tissue Cult. Lett.*, **2**, 74–75.

Galbraith, D. W. (1981). Mircrofluorimetric quantitation of cellulose biosynthesis by plant protoplasts using Calcofluor White. *Physiol. Plant.*, **53**, 111–116.

Gamborg, O. L. (1976). Plant protoplast isolation, culture and fusion. In *Cell Genetics in Higher Plants*, ed. D. Dudits, G. L. Farkas and P. Maliga, pp. 107–128. Akad. Kiado, Budapest.

Gamborg, O. L. (1977). Culture media for plant protoplasts. In *Handbook of Nutrition and Food*, ed. M. Rechcigl, pp. 415–422. CRC Press, Cleveland.

Gamborg. O. L., R. A. Miller and K. Ojima (1968). Nutrient requirements of suspension cultures of soybean root cells. *Exp. Cell Res.*, **50**, 151–158.

Gamborg, O. L., J. P. Shyluk and E. A. Shahin (1981). Isolation, fusion and culture of plant protoplasts. In *Plant Tissue Culture: Methods and Applications in Agriculture*, ed. T. A. Thorpe, pp. 115–153. Academic Press, New York.

Gilmour, D. M., M. R. Davey, E. C. Cocking and D. Pental (1987). Culture of low numbers of forage legume protoplasts in membrane chambers. *J. Plant. Physiol.*, **126**, 457–465.

Gleba, Yu. Yu. (1978). Microdroplet culture. *Naturwissenschaften*, **65**, 158–159.

Gleddie, S., W. A. Keller and G. Setterfield (1986). Somatic embryogenesis and plant regeneration from cell suspension derived protoplasts of *Solanum melogena* (eggplant). *Can. J. Bot.*, **64**, 355–361.

Glimelius, K. (1984). High growth and regeneration capacity of hypocotyl protoplasts in some *Brassicaceae*. *Physiol. Plant.*, **61**, 38–44.

Gosch, G., Y. P. S. Bajaj and J. Reinert (1975). Isolation, culture and induction of embryogenesis in protoplasts from cell suspensions of *Atropa belladonna*. *Protoplasma*, **86**, 405–410.

Griesbach, R. J. (1988). Recent advances in the protoplast biology of flower crops. *Sci. Hortic.* (*Amsterdam*), **37**, 247–256.

Grosser, J. M. and J. L. Chandler (1987). Aseptic isolation of leaf protoplasts from *Citrus, Poncirus, Citrus* × *Poncirus* hybrids and *Severinia* for use in somatic hybridization experiments. *Sci. Hortic.* (*Amsterdam*), **31**, 253–257.

Gunn, R. E. and P. R. Day (1986). *In vitro* culture in plant breeding. In *Tssue Culture and its Agricultural Applications*, ed. L. A. Withers and P. G. Alderson, pp. 331–336. Butterworths, London.

Gupta, P. K. and D. J. Durzan (1987). Somatic embryos from protoplasts of loblolly pine proembryonal cells. *Biotechnology*, **5**, 710–712.

Gupta, H. S., E. L. Rech, E. C. Cocking and M. R. Davey (1988). Electroporation and heat shock stimulate division protoplasts of *Pennisetum squamulatum*. *J. Plant Physiol.*, **133**, 457–459.

Guri, A. and S. Izhar (1984). Improved efficiency of plant regeneration from protoplasts of eggplant (*Solanum melongena* L.). *Plant Cell Rep.*, **3**, 247–249.

Haberlandt, G. (1902). Kulturversuche mit isolierten Pflanzenzellen. *S. B. Weisen Wien Mathnaturw.*, **111**, 69–92.

Hammatt, N., T. K. Ghose and M. R. Davey (1987). Regeneration in legumes. In *Cell Culture and Somatic Cell Genetics of Plants*, ed. I. K. Vasil, pp. 67–95. Academic Press, London.

Heyser, J. W. (1984). Callus and shoot regeneration from protoplasts of Proso millet (*Panicum miliaceum* L.). *Z. Pflanzenphysiol.*, **113**, 293–299.

Hidaka, T. and I. Kajiura (1988). Plantlet differentiation from callus protoplasts induced from *Citrus* embryo. *Sci. Hortic.* (*Amsterdam*), **34**, 85–92.

Horine, R. K. and A. W. Ruesink (1972). Cell wall regeneration around protoplasts isolated from *Convolvulus* tissue culture. *Plant Physiol.*, **50**, 438–445.

Horn, M. E., B. V. Conger and C. T. Harms (1988). Plant regeneration from protoplasts of embryogenic cell suspension cultures of orchard grass (*Dactylis glomerata* L.). *Plant Cell Rep.*, **7**, 371–374.

Horsch, R. B., R. T. Fraley, S. G. Rogers, H. J. Klee, J. E. Fry, M. A. W. Hinchee and D. S. Shah (1988). *Agrobacterium*-mediated gene transfer to plants: engineering tolerance to glyphosate. *Iowa State J. Res.*, **62**, 487–502.

Izhar, S. and J. B. Power (1977). Genetical studies with *Petunia* leaf protoplasts. 1. Genetic variation to specific growth hormones and possible genetic control on stages of protoplast development in culture. *Plant Sci. Lett.*, **8**, 375–383.

James, D. J. (1987). Cell and tissue culture technology for the genetic manipulation of temperate fruit trees. In *Biotechnology and Genetic Engineering Reviews*, vol. 5, pp. 33–79. Intercept, Dorset.

Jia, S. R., Y. Y. Fu and Y. Lin (1986). Embryogenesis and plant regeneration from cotyledon protoplast culture of cucumber *Cucumis sativus*. *J. Plant Physiol.*, **124**, 393–398.

Johnson, L. B., D. L. Stuteville, R. K. Higgins and D. Z. Skinner (1981). Regeneration of alfalfa plants from protoplasts of selected regenerated S clones. *Plant Sci. Lett.*, **20**, 297–304.

Jones, M. K. G. and P. J. Dale (1982). Reproducible regeneration of callus from suspension culture protoplasts of the grass *Lolium multiflorum*. *Z. Pflanzenphysiol.*, **105**, 267–274.

Kamo, K. K., K. L. Chang, M. E. Lynn and T. K. Hodges (1987). Embryogenic callus formation from maize protoplasts. *Planta*, **172**, 245–251.

Kanai, R. and G. E. Edwards (1973). Separation of mesophyll protoplasts and bundle sheath cells from maize leaves for photosynthetic studies. *Plant Physiol.*, **51**, 1133–1137.

Kao, K. N. (1977). Chromosomal behaviour in somatic hybrids of soybean — *Nicotiana glauca*. *Mol. Gen. Genet.*, **150**, 225–230.

Kao, K. N. and M. R. Michayluk (1974). A method for high-frequency intergeneric fusion of plant protoplasts. *Planta*, **115**, 355–367.

Kao, K. N. and M. Michayluk (1975). Nutritional requirements for growth of *Vicia hajastana* cells and protoplasts at very low population densities in liquid media. *Planta*, **126**, 105–110.

Kao, K. N. and M. Michayluk (1980). Plant regeneration from mesophyll protoplasts of alfalfa. *Z. Pflanzenphysiol.*, **96**, 135–141.

Kartha, K. K., M. R. Michayluk, K. N. Kao, O. L. Gamborg and F. Constabel (1974). Callus formation and plant regeneration from meosphyll protoplasts of rape plants (*Brassica napus* cv. Zephyr.). *Plant Sci. Lett.*, **3**, 265–271.

Khalid, N., C. H. Lee, J. B. Power and M. R. Davey (1988). Plant regeneration from electrofused protoplasts of *Chrysanthemum morifolium* and *Salpiglossis sinuata*. In *Progress in Plant Protoplast Research*, ed. K. J. Puite, J. J. M. Dons, H. J. Huizing, A. J. Kool, M. Kornneef and F. A. Krens, pp. 269–270. Kluwer, Dordrecht.

Kirby, E. G. (1980). Factors affecting proliferation of protoplasts and cell cutures of Douglas fir. In *Plant Cell Cultures: Results and Perspectives*, ed. F. Sala, B. Parisi, R. Cella and O. Ciferri, pp. 289–293. Elsevier, Amsterdam.

Kirby, E. G. (1988). Recent advances in protoplast culture of horticultural crops: conifers. *Sci. Hortic.* (*Amsterdam*), **37**, 267–276.

Klìmaszewska, K. and W. A. Keller (1986). Somatic embryogenesis in cell suspension and protoplast cultures of *Brassica nigra* L. Koch. *J. Plant Physiol.*, **122**, 251–260.

Kobayashi, S. (1987). Uniformity of plants regenerated from orange (*Citrus sinensis* Osbeck.) protoplasts. *Theor. Appl. Genet.*, **74**, 10–14.

Kobayashi, S., H. Uchimiya and I. Ikeda (1983). Plant regeneration from Trovita orange protoplasts. *Jpn. J. Breed.*, **33**, 119–122.

Kobayashi, S., I. Ikeda and H. Uchimiya (1985). Conditions for high frequency embryogenesis from orange (*Citrus sinensis* Osbeck.) protoplasts. *Plant Cell. Tissue Organ Cult.*, **4**, 249–259.

Koop, H. and H. G. Schweiger (1985). Regeneration of plants from individually cultivated protoplasts using an improved microculture system. *J. Plant Physiol.*, **121**, 245–257.

Krul, W. R. (1988). Recent advances in protoplast culture of horticultural crops: small fruits. *Sci. Hortic.* (*Amsterdam*), **37**, 231–245.

Kyozuka, J., Y. Hayashi and K. Shimamoto (1987). High frequency plant regeneration from rice protoplasts by novel nurse culture methods. *Mol. Gen. Genet.*, **206**, 408–413.

Larkin, P. J. (1986). Purification and viability determinations of plant protoplasts. *Planta*, **128**, 213–216.

Lee, C. H. and J. B. Power (1988). Interspecific gametosomatic hybridization in *Petunia hybrida*. *Plant Cell Rep.*, **7**, 17–18.

Lillo, C. and E. A. Shahin (1986). Rapid regeneration of plants from hypocotyl protoplasts and root segments of cabbage. *Hortic. Sci.*, **21**, 315–317.

Lillo, C. and J. E. Olsen (1989). Growth and shoot formation in protoplast-derived calli of *Brassica oleracea* ssp. *acephala* and ssp. *capitata*. *Plant Cell, Tissue Organ Cult.*, **17**, 91–100.

Lloyd, G. and B. H. McCown (1981). Commercially-feasible micropropagation of mountain laurel, *Kalmia latifolia*, by use of shoot tip culture. *Int. Plant Prop. Soc. Comb. Proc.*, **30**, 421–427.

Lu, C. Y., V. Vasil and I. K. Vasil (1981). Isolation and culture of protoplasts of *Panicum maximum* Jacq. (Guinea grass): somatic embryogenesis and plantlet formation. *Z. Pflanzenphysiol.*, **104**, 311–318.

McCown, B. H. and J. A. Russell (1987). Protoplasts culture of hardwoods. In *Cell and Tissue Culture in Forestry, vol. 2: Specific Principles and Methods: Growth and Developments*, ed. J. M. Bonga and D. J. Durzan, pp. 16–30. Nijhoff, Dordrecht.

Menczel, L., G. Lazar and P. Maliga (1978). Isolation of somatic hybrids by cloning *Nicotiana* heterokaryons in nurse cultures. *Planta*, **143**, 29–32.

Merkle, S. A. and H. E. Sommer (1987). Regeneration of *Liriodendron* tulipifera (family *Magnoliaceae*) from protoplast culture. *Am. J. Bot.*, **74**, 1317–1321.

Meyer, Y. and W. Herth (1978). Chemical inhibition of cell wall formation and cytokinesis, but not of nuclear division in protoplasts of *Nicotiana tabacum* L. cultivated *in vitro*. *Planta*, **142**, 253–262.

Michayluk, M. and K. N. Kao (1975). A comparative study of sugars and sugar alcohols on cell regeneration and sustained cell division in plant protoplasts. *Z. Pflanzenphysiol.*, **75**, 181–185.

Miura, Y. and M. Tabata (1986). Direct somatic embryogenesis from protoplasts of *Foeniculum vulgare*. *Plant Cell Rep.*, **5**, 310–313.

Morgan. A. and E. C. Cocking (1985). Plant regeneration from protoplasts of *Lycopersicon esculentum* Mill. *Z. Pflanzenphysiol.*, **106**, 97–104.

Muhlbach, H. P. (1980). Different regeneration potentials of mesophyll protoplasts from cultivated and a wild species of tomato. *Planta*, **148**, 89–96.

Murashige, T. and F. Skoog (1962). A revised medium for rapid growth and bioassays with tobacco tissue cultures. *Physiol. Plant.*, **15**, 473–497.

Nagata, T. and I. Takebe (1970). Cell wall regeneration and cell division in isolated tobacco mesophyll protoplasts. *Planta*, **92**, 301–308.

Nagata, T. and I. Takebe (1971). Plating of isolated tobacco mesophyll protoplasts on agar medium. *Planta*, **99**, 12–20.

Nagata, T. and T. Yamaki (1973). Electron microscopy of isolated tobacco mesophyll protoplasts cultured *in vitro*. *Z. Pflanzenphysiol.*, **70**, 452–459.

Niedz, R. P., S. M. Rutter, L. W. Handley and K. Smik (1985). Plant regeneration from leaf protoplasts of six tomato cultivars. *Plant Sci.*, **39**, 19–204.

Nishio, T. H. Yamagishi and K. Takayanagi (1987). Shoot regeneration capacity destined at an early stage of protoplast culture in cabbage. *Jpn. J. Breed.*, **37**, 22–28.

Nyman, M. and A. Wallin (1988). Plant regeneration from strawberry (*Fragaria* × *ananassa*) mesophyll protoplasts. In *Progress in Plant Protoplast Research*, ed. K. J. Puite, J. J. M. Dons, H. J. Huizing, A. J. Kool, M. Koornneef and F. A. Krens, pp. 101–102. Kluwer, Dordrecht.

Ochatt, S. J. (1987). Coltura di protoplasti come metodo per il miglioramento genetico nelle piante da frutto. *Frutticoltura*, **49**, 58–60.

Ochatt, S. J. (1990). Protoplast technology and top-fruit tree breeding. *Acta Hortic.*, **280**, 215–226.

Ochatt, S. J. and O. H. Caso (1986). Shoot regeneration from leaf mesophyll protoplasts of wild pear (*Pyrus communis* var. *pyraster* L.). *J. Plant Physiol.*, **122**, 243–249.

Ochatt, S. J. and J. B. Power (1988a). Rhizogenesis in callus from Conference pear (*Pyrus communis* L.) protoplasts. *Plant Cell, Tissue Organ Cult.*, **13**, 159–164.

Ochatt, S. J. and J. B. Power (1988b). An alternative approach to plant regeneration from protoplasts of sour cherry (*Prunus cerasus* L.). *Plant Sci.*, **56**, 75–79.

Ochatt, S. J. and J. B. Power (1988c). Plant regeneration from leaf mesophyll protoplasts of Williams' Bon Chretien (Syn. Bartlett) pear (*Pyrus communis* L.). *Plant Cell Rep.*, **7**, 587–589.

Ochatt, S. J. and J. B. Power (1989a). Selection for salt/drought tolerance using protoplast- and explant-derived tissue cultures of Colt cherry (*Prunus avium* × *pseudocerasus*). *Tree Physiol.*, **5**, 259–266.

Ochatt, S. J. and J. B. Power (1989b). Cell wall synthesis and salt (saline) sensitivity in cultured Colt cherry (*Prunus avium* × *pseudocerasus*) protoplasts. *Plant Cell Rep.*, **8**, 365–367.

Ochatt, S. J., E. C. Cocking and J. B. Power (1987). Isolation, culture and plant regeneration of Colt cherry (*Prunus avium* × *pseudocerasus*) protoplasts. *Plant Sci.*, **50**, 139–143.

Ochatt, S. J., P. K. Chand, E. L. Rech, M. R. Davey and J. B. Power (1988a). Electroporation-mediated improvement of plant regeneration from Colt cherry (*Prunus avium* × *pseudocerasus*) protoplasts. *Plant Sci.*, **54**, 165–169.

Ochatt, S. J., E. L. Rech, M. R. Davey and J. B. Power (1988b). Long-term effect of electroporation on enhancement of growth and plant regeneration of Colt cherry (*Prunus avium* × *pseudocerasus*) protoplasts. *Plant Cell Rep.*, **7**, 393–395.

Ochatt, S. J., M. R. Davey and J. B. Power (1989a). Tissue culture and top-fruit tree species. In *Methods in Molecular Biology*, vol. 6: *Plant Cell Culture*, ed. J. M. Walker and J. Pollard. The Humana Press, Clifton, New Jersey, in press.

Ochatt, S. J., E. M. Patat-Ochatt, E. L. Rech, M. R. Davey and J. B. Power (1988b). Somatic hybridization of sexually incompatible top-fruit tree rootstocks, wild pear (*Pyrus communis* var *pyraster* L.) and Colt cherry (*Prunus avium* × *pseudocerasus*). *Theor. Appl. Genet.*, **78**, 35–41.

Ochatt, S. J., A. S. Escandon and A. J. Martinez (1989c). Isolation, culture and plant regeneration from protoplasts of shrubby *Oxalis* species from South America. *J. Exp. Bot.*, **40**, 493–496.

Oka, S. and K. Ohyama (1985). Plant regeneration from leaf mesophyll protoplasts of *Broussonetia kazinoki* Sieb (paper mulberry). *J. Plant Physiol.*, **119**, 455–460.

Otsuka, H., N. Suematsu and M. Tode (1985). Plant regeneration from protoplasts of *Chrysanthemum morifolium* (Japanese). *Bull. Shizuoka Agric. Exp. Stn.*, **30**, 25–33.

Patat-Ochatt, E. M., S. J. Ochatt and J. B. Power (1988). Plant regeneration from protoplasts of apple rootstocks and scion varieties (*Malus* × *domestica* Borkh.). *J. Plant Physiol.*, **133**, 460–465.

Patat-Ochatt, E. M. and J. B. Power (1989). Advances in plant regeneration from apple protoplasts. *Acta Hortic.*, **280**, 285–288.

Pearce, R. S. and E. C. Cocking (1973). Behaviour in culture of isolated protoplasts of 'Paul's Scarlet Rose' suspension culture cells. *Protoplasma*, **77**, 165–180.

Pelcher, L. E., O. L. Gamborg and K. N. Kao (1974). Bean mesophyll protoplasts: production, culture and callus formation. *Plant Sci. Lett.*, **4**, 107–111.

Pental, D. and E. C. Cocking (1985). Some theoretical and practical possibilities of plant genetic manipulation using protoplasts. *Hereditas (Suppl.)*, **3**, 83–92.

Pirrie, A. and J. B. Power (1986). The production of fertile, triploid somatic hybrid plants (*Nicotiana glutinosa* [*n*] + *N. tabacum* [2*n*]) *via* gametic: somatic protoplast fusion. *Theor. Appl. Genet.*, **72**, 48–52.

Power, J. B., M. R. Davey, M. S. McLellan and D. Wilson (1989). Laboratory Manual: Plant Tissue Culture. University of Nottingham, UK.

Potrykus, I., C. T. Harms and H. Lorz (1976). Problems in culturing cereal protoplasts. In *Cell Genetics in Higher Plants*, ed. D. Dudits, G. L. Farakus and P. Maliga, pp. 129–140. Akad. Kiado, Budapest.

Prioli, L. M. and M. R. Sondhal (1989). Plant regeneration and recovery of fertile plants from protoplasts of maize (*Zea mays* L.). *Biotechnology*, **7**, 589–594.

Pua, E. C. (1987). Plant regeneration from stem-derived protoplasts of *Brassica algoglabra* Bailey. *Plant Sci.*, **50**, 153–160.

Ranch, J. and G. M. Pace (1988). Science in the art of plant regeneration from cultured cells: an essay and a proposal for a conceptual framework. *Iowa State J. Res.*, **62**, 537–569.

Rao, I. V. R., U. Mehta and H. Y. M. Ram (1982). Whole plant regeneration from cotyledonary protoplasts of *Crotalaria juncea*. In *Plant Tissue Culture 1982, Proc. 5th Int. Congress Plant Tissue Cell Culture*, ed. A. Fujiwara, pp. 595–596. Jpn. Assoc. Plant Tissue Culture, Tokyo.

Rao, P. S. and P. Ozias-Akins (1985). Plant regeneration through somatic embryogenesis in protoplast cultures of sandalwood (*Santalum album* L.). *Protoplasma*, **124**, 80–86.

Ratushnyak, Y. I., N. M. Piven and V. A. Rudas (1989). Protoplast culture and plant regeneration in *Lycium barbarum* L. *Plant Cell, Tissue Organ Cult.*, **17**, 183–190.

Raveh, D. and E. Galun (1975). Rapid regeneration of plants from tobacco protoplasts plated at low densities. *Z. Pflanzenphysiol.*, **76**, 76–79.

Raveh, D., E. Huberman and E. Galun (1973). *In vitro* culture of tobacco protoplasts: use of feeder techniques to support division of cells plated at low densities. *In Vitro*, **9**, 216–222.

Rech, E. L., S. J. Ochatt, P. K. Chand, J. B. Power and M. R. Davey (1987). Electroenhancement of division of plant protoplast-derived cells. *Protoplasma*, **141**, 169–176.

Rech, E. L., S. J. Ochatt, P. K. Chand, B. J. Mulligan, M. R. Davey and J. B. Power (1988). Electroporation increases DNA synthesis in cultured plant protoplasts. *Biotechnology*, **6**, 1091–1093.

Redenbaugh, M. K., R. D. Westfall and D. F. Karnosky (1980). Protoplast isolation from *Ulmus americana* L. pollen mother cells, tetrads and microspores. *Can. J. For. Res.*, **10**, 284–289.

Reinert, J. (1959). Uber die Kontrolle der Morphogenese und die Induktion von Adventivembryonen an Gewebekulturen aus Karotten. *Planta*, **53**, 318–333.

Revilla, M. A., S. J. Ochatt, S. Doughty and J. B. Power (1987). A general strategy for the isolation of leaf mesophyll protoplasts from deciduous fruit and nut tree species. *Plant Sci.*, **50**, 133–137.

Rhodes, C. A., K. S. Lowe and K. L. Ruby (1988). Plant regeneration from protoplasts isolated from embryogenic maize cell cultures. *Biotechnology.*, **6**, 56–60.

Robertson, D. and E. D. Earle (1986). Plant regeneration from leaf protoplasts of *Brassica oleracea* var *italica* cv Green Comet broccoli. *Plant Cell Rep.*, **5**, 61–64.

Robertson, D., E. D. Earle and M. A Mutschler (1988). Increased totipotency of protoplasts from *Brassica oleracea* plants previously regenerated in tissue culture. *Plant Cell, Tissue Organ Cult.*, **14**, 15–24.

Ruesink, A. W. (1978). Leucine uptake and incorporation by *Convolvulus* tissue culture and protoplasts under severe osmotic stress. *Physiol. Plant.*, **44**, 48–56.

Russell, J. A. and B. H. McCown (1986). Culture and regeneration of *Populus* leaf protoplasts isolated from nonseedling tissue. *Plant Sci.*, **46**, 133–142.

Russell, J. A. and B. H. McCown (1988). Recovery of plants from leaf protoplasts of hybrid poplar and aspen clones. *Plant Cell Rep.*, **7**, 59–62.

Saxena, P. K., R. Gill, A. Rashid and S. C. Maheshwari (1981). Plantlet formation from isolated protoplasts of *Solanum melongena* L. *Protoplasma*, **106**, 355–359.

Saxena, P. K. and R. Gill (1987). Plant regeneration from mesophyll protoplasts of the tree legume *Pithecelobium dulce* Benth. *Plant Sci.*, **53**, 257–262.

Schopke, C., L. E. Muller and H. Kohlenbach (1987). Somatic embryogenesis and regeneration of plantlets in protoplast cultures from somatic embryos of coffee (*Coffea canephora* P. ex Fr.). *Plant Cell. Tissue Organ Cult.*, **8**, 243–248.

Schweiger, H. G., J. Dirk, H. O. Koop, E. Kranz, G. Neuhaus, G. Spangenberg and D. Wolff (1987). Individual selection, culture and manipulation of higher plant cells. *Theor. Appl. Genet.*, **73**, 769–783.

Shahin, E. A. and J. F. Shepard (1980). Cassava mesophyll protoplasts isolation, proliferation and shoot formation. *Plant Sci. Lett.*, **17**, 459–465.

Shahin, E. A. (1985). Totipotency of tomato protoplasts. *Theor. Appl. Genet.*, **69**, 235–240.

Shekhawat, N. S. and A. W. Galston (1983). Isolation, culture, and regeneration of moth bean *Vigna aconitifolia* leaf protoplasts. *Plant Sci. Lett.*, **32**, 43–51.

Shepard, J. F. and R. E. Totten (1977). Mesophyll cell protoplasts of potato: isolation, proliferation and plant regeneration. *Plant Physiol.*, **60**, 313–316.

Shillito, R. D., J. Paszkowski and I. Potrykus (1983). Agarose plating and a bead type culture technique enable and stimulate development of protoplast-derived colonies in a number of plant species. *Plant Cell Rep.*, **2**, 244–247.

Shillito, R. D., M. W. Saul, J. Paszkowski, M. Muller and I. Potrykus (1985). High efficiency direct gene transfer to plants. *Biotechnology*, **3**, 1099–1103.

Shillito, R. D., G. K. Carswell, C. M. Johnson, J. J. DiMaio and C. T. Harms (1989). Regeneration of fertile plants from protoplasts of elite inbred maize. *Biotechnology.*, **7**, 581–587.

Sihachakr, D. and G. Ducreux (1987). Plant regeneration from protoplast culture of sweet potato (*Ipomoea batatas* Lam.). *Plant Cell Rep.*, **6**, 326–328.

Sim, G. E., C. S. Loh and C. J. Goh (1988). Direct somatic embryogenesis from protoplasts of *Citrus miti* Blanco. *Plant Cell Rep.*, **7**, 418–420.

Sommer, H. E. and H. Y. Wetztein (1984). Hardwoods. In *Handbook of Plant Cell Culture Vol. 3: Plant Species*, ed. P. V. Ammirato, D. A. Evans. W. R. Sharp and Y. Yamada, pp. 511–540. MacMillan, New York.

Spangenberg, G., H. U. Koop, R. Lichter and H. G. Schweiger (1986). Microculture of single protoplasts of *Brassica napus. Physiol. Plant.*, **66**, 1–8.

Srinivasan, C. and I. K.Vasil (1986). Plant regeneration from protoplasts of sugarcane. *J. Plant Physiol.*, **126**, 41–48.

Steward, F. C., M. O. Mapes and K. Mears (1958). Growth and organized development of cultured cells. II. Organization in cultures grown from freely suspended cells. *Am. J. Bot.*, **45**, 653–704.

Sticklen, M. B., S. C. Domir and R. D. Lineberger (1986). Shoot regeneration from protoplasts of *Ulmus* × Poineer. *Plant Sci.*, **47**, 29–34.

Tan, M. L. M. C., E. M. Rietveld, G. A. M. van Marrewijk and A. J. Kool (1987). Regeneration of leaf mesophyll protoplasts of tomato cultivars (*L. esculentum*): factors important for efficient protoplast culture and plant regeneration. *Plant Cell Rep.*, **6**, 172–175.

Thompson, J. A., R. Abdullah and E. C. Cocking (1989). Protoplast culture of rice using media solidified with agarose. *Plant Sci.*, **47**, 123–133.

Thompson, J. A., R. Abdullah, W. H. Chen and K. M. A. Gartland (1987). Enhanced protoplast division in rice (*Oryza sativa* L.) following heat shock treatment. *J. Plant Physiol.*, **127**, 367–370.

Tsai, C. K. (1988). Plant regeneration from leaf protoplasts callus of *Actinidia chinensis* var *chinensis*. *Plant Sci.*, **54**, 231–235.

Uchimiya, H. and T. Murashige (1976). Influence of the nutrient medium on the recovery of dividing cells from tobacco protoplasts. *Plant Physiol.*, **57**, 424–429.

Upadhya, M. D. (1975). Isolation and culture of mesophyll protoplasts of potato (*Solanum tuberosum* L.). *Potato Res.*, **18**, 438–445.

Van der Valk, P. and M. A. M. Zaal (1988). Regenerätion of protoplasts from protoplasts of *Poa pratensis* L. (Kentucky bluegrass). In *Progress in Plant Protoplast Research*, ed. K. J. Puite, J. J. M. Dons, H. J. Huizing, A. J. Kool, M. Koornneef and F. A. Krens, pp. 56–60. Kluwer, Dordrecht.

Vardi, A (1977). Isolation of protoplasts in *Citrus. Proc. Int. Soc. Citric.*, **2**, 575–578.

Vardi, A., P. Spiegel-Roy and E. Galun (1975). *Citrus* cell culture: isolation of protoplasts, plating densities, effect of mutagens and regeneration of embryos. *Plant Sci. Lett.*, **4**, 231–236.

Vardi, A., P. Spiegel-Roy and E. Galun (1982). Plant regeneration from *Citrus* protoplasts: variability in methodological requirements among cultivars and species. *Theor. Appl. Genet.*, **62**, 171–176.

Vardi, A., D. J. Hutchinson and E. Galun (1986). A protoplast to tree system in *Microcitrus* based on protoplasts derived from sustained embryogenic cells. *Plant Cell Rep.*, **5**, 412–414.

Vardi, A. and E. Galun (1988). Recent advances in protoplast culture of horticultural crops: *Citrus. Sci. Hortic.*, **37**, 217–230.

Vasil, I. K. (1987). Developing cell and tissue culture systems for the improvement of cereal and grass crops. *J. Plant Physiol.*, **128**, 193–218.

Vasil, I. K. (1988). Progress in the regeneration and genetic manipulation of cereal crops. *Biotechnology*, **6**, 397–402.

Vasil, V. and I. K. Vasil (1980). Isolation and culture of cereal protoplasts. Part 2: embryogenesis and plantlet formation from protoplasts of *Pennisetum americanum. Theor. Appl. Genet.*, **56**, 97–99.

Vasil, V., D. Wang and I. K. Vasil (1983). Plant regeneration from protoplasts of *Pennisetum purpureum* Schum. (Napier grass). *Z. Pflanzenphysiol.*, **111**, 319–325.

Vasil, V. and I. K. Vasil (1987). Formation of callus and somatic embryos from protoplasts of a commercial hybrid of maize (*Zea mays* L.). *Theor. Appl. Genet.*, **73**, 793–798.

Vatsya, V. and S. Bhaskaran (1982). Plant regeneration from cotyledonary protoplasts of cauliflower (*Brassica oleracea* L. var *botrytis* L.). *Protoplasma*, **113**, 161–163.

Von Arnold S. and T. Eriksson (1977). A revised medium for growth of pea mesophyll protoplasts. *Physiol. Plant.*, **39**, 257–260.

Wallin, A., K. Glimelius and T. Eriksson (1977). Pretreatment of cell suspensions as a method to increase the protoplast yield of *Haplopappus gracillis. Physiol. Plant.*, **40**, 307–311.

Watts, J. M., F. Motoyoshi and J. M. King (1974). Problems associated with the production of stable protoplasts of cells of tobacco mesophyll. *Ann. Bot. (London)*, **38**, 667–671.

Webb, K. J. (1988). Recent developments in the regeneration of agronomically important crops from protoplasts. In *Progress in Plant Protoplast Research*, ed. K. J. Puite. J. J. M. Dons, H. J. Huizing, A. J. Kool, M. Koornneef and F. A. Krens, pp. 27–32. Kluwer, Dordrecht.

Wei, Z.-M. and A.-H. Xu (1988). Plant regeneration from protoplasts of soybean (*Glycine max* L.). *Plant Cell Rep.*, **7**, 348–351.

Widholm, J. M. (1972). The use of fluorescein diacetate and phenosafranine for determining viability of cultured plant cells. *Stain Technol.*, **47**, 189–194.

Wilson, V. M., N. Haq and P. K. Evans (1985). Protoplast isolation, culture and plant regeneration in the winged bean, *Psophocarpus tetragonolobus* (L.) D.C. *Plant Sci.*, **41**, 61–68.

Xu, Z. H., M. R. Davey and E. C. Cocking (1982). Plant regeneration from root protoplasts of *Brassica. Plant Sci. Lett.*, **24**, 117–121.

Yamada, Y., Z. Yang and D. Tang (1986). Plant regeneration from protoplast-derived callus of rice (*Oryza sativa* L.). *Plant Cell Rep.*, **5**, 85–88.

Yarrow, S. A. and T. L. Barsby (1988). Recent advances in protoplast culture for horticultural crops: vegetable crops. *Sci. Hortic. (Amsterdam)*, **37**, 179–200.

Yarrow, S. A., E. C. Cocking and J. B. Power (1987). Plant regeneration from protoplasts of *Pelargonium aridum*, × *P. hortorum* and *P. peltatum. Plant Cell Rep.*, **6**, 102–104.

Zapata, F. J., P. K. Evans, J. B. Power and E. C. Cocking (1977). Effect of temperature on the division of leaf protoplasts of *Lycopersicon esculentum* and *L. peruvianum. Plant Sci. Lett.*, **8**, 119–124.

Zapata, F. J., K. C. Sink and E. C. Cocking (1981). Callus formation from leaf mesophyll protoplasts of three *Lycopersicon* species: *L. esculentum* cv Walter, *L. pimpinellifolium* and *L. hirsutum F. Glabratum. Plant Sci. Lett.*, **23**, 41–46.

# 8

# Micropropagation: Principles and Commercial Practice

ROBIN D. RICE
*Novalal plc, Suffolk, UK*
PETER G. ALDERSON
*Nottingham University, UK*
JOHN F. HALL
*Leicester Polytechnic, UK*
and
ASHOK RANCHHOD
*Southampton Institute, UK*

---

## 8.1 INTRODUCTION

Traditional horticultural skills required in the vegetative propagation of plant species are currently being accelerated and extended by the application of scientific methods. However, the use of tissue culture in the regeneration and commercial propagation of economically important plants is a comparatively recent and radical development.

The intrinsic property of excised shoots and roots to regenerate a whole plant has been recognized for several decades. *De novo* formation of roots and, in some cases, shoots remains central to vegetative propagation by cuttings. *In vitro* techniques considerably enhance this potential by the application of nutritional and hormonal regimes under aseptic conditions. Plant propagation by this method is termed micropropagation because miniature shoots or plantlets are initially derived. It has graduated from a research tool to become an important alternative to conventional horticultural procedures.

The discovery of micropropagation through shoot tip culture resulted from attempts to obtain virus-free plants by meristem isolation. The application of this procedure to *Cymbidium* orchids resulted in greatly enhanced rates of production. Subsequently a great diversity of plants have exhibited potential for vegetative propagation *in vitro*. It is estimated that European countries and the USA each produce over 30 million plants annually by this method.

Micropropagation has two primary uses: (i) rapid clonal propagation; and (ii) establishment, maintenance and distribution of pathogen-tested clones. A particular application is evident where species are conventionally slow to multiply, or where new hybrids derived from a plant breeding programme require rapid bulking for trials, or introduction to the trade. In parallel to propagation, potential exists through tissue culture for long distance shipment, or for long term storage of clonal material.

Micropropagation is analogous in many respects to conventional propagation and must compete for recognized horticultural markets. Its application on an enlarged scale has demanded a reappraisal of scientific and economic criteria. Although it has become integrated into the expanding ornamental trade and is widely accepted as a tool for plant breeders, the relative cost of micropropagation restricts its applications in other areas. The realization of the full potential of plant propagation by *in vitro* techniques across a broad spectrum of horticultural and agricultural crops now awaits the advent of new scientific and manufacturing technologies.

## 8.2 PATHWAYS OF REGENERATION

The regeneration of plants in aseptic culture is the focus of considerable effort on the part of the plant researcher, ranging from the need to rapidly propagate a particular cultivar to realizing the potential of systems with altered genotypes through recombinant DNA techniques. It is therefore encouraging that despite many inherent and at times insurmountable problems the number of species for which regeneration has become possible continues to rise rapidly. For example, in the important area of cereal crop tissue culture there are now reports of sustained regeneration from embryogenic cultures (Nabors *et al.*, 1983).

There are a number of interdigitating pathways for the regeneration of whole plants from excised plant parts. Three main pathways can be considered, *i.e.* regeneration of shoots from existing meristems, regeneration from *de novo* meristems and somatic embryogenesis. For some species all of the pathways may be appropriate, whereas in others only one or two may be possible given the current level of understanding. Clearly the predominant factor determining the ease with which any species can be cultured *in vitro* is the genetic complement of the plant, the importance of which cannot be overemphasized. The epigenetic state of plants may also be important, for example in woody plants the ability to regenerate shoots (axillary or adventitious) and embryos is very much dependent on the juvenile state of the tissues (Alderson, 1987).

### 8.2.1 Regeneration from Existing Meristems

This may be referred to as axillary shoot proliferation and requires that an existing meristem such as a shoot tip or a nodal bud is placed into culture. On a medium containing cytokinin, such explants proliferate a cluster of axillary shoots due to the loss of apical dominance in the original tissue. Shoots formed may be separated and subcultured onto fresh medium for further proliferation of axillary shoots. This response has been demonstrated for a range of herbaceous and woody plants and the level of cytokinin required has to be determined for each plant subject in order to obtain an acceptable rate of proliferation of shoots which are of a manageable size. The choice of cytokinin used may determine the rate of shoot proliferation, as shown for ericaceous plants, which respond better to isopentenyladenine than to benzylaminopurine (Norton and Norton, 1985).

Shoot proliferation from axillary meristems can significantly enhance multiplication rates and is the technique which has most readily found commercial application. There are few of the genetic instability problems which may be associated with other methods, supported by the report that many species have been kept for up to 15 years with no ill effects (Hussey, 1983).

### 8.2.2 Regeneration from Adventitious Meristems

Shoot multiplication either directly or by callus formation can be obtained by inducing adventitious shoot production on mature plant organs such as leaves, stems and roots. This requires the initiation of *de novo* meristems, which may be controlled by the balance of auxin and cytokinin in the culture medium. In general shoots are formed when a high ratio of cytokinin to auxin is present and roots when the medium contains the reverse balance of growth regulators. In contrast, regeneration in most bulbous species, which is *via* adventitious meristems, is promoted by low concentrations of auxins. An exogenous supply of cytokinin may not be required (van Aartrijk and van der Linde, 1986).

Regeneration from adventitious meristems may significantly enhance the potential multiplication rate of a plant, *e.g.* for *Saintpaulia* it is possible to obtain 900 plants from one leaf (Vasquez *et al.*, 1977). Thus the presence of morphologically competent cells that under appropriate chemical and physical conditions give rise to shoots in culture means that high multiplication rates may be achieved and that mechanization of the handling of the cultures is a possiblity. However, there may be problems of genetic instability in that the progeny arising from single cells may not be true to type. Polyploidization is the most common occurrence and is attributed to selection of preexistent cells (Evans and Bravo, 1986).

The indirect production of adventitious shoots from explants *via* a callus stage may accentuate the problem of genetic instability of the progeny. In addition, although the large scale multiplication of callus may confer commercial advantages, the repeated subculture of callus from many plant species may reduce its morphogenic potential. Regeneration of adventitious shoots from callus cultures is usually linked to a lowering of the levels of plant growth regulators in the culture medium or adjusting the auxin/cytokinin balance. More recently, Goldsworthy (1988) has shown that organogenesis in callus cultures may be controlled by the application of electric currents to callus.

### 8.2.3 Regeneration by Embryogenesis

Somatic embryogenesis presents the most promising technique for the rapid multiplication of plants. Somatic embryos may arise in culture directly on explants, *e.g.* coffee leaf discs (Sondahl and Sharp, 1977), or after a multiplication stage involving callus formation or liquid suspension culture, *e.g.* carrot (Ammirato, 1986). The large numbers of adventitious embryos produced are bipolar structures which possess both shoot and root meristems. These embryos arise from cells which are switched on to a series of cell divisions comparable to the development of a zygotic embryo. The induction of embryos requires a high level of auxin in the culture medium, whereas the development of somatic embryos from such 'induced' cells is often achieved by their transfer to a conditioning medium containing a low ratio of auxin to cytokinin or no growth regulators, as shown for embryogenesis in carrot cell suspensions (Ammirato, 1986).

The potential for mechanization and the production of artificial seeds, which may be stored and planted directly like normal seeds, makes this method of regeneration a very attractive system. Although there are difficulties still to be overcome with this system, considerable encouragement may be taken from the reports of tissue which does produce stable regenerable callus, the successful use of encapsulated celery somatic embryos (Gebhart, 1984) and the uniformity of *Trifolium repens* plants regenerated from somatic embryos (Maheswaran and Williams, 1987).

## 8.3 STAGES IN MICROPROPAGATION

It should be noted at the outset that there are no inviolable rules for the successful propagation of a particular species by tissue culture, and it is often necessary to adjust and readjust the composition of the ambient medium and even the environment so that cultures may be induced to grow and to organize. For descriptive purposes the procedures involved in micropropagation can be divided into five basic steps. The scheme of George and Sherrington (1984) is adopted here, as in many species there is a postrooting weaning period crucial to the establishment of the plant.

Stage 0: Preparation and pretreatment of the plant.
Stage 1: Initiation of the explant.
Stage 2: Multiplication of the tissue.
Stage 3: Regeneration of whole plants.
Stage 4: Hardening for subsequent field planting.

### 8.3.1 Stage 0: Preparation and Pretreatment of the Plant

The choice and pretreatment of the tissue to be propagated is in many cases of crucial importance and determines the success or failure of the subsequent initiation phase. The following factors must be considered. (i) The degree of contamination of the starting material is important. Clearly field grown plants are likely to have a far higher degree of bacterial/viral contamination and are consequently far more difficult to sterilize prior to *in vitro* culture. This is particularly true when parts of mature trees are used as explants and extreme and complicated sterilization procedures are required. It is therefore desirable to initiate cuttings from plants which have been specifically grown as stock plants in controlled environments, such as a well-managed greenhouse, without any overhead watering. (ii) The initiation of explants is more successful from healthy, vigorously growing material. This can be achieved by a number of methods, including heavy pruning to encourage vigorous growth, high levels of fertilizer and grafting to seedlings. Such juvenile tissues display a far greater degree of plasticity and have a greater morphogenic potential (Tisserat, 1985). For some species the most successful method of obtaining explants capable of regeneration requires that the starting material is itself cultured *in vitro* before being used as an explant source. (iii) Certain species including bulbs and trees require special treatments to break dormancy. The effects of temperature on subsequent regenerative behaviour can be both stimulatory and inhibitory depending upon several factors such as storage time, auxin concentration and culture temperature. A pretreatment of 30 °C is stimulatory for tulip and iris (van Aartrijk and van der Linde, 1986).

### 8.3.2 Stage 1: Initiation of the Explant

The initiation of the explant is most universally achieved on MS medium with a number of variations, although this medium is less suitable for certain plant genera. Media formulations for a wide variety of plants can be found in the general references. Herbaceous species readily respond to culturing, enabling the use of a variety of plant parts (root, stem, leaf, *etc.*) in contrast to perennials and some cereals. The choice of explant is often the shoot meristem and axillary buds. The predominant reason for this is the far higher degree of stability in culture and greater plasticity. In cereals the material of choice is the immature embryo which in a number of species will produce embryogenic callus when cultured on MS medium, usually with a number of auxin additives. Of these 2,4-D has been the most popular but success has also been achieved with CPA, 2,4,5-T, picloram, NAA, dicamba and benazolin (Schaeffer *et al.*, 1984). It would appear that few other additions have convincing effects and that auxin is the next important factor after the genetic constitution of the material. Typical levels for herbaceous species would be 1–10 mg $L^{-1}$ auxin (IAA and NAA most commonly) and 0.1–10 mg $L^{-1}$ cytokinin (kinetin and BAP). For woody species 0.1–5 mg $L^{-1}$ auxin (IAA and IBA), although 57 mg $L^{-1}$ has been used for *Malus domestica* (*cv* McIntosh) seedlings, with BAP at 1–10 mg $L^{-1}$. The use of low levels of gibberellins may also be beneficial (0.3–3 mg $L^{-1}$). Low concentrations of auxin are routinely used for bulbous species at a concentration of 10 $\mu$M or less.

The protocols for small fruit and grape tissue culture again usually utilize MS media with IBA or NAA and BAP additives. For strawberries it is beneficial to use adenine sulfate in the absence of a cytokinin as this improves establishment and reduces the occurrence of off-types which may subsequently form as a result of too rapid multiplication on media containing BAP. Alterations in the media for particular species may be beneficial (Swartz and Lindstrom, 1986). Thus *Ribes* fares better on MS medium with a 90% reduction in the level of nitrate ions, and half strength MS may be preferable for strawberries and raspberries. Conifers are usually cultured on a high salt medium initially, such as MS or SH with an auxin level around 0.005–0.5 $\mu$M. Combinations of growth regulators may improve multiplication, in Douglas Fir for instance, where BAP, 2iP, IAA and IBA together are superior to other treatments. Woody shrubs, *e.g.* mountain laurel and azalea, are grown on Lloyd and McCown (1980) medium, which is characterized by a low nitrate and high calcium content.

The establishment of plants in tissue culture is also affected by factors such as the season when the cuttings are taken. Spring is often found to be the best time to take shoots as they possess considerable vigour and less infection. In some species the production of phenolic compounds by the explant has deleterious effects, resulting in browning of the tissues. This can be combated by a combination of factors such as treatment with ascorbate or soaking in water for 24 h prior to culture (Swartz and Lindstrom, 1986). Charcoal and PVPP (polyvinylpolypyrrolidone) additives are also beneficial. Although not routinely used, antibiotics may be of some value in eliminating microorganisms, especially if it is unavoidable that mature and hence more heavily infested tissues, as may be required for clonal propagation of a particular characteristic, are used as explants.

### 8.3.3 Stage 2: Multiplication of the Tissue

The multiplication of the explants is a crucial stage in the propagation of any species for commercial exploitation, and the most rapid rates of multiplication are required. The most common additives to the standard media are cytokinins, usually as BAP or kinetin. However, there is a trade-off between a very rapid rate of multiplication which may be achieved with high levels of applied cytokinins and possible deleterious effects. Generally the faster the rate of multiplication (with high levels of BAP), the more problems arise as a result of cytokinin habituation *etc.* (see Section 8.4.2). There is possibly some benefit in using alternative cytokinins such as thidiazuron which is superior to BAP for *Malus* species giving higher rates of multiplication and fewer problems in later stages (Hammerschlag, 1986). Other factors affecting the rate of multiplication include light intensity, which is well documented for gooseberry, *i.e.* the rate increases up to a maximum at 18 $W\,m^{-2}$ (Wainwright and Flegmann, 1985). The choice of agar can also have important consequences. It can affect vitrification at later stages (see Section 8.4.3), and increased concentrations improve the multiplication rate but, in *Gebera* for instance, 10 $g\,L^{-1}$ becomes toxic after many subcultures and it is found that 3–5 $g\,L^{-1}$ is optimal. The length of the subculture passage is also significant and it is often found that several shorter passages are preferable to fewer long ones. Orientation of the explant on the induction medium can also influence the establishment and the rate of multiplication. A particular example is found in the use of lily bulb scales which possess basal polarity (van Aartrijk and Blom-Barnhoorn, 1983), an effect which is also found in other bulbous species. In oats somatic embryos are more readily obtained from immature embryos if they are placed scutellum uppermost, but for mature embryos the reverse is true (Kaur-Sawhney and Galston, 1984).

### 8.3.4 Stage 3: Regeneration of Whole Plants

The regeneration of whole plants for many species which are propagated through tissue culture normally means the production of roots, as axillary shoot proliferation remains the most common and reliable method for rapid multiplication. The production of roots is easily achieved in some species by reducing the cytokinin level with or without the addition of extra root-promoting auxins, especially IBA (0.2–5 $mg\,L^{-1}$). This treatment may be used fruitfully in tandem with reduced ionic strength of the medium, *e.g.* half strength MS with low nitrate. Fog chambers are recommended for soft fruit with a maximum initial light intensity of 350 $\mu mol\,m^{-2}\,s^{-1}$ at 15 °C. Charcoal and ascorbate may be used in the rooting medium but this is not routine. In woody species such as temperate fruit trees and nuts, phenolic cofactors often enhance rooting (Hammerschlag, 1986). Phloroglucinol has been reported to stimulate rooting and reduce the period of IBA treatment needed (James, 1983). In plums the addition of $GA_3$ is also effective, especially when combined with the use of a lower temperature and phloroglucinol. Most herbaceous species will respond readily to a reduction in the cytokinins and an increase in an appropriate auxin.

In contrast, regeneration from somatic embryos clearly requires the formation of the whole plant. This can be readily achieved in carrot and other members of the Umbelliferae by omitting the auxin from the liquid medium. The research on this system is well advanced and highly synchronized long term embryogenic cultures have been obtained for carrot (Nomura and Komamine, 1986). The position with other systems is less well advanced but in recent years progress has been made, particularly in the area of cereal crop production *via* somatic embryogenesis. The factors which have enabled the production of regenerants from cereal tissue are mainly the adjustment of the osmoticum, which in rice can prolong the retention of regenerative ability from 300 d (12 passages) to 1500 d (60 passages). Sorbitol was added at 3% (total medium osmolarity = 299 mosmol) which resulted in a regeneration frequency of 50–60%. Furthermore, regenerative potential was regained after 50 d on the sorbitol-containing medium (Kavi Kishor and Reddy, 1986). ABA has an influence on the ability of embryos to regenerate. The endogenous levels of ABA increase as embryogenesis occurs *in vitro* (Rajasekeran *et al.*, 1982) and levels of ABA correlate with the ability of the plant part to undergo embryogenesis, the levels being higher in the more responsive tissues (Rajasekeran *et al.*, 1987). It is not surprising to find therefore that embryogenesis is stimulated and maintained by ABA in a number of species. The addition of proline to the induction medium is also beneficial particularly in maize, although as with most treatments the response is genotype specific (Armstrong and Green, 1985). This effect is undoubtedly linked to the stress response of the tissue in culture, one consequence of which is the elevation of proline levels. Salt and other factors also stimulate this response. It has been generally established that an oxidizing environment favours embryogenesis and a reducing environment favours regeneration.

### 8.3.5 Stage 4: Hardening for Subsequent Field Planting

Plants grown under *in vitro* conditions adapt to the altered environment in several ways. Perhaps the most extreme form is the vitrification of the plant tissue, which makes it unsuitable for direct transfer to *in vivo* conditions. A postregenerative acclimatization period is required, therefore, to enable the survival of the tissue cultured plants *in vivo*. The requirements are broadly the same for most species. Plants in tissue culture are in an environment of high humidity, consequently mist chambers or fogging systems are used routinely. The relative humidity is kept high, almost at maximum during the day and at 80% at night. A prewash of the rooted shoots to remove any residual agar (an excellent food source for bacterial contamination) is essential and the use of a fertilizer relatively high in phosphate may be beneficial. It is necessary to use fungicides to avoid infection, particularly in the early stages when the humidity is so high before the plants are gradually weaned. The stomatal response of tissue cultured plants is defective and the wax layer is also drastically reduced on plants maintained *in vitro*, therefore transpiration losses will be dramatically high thereby reducing the chances of survival. It has been established that a reduction of the relative humidity to 80% is sufficient to enable the formation of functional stomata and a sufficiently high wax covering to reduce water loss to close to the value for seedling material (Short *et al.*, 1987). This reduction in relative humidity is most readily achieved *in vitro* by loosening the vessel closure and/or cooling the base of the vessel to 2–3 °C below the air temperature. Plants treated in this way and gradually weaned in a relatively high humidity with shading have a far higher survival rate.

## 8.4 *IN VITRO* PHENOMENA IN MASS PROPAGATION

The limited understanding of *in vitro* processes often necessitates an empirical approach to the micropropagation of a novel species. Morphogenic capacity is determined in part by the ambient physiological state of the donor plant (Marks, 1986; Rugini, 1984). More recently it has been recognized that the development of a satisfactory culture protocol does not necessarily guarantee a stable or predictive model of *in vitro* behaviour. Genetic, physiological and even pathological components of aseptic material interact with the culture environment during serial subculturing and may induce variable multiplication rates and habits of growth. For clonal propagation, this has important implications for the integrity and viability of this production system.

### 8.4.1 Genetic Instability

Within the whole plant there is considerable diversity of tissue type, degree of differentiation and genetic composition. The genetic stability of the explant *in vitro* is dependent both on its degree of organized structure and the influence of the aseptic environment in allowing expression of potential variability. Callus and cell cultures may have a diversity of origins and are genetically unstable, generating polyploidy, aneuploidy and chromosomal structural changes. Whilst such systems are recognized as potentially useful sources of somaclonal variants for crop improvement (Larkin and Scowcroft, 1981), they are clearly to be avoided by the clonal propagator.

Instability may derive from a genetic, ontogenetic or phenotypic base (Mantell, 1986). Phenotype refers to the appearance of the plant and is expressed by the interaction of genotype and environment. If the genotype is stable, phenotypic change is termed 'epigenetic' and is temporal in nature. This is discussed in connection with cytokinin-associated behaviour in Section 8.4.2.

#### *8.4.1.1 Genetic variation*

Axillary or apical culture systems are recognized as the most genetically conservative propagation methods. Genetically abnormal cells within the meristem are usually eliminated through competition (D'Amato, 1977) and consequently rates of variation are extremely low. By contrast, adventitious regeneration originates from single cells or small groups of cells and therefore the derepression of one somatic variant may result in a genetically abnormal population. Somatic variation may be confined to early divisions or arise continuously during serial subculturing (Cassells *et al.*, 1983). Hence, clonal fidelity may be affected in some systems of direct adventitious regeneration, or in axillary systems tending toward adventitious caulogenesis. The occurrence of off-types in micropropagated soft fruits ranged from 0.01–1%, and was variously attributed to frequency of subculture, unstable cultivars and supraoptimal cytokinin concentrations (Swartz and Lindstrom, 1986). This suggests that the avoidance of instability is at least partially a function of good management.

Differentiation between genetic and epigenetic flux can be problematic. Unstable genes may be readily expressed or may exhibit reversion, and by contrast epigenetic change itself may be relatively stable. In *Saintpaulia*, the frequency of an induced fluted leaf variant exceeded 70% in recycled adventitious leaf tissue, and was expressed in more than 30% of cultured progeny derived from variant plants grown to maturity *in vivo* (Cassells and Plunkett, 1986). This was provisionally attributed to a stable epigenetic adaptation to the *in vitro* environment. Genetic analysis of seed progeny is generally required for confirmation, as variation of an epigenetic nature will not be sexually transmitted.

The potential instability of adventitious regeneration has not precluded its successful application for clonal propagation, in particular for some monocotyledonous genera where the incidence of axillary meristems is relatively low.

#### *8.4.1.2 Ontogenetic variation*

Disorganization of shoot ontogeny can occur during adventitious meristem formation by mixing of cell or tissue arrangements. Ontogenetic variation is particularly pronounced in chimeras, where the constitutive layers of the shoot apex are genetically distinct. Leaf chimeras comprise a horticulturally important group due to the distinctive and marketable patterns of variegation. Chimeric breakdown *in vitro* may produce histogenically pure lines at frequencies exceeding 50% (McPheeters and Skirvin, 1984; Swartz and Lindstrom, 1986). Conversely, leaf disc cultures of some *Saintpaulia* forms have exhibited stable variegation, indicating organogenesis from multiple cell associations (Norris *et al.*, 1983). In general, strict adherence to axillary proliferation is a prerequisite to micropropagation, but does not guarantee stability. When in contact with exogenous cytokinins, highly regenerative genera may retain a chimeral association but with some tissue rearrangement, producing variable sectoral chimeras (Cassells and Plunkett, 1986). Unstable variegation, as in some micropropagated *Hosta sieboldiana*, is unacceptable to the trade. However, tissue culture of the green and yellow *Hosta* 'Frances Williams' was proved to be feasible as reversion to the single colour forms produced lines marketable in their own right (Meyer, 1980).

### 8.4.2 Cytokinin Habituation

Cytokinins are often essential for the multiplication stage of tissue culture. However, extended exposure to relatively low concentrations and increases in levels to enhance multiplication rates can have undesirable consequences. These include a tendency to cause rejuvenation of the tissue and the habituation to a growth condition where an exogenous supply of cytokinin is not required for continued multiplication. Rejuvenation of the tissues may not pose problems since such plants usually revert to the mature culture after a few generations. Thus in strawberry plants from tissue culture there is a greater tendency to produce runners than in the normally propagated plants (Marcotrigiano *et al.*, 1984). However, they revert after the second and subsequent generations. Other common traits, which may be a consequence of cytokinin treatment, are earlier flowering, increased thorniness, juvenile leaf forms and an increase of yield. There may be an overall decrease in the actual size of each fruit but the total number increases (Swartz and Lindstrom, 1986).

The underlying cause of cytokinin habituation lies in the establishment of a positive feedback loop which is activated by cytokinin itself (Meins, 1987). The characteristics of habituated cultures are often an inhibition of the flowering response, which can be a particular problem in rosette ornamentals, and a lack of root formation. These effects clearly have important commercial consequences and simple, effective treatments are required to normalize the affected plants. The results of a study commissioned by Microplants Ltd. in order to induce rooting in habituated *Kalmia latifolia* (mountain laurel) shoot cultures demonstrated that the addition of relatively high levels of auxin in the medium produced a high proportion of normal plantlets stimulating root production and reducing multiapexing (Table 1). The recommended method, *i.e* treatment with 25 mg $L^{-1}$IBA for 4 weeks, increased the percentage of normal shoots that rooted from 14% in the controls to 85%. It may thus be possible to use higher cytokinin levels, achieving faster multiplication rates, without jeopardizing the number of regenerable plants.

**Table 1** Rooting of Cytokinin Habituated Shoot Cultures of *Kalmia latifolia*[a]

| *Rooting medium*[b] | *Number of shoots with roots* | *Number of shoots without roots* |
|---|---|---|
| WPM + 5 μM 2iP (control) | 453 | 2726 |
| WPM + 100 μM IBA | 1375 | 241 |

[a]From Soe Yien Ang, Xing Ti and J. F. Hall (unpublished data). [b]Culture multiplied on Woody Plant Medium (WPM) + 35 μM 2ip for 10 weeks before transfer to rooting medium.

### 8.4.3 Vitrification

The term vitrification refers to the appearance of glossy translucent shoots *in vitro*. Clonal production may be completely retarded by such abnormal development, as vitrified shoots are usually difficult to proliferate and will not establish *in vivo*. Although it is a specific morphogenic phenomenon, vitrification is causally related to several environmental regimes and may be symptomatic of more than one physiological disorder. Excess cytokinins, matrix potential and ethylene have been implicated in the induction of vitreous tissue (Debergh and Maene, 1984; Kevers *et al.*, 1984). Increased agar concentrations may alleviate the condition by reducing water availability and impeding uptake of cytokinins (Vieitez *et al.*, 1985), although growth rates may be correspondingly reduced. Ventilation of the culture vessel is similarly influential to culture development by promoting diffusion of water vapour and secondary metabolites such as ethylene away from the culture atmosphere.

High levels of ammonium ions in MS medium have been associated with vitrification. The use of Hellers macronutrients prevented vitrification in *Castanea* (Vieitez *et al.*, 1985), and it was concluded that excessive uptake of ammonium from MS medium reduced the C:N ratio and decreased synthesis of cellulose and lignin, thereby increasing water uptake. In willow, concentrations of exogenous ammonium exceeding specific threshold levels induced a detoxification reaction, involving glutamate dehydrogenase, that resulted in a vitreous state (Daguin and Letouze, 1986).

The availability of water appears to be the central causal link. Consequently, water content of the gaseous phase has been proposed as the predominant element responsible for vitrification (Gaspar *et al.*, 1987). The temperature differential created by bottom heat from racks of fluorescent lighting, and aerial cooling by air conditioning exacerbates the high relative humidity within the culture container. Thus bottom cooling may be the most applicable method to control humidity and the incidence of vitreous degeneration (Gaspar *et al.*, 1987; Debergh, 1987a,b).

It is noteworthy that many genera do not commonly exhibit symptoms of vitrification during production and yet may still suffer physiological stress as a result of some of the factors outlined above. This may be manifested in fluctuating multiplication rates and suboptimal development, and in addition encourage greater proliferation of systemic contaminants.

### 8.4.4 Bacterial Contamination

Significant and sometimes total losses of clonal populations result from uncontrolled proliferation of bacteria and bacteria-like organisms. The incidence of bacterial infestation may be ubiquitous in plant species, and several genera of pathogens have been identified (See Bastiaens, 1983). Pathogenic, epiphytic and endophytic bacteria are expressed under aseptic conditions with the potential to overwhelm the plant culture (Bradbury, 1988). The high planting density in culture and rapid serial subculturing demanded by commercial micropropagation further promotes bacterial spread. Proliferation is often partially suppressed by cytokinins during stages I and II and subsequently becomes vigorously manifested during *in vitro* rooting, or gives rise during weaning to such symptoms as 'damping off' and basal rot. In addition, the potential for the spread of disease has been stressed in cultures carrying latent organisms across quarantine zones. Such organisms may become pathogenic in new environments (Cassells, 1986).

Bacterial contaminants may derive from four sources: (i) inadequate external disinfestation of explants; (ii) endogenous bacterial populations; (iii) inadequate subculture protocol; and (iv) operator error.

#### *8.4.4.1 Stock plant pretreatment*

Stock plants maintained under protection will contain less surface bacteria and possibly less endogenous infestation than plants grown in the field. Overhead watering should be avoided, and irrigation and nutrition should be provided by trickle or sand bed systems direct to the container (Debergh and Maene, 1981). It may be possible to harden stock plants against infection by regimes of calcium nutrition, or by pretreatment with appropriate bacterial and fungal sprays (Debergh and Maene, 1984). Appropriate pruning will yield fresh, juvenile growth of improved sanitary and physiological characteristics.

#### *8.4.4.2 Initiation*

Apical and nodal cuttings have commonly been used for culture establishment. Hypochlorite solutions are popular for surface disinfestation but appear only partially effective against exogenous bacteria (Mathias *et al.*, 1987). Alternatives include the use of combinations of antibiotics, novel bacteriocidal agents such as Alcide (Mathias *et al.*, 1987), Incyte (Podwyszynska and Hempel, 1987) and solutions of mercury(II) chloride. Of more relevance has been the realization that increasing explant size and physiological age is correlated with the incidence of endogenous contaminants. For example, in geranium, Reuther (1988) demonstrated a decreasing acropetal gradient of bacteria. Meristems are generally free of phloem- and xylem-restricted pathogens due to the absence of vascular connections. Meristem culture is considered the only appropriate method of initiating truly axenic cultures, and is an integral part of procedures recommended to produce certified seed potato stocks (Johansen *et al.*, 1984). A more readily applicable method to obtain satisfactory sanitary levels is by sterile dissection of shoot tips 1 mm in size or less. The technique does not require surface sterilization and can be applied to a wide diversity of genera including open rosette forms of growth. Alternatively, Zimmerman (1985) has suggested rapid etoliation *in vitro* to obtain bacteria-free shoots for excision and further subculture.

#### 8.4.4.3 *Subculture protocol*

Bacteria may be introduced during subculture from contaminated utility surfaces or from indirect contact with the operator. Flaming instruments with alcohol is a long established method but will not totally eradicate pathogens unless the instruments are made red hot. Glass bead sterilizers heat instruments to 230–240 °C and are rapid and effective if the instruments are paired and used alternately. For disinfestation of surfaces, sterilizing products such as 'Milton' are superior to accepted treatments like 70% alcohol. With respect to the operator, protective clothing should include surgical gloves, and the sterile working area should be screened from facial contact.

#### 8.4.4.4 *Detection of bacteria*

During early postinitiation stages cultures are grown singly in containers and can be inspected and sorted for contamination. Visual symptoms may include inhibition of growth, areas of tissue necrosis, death of cultures and characteristic growth of bacteria on the medium surface or as a 'halo' around the tissues in the medium.

At this stage it is also appropriate to remove samples for culture indexing procedures. The most commonly utilized are diagnostic media that promote bacterial growth, for example, potato dextrose agar, peptone yeast agar, yeast glucose calcium carbonate agar or nutrient glucose agar (Cassells, 1986; Cassells *et al.*, 1988). In a study of *Xanthomonas* infection of regal pelargoniums, Cassells *et al.* (1988) compared the relative sensitivities of bacteriological media, a specific ELISA preparation and a nonspecific cDNA probe. Positive identification was noted in 38%, 45% and 92% of samples respectively. By implication, general diagnostic media are fairly insensitive and may serve only to encourage acceptable sanitary levels. Debergh and Vanderschaeghe (1988) have obtained improved results using a single highly enriched medium (Saglio's) and several successive indexing stages as a routine health scheme.

However, in general it is likely that unless highly specific initiation and diagnostic procedures are utilized, micropropagation stocks will invariably possess some endogenous contamination and are in no way bacteria-free.

#### 8.4.4.5 *Antibiotics*

The potential application of antibiotics to control or eliminate bacterial pathogens and endophytes in plant cell tissue cultures has been reviewed (Bastiaens, 1983; Pollock *et al.*, 1983; Falkiner, 1988). To be effective, antibiotics must be soluble, unaffected by pH or media, possess minimal side effects, have a broad spectrum of activity, be bacteriocidal, offer little chance of resistance and be inexpensive. Broad range antibiotics without excessive phytotoxicity include the $\beta$-lactams, rifampicin, trimethoprim, erythromycin and tetracycline. Rifampicin may have the strongest bacteriostatic effect, although at high concentrations significant and residual phytotoxicity was noted on nectarine explants (Scortichini and Chiariotti, 1988). Combinations of antibiotics enhance effectiveness, *e.g.* Young *et al.* (1984) eliminated bacteria in shoot cultures of apple and rhododendron after treatment with cefatotaxine, tetracycline, rifampicin and polymixin B.

From a commercial perspective antibiotics are of limited use, and at best may have only a bacteriostatic effect (Debergh, 1987b). This may be due to: (i) phytotoxicity of concentrations of antibiotics required to eradicate bacteria populations in densely planted stocks of cultures; (ii) induced mutagenesis and selection of resistant pathogen strains with repeated application of antibiotics; and (iii) insufficient uptake of antibiotics to eradicate endogenous bacteria.

Consequently the use of antibiotics should be avoided and emphasis placed on rigorous initiation procedures and repeated culture indexing.

## 8.5 MICROPROPAGATION IN COMMERCIAL PERSPECTIVE

The production of laboratory microcuttings and their passage through the acclimatizing nursery environment for growing on in retail or wholesale outlets demonstrates a continuum with conventional techniques, and the requirement for integration into and expansion of horticultural market structures.

Worldwide there are variously estimated to be over 300 commercial laboratories, with the vast majority ancillary to parent nursery or biotechnology enterprises. Principal commodities have centred on tropical foliage or flowering ornamentals, and the production of early generations of disease-free plants. Very often the product is in the form of 'liners' requiring growing on by specialist nurseries.

Micropropagated plantlets have inherent advantages of increased health and vigour, and the potential for scheduled production of large numbers. However, the high labour and fixed cost elements increase unit costs and greatly limit market penetration. Technological advancement in the field of automation may realize the potential applications of plant tissue culture across a broad spectrum of market categories.

### 8.5.1 Advantages of Micropropagation

Plant propagation by tissue culture is considered by a corporate or entrepreneurial enterprise where: (i) existing developed markets are not filled or can be filled more effectively; and (ii) plants of improved quality or phenotype result from the micropropagation process to create new market opportunities.

#### *8.5.1.1 Rapidity of multiplication*

Orchids remain the classical example of a formerly commercially impracticable crop now widely available due to *in vitro* techniques of clonal and seed propagation. Early candidates were also derived from the fern and herbaceous ornamental market categories, notably *Nephrolepis* and *Saintpaulia.* Micropropagation has been most markedly predominant in the foliage plant industry such that the majority of Boston ferns and popular aroids are now tissue culture derived. Similarly, *in vitro* techniques have been more economical in several bulbous genera, particularly members of the *Liliaceae* and *Iridaceae* (see review by Read and Hosier, 1986).

The use of tissue culture in this context has prompted many nurserymen to perceive it solely as a rapid propagation system in direct competition with conventional nursery practice.

#### *8.5.1.2 Selection and introduction of new varieties*

Micropropagation companies are often utilized by growers wishing to enhance the competitive advantage of a new variety by rapid bulking to marketable quantities. The creation of foundation stocks in this manner allows production at premium prices, *e.g.* of strawberry, where otherwise micropropagation would be noncompetitive. Similarly, cut flower and pot plant markets may be encroached upon where conventional seed propagation of new heterozygous species results in a variable crop, *e.g.* gerbera, or reversion to the original phenotype, *e.g.* double flowered primulas.

Integration into existing plant breeding programmes allows rapid clonal reproduction of selected controlled crosses for trialling. This appears most advantageous where micropropagation competes with hand propagation or pollination techniques for F1 hybrid seed propagation in male sterile, self-incompatible, gynoecious or poor yielding varieties (Ng, 1986). The accelerated cycling of generation times may allow the breeder to introduce varieties in 3–4 years, thereby substantially reducing an investment programme spanning 7–15 years (Holdgate, 1977).

#### *8.5.1.3 Disease indexing and eradication*

Bacterial and viral pathogens endemic to plant populations can be reduced or eliminated by *in vitro* methods (Walkey, 1978). Bacterial indexing on diagnostic media is currently a routine and essential function of many micropropagation companies to prevent spread and spoilage *in vitro* through systemic contamination. Although it is expected that total absence of microorganisms may be unobtainable, it is generally accepted that the superior vigour of tissue cultured plantlets is due in part to the greatly increased hygiene of the growing environment. Consequently micropropagation has become the preferred method of production for several ornamentals with severe systemic diseases, such as *Gypsophila, Syngonium* (Chu, 1986) and hybrid *Philodendron* (Hartmann and Zettler, 1986).

Viruses have become widely disseminated throughout many agricultural and horticultural crops, in particular through the generalized use of vegetative propagation techniques, such as rootstock grafting by fruit growers, and the widespread monoculture of crops. Virus-indexed clones are therefore likely to assume a dominant role in many market categories. At present, this is chiefly manifested by small fruit producers for indexed mother plant production of strawberries and raspberries, and by the widespread micropropagation of indexed early potato stocks in line with nationally adopted certification and limited generation programmes (Harper, 1986; Jeffries, 1986).

The improved health status of aseptic cultures has greatly facilitated international dissemination of germplasm with respect to quarantine regulations (Kahn, 1986). *In vitro* programmes for the propagation and distribution of indexed *Citrus*, cassava, yam and sweet potato have been successfully implemented (Litz *et al.*, 1986).

#### *8.5.1.4 Improvement of plant morphology and physiology*

A consequence of prolonged and intensive culture is the induction of juvenile characteristics, *i.e* increased rootability, vigour and appearance of a juvenile leaf form (Hussey, 1983). This may be carried over to the *in vivo* state and in certain genera provide desirable characteristics of sufficient durability to create superior parent stocks. Increased fruit and runner production from aseptic culture has been recognized and widely exploited in strawberry (Swartz *et al.*, 1981), and greater vigour and precocity of fruiting noted in orchard trees (Jones, 1983; Webster *et al.*, 1985). In addition, a multibranched habit from micropropagated sources has created superior stocks of several herbaceous ornamentals. The retention of juvenile characteristics has led Marks (1986) to propose an *in vitro* induction phase prior to the establishment of conventionally hard-to-root fruit tree rootstocks. The resulting stockplants might be further utilized as a source of field cuttings with enhanced rootability.

### 8.5.2 Economics of Micropropagation

Micropropagated plantlets are often more expensive than the equivalent plantlets produced by conventional means. The individual cost components of such a system and related in-house business operations can be summarized as direct labour, materials, laboratory and nursery overheads, research and development, sales and administration.

From the above, three broad areas of cost can be defined: (i) fixed costs comprising utilities, supervision, research, general administrative expenses, lease, terms, *etc*; (ii) variable costs including labour, media, production supplies and energy consumption; and (iii) capital costs.

Published reports have commonly emphasized the predominance of the labour component (Anderson *et al.*, 1977; Dixon, 1986; Rajeevan and Pandey, 1986; Theiler-Hedtrich and Theiler-Hedtrich, 1983). A typical proportion could be given as 30% labour against 25% overhead (Holdgate and Aynsley, 1977). However, micropropagation companies must endure high fixed costs from start-up to the time the products enter the market and fulfil customer requirements. It has been estimated that fixed costs before profit range between 75% to 110% of direct labour and materials cost (Sluis and Walker, 1985).

#### *8.5.2.1 Fixed and capital costs*

Three areas can be distinguished. (i) Fixed costs pertaining to the administration of the enterprise. Each of the functional areas of administration, research and development, marketing and production requires qualified personnel due to the necessarily complex feedback of information. (ii) Fixed costs relating to maintenance requirements, particularly where this encroaches on research and development, for example where cultures are kept in a 'steady state' prior to production. (iii) Capitalization, depreciation and systems improvements incurred from capital items.

A laboratory with the capacity to produce approximately 1 million plants may realistically create £ 100 000–150 000 fixed costs per annum. Over a 2 year start-up period initial equipment and facility purchases may similarly total £ 100 000–150 000. With the addition of the variable cost element, such a laboratory would require approximately £ 350 000 of capital from day one.

#### 8.5.2.2 *Variable costs (operational)*

Scaled-up production in a micropropagation laboratory incurs variable costs that are influenced by the following factors: (i) multiplication factor of the plant; (ii) labour; (iii) space requirements in the growth room [time related and therefore dependent on (i)]; (iv) response to scaled-up production; (v) ease of rooting and establishment *in vitro/extra vitrum*; (vi) nursery costs; (vii) energy consumption; (viii) material costs; (ix) contamination/spoilage; (x) quality of plant product.

Simplistic calculations based on labour and multiplication rates have often resulted in vastly inflated expectations and excessively low costs (Zimberoff, 1984). For example, an idealized scenario may involve the production of 250 000 plantlets in the first year operating on a × 2 multiplication every four weeks. The number of transfer operations required will be 500 000. Cited operator rates vary from 60–300 transfers $h^{-1}$ (Anderson *et al.*, 1977; Sluis and Walker, 1985) and will be partially dependent on plant type and culture system. A transfer rate of 150 $h^{-1}$ can be assumed. At a minimum labour charge of £3 $h^{-1}$ the cost per operation will therefore be 2 p with a likely unit cost ex-laboratory culture of 4 p. Again, where a × 1.5 multiplication rate is assumed, three operations per culture will be required in addition to extra materials and shelf space, thereby increasing the cost per plant to 6 p.

The use of *in vitro* rooting, and nursery and weaning costs, involve an additional increment of 2 p and between 3 and 5 p respectively, depending on plant characteristics and operator efficiency.

In summary, this brings the overall variable costs of an idealized micropropagation production system, ex-nursery, at between 11 and 25 p per plantlet. Significantly, this is not dissimilar to general estimates for traditional systems of vegetative propagation. The fact that few, if any, micropropagation lines are available within this bracket indicates a series of mitigating factors affecting cost-effectiveness, *viz*: (i) variable multiplication rates throughout the year; (ii) culture stress, often manifested as vitrification (see Section 13.4.3); (iii) operator error, resulting in introduced contamination or poor quality harvesting; (iv) fungal/systemic contamination; (v) incorrectly applied production planning, resulting in irregular cycling and spoilage; and (vi) nursery losses, often linked to the above.

The consequences of any of the above will conspire to escalate costs, possibly doubling above the idealized norm (Sluis and Walker, 1985). In practice, a minimum realized multiplication rate of 2.5 is required (Constantine, 1986), and the plants selected should be fairly readily applicable to the micropropagation process, whilst problematic in conventional propagation systems.

#### 8.5.2.3 *Breakeven situations*

Where plantlets are manufactured in direct competition with conventional micropropagation, *i.e.* offering a crop of similar intrinsic value, the breakeven point will be between 500 000 and 800 000 plants (Figure 1). Fixed costs may vary from company to company and variable costs are based on a

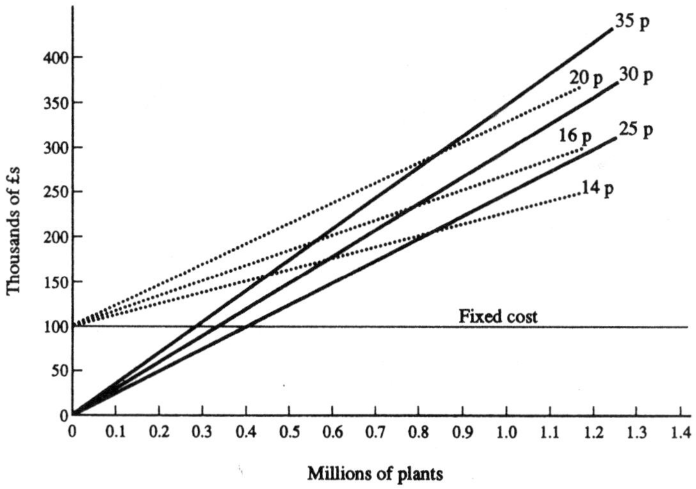


**Figure 1** Pricing of micropropagated plants: breakeven situations. Variable costs per plant are represented by dotted lines (- - -), selling prices by continuous lines (——)

realistic productivity rate, as discussed in the preceding section. The simple breakeven chart does not take into account hidden development costs for future productions, but gives some idea of options available for pricing decisions. Plants which are difficult to produce (variable costs $\geq$ 16 p) are often loss-making at conventional selling prices. In general, a unit that can produce and sell all of the one million plants that it produces in a season will function close to breakeven, and make a small profit. The fact that few companies actually reach the breakeven situation indicates the difficulties facing plant production and in selecting the correct product mix.

### 8.5.3 Market Development

The increase in leisure time and the consumer market as a whole has been paralleled by the tremendous expansion in the ornamental sector of the horticultural industry. The deployment of micropropagation has been correspondingly rapid and concentrates its production in the categories given in Table 2. Market penetration is determined by plants capable of carrying a $\leq$25 p price bracket. In many respects this is unsatisfactory as the industry needs to be market led rather than product led. The rapid adoption of readily micropropagated ferns and foliage plants in the predominantly ornamental sector has led to market saturation. Hartmann and Zettler (1986) point out that although production levels meet less than 10% of a 100 million plant market in the US, this still exceeded the developed market resulting in gluts and underpricing. Conversely, marketing input may seek to diversify the product base without due consideration of the technical limitations of efficient production.

Market development is therefore a contentious issue, influenced by manufacturing capability, geographical location, plant type and trends in consumer demand. The amenity plant trade continues to be the primary growth sector in both house and garden planting material. Successful *in vitro* scale-up of landscape ornamentals, particularly rhododendrons, azaleas and roses (Briggs and McCulloch, 1983) has created current growth in both the European and US markets. For florist crops there appears to be considerable potential for diversification in groups such as *Compositae*, *Liliaceae*, *Primulaceae* and *Ranunculaceae* (see review by Stimart, 1986). Competitiveness of fruit production may rely increasingly on self-rooted scions if field trials are successful (Jones, 1986), and on developmental work with stone fruits and nut crops. *In vitro* propagation of plantation crops remains commercially restricted to herbaceous genera, *i.e.* banana, plantain, pineapple and papaya. Although considerable regenerative potential has been identified in other genera (Litz *et al.*, 1985), cost effectiveness may be limited. The general use of seed propagation for vegetable crops again restricts *in vitro* applications, with the exception of potatoes, selected clones of asparagus and F1 hybrid seed production in crops such as cucumber (Walkey, 1986).

In each of the above categories disease indexing will allow increased premiums and enhance diversification. The ornamental sector demands a product of near perfection to be successfully marketed. Grafted rootstocks, root and tuber crops are highly susceptible to disease pressure; production of *Citrus*, sweet potato, cassava and various edible aroids is currently limited by bacterial and viral pathogens (Litz *et al.*, 1985). Consequently micropropagation companies are developing in-house pathology units to certify the health status of their products. The introduction of new varieties is also encouraging organizations to adopt trademarks and closer affiliations with plant breeding companies or specialist nurseries.

Particular opportunities exist for the export of novel and/or high health crops to developing countries, in conjunction with the introduction of tissue culture technology. The low cost of domestic labour considerably reduces production costs and may create independence from foreign imports. Eventually the predominance in micropropagation may shift to the third world, thereby mirroring trends in many manufacturing industries.

**Table 2** Percentage of World Tissue Culture Laboratories Producing Specific Crops[a]

| *Crop* | *Percentage of laboratories (%)* | *Crop* | *Percentage of laboratories (%)* |
|---|---|---|---|
| Orchids | 58 | Fruit crops | 15 |
| Foliage and flower | 33 | Vegetable crops | 10 |
| Ferns | 19 | Tropical plantation crops | 6 |
| Woody crops | 16 | | |

[a]Taken from Chu (1986).

To remain competitive, micropropagation companies must instigate fairly radical changes in both the biological system and its mode of handling. New technologies will allow penetration of markets with the highest unit volume, such as forestry and agriculture, which at present utilize seed-derived transplants at the 1 p level or less.

## 8.6 SYSTEMS ADVANCEMENT

In seeking to increase the competitiveness of an operation the technologist must analyze separate components affecting cost, productivity and quality of the end product. Short term considerations may include wastage reduction, increased yield, cycling rates and homogeneity in addition to a diversity of other factors involved in systems improvement. The longer term perspective must consider new manufacturing technologies.

The processes involved in micropropagation have not altered significantly in over 25 years. Separation and transfer of microcuttings take place at aseptic flow cabinet positions with subsequent growth in sealed containers. The use of fog or mist in nursery acclimatization is comparatively recent but planting is still a manual operation. Direct labour is one of the strongest cost elements, and consequently mechanization and automation are generally regarded as essential prerequisites to a significant lowering of unit costs. However, the magnitude of labour is determined in part by the nature of the propagules produced and the requirement for regular cycling. In association there must be a complex inventory control, production scheduling and supervisory input. Future advancement may therefore hinge as much on developing the biological basis of micropropagation as on applying novel solutions in the engineering field.

### 8.6.1 Systems Improvement

The stages in the micropropagation process may involve different objectives in terms of initiation, multiplication and establishment, but have similar priorities of reproducibility and yield. Ongoing research invariably aims to optimize media constituents and environmental factors influencing growth in culture and in the field. Similarly, a reduction is sought in wastage that derives from biological and human elements including systemic and introduced contaminants, poor quality harvesting and suboptimal work rates. Quality control by thorough and frequent inspection is essential.

Container type can significantly affect economies. A narrow necked tube or conical flask limits accessibility and thereby reduces work rates (Sluis and Walker, 1985). Enlarged vessels increase productivity by enabling larger volumes to be transferred at one time between opening and sealing the container. However, greater losses per container will result through contamination, and the dimensions of the vessel may affect *in vitro* development (George and Sherrington, 1984). The use of presterilized disposable vessels has been adopted by some companies to reduce labour and space requirements during medium preparation and dishwashing. Some plastics may be detrimental to the growth of certain genera.

The requirement for agar, a relatively expensive medium component, has been investigated and the merits of liquid-based media for shoot growth compared. In general, growth rates have increased on liquid media (George and Sherrington, 1984), probably due to the facilitation of nutrient uptake and prevention of localized accumulation of toxic metabolites. By contrast, gelled media do not require additional aeration and allow more organized growth and distribution of plant tissues within the container. The suitability of either system with respect to survival, multiplication and establishment will be influenced by genotype (Gabryzewska *et al.*, 1986). Agitated liquid medium has been proposed as the basis of mass propagation schemes for a diversity of plants including *Begonia* (Takayama and Misawa, 1982a) and *Lilium* species (Takayama and Misawa, 1982b). Rapid doubling times in the order of 20 d have been cited. However, such systems may be prone to *in vitro* generated abnormalities, such as vitrification, after serial subculturing.

With respect to nursery acclimation, development continues on varying treatments applied at Stages 3 and 4 with the dual intentions of hardening plantlets *in vitro* and optimizing nursery transplanting conditions in relation to the culture environment. *In vitro* conditions inhibit control of water loss and photoautotrophism, with little deposition of epicuticular wax or stomatal function. These physiological regimes are related (Dhawan and Bhojwani, 1987), and consequently treatments have included increased agar concentrations to create a drier culture environment, and increased light intensities (Conner and Thomas, 1981). Restoration of surface wax and stomatal operation has been achieved in carnation cultures by reducing the relative humidity to 80% (Short *et al.*, 1987), and photoautotrophism has been promoted in roses by lowering the concentration of carbohydrate (Langford and Wainwright, 1986). Commercially, methods involving additional materials and labour may negate the benefits of enhanced survival. Hartney (1986) proposed a simplified method of *Eucalyptus* micropropagation using a standardized potting mixture *in vitro* and gradual acclimation under permeable plastic film. Another cost-effective approach may be the application of antitranspirants immediately after transplanting (McComb and Newton, 1981).

Conditions of humidity during weaning have been studied, and fogging systems or polythene tents appear to provide less waterlogging than intermittent misting (Conner and Thomas, 1981). Photoautotrophism in weaned plants may be significantly enhanced by $CO_2$ enrichment and supplementary lighting (Desjardins *et al.*, 1987). An important development has been the evolution of plug technology for seed-derived transplants from the vegetable, bedding plant and forestry sectors. High density planting and ease of handling has promoted nursery and field mechanization (McCown, 1986). This suggests that suitable delivery systems may exist for prospective automated micropropagation laboratories.

### 8.6.2 Reduction of Cycling Requirements

Turnover of cultures is determined jointly by the requirement for plant productivity, and by the onset of deterioration that may occur very rapidly once accumulation of toxic metabolites has passed a critical level within the culture vessel. Particular attention has been paid to Stage 3, where *in vitro* rooting attracts the costs to each plant and may incur over 35% of the total unit price (Debergh and Maene, 1981). In addition, *in vitro* generated roots may be damaged during transfer, or be unsuitable for soil growth (Debergh and Maene, 1981). Successful establishment using unrooted cuttings direct from multiplying stocks has been carried out with ericaceous genera (Anderson, 1978; Lloyd and McCown, 1980), *Cordyline*, *Philodendron*, *Cryptanthus* (Debergh and Maene, 1981), *Dahlia*, *Alnus* and *Vaccinium* (Read and Fellman, 1985). However, the compacted and multi-branched habit of many Stage 2 cultures necessitates a pretreatment to obtain suitable propagules for the next stage. To meet this Maene and Debergh (1985) have developed a method of applying liquid additives directly to exhausted multiplication media to induce the phases of shoot elongation and root induction.

Minimal growth storage has potential for facilitating stock maintenance and scheduling, although at present it is principally associated with storage of elite or trial clones. Domestic markets usually necessitate seasonally biased activity with peak demand in a fairly narrow period. Consequently a significant size of stock has to be carried during slack periods, incurring additional labour and overhead costs. Prolongation of shelf life of cultures would therefore be highly beneficial. Cool storage is the most commonly applied regime but minimum critical temperatures of tropical genera may exceed 15 °C (Henshaw *et al.*, 1980). Dormancy may also be induced by osmotic or chemically based growth retardants. Potato has been successfully stored through combinations of low temperature (*ca.* 4 °C), mannitol, sucrose and abscisic acid (Henshaw *et al.*, 1980); however, individual application of similar treatments to soft fruit and herbaceous varieties gave variable responses (Gunning and Lagerstedt, 1985). Clearly, there is scope for future research in this area.

### 8.6.3 Alternative Biological Systems

Microcutting production by precocious axillary development is initiated and maintained by applied cytokinins. An increasing strength or concentration of cytokinin might appear to provide increased productivity; however, realizable yields are often close to or below minimum required rates. In practice the response is related to genotype and is optimal within a comparatively narrow concentration range. Supraoptimal cytokinin concentrations result in highly compacted shoot clusters tending toward adventitious caulogenesis. Such propagules display reduced rooting percentages, poor field survival and abnormal physiology (Debergh and Maene, 1984). The appropriate propagation ratio is therefore often one that will not provide the highest yield (Debergh, 1987a).

Adventitiously derived shoot production can be prodigious. Vasquez *et al.* (1977) obtained yields from *Saintpaulia* leaf segments totalling over 900 shoots per leaf. Similar methods have been commercially applied to other members of the Gesneriaceae, and to ornamentals such as *Kalanchoe*, *Lilium*, *Hosta* and *Petunia*. As regeneration can occur from tissue fragments, quality of handling is noncritical and has given rise to methods of rapid tissue maceration. Cooke (1979) ground *in vitro* plantlets of *Platycerium* and *Davallia* in sterile blenders, obtaining successful growth after plating the homogenate. Tideman and Hawker (1982) used normal fragmentation of *Euphorbia* and *Asclepias* with regeneration potential dependent on species, medium and explant source. Although these techniques are amenable to automated bulk handling and potentially provide highly economic rates of multiplication they have not been widely adopted by the industry. The majority of genotypes cannot survive or regenerate satisfactorily following such treatment.

Somatic embryogenesis has been reported in over 100 plant species. Commercial application remains limited to very few, principally of the Umbelliferae (Ammirato, 1983). In the majority, regenerated embryos remain attached to associated tissues and require manual separation. Productivity can be limited in such cases but still provides a feasible method of propagation where conventional methods are unsatisfactory. For example, turnover of oil palm callus may average 40 d, with several months subsequently required for embryo and shoot development (Paranjothy, 1984). However, cost effectiveness is achieved by the premium value of clones with high oil yields, the characteristics of which are lost through variable seed progeny.

The greatest potential yields are obtainable where free embryoids are derived from proembryonic masses in liquid culture. The relative ease of handling liquid volumes containing high densities of embryoids makes this system particularly attractive for scaling up. The self-singulating nature of the individuals produced aids both nursery and field delivery. Somatic embryogenesis induces both root and shoot meristems in a bipolar structure, thereby obviating the equivalent steps of conventional micropropagation.

Lutz *et al.* (1985) have reviewed the feasibility of a suspension-based cloning scheme using carrot. Variation between culture tissues and cultivars, lack of developmental synchrony and phenotypic and genotypic abnormalities were various factors threatening system integrity. The high yield (5000 cells $mL^{-1}$ was reduced by 50% after grading to obtain a uniform stock. Potential for variation and loss of regenerative ability *in vitro* was time related, indicating strict constraints to the duration of the regenerative period (Lutz *et al.*, 1985). The relatively high frequency of variants manifested in liquid culture systems is perhaps the primary question mark over viability as a clonal propagation system. The stability of occasional genera such as *Hemerocallis* (Krikorian, 1982) suggests the possibility of regulation and control. An alternative may be direct and secondary embryogenic schemes that minimize variation and are adaptable to mechanization (Williams and Maheswaran, 1985), although in terms of their utility they may not differ appreciably from adventitious shoot production.

Transplantation presents additional problems. The naked embryo is not equivalent to seed and requires nutrition, and protection from pathogens, dessication and physical damage. Three delivery systems have been proposed, comprising fluid drilling, seed tapes and encapsulation. In general, each method involves surrounding the embryo with a protective, cohesive material containing suitable additives. Currently encapsulation appears to hold the greatest promise and may be adaptable to conventional seeding machines. Such techniques have given rise to the concept of synthetic seeds, and the potential of a transplant unit price considerably below the 1 p level (Redenbaugh *et al.*, 1987).

### 8.6.4 Robotics and Automation

Comparatively few cited reports exist on the application of mechanization to laboratory-based plant propagation. This may reflect a paucity of university-sponsored research, and an emphasis in the private sector toward more profitable manufacturing industries.

The Centre for Research on Intelligent Systems at Deakin University, Australia, has developed a microprocessor-controlled prototype capable of producing 1 plant $s^{-1}$ or 85 000 per day (Hartney and Kabay, 1984). If successful, this can be applied to high volumes of forest trees during restricted planting periods (Hartney and Kabay, 1984). Research from a French group has also been cited (De Bry, 1986) on an intelligent robot capable of a high degree of image processing and a production potential exceeding 1000 operations per hour.

One of the principal obstacles to mechanization of an axillary shoot system is the discrimination required to identify and select suitable explants. Herbaceous and woody plants tend to assume rosette and coppice formations of shoots respectively *in vitro*. The configuration of growth can be complex and exhibit considerable random variation in form. Consequently, standards of quality control of mechanically derived propagules may be difficult to maintain.

The concept of robotics and its application to plant shoot cultures has been reviewed by De Bry (1986). The primary areas of research have included both visual and tactile sensory systems, and computer-associated software for both decision making and interpretation. The successful development of fruit-harvesting robots capable of the myriad functions involved in efficiently selecting mature fruit from an orchard suggests that shoot culture systems might similarly be solvable.

However, it is important to question the basis whereby an advanced engineering approach is required. Current processes of micropropagation are primitive, and constraints are defined not by labour *per se* but by the biological and manufacturing components that create the labour requirement. *In vitro* systems can be developed that faciliate mechanization and automation. Aitken-Christie (1986) has devised a shoot harvesting scheme for *Pinus radiata* that is being automated. In this system, shoot hedges are suspended on agar and fed by a nutrient solution; the combined functions of cropping and liquid replenishment greatly reduce the need for costly transfer and result in enhanced growth. Embryogenic suspensions are highly amenable to batch handling, and laboratory and field mechanisms may already exist for scaled-up production. Bioreactor designs are adaptable and have been used for large scale multiplication of plant cells (Styer, 1985; Fowler, 1987). Fermentation techniques of this kind have also been applied to plantlets and plant organs for mass propagation (Takayama *et al.*, 1986). Most recently Levin *et al.* (1988) have reported the commercial implementation of a fully automated suspension-based system, possibly the first of its kind. The process comprises three stages involving organogenic or embryogenic growth in bioreactors, sorting of propagules in a bioprocessor to culture containers and automated transplanting of developed plantlets to soil. Work rates of 12 500 and 8 000 plantlets per hour have been cited for propagule processing and transplanting respectively, and production savings of up to 60% over conventional tissue culture systems have been claimed.

The prospect of automated laboratories each capable of annually producing several million plants raises doubts concerning whether the level of market development will be equal to the efficiency of its manufacturing base. Although market penetration in novel sectors will be achieved and greatly reduced manufacturing costs stimulate consumer demand, there will be the danger of market saturation in many sectors.

## 8.7 REFERENCES


Aitken-Christie, J. and C. Jones (1987). Toward automation: radiata pine shoot hedges *in vitro*. *Plant Cell Tissue Org. Cult.*, **8**, 185–196.

Alderson, P. G. (1987). Micropropagation of woody plants. In *Micropropagation in Horticulture–Practice and Commercial Problems*, ed. P. G. Alderson and W. M. Dullforce, pp. 37–52. Univ. Nottingham Trent Print Unit.

Ammirato, P. V. (1983). Embryogenesis. In *Handbook of Plant Cell Culture*, ed. D. A. Evans, W. R. Sharp. P. V. Ammirato and Y. Yamada, vol. 1, pp. 82–123. Macmillan, New York.

Ammirato, P. V. (1986). Control and expression of morphogenesis in cultures. In *Plant Tissue Culture and its Agricultural Applications*, ed. L. A. Withers and P. G. Alderson, pp. 23–45. Butterworths, London.

Anderson, W. C. (1978). Rooting of tissue-cultured rhododendrons. *Comb. Proc. Int. Plant Propag. Soc.*, **28**, 135–139.

Anderson, W. C., G. W. Meagher and A. G. Nelson (1977). Cost of propagating broccoli plants through tissue culture. *HortScience*, **12**, 543–544.

Armstrong, C. L. and C. E. Green (1985). Establishment and maintenance of friable, embryogenic maize callus and the involvement of L-proline. *Planta*, **164**, 207–214.

Bastiaens, L. (1983). Endogenous bacteria in plants and their implication in tissue culture — a review. *Meded. Fac. Landbauwwet., Rijksunir. Gent.*, **48**, 1–11.

Bradbury, J. F. (1988). Identification of cultivatable bacteria from plants and plant tissue cultures by use of simple classical methods, *Acta Hortic.*, **225**, 27–37.

Briggs, B. A. and S. M. McCulloch (1983). Progress in micropropagation of woody plants in the United States and Western Canada. *Comb. Proc. Int. Plant Propag. Soc.*, **33**, 239–248.

Cassells, A. C. (1986). Production of healthy plants. In *Micropropagation in Horticulture — Practice and Commercial Problems*, ed. P. G. Alderson and W. M. Dullforce, pp. 53–70. Univ. Nottingham Trent Print Unit.

Cassells, A. C., E. M. Goetz and S. Austin (1983). Phenotypic variation in plants produced from lateral buds, stem explants and single cell derived callus of potato. *Potato Res.*, **26**, 367–372.

Cassells, A. C., M. A. Harmey, B. F. Carney, E. McCarthy and A. McHugh (1988). Problems posed by cultivable bacterial endophytes in the establishment of axenic cultures of *Pelargonium* x *domesticum*: the use of *Xanthomonas pelargonii* — specific ELISA, DNA probes and culture indexing in the screening of antibiotic treated and untreated donor plants. *Acta Hortic.*, **225**, 153–162.

Cassells, A. C. and A. Plunkett (1986). Habit difference in African violets produced from leaf cuttings, and *in vitro* from leaf discs and recycled axenic leaves. In *Plant Tissue Culture and its Agricultural Applications*, ed. L. A. Withers and P. G. Alderson, pp. 105–111. Butterworths, London.

Chu, I. Y. E. (1986). The application of tissue culture to plant improvement and propagation in the ornamental horticulture industry. In *Tissue Culture as a Plant Production System for Horticultural Crops*, ed. R. H. Zimmerman, R. J. Griesbach, F. A. Hammerschlag and R. H. Lawson, pp. 15–34. Nijhoff, The Hague.

Conner, A. J. and M. D. Thomas (1981). Reestablishing plantlets from tissue culture: a review. *Comb. Proc. Int. Plant Propag. Soc.*, **31**, 342–357.

Constantine, D. R. (1986). Micropropagation in the commercial environment. In *Plant Tissue Culture and its Agriculture Application*, ed. L. A. Withers and P. G. Alderson, pp. 175–186. Butterworths, London.

Cooke, R. C. (1979). Homogenization as an aid in tissue culture propagation of *Platycerium* and *Davallia*. *HortScience*, **14**, 21–22.

Daguin, F. and R. Letouze (1986). Ammonium-induced vitrification in cultured tissues. *Physiol. Plant.*, **66**, 94–98.

D'Amato, F. (1977). Cytogenetics of differetiation in tissue and cell cultures. In *Applied and Fundamental Aspects of Plant Cell, Tissue and Organ Culture*, ed. J. Reinert and Y. P. S. Bajaj, pp. 343–357. Springer-Verlag, Berlin.

Debergh, P. C. (1987a). Improving micropropagtion. *Int. Assoc. Plant Tssue Cult. Newsletter*, **51**, 2–10.

Debergh, P. C. (1987b). Recent trends in the application of tissue culture to ornamentals. In *Plant Tissue and Cell Culture*, ed. C. E. Green, D. A. Somers, W. P. Hackett and D. D. Biesboer, pp. 383–393. Liss, New York.

Debergh, P. C. and L. Maene (1981). A scheme for commercial propagation of ornamental plants by tissue culture. *Sci. Hortic.*, **14**, 335–345.

Debergh, P. C. and L. Maene (1984). Pathological and physiological problems related to the *in vitro* culture of plants. *Parasitica*, **40**, 69–75.

Debergh, P. C. and A. M. Vanderschaeghe (1988). Some symptoms indicating the presence of bacterial contaminants in plant tissue cultures. *Acta Hortic.*, **225**, 77–81.

De Bry, L. (1986). Robots in plant tissue culture: an insight. *Int. Assoc. Plant Tissue Cult. Newsletter*, **49**, 2–22.

Desjardins, Y., A. Gosselin and S. Yelle (1987). Acclimatization of *ex-vitro* strawberry plantlets in $CO_2$-enriched environments and supplementary lighting. *J. Am. Soc. Hortic. Sci.*, **112**, 846–851.

Dhawan, V. and S. S. Bhojwani (1987). Hardening *in vitro* and morphophysiological changes in the leaves during acclimatization of micropropagated plants of *Leucaena leucocephala* (Lam.) De Wit. *Plant Soc.*, **53**, 65–72.

Dixon, G. R. (1986). The practicalities and economics of micropropagation for the amenity plant trade. In *Micropropagation in Horticulture — Practice and Commercial Problems*, ed. P. G. Alderson and W. M. Dullforce, pp. 183–196. Univ. Nottingham Trent Print Unit.

Evans, D. A. and J. E. Bravo (1986). Phenotypic and genotypic stability of tissue cultured plants. In *Tissue Culture as a Plant Production System for Horticultural Crops*, ed. R. H. Zimmerman, R. J. Griesbach, F. A. Hammerschlag and R. H. Lawson, pp. 73–94. Nijhoff, The Hague.

Falkiner, F. R. (1988). Strategy for the selection of antibiotics for use against common bacterial pathogens and endophytes of plants. *Acta. Hortic.*, **225**, 53–56.

Fowler, M. W. (1987). Process systems and approaches for large scale plant cell culture. In *Plant Tissue and Cell Culture*, ed. C. E. Green, D. A. Somers, W. P. Hackett and D. D. Biesboer, pp. 459–471. Liss, New York.

Gabryszewska, E., M. Hempel and U. Sozzek (1986). Suitability of different culture methods for micropropagation of some ornamental plants. In *6th International Congress of Plant Tissue and Cell Culture* (*Abst.*), ed. D. A. Somers, B. G. Gengenbach, D. D. Biesboer, W. P. Hackett and C. E. Green, p. 433. Univ. of Minnesota, Minneapolis, USA.

Gaspar, Th., C. Kevers, P. C. Debergh, L. Maene, M. Paques and Ph. Boxus (1987). Vitrification: morphological, physiological and ecological aspects. In *Cell and Tissue Culture in Forestry*, ed. J. M. Bonga and D. J. Durzan, vol. 1, pp. 152–166. Nijhoff, The Hague.

Gebhart, F. (1984). Plants Genetics Inc. report successful vegetable harvest from 'synthetic seeds'. *Genet. Eng. News*, **4**, 31.

George, E.F. and P. D. Sherrington (1984). *Plant Propagation by Tissue Culture—Handbook of Commercial Laboratories.* Exegetics Ltd., Basingstoke, UK.

Goldsworthy, A. (1988). Growth control in plant tissue cultures. In *Advances in Biotechnological Processes*, ed. A. Mizrahi, vol. 9, pp. 35–52. Liss, New York.

Gunning, J. and H. B. Lagerstedt (1985). Long-term storage techniques for *in vitro* plant germplasm. *Comb. Proc. Int. Plant Propag. Soc.*, **35**, 199–205.

Hammerschlag, F. A. (1986). Temperate fruits and nuts. In *Tissue Culture as a Plant Production System for Horticultural Crop*, ed. R. H. Zimmerman, R. J. Griesbach, F. A. Hammerschlag and R. H. Lawson, pp. 221–236. Nijhoff, The Hague.

Harper, P. C. (1986). Introduction to the use of micropropagation in clean stock programmes. In *Healthy Planting Material: Strategies and technologies*, ed. D. Rudd-Jones and F. A. Langton, pp. 179–181. BCPC Monograph No. 33.

Hartmann, R. D. and F. W. Zettler (1986). Tissue culture as a plant production system for foliage plants. In *Tissue Culture as a Plant Production System for Horticultural Crops*, ed. R. H. Zimmerman, R. J. Griesbach, F. A. Hammerschlag and R. H. Lawson, pp. 293–299. Nijhoff, The Hague.

Hartney, V. J. (1986). Commercial aspects of micropropagating eucalypts. In *6th International Congress of Plant Tissue and Cell Culture* (*Abst.*), ed. D. A. Somers, B. G. Gengenbach, D. D. Biesboer, W. P. Hackett and C. E. Green, p. 14. Univ. of Minnesota, Minneapolis, USA.

Hartney, V. J. and E. D. Kabay (1984). From tissue culture to forest trees. *Comb. Proc. Int. Plant Propag. Soc.*, **34**, 93–99.

Henshaw, G. G., J. F. O'Hara and R. J. Westcott (1980). Tissue culture methods for the storage and utilization of potato germplasm. In *Tissue Culture Methods for Plant Pathologists*, ed. D. S. Ingram and J. P. Helgeson, pp. 71–76. Blackwell Scientific, Oxford.

Holdgate, D. P. (1977). Propagation of ornamentals by tissue culture. In *Plant Cell, Tissue and Organ Culture*, ed. J. Reinert and Y. P. S. Bajaj, pp. 18–43. Springer-Verlag, Berlin.

Holdgate, D. P. and J. S. Aynsley (1977). The development and establisment of a commercial tissue culture laboratory. *Acta Hortic.*, **78**, 31–36.

Hussey, G. (1983). *In vitro* propagation of horticultural and agricultural crops. In *Plant Biotechnology*, ed. S. H. Mantell and H. Smith, pp. 111–138. Soc. Exp. Biol. Seminar Series 18, Cambridge University Press.

James, D. J. (1983). Adventitious root formation *in vitro* in apple rootstock (*Malus pumila*). Factors affecting the length of the auxin sensitive phase in M.9. *Physiol. Plant.*, **57**, 149–153.

Jeffries, C. J. (1986). The Scottish seed potato classification scheme and the production of nucleus stocks using micropropagation. In *Healthy Planting Material: Strategies and Technologies*, ed. D. Rudd-Jones and F. A. Langton, pp. 239–247. BCPC Monograph No. 33.

Johansen, D. G., K. W. Kurtson, S. R. Slack and E. D. Jones (1984). Recommendations for standards and guidelines for acceptance of tissue cultured potato seed stocks into certification programmes. *Am. Potato J.*, **61**, 369–370.

Jones, O. P. (1983). *In vitro* propagation of tree crops. In *Plant Biotechnology.*, ed. S. H. Mantell and H. Smith, pp. 139–159. Soc. Exp. Biol. Seminar Series 18, Cambridge University Press.

Jones, O. P. (1986). Micropropagation of strawberry and temperate fruit trees. In *Micropropagation in Horticulture — Practice and Commercial Problems.*, ed. P. G. Alderson and W. M. Dullforce, pp. 71–83. Univ. Nottingham Trent Print Unit.

Kahn, R. P. (1986). Plant quarantine and international shipment of tissue culture plants. In *Tissue Culture as a Plant Production System for Horticultural Crops*, ed. R. H. Zimmerman, R. J. Griesbach, F. A. Hammerschlag and R. H. Lawson, pp. 147–164. Nijhoff, The Hague.

Kaur-Sawhney, R. and A. W. Galston (1984). Oats. In *Handbook of Plant Cell Culture*, ed. W. R. Sharp, D. A. Evans, P. V. Ammirato and Y. Yamada, vol. 2, pp. 92–107. Macmillan, New York.

Kavi Kishor, P. B. and G. M. Ruddy (1986). Retention and revival of regenerating ability by osmotic adjustment in long-term cultures of four varieties of rice. *J. Plant Physiol.*, **126**, 49–54.

Kevers, C., M. Coumans, M. F. Coumans-Gilles and Th. Gaspar (1984). Physiological and biochemical events leading to vitrification of plants cultured *in vitro*. *Physiol. Plant.*, **61**, 69–74.

Krikorian, A. D. (1982). Cloning higher plants from aseptically cultured tissues and cells. *Bot. Rev.*, **57**, 151–218.

Langford, P. J. and H. Wainwright (1986). Photosynthetic ability of *in vitro* grown rose shoots in relation to media components. In *6th International Congress of Plant Tissue and Cell Culture (Abst.)*, ed. D. A. Somers, B. G. Gengenbach, D. D. Biesboer, W. P. Hackett and C. E. Green, p. 433. Univ. of Minnesota, Minneapolis, USA.

Larkin, P. J. and J. M. Scowcroft (1981). Somaclonal variation — a novel source of variability from cell cultures for plant improvement. *Theor. Appl. Genet.*, **60**, 197–214.

Levin, R., V. Gaba, B. Tal, S. Hirsch, D. Denola and I. K. Vasil (1988). Automated plant tissue culture for mass propagation. *Bio/Technology*, **6**, 1035–1040.

Litz, R. E., R. L. Jarret and M. P. Asokan (1986). Tropical and subtropical fruits and vegetables. In *Tissue Culture as a Plant Production System for Horticultural Crops*, ed. R. H. Zimmerman, R. J. Griesbach, F. A. Hammerschlag and R. H. Lawson, pp. 237–251. Nijhoff. The Hague.

Litz, R. E., G. A. Moore and L. Srinivasan (1985). *In vitro* systems for propagation and improvement of tropical fruits and palms. *Hortic. Rev.*, **7**, 157–200.

Lloyd, G. B. and B. H. McCown (1980). Commercially feasible micropropagation of mountain laurel, *Kalmia latifolia*, by use of shoot tip culture. *Comb. Proc. Int. Plant Propag. Sci.*, **30**, 421–437.

Lutz, J. D., J. R. Wong and J. Rowe (1985). Somatic embryogenesis for mass cloning of crop plants. *Basic Life Sci.*, **32**, 105–116.

Maene, L. and P. C. Debergh (1985). Liquid medium additions to established tissue cultures to improve elongation and rooting *in vitro*. *Plant Cell Tissue Org. Cult.*, **5**, 1–11.

Maheswaran, G. and E. G. Williams (1987). Uniformity of plants generated by direct somatic embryogenesis from zygotic embryos of *Trifolium repens*. *Ann. Bot. (London)*, **59**, 93–97.

Mantell, S. H. (1986). Problems of stability in mass propagation. In *Micropropagation in Horticulture — Practice and Commercial Problems*, ed. P. G. Alderson and W. M. Dullforce, pp. 145–159. Univ. Nottingham Trent Print Unit.

Marcotrigiano, M., H. G. Swartz, S. E. Gray, D. Tokarcik and J. Popenoe (1984). The effect of benzylaminopurine on the *in vitro* multiplication rate and subsequent field performance of tissue culture propagated strawberry plants. *Adv. Strawberry Prod.*, **3**, 23–25.

Marks, T. R. (1986). Micropropagation of hardy ornamental nursery stock. In *Micropropagation in Horticulture — Practice and Commercial Problems*, ed. P. G. Alderson and W. M. Dullforce, pp. 71–83. Univ. Nottingham Trent Print Unit.

Mathias, P. J., P. G. Alderson and R. R. B. Leakey (1987). Bacterial contamination in tropical hardwood cultures. *Acta Hortic.*, **212**, 43–48.

Mathias, R. J. and L. A. Boyd (1986). Cefatoxime stimulates callus growth, embryogenesis and regeneration in hexaploid bread wheat (*Triticum aestivum* L. E. M. Thell.). *Plant Sci.*, **46**, 217–223.

McComb, J. A. and S. Newton (1981). Propagation of kangaroo paws using tissue culture. *J. Hortic. Sci.*, **56**, 181–183.

McCown, D. D. (1986). Plug systems for micropropagules. In *Tissue Culture as a Plant Production System for Horticultural Crops*, ed. R. H. Zimmerman, R. J. Griesbach, F. A. Hammerschlag and R. H. Lawson, pp. 53–60. Nijhoff, The Hague.

McPheeters, K. and R. M. Skirvin (1984). Field observations of micropropagated 'Thornless Evergreen' blackberry. *HortScience*, **19**, 554.

Meins, Jr., F. (1987). Hormones and the molecular basis of determination in plants. In *Advances in the Chemical Manipulation of Plant Tissue Cultures.*, ed. M. B. Jackson, S. H. Mantell and J. Blake, pp. 19–29. B. P. G. R. G., Monograph 16.

Meyer, M. M. (1980). *In vitro* propagation of *Hosta sieboldiana*, *HortScience*, **15**, 737–738.

Nabors, M. W., J. W. Heyser, T. A. Dykes and K. J. deMott (1983). Long duration high frequency plant regeneration from cereal tissue cultures. *Planta*, **157**, 385–391.

Ng, T. J. (1986). Use of tissue culture for micropropagation of vegetable crops. In *Tissue Culture as a Plant Production System for Horticultural Crops*, ed. R. H. Zimmerman, R. J. Griesbach, F. A. Hammerschlag and R. H. Lawson, pp. 259–270. Nijhoff, The Hague.

Nomura, K. and A. Komamine (1986). Somatic embryogenesis in cultured carrot cells. *Dev. Growth Differ.*, **28**, 511–517.

Norris, R., R. H. Smith and K. C. Vaughn (1983). Plant chimeras used to establish *de novo* origin of shoots. *Science (Washington D.C.)*, **220**, 17–24.

Norton, M. E., and C. R. Norton (1985). *In vitro* propagation of Ericaceae: a comparison of the activity of the cytokinins $N^6$-benzyladenine and $N^6$-isopentenyladenine in shoot proliferation. *Sci. Hortic.*, **27**, 335–340.

Paranjothy, K. (1984). Oil palm. In *Handbook of Plant Cell Culture*, ed. P. V. Ammirato, D. A. Evans, W. R. Sharp and Y. Yamada, vol. 3, pp. 591–605. Macmillan, New York.

Podwyszynska, M. and M. Hempel (1987). Identification and elimination of slowly growing bacteria from a micropropagated gerbera. *Acta Hortic.*, **212**, 112.

Pollock, K., D. G. Barfield and R. Shields (1983). The toxicity of antibiotics to plant cell cultures. *Plant Cell Rep.*, **2**, 36–39.

Rajasekeran, K., M. B. Hein, G. C. Davis, M. G. Carnes and I. K. Vasil (1987). Endogenous growth regulators in leaves and tissue cultures of *Pennisetum purpureum* Schum. *J. Plant Physiol.*, **130**, 13–25.

Rajasekeran, K., J. Vine and M. G. Mullins (1982). Dormancy in somatic embryos and seeds of *Vitis*: changes in endogenous abscisic acid during embryogeny and germination. *Planta*, **154**, 139–144.

Rajeevan, M. S. and R. M. Pandy (1986). Economics of mass propagation of Papaya through tissue culture. In *Plant Tissue Culture and its Agricultural Application*, ed. L. A. Withers and P. G. Alderson, pp. 211–216. Butterworths, London.

Read, P. E. and C. D. Fellmam (1985). Accelerating acclimation of *in vitro* propagated woody ornamentals. *Acta Hortic.*, **166**, 15–19.

Read, P. E. and M. A. Hosier (1986). Tissue culture propagation of ornamental crops: an overview. In *Tissue Culture as a Plant Production System for Horticultural Crops*, ed. R. H. Zimmerman, R. J. Griesbach, F. A. Hammerschlag and R. H. Lawson, pp. 283–292. Nijhoff, The Hague.

Redenbaugh, K., P. Viss, D. Slade and J. A. Fujii (1987). Scale up: artificial seeds. In *Plant Tissue and Cell Culture*, ed. C. E. Green, D. A. Somers, W. P. Hackett and D. D. Biesboer, pp. 473–494. Liss, New York.

Reuther, G. (1988). Problems of transmission and identification of bacteria in tissue culture propagated geraniums. *Acta Hortic.*, **225**, 139–152.
Rugini, E. (1984). *In vitro* propagation of some olive (*Olea europea sativa* L.) cultivars with different root-ability, and medium development using analytical data from developing shoots and embryos. *Sci. Hortic.*, **24**, 123–134.
Schaeffer, G. W., M. D. Lazar and P. S. Baenziger (1984). Wheat. In *Handbook of Plant Cell Culture*, vol. 2, ed. W. R. Sharp, D. A. Evans, P. V. Ammirato and Y. Yamada, pp. 108–136. Macmillan, New York.
Scortichini, M. and A. Chiariotti (1988). *In vitro* culture of *Prunus persica* var. Laevis Gray (nectarine): detection of bacterial contaminants and possibility of decontamination by means of antibiotics. *Acta Hortic.*, **225** 109–113.
Short, K. C., J. Warburton and A. V. Roberts (1987). *In vitro* hardening of cultured cauliflower and chrysanthemum plantlets to humidity. *Acta Hortic.*, **212**, 329–334.
Sluis, C. J. and K. A. Walker (1985). Commercialisation of plant tissue culture. *Int. Assoc. Plant Tissue Cult. Newsletter*, **47**, 2–12.
Sondahl, M. L. and W. R. Sharp (1977). High frequency induction of somatic embryos in cultured leaf explants of *Coffea arabica* L. *Z. Pflanzenphysiol.*, **81**, 395–408.
Stimart, D. P. (1986). Commercial micropropagation of florist flower crops. In *Tissue Culture as a Plant Production System for Horticultural Crops*, ed. R. H. Zimmerman, R. J. Griesbach, F. A. Hammerschlag and R. H. Lawson, pp. 301–315. Nijhoff, The Hague.
Styer, D. J. (1985). Bioreactor technology for plant propagation. *Basic Life Sci.*, **32**, 117–130.
Swartz, H. J., G. J. Galletta and R. H. Zimmerman (1981). Field performance and phenotypic stability of tissue culture propagated strawberries. *J. Am. Soc. Hortic. Sci.*, **106**, 667–673.
Swartz, H. J. and J. T. Lindstrom (1986). Small fruit and grape tissue culture from 1980 to 1985: commercialization of the technique. In *Tissue Culture as a Plant Production System for Horticultural Crops*, ed. R. H. Zimmerman, R. J. Griesbach, F. A Hammerschlag and R. H. Lawson, pp. 201–220. Martinus Nijhoff, The Netherlands.
Takayama, S., Y. Arima and M. Akita (1986). Mass propagation of plants by fermenter culture techniques. In *6th International Congress of Plant Tissue and Cell Culture (Abst.)*, ed. D. A. Somers, B. G. Gengenbach, D. D. Biesboer. W. P. Hackett and C. E. Green, p. 449. Univ. of Minnesota, Minneapolis, USA.
Takayama, S. and M. Misawa (1982a). Factors affecting differentation and growth *in vitro* and a mass propagation scheme for *Begonia* × *hiemalis*. *Sci. Hortic.*, **16**, 65–75.
Takayama, S. and M. Misawa (1982b). A scheme for mass propagation of *Lilium in vitro*. *Sci. Hortic.*, **18**, 353–362.
Theiler-Hedtrich, R. and C. M. Theiler-Hedtrich (1983). The use of tissue culture for the production of ornamental plants: some economical aspects. *Acta Hortic.*, **131**, 179–192.
Tideman, J. and J. S. Hawker (1982). *In vitro* propagation of latex-producing plants. *Ann. Bot.*, **49**, 273–279.
Tisserat, B. (1985). Embryogenesis organogenesis and plant regeneration. In *Plant Tissue Culture: A Practical Approach*, pp. 79–105. IRL Press, Oxford.
van Aartrijk, J. and G. J. Blom-Barnhoorn (1983). Adventitious bud formation from bulb scale explants of *Lilium speciosum* Thunb. *in vitro*. Effects of wounding, TIBA and temperature. *Z. Pflanzenphysiol.*, **110**, 355–363.
van Aartrijk, J. and P. C. G. van der Linde (1986). *In vitro* propagation of flower bulb crops. In *Tissue Culture as a Plant Production System for Horticultural Crops*, ed. R. H. Zimmerman, R. J. Griesbach, F. A. Hammerschlag and R. H. Lawson, pp. 317–331. Nijhoff, The Hague.
Vasquez, A. M., M. R. Davey and K. C. Short (1977). Organogenesis in cultures of *Saintpaulia ionantha*. *Acta Hortic.*, **78**, 249–258.
Vieitez, A. M., A. Ballester, M. C. San-Jose and E. Vieitez (1985). Anatomical and chemical studies of vitrified shoots of chestnut regenerated *in vitro*. *Physiol. Plant.*, **65**, 177–184.
Wainwright, H. and A. W. Flegmann (1985). The micropropagation of gooseberry: *in vitro* proliferation and *in vivo* establishment. *J. Hortic Sci.* **60**, 485–491.
Walkey, D. G. A. (1978). Elimination of bacterial and viral pathogens by tissue culture. In *Frontiers of Plant Tissue Culture*, ed. T. A. Thorpe, pp. 245–254. Univ. of Calgary, Canada.
Walkey, D. G. A. (1986). Micropropagation of vegetables. In *Micropropagation in Horticulture Practice and Commercial Problems*, ed. P. G. Alderson and W. M. Dullforce, pp. 97–112. Univ. Nottingham Trent Print Unit.
Webster, A. D., H. Oehl, J. E. Jackson and O. P. Jones (1985). The orchard establishment, growth and precocity of four micropropagated apple scion cultivars. *J. Hortic. Sci.*, **60**, 169–180.
Williams, E. G. and G. Maheswaran (1985). Embryo propagation by direct somatic embryogenesis and multiple shoot formation. *Basic Life Sci.*, **32**, 366–367.
Young, P. M., A. S. Hutchins and M. L. Canfield (1984). Use of antibiotics to control bacteria in shoot cultures of woody plants. *Plant Sci. Lett.*, **34**, 203–206.
Zimberoff, S. (1984). Cost parameters for choosing crops for micropropagation. *Comb. Proc. Int. Plant Propag. Soc.*, **34**, 165–167.
Zimmerman, R. H. (1985). Application of tissue culture propagation to woody plants. In *Tissue Culture in Forestry and Agriculture*, ed. R. R. Henke, K. W. Hughes, M. J. Constantin and A. Hollaender, pp. 165–177. Plenum Press, New York.

# 9

# Plant Genetic Transformation

ANTHONY G. DAY
*University of Cambridge, UK*
and
CONRAD P. LICHTENSTEIN
*Imperial College of Science, Technology and Medicine, London, UK*

## 9.1 INTRODUCTION

Plant genetic engineering is a very young science. It was only in 1983 that chimeric genes were first expressed in genetically transformed plant tissue (Bevan *et al.*, 1983; Herrera-Estrella *et al.*, 1983). Only after this development, and the concomitant availability of selectable and nonselectable marker genes which express in plant tissue, did work on plant genetic transformation systems begin in earnest. Since that time there has been an explosion in the literature of plant genetic transformation. We propose to review some of that literature and, we hope, give a feeling for the most active and interesting areas of research.

The *Agrobacterium* system was the first used to transform plant tissue; there have been many developments since then and we will try to cover them. We will also discuss direct gene transfer (DGT) of naked DNA to plant cells, a technique that has also been developed to a high degree of sophistication. Finally, we will review some of the techniques and approaches which have not yet begun (*e.g.* organelle transformation) or have only just begun (*e.g.* transposon and gemini virus vectors) to yield the fruits of their early promise.

## 9.2 *AGROBACTERIUM TUMEFACIENS*

### 9.2.1 A Natural Genetic Engineer

It could be said that the birth of modern plant genetic engineering stems from the recognition of the natural engineering ability of the bacterium *Agrobacterium tumefaciens* (Chilton *et al.*, 1977; Chilton *et al.*, 1980; Willmitzer *et al.*, 1980; Yadav *et al.*, 1982; Zambryski *et al.*, 1980). This opportunistic soil phytopathogen causes crown gall tumours in wounded gymnosperms and dicotyledonous angiosperms (dicots). Oncogenic strains contain a single copy of a large (150–250 kb) tumour-inducing (Ti) plasmid (Figure 1). Part of this plasmid DNA (the 'transfer' or T-DNA) is transferred to the wounded plant cell and stably intergrated into the genome.

The genes encoded on the T-DNA, whilst bacterial in origin, contain plant (*i.e.* eukaryotic) regulatory signals enabling expression in infected plant cells. The expression of these genes has the following two consequences: (i) the synthesis of phytohormones; and (ii) the synthesis of opines. The former are necessary for the neoplastic transformation of the infected tissue, which proliferates to produce the characteristic tumorous gall of crown gall disease. The latter are amino acid derivatives which diffuse from the tumour into the surrounding soil where they may serve as a sole carbon source for *Agrobacteria* harbouring a Ti plasmid; they also induce the *tra* operon which allows conjugal transfer of the Ti plasmid to other *Agrobacteria.*

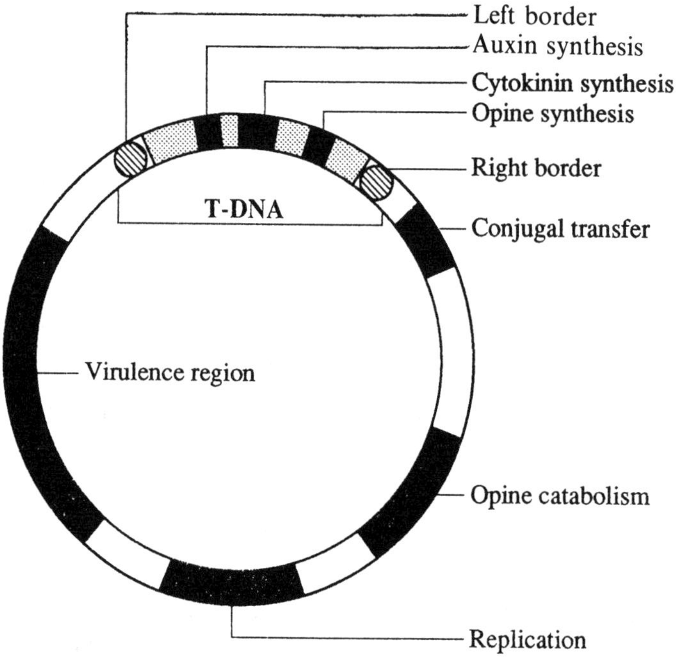


**Figure 1** Octopine-type Ti plasmid

Thus, the Ti plasmid of *A. tumefaciens* has evolved an elegant genetic parasitism, in which infection results in the hijacking of plant metabolic resources to produce food (opines) which are metabolizable only through bacterial products encoded by the Ti plasmid; this plasmid can spread only in the presence of infected tissue. Furthermore, infected tissue proliferates, thereby increasing the amount of opine available.

Ti plasmids can be adapted to make useful vectors for the genetic transformation of plant cells with recombinant DNA; in many cases these transgenic plant cells can be regenerated into whole plants. The construction of Ti-based vectors has been greatly facilitated by the observation that all of the material on the T-DNA, with the exception of two 25 bp border regions, is inessential for T-DNA transfer (Garfinkel *et al.*, 1981; Zambryski *et al.*, 1983), and that it can be deleted (phytohormone synthesis interferes with plant regeneration) and replaced with recombinant DNA.

### 9.2.2 *Agrobacterium* Biology

Before discussing in any detail the development of Ti-based vectors for plant transformation we will first give an overview of the biology of T-DNA transfer and crown gall disease as it is understood at present. Comprehensive reviews on the subject have recently been published (Hooykaas, 1988; Lichtenstein and Fuller, 1987; Melchers and Hooykaas, 1987; Memelink *et al.*, 1987; Stachel and Zambryski, 1986) and extensive literature references will be found therein.

For successful infection of a plant cell by oncogenic *Agrobacteria* the following four steps must take place (Figure 2). (i) The *Agrobacterium* recognizes some signal molecule exuded by the susceptible (wounded) plant cells and, in a chemotactic response, moves up the concentration gradient towards these plant cells. The *Agrobacterium* must then become attached to the cell. (ii) In response to the same, or different, signal molecules the *vir* regulon on the Ti plasmid is induced to express the gene necessary for T-DNA transfer to the plant cell. (iii) The T-DNA is integrated into the plant genome and the T-DNA genes are expressed in the plant cell. The expression of the *onc* genes gives rise to cell proliferation and expression of the opine genes whose products are responsible for the synthesis of opine amino acid derivatives. Ti plasmids are classified according to the opine for which they encode. At least six *Agrobacterium* strains have been distinguished on this basis (Figure 3). (iv) Finally, the opine(s) diffuses from the tumour tissue and into the surrounding *Agrobacteria*- containing soil, where it induces the two Ti plasmid encoded operons, *tra* and *oc*. The former encodes the genes necessary for conjugal transfer of the Ti plasmid to other *Agrobacteria*, while the latter encodes the genes required for opine catabolism.

#### *9.2.2.1 Attachment*

Initially, the *Agrobacterium* has a chemotactic response towards phenolic compounds released by wounded plant tissue and moves up the concentration gradient towards the wounded cell. There is some conflicting evidence in the literature at present regarding the molecular basis for this chemotaxis.

A phenolic compound, acetosyringone (Figure 4), has been shown to induce the *vir* genes at a concentration of $1.5–10 \times 10^{-6}$ M (Stachel *et al.*, 1985). At a lower concentration ($10^{-7}$ M) this same compound has been shown to induce a chemotactic response in *Agrobacterium* (Ashby *et al.*, 1987), and this response has been seen to be dependent on the constitutively expressed, Ti plasmid encoded *vir* genes, *virA* and *virG* (Shaw *et al.*, 1988). However, Parke *et al.* (1987) have obtained contradictory results. They found that acetosyringone produced no chemotaxis in *agrobacteria*; however, other *vir*-inducing phenolics which did produce a chemotactic response did so in strains which harboured no Ti plasmid. Different strains of *Agrobacteria* were used for the two sets of experiments and it is possible that this is the reason for the apparently contradictory results. Perhaps *in vivo*, some, as yet unelucidated, compound is the major chemotactic attractant signalling the presence of wounded plant cells to *Agrobacteria*, and the observed strain dependent responses to the compounds examined are, therefore, fortuitous.

Following the chemotactic response, *Agrobacterium* attaches to the plant cell (the bacterial plant cell recognition process has been reviewed by Halverson and Stacey, 1986). It has been shown (Douglas *et al.*, 1986), by deletion and complementation analysis, that two neighbouring, constitutively expressed, chromosomally encoded genes, *chvA* and *chvB*, are required for attachment. Dylan *et al.* (1986) have shown that *chvA* and *chvB* have functional equivalents in the *Rhizobium ndvA* and *ndvB* loci and can be complemented by them; these *Rhizobium* loci are required for normal nodulation and nitrogen fixation in the root nodules of legumes. It has similarly been shown by Thomashow *et al.*,

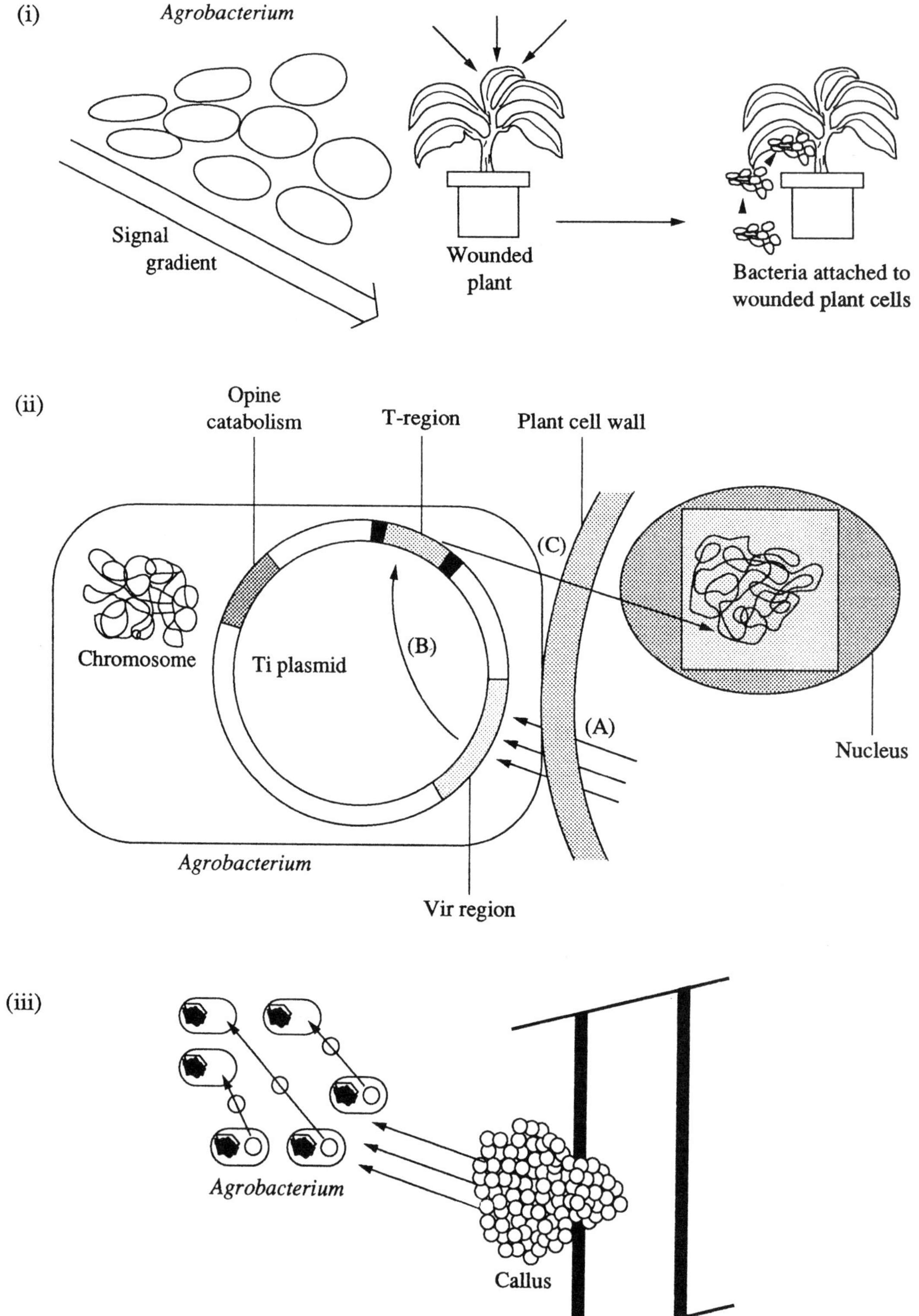


**Figure 2** Mechanism of *Agrobacterium* infection. (i) *Agrobacterium* move up a plant wound exudate concentration gradient towards wounded plant tissue. Once they reach the wounded plant tissue, they attach to the plant cells. (ii) Wound exudate (A) stimulates the vir region of the Ti-plasmid to (B) act upon the T-region and initiate (C) T-strand transfer and integration. (iii) Undifferentiated callus tissue proliferates from the wound site and synthesizes opines. The opines are catabolized by the *Agrobacterium* and induce conjugal transfer of the Ti plasmid from one *Agrobacterium* to another

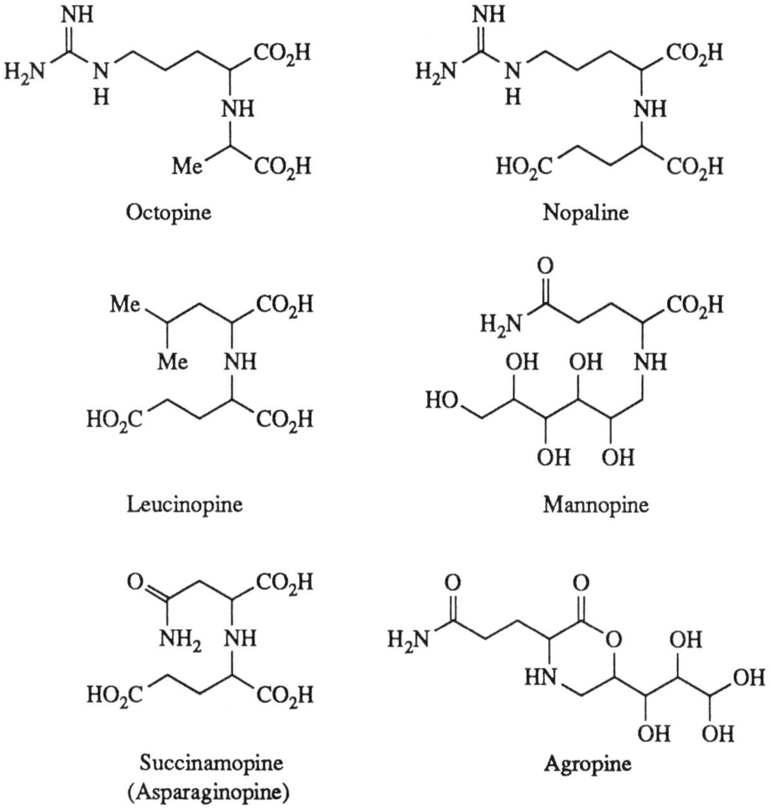


**Figure 3** Opines

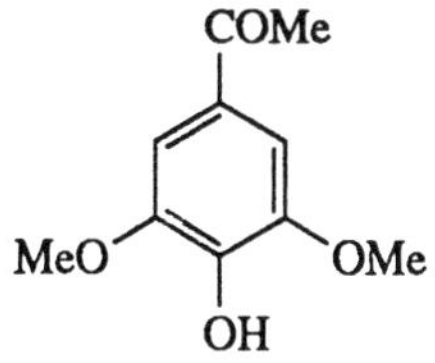

**Figure 4** Acetosyringone

(1987) that mutations in the chromosomally encoded *pscA* locus abolish *Agrobacterium* virulence by virtue of preventing attachment. Mutants in this locus produce none of the four species of exopolysaccharide normally synthesized by the wild-type strain.

Very little is known about the plant components necessary for attachment. It is difficult to distinguish between productive and nonproductive binding in the absence of a marker for productive binding during *in vitro* studies on plant cells and cellular components. It may be that the apparent immunity of many monocotyledonous plants (monocots) to *Agrobacterium* infection is due, in part, to the absence of suitable receptors for binding; however, Douglas *et al.* (1986) have some evidence that *Agrobacterium* can bind specifically to suspension cells of the monocot, bamboo. An alternate explanation is that monocots, once infected and transformed, fail to respond to the synthesized phytohormones.

### *9.2.2.2 T-DNA transfer and integration*

This work has recenly been reviewed in detail (Hooykaas, 1988; Koulolíková-Nicola *et al.*, 1987; Lichtenstein and Fuller, 1987; Melchers and Hooykaas, 1987), and the experimental bases for the conclusions presented here are not discussed in any detail.

In response to plant phenolics the constitutively expressed vir protein, virA, is thought to be a membrane protein which acts as a signal sensor and transmits the extracellular signal to the (also constitutively expressed) intracellular *virG* gene product, thus leading to its expression at an increased level. The virG protein acts as an inducer for the rest of the *vir* genes. Once activated the *vir* genes act on the T-DNA. T-DNA is defined by the presence of two 25 bp imperfect repeats (the left and right borders); transfer of T-DNA is completely independent of the sequence between the borders. The T-region transfer enhancer, the so-called 'overdrive' sequence, is adjacent to the right border and increases the efficiency of transfer (Perealta *et al.*, 1986; van Haaren *et al.*, 1987a, 1987b). The transfer is polar, going from right border to left border. If the right border is deleted, T-DNA is not transferred. If the left border is deleted, transfer takes place but the T-DNA transfer continues past the T-DNA to some pseudoborder, or in the case of smaller plasmids ( $< 50$ kb) perhaps all the way around back to the right border (Rubin, 1987).

Upon induction, the T-DNA is nicked precisely in the lower strand within the border repeats by virD protein(s) (Melchers and Hooykaas, 1987 and refs. therein; Yanofsky *et al.*, 1986). Following this event, single-stranded T-strand is produced, corresponding to the bottom strand of the T-DNA. Production of this strand is also thought to be associated with DNA replication or repair as the T-region of the Ti plasmids have partial single- and/or triple-stranded structures indicative of possible replication intermediates.

What is the mechanism of T-DNA transfer? It is likely to be similar to bacterial conjugation. Buchanan-Wollaston *et al.* (1987) have shown that plasmid *mob* and *oriT* functions together, can replace the requirement for T-DNA borders and allow conjugal transfer of plasmid DNA to plants. During bacterial conjugation *oriT* is the origin of transfer and *mob* supplies the mobilization functions, including nicking at *oriT* for transfer of single-stranded plasmid DNA into a recipient bacterium (reviewed by Willets and Wilkins, 1984). In *Agrobacterium* the following model has been proposed for transfer of T-DNA to plants (Lichtenstein and Fuller, 1987 and refs. therein): (i) a strand specific nick is made in each of the borders by virD protein(s); (ii) rolling circle DNA replication is initiated from the right border nick and (perhaps with the aid of a helicase activity) the T-strand is extruded in a 5′ to 3′ direction. The left border nick signals the end of transfer; (iii) in some manner, as yet unclear, the single-stranded DNA is transferred to the wounded plant cell, transported to the nucleus (do *vir*-encoded proteins allow passage through the nuclear membrane?) and integrated into the plant nuclear genome. There is some evidence for the role of a single-stranded DNA binding protein in T-strand transfer (Gietl *et al.*, 1987). It is probable that the integration of the T-DNA is carried out, at least in part, by plant-encoded functions which are expressed during replication and repair of the wounded plant cell. The sites of integration in the plant DNA appear to be random; and (iv) once integrated into the plant nuclear genome, the T-DNA-encoded functions are expressed as described above (Section 9.2.1).

### 9.2.3 Ti-Plasmid Based Vectors

This area has been reviewed by An (1987), Fraley *et al.* (1986), Klee *et al.* (1987), and Lichtenstein and Fuller (1987).

The development of Ti-plasmid-based vectors was facilitated by the observation that plant tissue infected by *Agrobacterium* is able to grow *in vitro* in the absence of phytohormones. This was used as a dominant selectable marker for transformation. Furthermore, recombinant DNA inserted into the T-region is expressed in plants transformed with that T-DNA if the recombinant DNA is attached to the correct plant regulatory signals. These signals were initially taken from T-DNA genes, notably the octopine synthase (*ocs*) and nopaline synthase (*nos*) genes. After unsuccessful attempts to express foreign genes in plants (with nonplant regulatory signals, *e.g.* Barton *et al.*, 1983; Shaw *et al.*, 1983), the first successes were reported by Bevan *et al.* (1983) and Herrera-Estrella *et al.* (1983). Herrera-Estrella *et al.* (1983) expressed the octopine synthase and chloramphenicol acetyl transferase (CAT) genes regulated by the nopaline synthase promoter in transgenic tumour tissue; they also reported the expression of the neomycin phosphotransferase II gene and dihydrofolate reductase gene, thus providing dominant selectable markers in the absence of oncogenicity. Bevan *et al.* (1983) reported the expression of neomycin phosphotransferase in transgenic tumours at about the same time.

The Ti plasmids themselves are too large (150–250 kb) for direct genetic manipulation and it is fortunate that the *vir* region of Ti plasmids will act *in trans* on T-DNA carried on another plasmid. This allows the use of small plasmids more amenable to the manipulation of recombinant DNA in *E. coli*.

All vectors require a selectable marker for introduction into *Agrobacterium*, a selectable marker functional in plants and borders for T-DNA transfer to plants.

There are two types of Ti-based vectors currently in use, *viz.* cointegrate vectors and binary vectors.

#### 9.2.3.1 Cointegrate or shuttle vectors

The cointegrating vectors have a region of homology between themselves and the accepting Ti plasmid. One of the main systems in use today is the pGV3850 system (Zambryski *et al.*, 1983). In this system the acceptor Ti plasmid pGV3850 has the *onc* genes replaced by pBR322; the rest of the plasmid, including the borders and the *nos* gene (used as a screen for transformation), are intact. Any plasmid containing pBR322 DNA (into which foreign genes for expression in plants can be cloned) can be used as the shuttle vector. This plasmid, having a ColE1 origin of replication, cannot be maintained in *Agrobacterium*. It can be mobilized to *Agrobacterium* by a 'triparental mating' (Figure 5) with a plasmid which supplies the *mob* and *tra* genes *in trans* (*e.g.*, pRK2013; Ditta *et al.*, 1980), but only the Ti plasmid will be maintained in *Agrobacterium*; selection for an antibiotic resistance gene from the shuttle vector will select for a single crossover event in which the shuttle vector has become integrated between the border sequences of pGV3850. A similar system has been developed by Fraley *et al.* (1985), in which all except a small region of T-DNA, 'the limited internal homology' (LIH) and the left border, has been deleted from the accepting Ti-plasmid. The shuttle vector contains the LIH region and the right border; upon integration, a functional T-DNA with intact left and right borders is created.

#### 9.2.3.2 Binary vectors

Binary vectors rely on the observation that the *vir* region is *trans* acting. The *vir* functions are supplied *in trans* from a coresident 'helper' Ti-plasmid, usually deleted for the T-DNA region. They contain a broad host range origin of replication and a bacterial antibiotic resistance gene, for selection and maintenance in *Agrobacterium*; they also have T-DNA borders between which they typically carry a selectable marker which expresses in plants and a polylinker region, containing multiple, unique restriction endonuclease sites for cloning genes of interest. Many binary vectors have been designed (*e.g.* An, 1986; Bevan, 1984; Hoekema *et al.*, 1985; Klee *et al.*, 1985; Matzke and Matzke, 1986; Simoens *et al.*, 1986; Töpfer *et al.*, 1987). As an example, we will describe the binary vector most often used in our laboratory, pGA482 (An, 1986) (Figure 6). This plasmid is 13.2 kb in size and contains the following features: (i) an origin of transfer (*oriT*) and a broad host range origin of

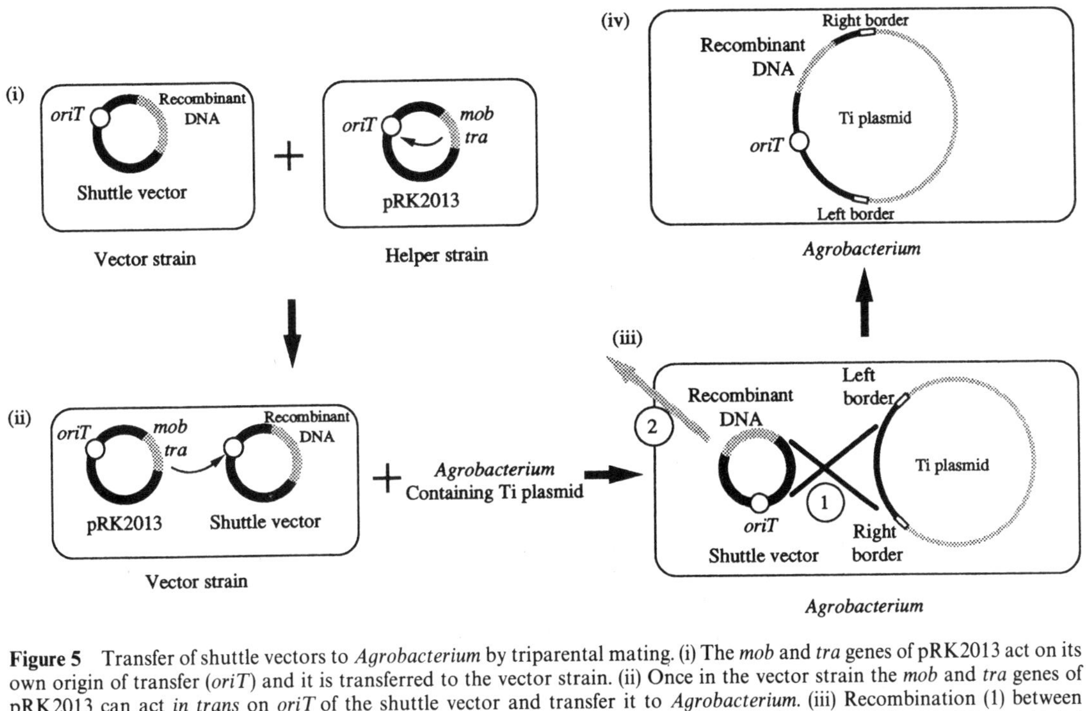


**Figure 5** Transfer of shuttle vectors to *Agrobacterium* by triparental mating. (i) The *mob* and *tra* genes of pRK2013 act on its own origin of transfer (*oriT*) and it is transferred to the vector strain. (ii) Once in the vector strain the *mob* and *tra* genes of pRK2013 can act *in trans* on *oriT* of the shuttle vector and transfer it to *Agrobacterium*. (iii) Recombination (1) between homologous regions on the shuttle and Ti plasmids gives rise to transfer of DNA from the shuttle vector into the Ti plasmid. (iv) The shuttle vector does not have an origin of replication which functions in *Agrobacterium* and is lost (2), leaving a recombinant Ti plasmid behind

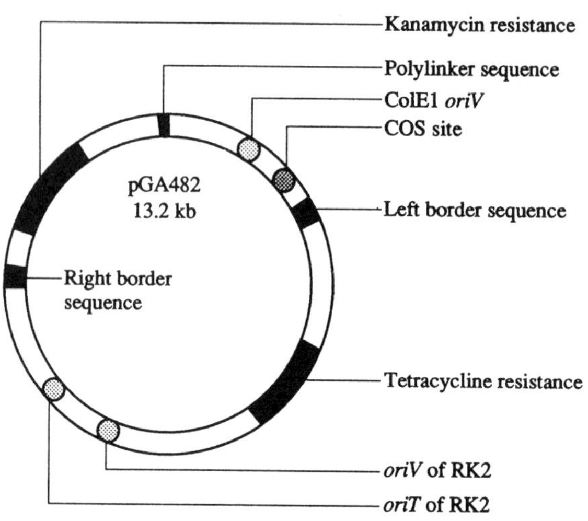


**Figure 6** An example of a binary vector, pGA482

replication (*oriV*) allowing conjugation to, and maintenance in, *Agrobacterium*. The mobilization (*mob*) and transfer (*tra*) functions must be supplied, *in trans*, by a helper plasmid, as described above for the shuttle vectors (Section 7.2.3.1). Alternatively, binary vector plasmid DNA can be introduced by transformation of 'competent' *Agrobacterium tumefaciens* cells; the transformation efficiencies are much lower than those to *E. coli.* (see Lichtenstein and Draper, 1985); (ii) the high copy number origin of replication from plasmid ColE1 (*oriV*) which, whilst being nonfunctional in *Agrobacterium*, facilitates recovery of high yields from *E. coli.*; (iii) a bacterial tetracycline resistance gene, for selection in *E. coli.* and *Agrobacterium*; (iv) a chimeric kanamycin resistance gene which expresses in plants, to allow for selection of transformed plant tissue. This gene has the coding region of neomycin phosphotransferase II (NPTII) from transposon Tn5 flanked by the nopaline (*nos*) promoter and polyadenylation signal; (v) a polylinker sequence for the cloning of foreign DNA; (vi) a bacteriophage $\lambda$-*cos* site. This, together with *oriV*, will allow the integrated DNA and flanking plant DNA to be cloned as a cosmid by $\lambda$-*in vitro* packaging of plant nuclear DNA fragments. As far as we are aware, no reports of this have yet been published; and (vii) the left and right T-DNA borders flanking (iv), (v) and (vi).

#### 9.2.3.3 *Relative merits of the shuttle and binary vector systems*

The shuttle vector systems have the disadvantage that the frequency of introduction into *Agrobacterium* is only about $10^{-5}$–$10^{-7}$, compared with $10^{0}$–$10^{-1}$ for the binary vector systems (Fraley *et al.*, 1986). Shuttle vectors are also dependent on a specific Ti plasmid, or set of Ti plasmids with which they have homology; there is evidence that the host range varies with the *Agrobacterium* strain (Byrne *et al.*, 1987; Owens and Cress, 1985). On the other hand, shuttle vectors are, in contrast to binary vectors, completely stable in *Agrobacterium* in the absence of drug selection and there is evidence that shuttle vectors can transform (tomato at least) at higher frequency than binary vectors (McCormick *et al.*, 1986). Perhaps the latter observation is due to the so called 'proximity effect', in which the *trans*-acting *vir* factors are at a higher local concentration when the T-DNA is on the same plasmid as the *vir* region.

### 9.2.4 Transformation with *Agrobacterium*

This topic has been reviewed by Rogers *et al.* (1986a) and Rogers and Klee (1987).

There are three major, well-established techniques for the transformation of plants with *Agrobacterium*, namely: (i) inoculation of whole or decapitated plants; (ii) cocultivation of protoplasts with *Agrobacterium*; and (iii) 'leaf disc transformation'. The first is useful only in the study of oncogenic strains and will not be discussed further here (the laboratory manual by Draper *et al.* (1988) covers this technique in some detail).

### 9.2.4.1 *Cocultivation*

In this procedure, two- or three-day old protoplasts (plant cells that have had their cell walls enzymatically removed; $10^{-4}$–$10^{5}$ cm$^{-3}$), undergoing their first round of cell division, are inoculated with *Agrobacterium* at $10^{7}$–$10^{8}$ cm$^{-3}$ and cultured for between 36 and 48 h. At this time the cells are washed and treated with antibiotics to kill the *Agrobacterium*, and microcalli are allowed to develop before transferal to media containing the appropriate regime to select for transformation (*e.g.* antibiotic resistance or phytohormone independent growth). Very large numbers of transformants can be obtained in this way; transformation efficiency can approach 50% of viable calli (An *et al.*, 1985). The level of gene expression on foreign DNA can vary according to the site of insertion, the so-called 'position effect'. An advantage of this technique for gene expression studies is that proteins or RNA can be extracted from a number of calli, arising from individual transformation events, to give an average independent of such position effects (Fraley *et al.*, 1986). The main drawbacks of this technique are: (i) the relative technical difficulty of preparing protoplasts; (ii) the long time required to regenerate reasonable amounts of callus or whole plants (3–5 months); and (iii) the susceptibility of the procedure to cell culture induced chromosomal abnormalities (Evans and Sharpe, 1983). Perhaps more important is the difficulty of regenerating whole plants from protoplasts of most cereal crops.

Cocultivation systems in which suspension culture cells (not protoplasts) are inoculated with *Agrobacterium* have been developed and in tobacco give similar transformation efficiencies to systems that use protoplasts (An, 1985). This may be a route for the efficient transformation and regeneration of whole plants from species not amenable to regeneration from protoplasts (*e.g.* embryogenic carrot suspension cells have been efficiently transformed and plants regenerated by somatic embryogenesis; Scott and Draper, 1987).

### 9.2.4.2 *Leaf disc transformation*

This term was originally applied to the use of surface sterilized tobacco, petunia or tomato leaf pieces (Horsch *et al.*, 1985), but the term has since been applied to other axenic explants, the choice of which is dependent on the ability to regenerate well in tissue culture, *e.g.* stem sections of *Brassica napus* (Fry *et al.*, 1987) and developing hypocotyl sections of flax (Basiran *et al.*, 1987). A typical protocol for tobacco leaf disc transformation would be as follows: surface sterilized leaves from greenhouse grown plants or from axenically grown leaves are cut into 1 cm$^{2}$ pieces and transferred to plates containing an appropriate agar-solidified plant medium, these plates may also contain feeder protoplast cells, although in our hands reasonable transformation can be obtained without them. The plates are incubated for two days to condition the medium (presumably with *vir*-inducing signal substances) and then the leaf pieces are dipped in a 50-fold dilution of an overnight *Agrobacterium* culture and returned to the plates. These are incubated for a further 48 h to allow bacteria to bind to the wounded plant cells and for T-DNA transfer to occur. At this time the leaf pieces are transferred to media containing carbanecillin (to kill the *Agrobacterium*), and the appropriate antibiotic to select for transformation and phytohormones to allow either shoot or callus regeneration. Callus or callus plus shoots (depending on the medium used) will start to appear at about one week. After a further two weeks callus can be excised and transferred to separate plates for further culture and shoots can be excised and transferred to a rooting medium. Once the transformed shoots have rooted they may be propagated in tissue culture or, when they reach a reasonable size, transferred to soil in the greenhouse or growth room. Most plants obtained in this way will be morphologically normal and will produce viable seed.

It should be noted at this point that not all of the shoots generated by this method will be transformed; this is perhaps due to cross-feeding from adjacent transformed tissue. The ability to root on media containing the selective drug, however, is almost always indicative of transformation.

The advantages of leaf disc transformation over cocultivation are its relative simplicity, the enhanced ability to regenerate whole plants from species which are not amenable to regeneration from protoplasts and the rapidity with which transgenic plants can be obtained, typically four to six weeks. Horsch and Klee (1986) have developed a method by which leaf discs can be assayed for opine synthesis, and thus transformation, within two days of inoculation!

### 9.2.5 The Structure and Stability of T-DNA Introduced into Plants

#### *9.2.5.1 Mendelian segregation*

It has been shown in a number of studies that once T-DNA is integrated into the plant genome it segregates stably in a Mendelian fashion (reviewed by Fraley *et al.*, 1986).

Budar *et al.* (1986) have carried out a series of genetic crosses with 44 independently transformed plants and analyzed the segregation of the phenotypic marker, kanamycin resistance. It was found that 40 of these plants were amenable to analysis by Mendelian genetics: 35 segregated as having a single kanamycin resistance locus and five as having two independent resistance loci. Interestingly, two of the single locus plants appeared to have the T-DNA inserted such that it caused a recessive lethal mutation; perhaps this will be a way into the use of T-DNA for insertional mutagenesis. Müller *et al.* (1987) have analyzed two transgenic tobacco plants carrying single copies of T-DNA. The marker was segregated according to Mendelian genetics. Extremely low levels of instability were observed: of 1800 seeds analyzed, none appeared to have lost kanamycin resistance, indicating stability through meiosis (seeds were produced by back crossing homozygotes to untransformed plants).

Peerbolte and coworkers (1987a) have studied a large number of transformed F1 seedlings at the DNA level and have also observed fidelity of T-DNA transfer; furthermore, this was true in three states of differentiation (shoot, callus and graft).

Although T-DNA appears to be faithfully inherited in F1 and F2 progeny, levels of gene expression vary. Czernilofsky *et al.* (1986a) examined the level of NPTII activity in eight independent transformants (NPTII gene activity segregated as a single genetic locus in the F1 generation in all cases) and their progeny (which was in each case shown to have T-DNA identical to that of the parent). There was an approximately 20-fold variation in the levels of expression between the eight primary transformants. While there was some correlation between parental and progeny levels of expression, the variation in expression between six F1 progeny (from the same parent) was up to 10-fold. This may be due to methylation (see Section 9.2.5.4). It was also shown that plants that were homozygous for NPTII expressed it at higher levels than heterozygous plants.

#### *9.2.5.2 The fate of simultaneously transferred unlinked T-DNAs*

Another factor that may be important in the design of experiments is the fate of two unlinked T-DNAs transferred to the same plant tissue following inoculation with an *Agrobacterium* strain or mixture of strains.

It has been found that cocultivation using *Agrobacterium* strains containing both an oncogenic helper Ti plasmid (*i.e.* with wild-type T-DNA) and a binary vector sometimes gives rise to higher efficiencies of transformation than equivalent strains containing nononcogenic helper plasmids (lacking T-DNA; An *et al.*, 1985). In these experiments it was also observed that, by phenotype at least, 20–60% of transformants (depending on whether phytohormone independence or kanamycin resistance was the initial selection) had integrated and were expressing both the wild-type and recombinant T-DNAs. In a similar system, de Framond *et al.* (1986) demonstrated that, in the single plant which they analyzed, two unlinked T-DNAs transferred to the same cell segregated in the F1 generation. This suggests the possibility of cotransforming a plant cell with a selectable and nonselectable gene, taking segregants which have lost the selectable gene in the F1 generation and cotransforming again with a second gene and the same selectable marker gene.

It was not possible in either of the above cases to distinguish between transfer of two T-DNAs from one bacterium and transfer of one T-DNA from each of two bacteria. To address this problem Depicker *et al.* (1985) compared the frequencies of cotransformation of protoplasts by T-DNAs carried in separate *Agrobacteria* with T-DNAs on the same Ti plasmid in a single *Agrobacterium*. The results suggest that the frequency of double transformation by two *Agrobacteria* is equal to the product of their independent transformation frequencies; for example, if a single transformation takes place at 40% efficiency, then a double transformation event will take place at 16% efficiency ($0.4 \times 0.4 = 0.16$). The number of double transformants arising from two T-DNAs resident on the Ti plasmid in a single *Agrobacterium* is about 60% of the total transferred, independent of the overall transformation efficiency. This suggests that the rate-limiting step in *Agrobacterium* infection is the formation of a successful plant cell/bacteria interaction and not DNA transfer and integration or plant cell competency (Depicker *et al.*, 1985).

From the point of view of experimental design, however, the important question is, 'Can we get reasonable levels of double transformation events from separate *Agrobacterium*'? If we can, it greatly reduces the difficulty of carrying out experiments which require more than one independent transformation of a single plant cell. Recently it was shown (McKnight *et al.*, 1987) that, when two *Agrobacterium* strains containing different T-DNAs (in the same genetic background) were used in a leaf disc inoculation, about 20% of the transformants contained both T-DNAs and the markers were shown to segregate independently in the F1 generation. When they used the related *Agrobacterium rhizogenes* (see Section 9.2.8) they obtained roots (a phenotype equivalent to crown gall in *A. tumefaciens*) containing one marker (in addition to rootiness itself) in 50% of the transformants and both markers in 11%. This indicates that in 11% of the transformants three(!) independent T-DNAs had been transferred to the plant cell. Segregation in the *A. rhizogenes* case was not examined. Work carried out by Petit *et al.* (1986) gave similar results.

The above work indicates that it is not necessary to put two unlinked T-DNAs in the same *Agrobacterium* in order to get reasonable frequencies of cotransformation.

### *9.2.5.3 Structure, stability and rearrangements in transferred T-DNAs*

A number of studies have been carried out on the structure of inserted T-DNA in plant genomes and the occurrence of truncated or rearranged integrations has been noted. Some of these studies are discussed below.

Van Lijsbettens *et al.* (1986) demonstrated that of nearly 1000 subclones from a *Nicotiana tobacum* callus produced by the wild-type *Agrobacterium* C-58 only six were morphologically different from the parental callus, and none of these differences could be correlated with gross rearrangements of T-DNA structure. The implication of this is that once the T-DNA has been integrated into the plant genome it remains stable through mitosis; this, together with the meiotic stability discussed in Section 9.2.5.1, confirms the potential utility of transgenic crop plants to commercial breeders. Inactivation of T-DNA in transformed tissue at the DNA level has, however, been shown to occur, albeit at very low frequency, in at least one instance. Peerbolte *et al.* (1987d) found a shooty crown gall line which reverted to wild-type after three years in tissue culture due to T-DNA rearrangements and deletions (so-called 'somaclonal variation'). In the light of the study of Van Lijsbettens *et al.* (1986) this is probably not important so far as the applications to commercial cultivars are concerned.

An attempt has been made (Czernilofsky *et al.*, 1986b) to amplify an NPTII gene in transformed tobacco callus by exposure to increasing levels of kanamycin, both in small steps (0–700 $\mu g\,cm^{-3}$ in 100 $\mu g$ steps) and in single jumps to high concentration. No amplification was found. In animal cells this technique works well (*e.g.* Schimke, 1984).

It has been found (Van Lijsbettens *et al.*, 1986) that in calli produced by cocultivation of *Agrobacterium* with protoplasts about 12% had morphological abnormalities associated with truncated T-DNA and it is thought that this is related to the presence of 'pseudoborders' within the T-DNA. In contrast, if calli are produced by the inoculation of whole plants, then the incidence of truncated T-DNA is considerably less (although expression of the T-DNA genes in these tumours is very variable, discussed below in Section 9.2.5.4). If this observation can be extended to recombinant T-DNA, it serves as another reason for choosing leaf disc transformation over cocultivation wherever possible.

The arrangement of T-DNA integrated into the plant chromosome seems to be partly dependent on the *Agrobacterium* strain and vector system (binary or cointegrate) used. Early work (Zambryski *et al.*, 1980) on a tumour line derived from the nopaline strain T37 showed that T-DNA was integrated as an inverted repeat and several direct repeats. More recently, Jorgensen *et al.* (1987) have shown, for the nopaline strain C58, that arrays of inverted repeats are the predominant structure in transformed plants, both head to head and tail to tail repeats being equally well represented. T-DNA junctions were also analyzed and right ends were considerably more homogeneous than left ends. This is consistent with the right to left end polarity of T-strand transfer if initiation of transfer is more precise than termination.

These results are in contrast to those found for octopine-type strains where inverted repeat structures are relatively uncommon (Jorgensen *et al.*, 1987 and refs. therein). An analysis of T-DNA structure in plants transformed by the disarmed octopine strain LBA4404 harbouring a binary vector has been carried out (Spielmann and Simpson, 1986). Between one and six copies of the T-DNA had been transferred to the nine plants analyzed and approximately 30% of the transferred T-DNAs were incomplete or rearranged. Seven of the nine plants had T-DNA in more than one genetic locus. A study of the literature revealed that insertion at multiple loci is quite unusual (Spielmann and

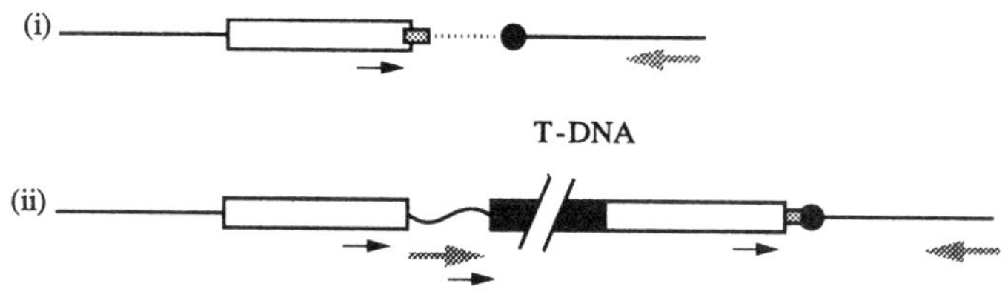


**Figure 7** Rearrangements made by T-DNA insertion (after Gheysen *et al.*, 1987). (i) The unoccupied target site. (ii) Diagram of the left and right border junctions of the T-DNA insertion. T-DNA is shown as an interrupted black box, and the 158 bp target duplication is shown as an empty box. On the right there is the 158 bp duplication (open box), an unchanged 7 bp sequence (black square), the deletion of target sequences (dotted line) and the C to T transition (filled circle). On the left, rearrangements include the 158 bp duplication and the 'filler' DNA (wavy line). Arrows indicate homology of 'filler' DNA to other sequences

Simpson, 1986 and refs. therein); it was suggested that it occurred because the binary vector, containing the T-DNA, is maintained at four to seven copies per *Agrobacterium* cell. In contrast, most of the literature examples cited used wild-type or cointegrate Ti plasmids, *i.e.* single copy plasmids. This hypothesis is entirely consistent with the high degree of cotransformation found when two different T-DNAs are present in the same *Agrobacterium* (Depicker *et al.*, 1985; see Section 9.2.5.2). If true, the difference in the number of integration loci obtained by use of either cointegrate or binary vectors is another factor to be considered in experimental design.

An extremely detailed analysis of T-DNA transferred from the octopine strain LBA1501 has been carried out by Peerbolte *et al.* (1987b, 1987c), and quite a high degree of infidelity of transfer was observed. It is recommended that the reader refer to the original papers as there is not room in this review to go into the details.

Only recently has a thorough investigation of the plant sequences involved in T-DNA integration been carried out with the cloning of a virgin target site (Gheysen *et al.*, 1987). This allowed duplications and rearrangements in both T-DNA and plant DNA to be elucidated. The tissue used was from a C58 nopaline strain derived tumour. Plant/T-DNA junction sequences were cloned and used as probes to screen a genomic library constructed from untransformed tissue. The results may be summarized as follows (Figure 7). At the right junction there is a direct transition from T-DNA to plant DNA; however, there is a deletion of 27 bp at a position 165 bp to the right of the junction followed by a single C to T transition. At the left end the transition is not direct: there is a 33 bp 'filler' sequence between the T-DNA and the plant DNA and this sequence is an imperfect, inverted repeat of two regions from the right end junction. Following this filler sequence is a duplication of 158 bp of target sequence (*i.e.* identical to the first 158 bp of plant DNA from the right junction). It is not possible to propose a mechanism for T-DNA integration from analysis of a single event, although from the degree of rearrangement it can be presumed that it is a multistep event. It was also noted that the target site was particularly AT rich around the point of integration, but whether this is coincidental will not be known until further analyses have been carried out. Clearly, in the light of this work, experiments based on the assumption of precise insertion of T-DNA into the plant genome (*cf.* bacterial transposons) will have to be approached cautiously.

### *9.2.5.4 The role of methylation in transgenic gene expression*

Although gene expression is discussed elsewhere (Chapter 2), it is pertinent at this point to discuss briefly the stability of levels of gene expression in transgenic plants.

It has been observed (An, 1986) that the activity of the nopaline synthase (*nos*) promoter can vary up to 200-fold between independent transgenic tissue and that the activity of closely linked genes under the control of identical *nos* promoters has no apparent correlation with this. This has been attributed by various workers to some sort of little understood 'position effect'. Although chromatin structure and position effects probably play a role in the level of transgenic gene expression, an equally, if not more, important factor is cytosine methylation. Methylation of T-DNA was first reported by Gelvin *et al.* (1983). Subsequently, T-DNA methylation has been correlated with aberrant tumour phenotype (suggesting underexpression of certain T-DNA genes) and with low levels of transcription (Amasino *et al.*, 1984; Hepburn *et al.*, 1983; Peerbolte *et al.*, 1986b; Van Lijsbettens *et al.*, 1986). In both instances, suppression of gene expression was found to be reversible by the action of the demethylating drug 5-azacytidine. We have regenerated plants from the leaf tissue of a previously transformed, CAT positive, kanamycin resistant plant. Despite selection for kanamycin resistance, approximately 50% of such regenerants failed to express the nonselected, but adjacent, CAT gene at a detectable level (Day and Lichtenstein, unpublished). One of the resultant CAT negative, kanamycin resistant plants was similarly taken through tissue culture and about 50% of these regenerants did express CAT. We assume in this case that the plants are all clonal copies of one another and that the variation in CAT expression is due to methylation, although this has yet to be confirmed. It appears that the *nos* promoter is particularly prone to these high levels of variation and the stronger promoters, such as the CaMV 35S and 19S promoters, have a much lesser degree of variation of expression (Negrutiu *et al.*, 1987a); perhaps the *nos* promoter directs its own methylation.

### 9.2.6. Transformation of Monocots

One of the major shortcomings of *Agrobacterium*-mediated plant transformation has been its limited host range. For a long time monocots, which includes the majority of important crop plants (maize, wheat and rice), had been considered refractive to *Agrobacterium*-mediated transformation (De Cleene and De Ley, 1976). However, it was reported (Hooykaas-Van Slogteren *et al.*, 1984) that inoculation of both *Chlorophytum* and *Narcissus* (Liliaceae and Amaralyllidaceae, respectively) with oncogenic *Agrobacterium* produced small tumours with which opines were associated. Similar results have been reported by Graves and Goldman for *Gladiolus* sp. (Iridaceae, 1987) and *Zea mays* (Graminaceae, 1986).

As yet, proof at the nucleic acid level of T-DNA integration into the monocot host genome has not been forthcoming in the above species, although it has been shown for the Liliaceae *Asparagus officinalis* (Bytebier *et al.*, 1987) and for *Dioscorea bulbifera* (yam; Schäfer *et al.*, 1987). In the latter case tumour formation only occurred if the *Agrobacteria* were preinduced by incubation with *Solanum tuberosum* (the common dicot, potato) wound exudate. It is possible that this preinduction is a general requirement for *Agrobacterium*-mediated transformation of the majority of monocots. If this turns out to be the case we can look forward to a deluge of reports on monocot transformation.

Opine-producing tumours have been obtained by inoculation of stem segments of *Asparagus officinalis* (Liliaceae; Hernalsteens *et al.*, 1984), and the T-DNA integration has been shown to occur normally (Bytebier *et al.*, 1987). In the same paper the regeneration of transformed asparagus plants, after transformation with disarmed Ti plasmid vectors, was reported. *Asparagus officinalis* is one of the few monocots amenable to regeneration from single cells.

The practical difficulties of detecting *Agrobacterium*-mediated monocot transformation have recently been reviewed by Hooykaas and Schilperoort (1987).

It is germane at this point to mention the 'agroinfection' of maize by the single-stranded DNA virus, maize streak virus (MSV, discussed more fully in the next section; Grimsley *et al.*, 1987). Normally, MSV can only be transmitted by its insect vector; naked DNA and even whole virus isolate are uninfectious by mechanical inoculation. However, if the virus is cloned between T-DNA borders, and the resulting *Agrobacterium* strain used to infect maize, a systemic viral infection results. From these results it is clear that the *Agrobacterium* must be binding to the plant cells and that T-DNA transfer is taking place; however, the fate of the T- strand after transfer is not clear.

It may be seen from the results discussed above that the problem of transformation of monocots by *Agrobacterium* is gradually being solved: Liliaceae, Amaralyllidaceae, Graminaceae and Dioscorea all show some degree of susceptibility to Ti plasmid transformation and it is our opinion that it will not be too long before these transformations become routine. Perhaps a more intractable problem in the transformation of the important monocot crop plants will be that of whole plant regeneration from single cells or protoplasts. The first reports of transformed, regenerated maize plants from protoplasts have recently been published (Rhodes *et al.*, 1988).

### 9.2.7 Viral Infection of Plants by *Agrobacterium*: 'Agroinfection'

The use of *Agrobacterium* as a vector to facilitate the virus infection of plants has been termed agroninfection (Grimsley *et al.*, 1986) and has been reviewed elsewhere (Grimsley and Bisaro, 1987).

The first experiments were carried out using the double-stranded DNA virus, cauliflower mosaic virus (CaMV; Grimsley *et al.*, 1986; see Section 9.4.1). It had previously been observed that cloned CaMV was not infectious by mechanical inoculation unless the viral DNA was first excised from the bacterial vector or a cloned tandem duplication of CaMV was used. This implied that recombination between the duplications of the viral genome allowed it to escape from the vector and replicate. Ti plasmids were constructed containing 1.4 and 2 tandem genomes of CaMV and these were both found to give rise to systemic viral infection in *Brassica rapa* (turnip). Furthermore, the role of the

*Agrobacterium* was confirmed by the observation that, while inoculation with less than 1 μg of plasmid DNA rarely led to infection, agroinfection with bacteria containing the equivalent of 10 ng of CaMV DNA consistently caused systemic infection.

As mentioned above (Section 9.2.6), agroinfection has also been used to infect the graminaceous monocot maize with the single-stranded DNA gemini virus, maize streak virus (MSV; Grimsley *et al.*, 1987); again, a tandem repeat was used. This virus is very recalcitrant to infection by mechanical inoculation and previously the only reports of successful infection used an insect vector. The same technique has also been applied to the related (and similarly recalcitrant) digitaria streak virus (DSV; Donson *et al.*, 1988), to wheat dwarf virus (WDV; Hayes *et al.*, 1988a) and also to the more distantly related virus, tomato golden mosaic virus (TGMV; Elmer *et al.*, 1988; Hayes *et al.*, 1988b).

TGMV has a bipartite genome of two circular, single-stranded DNA molecules approximately 2.5 kb in length, both of which are required for systemic infection. The only homology between the two molecules is the so-called 'common region' of about 200 bp. It has previously been shown that when dimers of TGMV DNA A are cloned into the chromosome of petunia plants (*via Agrobacterium*-mediated transformation), a small number of unit length DNA A molecules is produced (Rogers *et al.*, 1986b), suggesting that replication takes place. In the same work, trimers of TGMV B were cloned into the chromosome of petunia, but no unit length DNA molecules were found. However, when the A2 and B3 plants were crossed, a systemic infection resulted. This suggests that the A particle encodes the functions necessary for replication and the B particle those necessary for cell to cell spread and symptom development. Hayes *et al.* (1988b) have found that when wild-type tobacco plants are agroinfected with bacteria containing (within the same T-DNA) either dimers of DNA A and DNA B, or a partial dimer (1.6 mer) of A and a dimer of B or a dimer of B and a partial dimer of A with part of the coat protein gene missing, systemic infection is readily obtained. Systemic infection was also observed when plants were agroinfected with a mixture of a strain containing a dimer of A and a strain containing a dimer of B. The latter strain was also found to complement A2 transgenic plants (containing dimers of A in the genome) and the former B3 plants (containing trimers of B in the genome), giving rise to systemic infection in both cases. Similar work has been carried out by Elmer *et al.* (1988). The results were almost the same, with one exception: they found that A2 plants were not complemented by agroinfection with the B component.

The mechanism of agroinfection is not clear from these experiments and further work will help to elucidate whether infections arise by a replicative mechanism or by excision of viral DNA *via* homologous recombination between repeated sequences. Whether integration of the T-DNA into the plant chromosomal DNA occurs and whether or not it is necessary during infection remain open questions. Mechanism notwithstanding, agroinfection provides an efficient and reproducible means of infecting plants with virus. Already, it has shown its utility with regards to viral vectors (Section 9.4.2) and may one day help to answer questions about interactions between *Agrobacteria* and monocots.

### 9.2.8 *Agrobacterium rhizogenes*

*Agrobacterium rhizogenes* is a species very closely related to *A. tumefaciens*. The only major difference between the two is that while *A. tumefaciens* harbours a Ti plasmid (which causes a crown gall on infection), *A. rhizogenes* harbours an Ri plasmid which causes a rooty phenotype on infection (reviewed in Hooykaas, 1988; Melchers and Hooykaas, 1987; Figure 8).

Ri plasmids (in common with octopine-type Ti plasmids) have two T-regions, the $T_L$ and $T_R$ regions, either or both of which may be transferred to plants on infection. Integration of T-DNA from Ri plasmids gives rise to a characteristic 'hairy root' phenotype; adventitious roots grow from the infected wound site whether it is on stem or leaf. Several genes responsible for this neoplastic transformation have been identified, they have no homology with the three *onc* genes present in the T-DNA of Ti plasmids, and their functions are, as yet, unknown (Hooykaas, 1988).

Ri plasmids have been used to transform plants. Whole plants may be regenerated from the adventitious roots arising from integration of Ri plasmid T-DNA under the appropriate conditions. Tepfer (1984) showed that plants with altered morphology were produced and that the altered phenotype and genotype is heritable. In all species studied, the altered phenotype consisted of high root growth rate in culture, reduced apical dominance in roots and stems, wrinkled leaves and increased width to length ratio in leaves; other (species dependent) alterations in phenotype were also observed. This altered morphology could be used as a marker for transformation.

Simpson *et al.* (1986) have shown that a binary vector resident in an Ri strain of *Agrobacterium* successfully cotransfers T-DNA to hairy roots. The frequency to contransformation was dependent on

**Figure 8** A: Crown gall tumour on sugar beet stem. B: Rooty (Ri) tumour on carrot discs

both the plant species and the marker/selection. When nopaline synthase was the marker, cotransfer took place in 33–55% of cases in alfalfa, 33% in tomato and 2% in soybean; when kanamycin resistance was the selection, the frequencies of cotransfer were 4% in alfalfa, 19% in tomato, 18% in tobacco and 0% in soybean. The lower frequency of cotransformation observed with selection on kanamycin may have been due to the use of too high levels of the drug. However, in three of the four plant species examined, the frequency of cotransfer of a nonselected marker was reasonable. Similar results have been obtained by, for example, Hoekema *et al.* (1984) and Shahin *et al.* (1986).

A number of different workers have constructed shuttle vectors (see Section 9.2.3.1) based on Ri plasmids (*e.g.* Comai *et al.*, 1985; Jensen *et al.*, 1986; Morgan *et al.*, 1987; Stougaard *et al.*, 1987; Tepfer and Casse-Delbert, 1987). We will take the work of Stougaard *et al.* (1987) as an example. Replacement mutagenesis was used to insert pBR322 DNA into the $T_L$ region of an Ri plasmid while retaining the wild-type root-inducing phenotype. Cointegrates could then be formed by homologous recombination with pBR322 vectors containing DNA to be cloned into plants. Transformed 'roots' (and regenerated plants) were shown to contain the cloned DNA in addition to the Ri DNA. When a binary Ti plasmid vector was used in conjunction with the cointegrate Ri plasmid victor, the frequency of cotransformation (as assayed by selection on kanamycin) was 60% of the rooty lines tested. Thus, Ri plasmid *vir* functions must recognize Ti plasmid border sequences for T-DNA transfer.

The combination of the Ri system, enabling rapid and easy regeneration of transformed roots into plants (albeit with aberrant morphology), together with the better-developed Ti based system holds great potential for work involving plant genetic transformation.

### 9.2.9 Conclusion

The *Agrobacterium* system for plant transformation has been, and will continue to be, of great utility in the analysis of plant gene function and in the engineering of crop plants, particularly as the problems of its use for monocots are being solved. It has already been shown to be possible to rescue a single gene from a plant by shotgun cloning using leaf disc transformation (DNA from a plant having a single copy of an NPT II gene was shotgun cloned into a Ti plasmid which was then used to transform leaf discs: one disc in 20 gave rise to kanamycin resistant tissue; Klee *et al.*, 1987), and plants have been engineered which are resistant to herbicide (Comai *et al.*, 1985; De Block *et al.*, 1987), virus (*e.g.* Abel *et al.*, 1986; Hemenway *et al.*, 1988) and insect attack (Vaeck *et al.*, 1987).

## 9.3 DIRECT GENE TRANSFER

### 9.3.1 A Physicochemical Method of Genetic Transformation

Direct gene transfer (DGT) involves the application of DNA containing the gene(s) of interest directly to the tissue to be transformed. The uptake and/or integration and/or expression of the foreign genes is mediated by the physical and chemical conditions of application. These techniques have been applied to yeast and mammalian systems for some time (Chu and Sharp, 1981; Hinnen *et al.*, 1978). However, it is only recently that DGT has been applied to plants (for recent reviews, see Lichtenstein and Fuller, 1987; Negrutiu *et al.*, 1987a; Potrykus *et al.*, 1985a, 1987a, 1987b; Saul *et al.*, 1987; for a review of the techniques, see Fromm *et al.*, 1987; Langbridge *et al.*, 1987; Shillito and Potrykus, 1987).

There are a number of advantages to DGT over *Agrobacterium*- mediated gene transfer. One is the general applicability of DGT to plant species: whereas *Agrobacteria* have a limited host range, particularly with respect to monocots, DGT is applicable to any species from which it is possible to make protoplasts. A further advantage is the utility of DGT in transient expression systems where it is possible to check the expression of DNA or RNA transferred to plant protoplasts within two days of initiating the experiment. The main disadvantage of DGT is the necessity (in most cases) of using protoplasts; this is technically more demanding and most species of cereal cannot yet be regenerated from protoplasts. However, the regeneration of genetically transformed, whole plants from maize protoplasts has recently been reported by Rhodes *et al.* (1988). In this study, all of the transgenic plants were sterile; it is not clear at present if this is intrinsic to regeneration, or owing to some other experimental parameter (*e.g.* culture line or age). Another disadvantage of DGT is the complex patterns of integration often observed (Section 9.3.3).

The two techniques most often associated with DGT in plants are electroporation and poly(ethylene glycol) (PEG) mediated DNA transfer. Both of these techniques have been applied to transient and stable transformations of plant cells. In both, the membrane is temporarily made permeable and DNA uptake by the cell is possible. In the latter, the chemical conditions allow the uptake, whilst in the former it is an electrical impulse of high field strength that reversibly permeabilizes the cell membrane (Zimmermann *et al.*, 1984) and allows DNA uptake by the plant protoplasts.

### 9.3.2 Transformation of Plant Protoplasts by Electroporation and Chemically Stimulated DNA Uptake

Early experiments, in which naked DNA was incubated with plant protoplasts in the presence of various transforming agents, gave rise to transformants at a low frequency of $\sim 10^{-6}$ (Davey *et al.*, 1980; Draper *et al.*, 1982; Krens *et al.*, 1982; Lurqin and Kado, 1977). When naked Ti plasmids were used as the transforming DNA, the function of the T-DNA borders was not retained and the *vir* region was not required for the transformation–integration event (Krens *et al.*, 1985). Since this time the procedure has been optimized and transformation efficiencies of $10^{-2}$–$10^{-4}$ are routinely obtained (Fromm *et al.*, 1986; Negrutiu *et al.*, 1987b; Potrykus *et al.*, 1985a).

A number of factors appear to affect the efficiency of DGT by naked DNA. Negrutiu *et al.* (1987a, 1987b) systematically studied 17 parameters affecting direct gene transfer in over 100 combinations for two the species, *Nicotiana tobacum* and *N. plumbaginifolia*. Factors investigated included: (i) plant species; (ii) growth conditions; (iii) protoplast source and method of isolation; (iv) bathing medium for transformation; (v) state, and size concentration of plasmid and carrier DNA; (vi) PEG molecular weight, concentration and pH; (vii) timing and order of steps; (viii) heat shock; (ix) electroporation; (x) cell cycle stage; and (xi) irradiation and other factors (Negrutiu *et al.*, 1987a, 1987b). The major parameters found to be important for efficient transformation are: the plant species; the concentration and state of the plasmid DNA; the presence and concentration of carrier DNA; cations and PEG concentration; and electroportion (Negrutiu *et al.*, 1987a, 1987b). Synchronization of the cell cycle by aphidicolin, which inhibits α-like DNA polymerase, has been claimed to increase efficiency in both stable (Meyer *et al.*, 1985) and transient (Okada *et al.*, 1986a) transformation systems. However, Negrutiu *et al.*, (1987b) found that the use of this compound gave a 10-fold drop in efficiency. It is probable that other differences in protocol led to these apparently contradictory results.

### *9.3.2.1 Stable transformation*

A typical experiment to obtain stable transformation of plants by PEG-mediated DNA uptake might be as follows (Shillito and Potrykus, 1987): 1 $cm^3$ of *N. tobacum* protoplasts at $2 \times 10^6 cm^{-3}$ in an appropriate medium is placed in a 10 $cm^3$ sterile plastic tube and heat shocked for 5 min at 45 °C, followed by cooling on ice. Linearized plasmid DNA is added (10 $\mu$g) together with calf thymus carrier DNA (50 $\mu$g) in (0.05 $cm^3$) distilled water; after mixing, 0.5 $cm^3$ of 40% PEG 6000 is added. The mixture is incubated for 30 min at room temperature and, following dilution by give successive additions of medium (2 $cm^3$) the protoplasts are centrifuged; cells are then resuspended and plated in a medium containing agarose. After a number of days the recovering protoplasts may be transferred to media containing selective agents where appropriate. Transformation efficiencies are reported to be of the order $10^{-4}$–$10^{-3}$ of surviving protoplasts.

A protocol for the transformation of protoplasts by electroporation is given in the same reference (Shillito and Potrykus, 1987). Protoplasts are resuspended in buffered 0.4 M mannitol (to maintain the correct osmoticum) containing 6 mM magnesium chloride at a concentration of $1.6 \times 10^6$ cells $cm^{-3}$. Aliquots of 0.37 $cm^3$ are transferred to the electroporator chamber and magnesium chloride is added until the resistance is 1–1. 1 kΩ. Heat shock is carried out as before for 5 min at 45°C. 0.25 $cm^3$ aliqots of the suspension are removed to polycarbonate tubes, 4 $\mu$g of linearized plasmid DNA and 20 $\mu$g of calf thymus DNA are added in 0.02 $cm^3$ of water and half volume of 24% PEG 6000 (previously adjusted to 1.2k Ω with magnesium chloride) is added. After 10 min the samples are transferred back to the electroporation chamber and pulsed three times at 10 min intervals with a field strength of 1.4–15 kV $cm^{-1}$. Following this step, cells are plated as described above. This protocol is reported to routinely yield transformants at an efficiency of 1–3% of surviving protoplasts.

Negrutiu *et al.* (1987b) have reported a method of transforming *Nicotiana* protoplasts by the synergistic interaction of $Mg^{2+}$, PEG and possibly $Ca^{2+}$. The optimal concentration of $Mg^{2+}$ was different for *N. tobacum* and *N. plumbaginifolia*. This protocol was reported to give transformation efficiencies as high, or higher, than those obtained under optimum electroporation conditions (up to 4.8% for *N. tobacum* protoplasts); this is 100-fold higher than anything previously reported for a chemical salt and PEG transformation method.

A number of points should be made with regard to the above described experimental protocols. First of all, it should be stressed that conditions described for one species will not necessarily give as high efficiencies in other species, and when identical conditions were used, efficiencies varied by as much as an order of magnitude (Negrutiu *et al.*, 1987a, 1987b). At present, however, there have been no reports of protoplasts from any species which are completely refractory to transformation, and species from which protoplasts have been transformed include: *Nicotiana tobacum, Nicotiana plumbaginifolia, Lolium multiflorum, Triticum monococcum, Oryza sativa, Petunia hybrida, Hyoscyamus muticus, Brassica compestris* and *Zea mays* (Negrutiu *et al.*, 1987a; Potrykus *et al.*, 1987a and refs. therein). Electroporation is in many respects more of an art than a science, it being difficult to compare results from different laboratories or to use published protocols without further optimization. Electroporators may differ in either giving an exponentially decaying pulse discharged through a capacitor (usually home made) or a step wave pulse (commercial electroporators). Voltages, discharge time, number of pulses and pre-/post-treatments can all vary. The construction of the chambers varies both in terms of shape and material. The intensity of the electrical pulse required for efficient transformation is inversely proportional to the diameter of the protoplasts (Zimmermann *et al.*, 1982), which varies from species to species. Finally, even such things as trace amounts of ions in the water used to make up solutions can affect results (purer not necessarily being better). Clearly, protocols have to be optimized for each laboratory and, once good results have been obtained, it is a good idea to use materials from the same batch for subsequent experiments (particularly for crude substances such as enzymes used in the preparation of protoplasts).

#### *9.3.2.2 Transient transformation*

Transient transformation has been reviewed by Fromm and Walbot (1987).

Different conditions are required for transient expression of foreign DNA and RNA (*cf.* stable transformation). Circular plasmids and metabolically active, nondividing cells seem to represent optimal conditions for transient expression (Negrutiu *et al.*, 1987; Okada *et al.*, 1986a); conditions harsh enough to reduce the ability of a cell to divide may allow ingress of larger amounts of nucleic acid and leave transcriptional and translational ability unimpaired.

Conditions for the transient expression of DNA in both dicot and monocot protoplasts, mediated by electroporation, have been investigated using a number of reporter genes (neomycin phosphotransferase, NPT II; nopaline synthase, NOS; and chloramphenicol acetyl transferase, CAT) and optimal activity could be monitored 36 h after electroporation (after which activity decayed exponentially: Fromm *et al.*, 1985; Hauptmann *et al.*, 1987; Ou-Lee *et al.*, 1986).

The expression of mRNA electroporated into both plant and animal cells has been studied by Callis *et al.* (1986). It was found that in monkey fibroplasts, and in maize and carrot protoplasts, CAT mRNA is only efficiently expressed if it is both 5′ -capped and 3′-polyadenylated; expression *in vitro* (rabbit reticulocyte lysate) is much less affected by these parameters.

Electroporation has also been successfully used to mediate infection of protoplasts with viral RNA (Nishiguchi *et al.*, 1985; Okada *et al.*, 1986; Watts *et al.*, 1987). Attempts to infect protoplasts with whole virus particles by this method have been less successful (Watts *et al.*, 1987) unless large amounts of virus are used (500 $\mu$g cm$^{-3}$ *cf.* 10 $\mu$g cm$^{-3}$ for RNA; Okada *et al.*, 1986b).

Detailed methods for the electroporation and transient expression of DNA and RNA into plant protoplasts, including instructions for the building of simple electroporators, are given by Fromm *et al.* (1987) and Langbridge *et al.* (1987).

### 9.3.3 The Structure and Stability of DNA Stably Introduced into Plants by DGT

A number of workers have analyzed the fate of DNA transferred to plant tissue by DGT. Hain *et al.* (1985) obtained kanamycin resistant tobacco callus by means of calcium phosphate mediated protoplast transformation. Southern blot analysis of two of these clones demonstrated that the DNA had integrated in a complex pattern (*cf. Agrobacterium*-mediated transfer of the same gene), possibly including concatenation of the input DNA. (Southern blot analysis is a technique in which DNA is cut by restriction endonucleases, the fragments separated electrophoretically, and bands (fragments) with homology to a specific probe visualized. Northern blot analysis is the analogous technique applied to RNA. A fuller explanation will be found in any introductory text on molecular biology.) One regenerated plant was analyzed and kanamycin resistance was shown to be inherited in Mendelian fashion as a single dominant locus. Similar work was carried out by Potrykus *et al.* (1985b): two transformed tobacco cell lines were subjected to a rigorous genetic analysis, in addition to clonal (*i.e.* propagation of plants in tissue culture) and Southern blot analyses. In both cases, kanamycin resistance was inherited as a single dominant Mendelian trait, and the gene was stably maintained during clonal propagation. Southern analysis demonstrated that all kanamycin resistant tissue had the expected band as well as extra bands corresponding to rearranged DNA. The additional bands were shown to be genetically linked to the expected band. Instability was very occasionally observed in clonally propagated plants; in one case, all of the rearranged bands had disappeared (but not the expected band) and in another, the expected band had disappeared and the other bands remained. In the latter case, the kanamycin resistant phenotype was also lost. This instability has not been observed in *Agrobacterium*-transformed plants (Section 9.2.5) and may be due to homologous recombination between rearranged (inactive) and nonrearranged foreign DNA sequences. Detailed analyses of tobacco transformed by direct gene transfer have also been carried out by Czernilofsky *et al.* (1986a, 1986b) and by Riggs and Bates (1986). The latter analyzed 26 plants. Most plants transmitted kanamycin resistance as a single dominant marker, although three exhibited patterns consistent with the inheritance of homozygous lethal mutations. Southern blot analysis of nine plants suggested that one of these plants had integrated the transforming DNA at more than one site. Evidence for plasmid recircularization or concatemer formation was found in most of the plants and in every case the EcoRI site used to linearize the plasmid prior to transformation was lost; head to head and head to tail (incomplete) concatemers were detected. In six of the nine plants, an unexpected 6.5 kb fragment was observed suggesting that, in addition to ligation, the plasmid had undergone some sort of regular rearrangement. It was suggested that the observed plasmid ligations were more likely to have occurred prior to integration. Similar results were obtained by Czernilofsky *et al.* (1986a, 1986b). The integrated DNA occurred in multiple arrays of 10–25, mostly identically altered units of donor DNA.

Potrykus *et al.* (1985c) also obtained G-418 resistant colonies after transformation of the cereal monocot *Lolium multiflorum* by PEG treatment. (G418, or Geneticin, is, like kanamycin, an aminoglycoside antibiotic which is inactivated by NPTII.) The two colonies analyzed by southern blot both gave bands of the expected size (*i.e.* undeleted) at copy number greater than five per haploid genome when probed with an internal fragment with homology to the NPTII gene. This again suggested concatenation (although heredity could not be investigated as only callus was produced). Additional bands were not seen. Similarly, Fromm *et al.* (1986) transformed maize by electroporation and analyzed four of the kanamycin resistant calli by Southern blot. All four contained between one and five copies of the intact gene; two of the calli also had rearranged copies of the gene present.

In 220 independent transformants genetically analyzed by Potrykus *et al.* (1987a; see also Saul *et al.*, 1987), 77% passed on the kanamycin resistant phenotype as a single dominant locus, 8% as two independent loci, 2% as two linked loci, 2% as three or more linked or unlinked loci, 3% passed on the marker only when crossed as the female parent and 8% had patterns difficult to assign.

In the experiments described above, the foreign DNA segregated mostly as a single dominant genetic locus. The observation of primarily single loci inheritance has been confirmed at the molecular level in an elegant experiment in which metaphase plates from two independent transformants were examined by *in situ* hybridization using a $^{3}$H-labelled probe (Mouras *et al.*, 1987). After 12 weeks exposure it was possible to see in both cases a signal localized on (different) homologous chromosome pairs. This confirms that the transforming DNA had integrated into the plant chromosomes in single blocks.

It had previously been observed that calf thymus carrier DNA was integrated along with the DNA containing the selectable marker (Krens *et al.*, 1985) and this DNA was shown to be genetically linked to the selected DNA (Peerbolte *et al.*, 1985). This suggested that it should be possible to cotransform protoplasts with nonlinked, selected and nonselected DNA's; this is indeed the case. Czernilofsky *et al.* (1986a) have successfully cotransformed tobacco (using calcium phosphate/PEG) with the selectable NPTII gene and the nonselected maize transposable element *Ac*. The frequencies of co-transformation are not given. Schocher *et al.* (1986) carried out similar experiments with tobacco protoplasts (transformed by electroporation) in which a zein genomic clone was cotransformed with an NPTII gene, and a nopaline synthase (*nos*) gene with an NPTII gene. They found sequences from the zein clone in 88% of kanamycin resistant calli; the zein clone was full length in 33% of the cases. Of the kanamycin resistant calli cotransformed with the *nos* gene, 47% were found to have a complete and active *nos* gene present. These results further extend the potential utility of chemical- and electroporation-mediated DGT for the investigation of plant gene expression and crop improvement, particularly in species which are not amenable to *Agrobacterium*-mediated transformation.

A cautionary note about DGT has been sounded by Negrutiu *et al.* (1987a), who pointed out a possible mutagenic effect of exogenous DNA; reversion frequencies in nitrate reductase-less plant cells were an order of magnitude higher in cells treated with DNA than the spontaneous reversion frequencies, and indeed even higher than the reversion frequencies induced by UV or gamma radiation. It is suggested that this may pose a problem when attempts are made to transfer native genes (particularly those for which there is no tight selection or specific assay) by means of transformation with total genomic DNA.

### 9.3.4 DGT by Other Means

#### *9.3.4.1 Liposome- and spheroplast-mediated DGT*

Artificial lipid vesicles (liposomes) and bacterial spheroplasts (bacteria from which the cell wall has been enzymatically removed) harbouring plasmids have both been used to transform plant protoplasts.

Expression of RNA introduced into tobacco protoplasts by liposome-mediated transformation has been observed (Fraley *et al.*, 1982; Rouze *et al.*, 1983) at efficiencies of between 20% and 70%. Tonaka *et al.* (1984) used *E. coli* spheroplasts to successfully infect *Brassica chinensis* mesophyll protoplasts with cloned cauliflower mosaic virus at an efficiency of between 1% and 5%.

Perhaps more interesting are the liposome-mediated stable transformations obtained by Deshayes *et al.* (1985) and the *E. coli* spheroplast-mediated gene transfer observed by Hain *et al.* (1984). In the former case a plasmid carrying a kanamycin resistance gene was successfully transferred to tobacco protoplasts and stably integrated at a frequency of $\sim 10^{-5}$. Three transformants analyzed further all passed on kanamycin resistance as a single dominant trait; Southern blot analysis suggested that the transferred DNA was present in four to five tandemly repeated, essentially unrearranged copies.

The potential advantages of liposome- and spheroplast-mediated DGT are: the protection of sequestered nucleic acids from nuclease digestion during introduction into the cellular environment; no carrier DNA is required; and it is possible that less rearrangement takes place. In a transient expression system, Rosenberg *et al.* (1988) found that if liposome-mediated transformation was used to transfer CAT DNA to tobacco protoplasts, then CAT activity could be detected up to nine days after transformation (*cf.* 48 h for naked DNA transformation).

On the other hand, simpler protocols (electroporation- and PEG-mediated DNA uptake) give higher transformation efficiencies. The preparation of liposomes requires sonication which may well shear DNA larger than 25 kb (Deshayes *et al.*, 1985), and there is always the possibility that the components of the liposome system are toxic and/or mutagenic to plant cells.

### *9.3.4.2 Microinjection*

The physical microinjection of DNA directly into the cytoplasm or nucleus of plant protoplasts has been successfully attempted a number of times (Crossway *et al.*, 1986; Reich *et al.*, 1985; Steinbiss *et al.*, 1985). The topic has been reviewed (Miki *et al.*,1987).

The technique has progressed from the production of one transformant per scientist per year of intensive work (Steinbiss *et al.*, 1985) to the point where up to 10% of injected protoplasts give rise to transformants (Crossway *et al.*, 1986). The protoplasts are held on a holding pipette by slight suction while the DNA is introduced selectively into either the cytoplasmic or nuclear space using a microcapillary pipette (Figure 9). Following injection, individually injected protoplasts can be cultured in a hanging drop of media. A 30 kb T-DNA-containing plasmid was used and one or two copies of foreign DNA were integrated. In all cases the integrated DNA was less than full length and in only four of the 11 transformants obtained were sequences homologous to the NPTII gene found. It was suggested that microinjection through a small capillary shears the DNA.

Considering the extreme technical difficulty, the high cost of micromanipulators and other equipment needed to carry out the experiments and relatively poor transformation efficiencies (per man hour at least) compared to DGT mediated by PEG and electroporation, microinjection is unlikely to be of general use. This situation may change if technical difficulty decreases, or if it becomes possible to microinject plastids, nuclei or whole chromosomes, or if whole cells rather than protoplasts can be used for the microinjection experiments.

### *9.3.4.3 Electroinjection into whole cells*

Recently, Morikawa *et al.* (1986) introduced TMV RNA into intact tobacco mesophyll cells by electroporation in the absence of PEG. The RNA was shown to be expressed and a homogenate of the electroporated cells was shown to be infectious when inoculated onto tobacco leaves. The applicablity of this technique to transient expression or stable transformation has been questioned by Negrutiu *et al.* (1987a); they asked whether the technique was applicable to cells that had been isolated by nonenzymatic treatment and observed that the electroporation conditions used would give extremely low transformation frequencies with tobacco protoplasts. However, Lindsey and Jones (1987) obtained a low level of CAT activity after transient transformation of intact sugar beet suspension cells by electroporation.

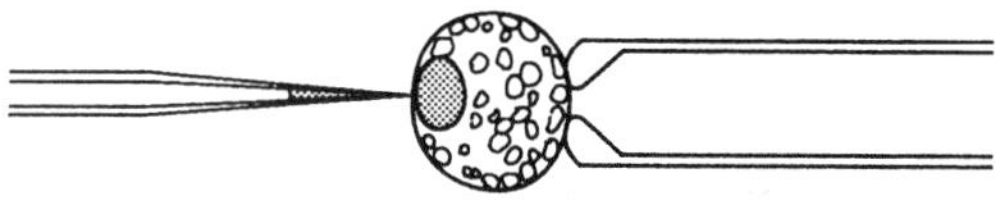

**Figure 9** Microinjection. The protoplast is held with gentle suction by a holding pipette and DNA is injected directly into the nucleus with a microcapilliary pipette

Further work will no doubt tell if this method is generally applicable; if it is, it will be a step towards the transformation of plants which at present cannot be regenerated from protoplasts.

#### *9.3.4.4 DGT into whole plants*

Ohta (1986) reported efficient gene transfer to maize by the application of a viscous mixture of (foreign) total genomic DNA and ungerminated pollen onto silks. However, the process led to mutations at a high level ( ~ 9%) and there was no proof of transformation at the DNA or RNA level. Also, transmission of the marker gene was unstable and crosses to determine the Mendelian (or otherwise) segregation of the marker gene in progeny were not reported. We await further reports on this technique with interest.

Work by de la Peña *et al.* (1987) has elaborated the transformation of rye plants by injection of young floral tillers with DNA. Their previous work had demonstrated that, 14 days prior to the first meiotic metaphase, archesporial cells are highly sensitive to exogenously applied chemicals injected into the developing floral tillers. They injected the developing floral tillers with an aqueous solution containing ~ 30 $\mu$g of plasmid DNA carrying the NPTII gene. Seeds from the injected tillers were obtained by cross pollination with other injected tillers (rye is self incompatible): 3023 seeds from the cross pollination of 98 plants were plated out on media containing kanamycin and of these seven seedlings appeared to be resistant to kanamycin. From these seven seeds two were shown to express NPTII enzyme activity and the presence of the gene was confirmed by Southern blot analysis, where it was found to be integrated in multiple copies, some rearranged. This transformation at about 0.1% was the first confirmed report of a transgenic, graminaceous, monocotyledonous *plant*. It may turn out to be the method of choice for transforming monocots, avoiding, as it does, the need for tissue culture.

#### *9.3.4.5 Microprojectile bombardment*

An unusual approach to plant genetic transformation is that developed by Klein *et al.* (1987). Microscopic ( ~ 1 $\mu$m) tungsten particles, coated with plasmid DNA, were accelerated to 400 m $s^{-1}$ in a 'particle gun' apparatus. This speed allowed the particles to penetrate intact cell walls.

A chimeric CAT gene has been delivered to and expressed in onion epidermis cells. Preliminary work has been reported in the application of this system to maize (Klein *et al.*, 1988).

### 9.3.5 Conclusion

From the experiments described above it is clear that DGT complements *Agrobacterium*-mediated transformation. It is unlimited by host range and allows the transient expression of introduced DNA and RNA; this is a great advantage in the investigation of plant gene function when stable transformation experiments take a minimum of months. It is possible that, due to the different mechanism of DNA uptake and integration (*cf. Agrobacterium*-mediated transformation), some experiments, such as gene targetting by homologous recombination, will be aided by the DGT approach.

## 9.4 VIRUS VECTORS

Plant viruses have not yet fulfilled their promise as vectors for the genetic engineering of plants. They have the potential to complement the existing technologies of DGT and *Agrobacterium* effectively. Whole plants may be systemically infected and thus gene expression can be investigated in differentiated tissue as opposed to protoplasts or callus (particularly in plants recalcitrant to regeneration). Systemic infection typically takes place over days or weeks, not months, and viral particles accumulate to high levels leading to gene amplification and thus the possibility of expression at high levels (see Section 9.4.2). However, no plant virus so far discovered integrates into the host genome, and transmission through the germline has not been possible. Transmission by vegetative propagation (*cf*. the abutilon mosaic virus which is passed on in vegetatively propagated *Abutilon sellovianum*, and gives rise to the decorative variegation of the leaves) is possible and may even be an advantage to commercial plant breeding and genetic engineering companies. There are also problems with regard to limited host range and packaging constaints on the size of the cloned insert in viral vectors.

There are two broad groups of plant viruses, the DNA and the RNA viruses. The DNA viruses make up only 1–2% of known plant viruses (Lebeurier, 1986). They may be subdivided into two groups, the double-stranded caulimoviruses and the single-stranded gemini viruses. Because it is easier to work with DNA viruses, most of the work (although not all) on vector development has been carried out on these viruses, and we shall consider this group first.

### 9.4.1 Caulimoviruses

There are at least 12 members of the caulimovirus family (Hull and Davies, 1983), and of these the cauliflower mosaic virus (CaMV) is the best characterized. The biology of this virus has been recently well reviewed (Gronenborn, 1987; Lichtenstein and Fuller, 1987) and extensive literature references can be found. Reviews have also been published on the potential of this virus for vector development (Fraley *et al.*, 1986; Hull and Davies, 1983; Lebeurier, 1986; Lichtenstein and Fuller, 1987). We shall only describe the biology as it pertains to vector development and the evidence for the described biology will be found in the reviews cited above.

CaMV has a double-stranded DNA genome of approximately 8 kb in length with a host range limited, in the main, to cruciferous plants such as cauliflower and turnip. Aphids are the natural transmission vectors; however, it is also infectious by mechanical inoculation and agroinfection (see Section 9.2.7).

The viral DNA is transcribed in the nucleus of infected cells by host-encoded RNA polymerase II. There are two major transcripts produced (Figure 10), a 35S RNA corresponding to the entire viral DNA plus a terminal repeat of 180 bp, and a subgenomic messenger 19S RNA terminating at the same point as the 35S RNA. There are eight potential open-reading frames (ORFs; Figure 10). The 19S transcript probably encodes ORFVI, responsible for symptom development (Baughman *et al.*, 1988). It has been proposed that the other ORFs are translated from the polycistronic 35S transcript by a 'relay race' mechanism, in which the ribosomes pause after passing the termination codon of one transcript and reinitiates at the next initiation codon on the transcript. It has also been suggested that ORFs IV and V could be expressed as a long precursor protein which is subsequently cleaved to produce two functional proteins. In support of the relay race mechanism, and of importance to vector design, is the observation that a nonsense mutation in ORFII affects the translation of the downstream ORFIII in a polar fashion (there is other evidence for the relay race mechanism, see Lichtenstein and Fuller, 1987); clearly, any foreign DNA cloned CaMV must take account of the polycistronic nature of the viral 35S transcript.

The virus replicates by priming of the 35S transcript with a host-encoded t-RNA-met, followed by minus strand DNA synthesis catalyzed by a virus-encoded reverse transcriptase.

ORFII has been shown to be an aphid transmission factor (Daubert *et al.*, 1983) and is not essential

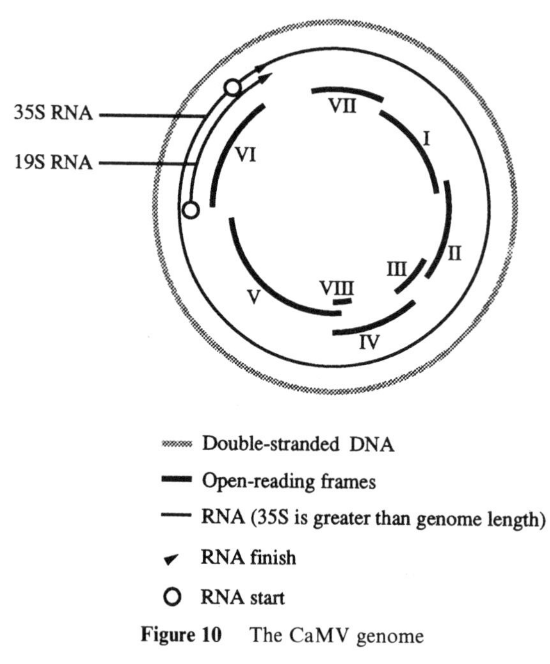


**Figure 10** The CaMV genome

for viral infectivity as a naturally occurring, severely deleted ORFII mutant has been isolated; ORFVII has also been shown to be dispensible (Howarth *et al.*, 1981; Howell *et al.*, 1981). Gronenborn *et al.* (1981) successfully cloned foreign DNA into ORFII and found that the upper limit for foreign DNA propagated by CaMV in this way was ~250 bp. Brisson *et al.* (1984) replaced ORFII with a small (234 bp) bacterial dihydrofolate reductase (*dhfr*) gene and it was expressed and stably propagated in turnip plants. This gene confers resistance to the compound methotrexate, and turnip plants infected with CaMV containing the *dhfr* gene were resistant to it. It was noted that when one of the constructs used had a slightly longer noncoding region (47 bp *cf.* 9 bp) between the end of ORFI and the start of the *dhfr* gene, the former construct was unstable to passage through the plant and expressed the *dhfr* gene at a reduced level compared to the latter. This sensitivity to the length of intergenic sequences found in CaMV shows the care that has to be taken when designing vectors of this kind.

In the simple CaMV vector approach described above the limit for cloned inserts is probably about 800 bp (by deletion and addition). A different approach is to construct complementing defective mutants, such that the 'vector' mutant carries the foreign DNA and only a few CaMV functions, and the other mutant carries all of the complementing CaMV ORFs (but is deficient in the ORFs carried on the 'vector'). This approach has been tested (Choe *et al.*, 1985); pairs of complementing, noninfective CaMV mutants were constructed and used in a mixed infection. Successful infection occurred but wild-type virus was rescued in all cases; recombination events had deleted the foreign DNA. This high frequency of recombination in CaMV has been observed before (Walden and Howell, 1982). Perhaps, if complementary defective mutants can be constructed with little or no homology, this approach to vector design will bear fruit. Pazkowski *et al.* (1986) have also attempted to use complementation as a route to using CaMV as a vector; they replaced the essential ORFVI (translated from the 19S transcript) with the coding region from NPTII. Unfortunately, this mutant was noninfective both on its own and when complemented by the wild-type virus. The NPTII gene was expressed from the viral promoter when the mutant virus was transferred to turnip protoplasts by direct gene transfer, but there was no escape of the virus and viral sequences were only associated with high molecular weight genomic DNA.

Another approach to complementation in CaMV vectors would be to put a complementing defective virus genome into the plant genome by means of *Agrobacterium*-mediated transformation. This would give a system equivalent to the COS cell system for the propagation of defective mammalian SV40 virus-based vectors (Gluzman, 1981 see Section 9.4.2 for a successful use of this approach in plants). The feasibility of this approach in CaMV-based vectors is beginning to be demonstrated. The work of Pazkowski *et al.* (1986) described above demonstrates that gene products of mutant CaMV genomes cloned into plants can be expressed. Shewmaker *et al.* (1985) have cloned a full length copy of CaMV into a variety of plant species and shown that both the 19S and 35S transcripts are expressed (at varying levels in different species). Futhermore, Walden (personal communication) has made transgenic tobacco plants (outside the normal host range of CaMV) containing dimers of CaMV and shown that leaf homogenate from these plants is infectious when inoculated onto turnip; it is not yet clear if the virus replicates in tobacco, or if the infection turnip is mediated by the 19S and 35S transcripts. Perhaps, tobacco, which is more amenable to genetic transformation, can be used as a model system for this work. Clearly, much still needs to be done before CaMV becomes the basis of a useful vector system; in the light of the exciting work on gemini virus based vectors recently reported (see below), we should seriously consider whether it is worth expending such effort.

### 9.4.2 Gemini Viruses

Gemini viruses (reviewed by Davies *et al.*, 1987; Harrison, 1985; Stanley, 1985) are single-stranded DNA viruses that replicate in the nuclei of cells *via* a double-stranted DNA intermediate. The name derives from the bisegmented nature of the virions as they appear in electron micrographs. Cloned copies of several gemini virus members are, between them, transmissible to a wide host range including both monocots and dicots (Harrison, 1985). Gemini viruses may be subdivided into whitefly- and leafhopper-transmitted virus; leafhopper-transmitted viruses in general have monopartite genomes, whilst whitefly-transmitted viruses have bipartite genomes, as described in Section 9.2.7 for TGMV (genome lengths for both groups are ~2.5 kb). Members of this latter class have recently been reported as plant vectors.

It has previously been shown that the bipartite cassava latent virus (CLV) DNA 1 encodes functions essential for replication but not systemic infection or cell to cell transmission (Townsend *et al.*, 1986);

it has also been shown that DNA 1 encodes the coat protein which is not essential for infectivity (Stanley and Townsend, 1986). Ward *et al.* (1988) have shown that although deletion of the coat protein gene abolishes infectivity by manual inoculation with linear DNA 1 and DNA 2 molecules, mutants in which it is replaced by sequences of similar size are fully infectious. When the coat protein gene was replaced by the bacterial CAT gene and the engineered DNA 1 inoculated onto *Nicotiana benthamiana* together with wild-type DNA 2, the plants developed normal symptoms of infection and CAT was expressed systemically at high levels (80 units $mg^{-1}$ soluble protein) 10 days after inoculation. The level of CAT activity was monitored over a period of four weeks and found to be maintained. Complementary defective mutants of CLV undergo frequent recombination to give wild-type virus (Ward *et al.*, 1988). However, no instability was observed in the virus expressing the CAT gene, presumably because there is no selective pressure for deletion of this gene.

Similar work has also been carried out on the closely related tomato golden mosaic virus (TGMV) by Hayes *et al.* (1988c). It is useful to restate some of the work on this virus, described in Section 9.2.7, before describing the results.

When dimers of TGMV DNA A were cloned into the chromosomes of petunia plants (*via Agrobacterium*-mediated transformation), freely replicating DNA A components were found (Roger *et al.*, 1986b). Trimers of TGMV B were cloned into the chromosomes of petunia, but no freely replicating DNA B molecules were found. However, if these A2 and B2 plants were crossed, a systemic infection resulted. This suggests that perhaps the A particle encodes the functions necessary for replication and the B particle those necessary for cell to cell spread and symptom development. *Agrobacterium* strains containing B dimers complement A2 plants (containing dimers of A in the genome), and *Agrobacterium* strains containing A dimers complement the B3 plants (containing trimers of B in the genome); systemic infection was seen in both cases. The virus coat protein is not required for systemic infection of plants (Brough *et al.*, 1988; Hayes *et al.*, 1988b).

Hayes *et al.* (1988c) constructed vectors in which the NPTII coding region replaced the coat protein coding region. These vectors contained either 1.6 copies of tandemly repeated NPTII encoding TGMV A DNA (pBIN19A1.6*neo*) or 1.6 copies of tandemly repeated NPTII encoding TGMV A DNA plus 2 copies of TGMV B DNA (pBIN19A1.6*neo*B2) between T-DNA borders, analogous to those described in the agroinfection experiments (see Section 9.2.7; Hayes *et al.*, 1988b). Wild-type tobacco plants were agroinfected with pBIN19A1.6*neo* B2, and B2 transgenic plants (containing TGMV B DNA integrated into the genome) were agroinfected with pBIN191.6*neo*; four out of 20 of the former and 18 out of 20 of the latter were systemically infected as determined by DNA dot blot analysis using probes specific for TGMV DNA A and DNA B. None of the *neo*-infected plants showed any symptoms. The infections were confirmed by Southern blot; single-stranded, supercoiled and open circular forms of viral DNA of the expected size were found. When the 'wild-type' *neo*-less pBIN19A1.6B2 construct was used to infect tobacco plants, infection was more efficient 19 out of 20 plants developed symptoms.

Plants containing A1.6*neo* DNA integrated into the chromosome were also constructed. On Southern DNA blotting the expected monomeric forms of TGMV A *neo* DNA (with the NPTII coding region replacing the coat protein gene) were found in the plant DNA. NPTII activity was assayed and the relative activities were as follows: a transgenic plant containing seven copies of NPTII regulated by the *nos*-promoter, 1: a transgenic plant containing A 1.6 *neo* DNA, 40; a wild-type plant agro-infected with pBIN19A 1.6 *neo* B2, 80; and a transgenic B2 plant agroinfected with pBIN19A 1.6 *neo*, 240!

The size of the NPTII expressing TGMV A DNA is 2708 bp compared to the wild- type DNA A which is 2588 bp. The size limit of this vector has since been extended by inclusion of the coding region from the glucuronidase (GUS) gene in place of the NPTII coding region (Hayes, personal communication); this is an increase of 600 bp.

There are some interesting differences between the work on CLV and that on TGMV. The loss of infectivity observed in CLV coat protein deletion mutants was not observed in TGMV coat protein deletion mutants, suggesting that perhaps the size constraints in CLV are tighter than those in TGMV. Also, the TGMV construct produced symptomless plants while the CLV construct did not. As the marker gene, host, and method of infection differed in the two experiments, it is not clear at this stage whether the differences observed are due to the viruses or the way in which the experiments were carried out.

Gemini virus based vectors appear to hold great promise. They seem to provide a useful means of gene amplification in plants, either by systemic infection of plants with the vector to provide a rapid assay of gene expression, or by producing plants transgenic for a partial dimer of the vector to provide inheritable gene amplification. The upper limit of foreign DNA that can be replicated by a gemini virus vector has not yet been determined, but it is not limited by packaging constraints as it is in the caulimovirus vectors. Furthermore, gemini viruses infect a wider range of plant species.

### 9.4.3 RNA Viruses

The potential general utility of RNA virus based vectors is uncertain. To our knowledge, there have been no demonstrations of directly infections cDNA forms of RNA viruses; thus infection by *in vitro* synthesized RNA transcripts is the only obvious route to engineered RNA viruses (Fraley *et al.*, 1986 and refs. therein). It has also been suggested that the high error frequency during viral RNA synthesis will cause instability in foreign genes (van Vloten-Doting *et al.*, 1985). There is no selective pressure to correct errors induced in foreign genes, in contrast to mutations in endogenous viral genes. This suggestion has been disputed (Siegel, 1985).

There have been two recent reports of foreign genes expressed in RNA viral vectors. French *et al.* (1986) have reported the expression of CAT mediated by RNA transcripts of a coat protein fusion of bromo mosaic virus in barley protoplasts. Since protoplasts were used, it is difficult to compare the results to those obtained with other viral vectors. Takamatsu *et al.* (1987) made similar constructs in tobacco mosaic virus which produced local lesions when the RNA transcripts were inoculated on tobacco leaves. CAT was detected (0.1 unit $mg^{-1}$ of soluble protein) at 0.125% of the activity obtained with the gemini virus CLV-based vector. Also, the coat protein is necessary in TMV for systemic infection and no CAT or virus was detected on leaves distal to the site of inoculation.

These results suggest that a great deal more work must be done if RNA virus based vectors are to have any realistic future.

## 9.5 TRANSPOSON VECTORS

Transposons are pieces of DNA (often encoding drug resistance in bacteria) that can jump from one piece of DNA to another *in vivo*. Evidence for the existence of transposons was first observed in maize by McClintock (1951). Much later, recombinant DNA technology allowed confirmation of her intuition, and for her early work McClintock won the Nobel prize. The transposon she discovered, the *Ac* element, creates an 8 bp duplication at the site of integration, and it can excise itself precisely from one part of the genome and integrate in another, distal position. For a review, see Wienand and Saedler (1987).

This pioneering work by McClintock has recently led to the possibility of transposon vectors. The *Ac* element from maize has been transferred to tobacco by *Agrobacterium*-mediated transformation where it was shown to excise and reintegrate elsewhere as in its natural 'host', maize (Baker *et al.*, 1986). The same group had developed a phenotype assay for excision of the *Ac* element in tobacco (Baker *et al.*, 1987). The *Ac* element has been cloned between the promoter and coding region of the NPTII gene so as to inactivate it; upon excision of the *Ac* element, a functional NPTII gene is recreated. After *Agrobacterium*-mediated transfer to tobacco, excision events could be monitored by the appearance of kanamycin resistant colonies. If selection was applied 10–12 days after cocultivation kanamycin resistant calli were detected in the *Ac*-transformed clones at between 13% and 40% of the number found with a wild-type NPTII gene. If, however, selection was not applied until three to four months following cocultivation, the number of kanamycin resistant calli went up to 70%, suggesting that *Ac* excision continues for more than two weeks after its introduction into tobacco cells. The results were confirmed by Southern blot analysis.

Several plant genes have been cloned by gene tagging using transposons. A mutated gene carrying a transposon insert is first isolated using a previously cloned transposon as a probe to find the element and flanking DNA; the flanking DNA is then used as a probe to isolate the gene from the wild-type line (refs. in Baker *et al.*, 1987). Gene tagging has previously been limited to the few plant species which possess well-characterized transposable elements. In principle, this work could extend the usefulness of gene tagging to any plant species which is amenable to genetic transformation (by *Agrobacterium* or DGT) and whole plant regeneration, so long as the *Ac* element functions in that species.

## 9.6 ORGANELLE TRANSFORMATION

Organelle transformation is of interest, both because such commercially important properties as cytoplasmic male sterility and resistance to the triazine herbicides are encoded there, and also because much of the fundamental biochemistry of plants (*e.g.* photosynthesis) occurs in organelles.

Transfer and expression of foreign genes in plant organelles is difficult; much of the unpublished (and unsuccessful) work on organelle transformation carried out in the last few years has been reviewed by Cornelissen *et al.* (1987). There are up to 1000 organelle genome copies in a single plant cell (Cornelissen *et al.*, 1987); by comparison with bacteria, we can see that there are strong possibilities that selective replicative disadvantage and recombination events will quickly remove a recombinant plastid or mitochondrial genome. Furthermore, differential codon usage in mitochondria poses a potential problem.

De Block *et al.* (1985) have reported a successful chloroplast transformation mediated by *Agrobacterium* cocultivation. A chimeric gene consisting of the nopaline synthase gene promoter ($P_{nos}$) is fused to the coding region of the CAT gene. Chloramphenicol resistant plants obtained from chloramphenicol resistant calli were unable to transmit the phenotype through pollen and a typically maternal inheritance pattern was observed. The CAT activity was shown to be localized exclusively in the chloroplast fraction by Southern blot analysis and CAT assay on isolated nuclei, mitochondria and chloroplasts. The CAT activity and associated DNA was lost from plants not grown under chloramphenicol selection.

Numerous attempts to extend this work using chloroplast specific promoters (Cornelissen *et al.*, 1987) have been unsuccessful. It has been suggested that the promoters used were too strong and the success enjoyed by De Block *et al.* (1985) was due to the use of the Ti plasmid-derived *nos* promoter (Cornelissen *et al.*, 1987); this eukaryotic promoter is weakly functional in bacteria (and thus presumably, plant organelles). Progress in organelle transformation is probably going to be slow and difficult.

An alternative approach is to target nuclear gene products into chloroplasts (reviewed by Szabo and Cashmore, 1987). A particularly clever example is given by the work of Cheung *et al.* (1988). The coding region from a chloroplast gene *psbA* (encoding resistance to the herbicide, atrazine) was fused to the transcriptional and transit peptide sequences of the nuclear, rubisco small subunit gene. This chimera was transferred to the nuclear genome to tobacco by *Agrobacterium*-mediated transformation. The foreign protein was synthesized and transported to the chloroplasts: atrazine tolerant tobacco plants resulted.

Another type of organelle transformation, which is discussed in Chapter 11, is the transfer of organelles from one species to another by protoplast fusion. A successful example is the transfer of terbutryn (herbicide) resistant plastids from *Nicotiana plumbaginifolia* to *N. tabacum* without transfer of the accompanying mitochondrially encoded cytoplasmic sterility (Menczel *et al.*, 1986). This could not have been accomplished by standard plant-breeding techniques.

## 9.7 CONCLUDING REMARKS

As stated in the introduction, plant genetic engineering is yet a young science. Although it has progressed by leaps and bounds, it is still in its infancy. The development of techniques for plant genetic transformation has been rapid, and continues to be so. The complementary technique of plant tissue culture has to catch up before the full commercial and scientific potential of plant genetic engineering can be realized.

Our knowledge of plant biochemistry and development is, as yet, not as advanced as our knowledge of bacterial, fungal, *Drosophila* and mammalian systems. As the technology of plant genetic manipulation becomes as versatile and sophisticated as that available in other systems, so our understanding of plants at the molecular level will increase accordingly. As the function of new genes is elucidated, so our ability to engineer plants for specific traits will increase. In the study of developmental biology, plants have a number of advantages over other systems: they are totipotent, and differentiated cells can be de- and re-differentiated, sometimes into whole plants or isolated tissue types. Furthermore, fully differentiated transgenic plants can be produced from a single cell in a matter of only weeks.

We feel that the future of plant genetic engineering will be both exciting and intellectually challenging.

## ACKNOWLEDGEMENTS

We wish to thank all of the people who kindly sent preprints and communicated unpublished data; Alison Davis for help in the preparation of figures; and Jorge Taver-Torres, Carlos Flores, and Jyoti Lakhanpal for constructive comments on the manuscript. We also thank Jane Clark for surrendering the MacintoshPlus on which this review was written. Finally, we apologize to all those people whose work was omitted through oversight or space constraints.

## 9.8 REFERENCES


Abel, P., R. S. Nelson, B. De, N. Hoffman, S. G. Rogers, R. T. Fraley and R. N. Beachy (1986). Delay of disease development in transgenic plants that express the tobacco mosaic virus coat protein gene. *Science (Washington D.C.)*, **232**, 738–743.

Amasino, R. M., A. L. T. Powell and M. P. Gordon (1984). Changes in T-DNA methylation and expression are associated with phenotypic variation and plant regeneration in a crown gall tumour. *Mol. Gen. Genet.*, **197**, 437–446.

An, G. (1985). High efficiency transformation of cultured tobacco cells. *Plant Physiol.*, **79**, 568–570.

An, G., B. D. Watson, S. Stachel, M. P. Gordon and E. W. Nebster (1985). New cloning vehicles for the transformation of higher plants. *EMBO J.*, **4**, 227–284.

An, G. (1986). Development of plant promoter expression vectors and their use for analysis of differential activity of nopaline synthase promoter in transformed tobacco cells. *Plant Physiol.*, **81**, 86–91.

An, G. (1987). Binary Ti vectors for plant transformation and promoter analysis. *Methods Enzymol.*, **153**, 292–305.

Ashby, A. M., M. D. Watson and C. H. Shaw (1987). A Ti plasmid determined function is responsible for chemotaxis of *Agrobacterium tumefaciens* towards the plant wound product acetosyringone. *FEMS Microbiol. Lett.*, **41**, 189–192.

Baker, B., J. Schell, H. Lörz and N. Fedoroff (1986). Transposition of the maize-controlling element 'activator' in tobacco. *Proc. Natl. Acad. Sci. USA*, **83**, 4844–4848.

Baker, B., G. Coupland, N. Fedoroff, P. Starlinger and J. Schell (1987). Phenotypic assay for excision of the maize-controlling element *Ac* in tobacco. *EMBO J.*, **6**, 1547–1554.

Barton, K. A., A. N. Binns, A. J. M. Matzke and M. Chilton (1983). Regeneration of intact tobacco plants containing full length copies of gentically engineered T-DNA, and transmission of T-DNA of F1 progeny. *Cell*, **32**, 1033–1048.

Basiran, N., P. Armitage, R. Scott and J. Draper (1987). Genetic transformation of flax by *Agrobacterium tumefaciens*: regeneration of transformed shoots *via* a callus phase. *Plant Cell Rep.*, **6**, 396–399.

Baughman, G. A., J. D. Jacobs and S. H. Howell (1988). Cauliflower mosaic virus gene VI produces a symptomatic phenotype in transgenic tobacco plants. *Proc. Natl. Acad. Sci. USA*, **85**, 733–737.

Bevan, M. W. (1984). *Agrobacterium* vectors for plant transformation. *Nucleic Acids. Res.*, **12**, 8711–8721.

Bevan, M. W., R. B. Flavell and M. D. Chilton (1983). A chimeric antibiotic resistance gene as a selectable marker for plant cell transformation. *Nature (London)*, **304**, 184–187.

Brisson, N., J. Pazkowski, J. R. Penswick, B. Gronenborn, I. Potrykus and T. Hohn (1984). Expression of a bacterial gene using a viral vector. *Nature (London)*, **310**, 511–514.

Brough, C. L., R. J. Hayes, A. J. Morgan, R. H. A. Coutts and K. W. Buck (1988). Effects of mutagenesis *in vitro* on the ability of cloned tomato golden mosaic virus DNA to infect *Nicotiana benthamiana* plants. *J. Gen. Virol.*, **69**, 503–514.

Buchanan-Wollaston, V., J. E. Passiatore and F. Cannon (1987). The *mob* and *oriT* mobilization functions of a bacterial plasmid promote its transfer to plants. *Nature (London)*, **328**, 172–175.

Budar, F., L. Thia-Toong, M. Van Montagu and J. P. Hernalsteens (1986). *Agrobacterium*-mediated gene transfer results in transgenic plants transmitting T-DNA as a single Mendelian factor. *Genetics*, **114**, 303–313.

Byrne, M., R. McDonnell, M. Wright and M. Carnes (1987). Strain and genotype specificity in *Agrobacterium*-soybean interaction. *Plant Cell, Tissue Organ Cult.*, **8**, 3–15.

Bytebier, B., F. Deboeck, H. De Greve, M. Ban Montagu and J. P. Hernalsteens (1987). T-DNA organization in tumour cultures and transgenic plants of the monocot *Asparagus officinalis*. *Proc. Natl. Acad. Sci. USA*, **84**, 5345–5349.

Callis, J., M. Fromm and V. Walbot (1987). Expression of mRNA electroporated into plant and animal cells. *Nucleic Acids Res.*, **15**, 5823–5831.

Cheung, A., L. Bogorad, M. Van Montagu and J. Schell (1988). Relocating a gene for herbicide tolerance: a chloroplast gene is converted into a nuclear gene. *Proc. Natl. Acad. Sci. USA*, **85**, 391–395.

Chilton, M. D., M. H. Drummond, D. J. Merlo, D. Sciaky, A. L. Montoya, M. P. Gordon and E. W. Nester (1977). Stable incorporation of plasmid DNA into higher plant cells: the molecular basis of crown gall tumorigenesis. *Cell*, **11**, 263–271.

Chilton, M. D., R. K. Saiki, N. Yadav, M. P. Gordon and F. Quetier (1980). T-DNA from *Agrobacterium* Ti plasmid is in the nuclear DNA of crown gall tumour cells. *Proc. Natl. Acad. Sci. USA*, **77**, 4060–4064.

Choe, I. S., U. Melcher, K. Richards, G. Lebeurier and R. C. Essenberg (1985). Recombination between mutant cauliflower mosiac virus DNAs. *Plant Mol. Biol.*, **5**, 281–289.

Chu, G. and P. Sharp (1981). SV40 DNA transformation of cells in suspension: analysis of the efficiency of transcription and translation of T-antigen. *Gene*, **13**, 197–202.

Comai, L., D. Facciotti, W. R. Hiatt, G. Thompson, R. E. Rose and D. M. Stalker (1985). Expression in plants of a mutant *aroA* gene from *Salmonella typhimurium* confers tolerance to glyphosate. *Nature (London)*, **317**, 741–744.

Cornelissen, M. J., M. De Block, M. Van Montagu, J. Leemans, P. H. Schreier and J. Schell (1987). Plastid transformation: a progress report. In *Plant DNA Infectious Agents (Plant Gene Research)*, T. Hohn and J. Schell, pp. 311–320, Springer, Wien.

Crossway, A., J. V. Oakes, J. M. Irving, B. Ward, V. C. Knauf and C. K. Shewmaker (1986). Integration of foreign DNA following microinjection of tobacco mesophyll protoplasts. *Mol. Gen. Genet.*, **202**, 179–185.

Czernilofsky, A. P., R. Hain, B. Baker and U. Wirtz (1986a). Studies of the structure and functional organization of foreign DNA integrated into the genome of *Nicotiana tobacum*. *DNA*, **5**, 473–482.

Czernilofsky, A. P., R. Hain, L. Herrera-Estrella, H. Lörz, E. Goyvaerts, B. Baker and J. Schell (1986b). Fate of selectable marker DNA integrated into the genome of *Nicotiana tobacum*. *DNA*, **5**, 101–113.

Daubert, S., R. J. Shepherd and R. C. Gardner (1983). Insertional mutagenesis of the cauliflower mosaic virus genome. *Gene*, **25**, 21–28.

Davey, M. R., E. C. Cocking, J. Freeman, N. Pearce and I. Tudor (1980). Transformation of petunia protoplasts by isolated *Agrobacterium* plasmids. *Plant Sci. Lett.*, **18**, 307–313.

Davies, J. W., R. Townsend and J. Stanley (1987). The structure, expression, functions and possible exploitation of gemini virus genomes. In *Plant DNA Infectious Agents (Plant Gene Research)*, ed. T. Hohn and J. Schell, pp. 31–53. Springer, Wien.

De Block, M., J. Schell and M. Van Montagu (1985). Chloroplast transformation by *Agrobacterium tumefaciens*. *EMBO J.*, **4**, 1367–1372.

De Block, M., J. Botterman, M. Vaendewiele, J. Dockx, C. Thoen, V. Thoen, V. Gosselé, N. Rao Movva, C. Thompson, M. Van Montagu and J. Leemans (1987). Engineering herbicide resistance in plants by expression of a detoxifying enzyme. *EMBO J.*, **6**, 2513–2518.

De Cleene, M. and J. De Ley (1976). The host range of crown gall. *Bot. Rev.*, **42**, 389–466.

de Framond, A. J., E. W. Back, W. S. Chilton, L. Keyes and M. Chilton (1986). Two unlinked T-DNAs can transform the same tobacco plant and segregate in the F1 generation. *Mol Gen. Genet.*, **202**, 125–131.
de la Peña, A., H. Lörz and J. Schell (1987). Transgenic rye plants obtained by injecting DNA into young floral tillers. *Nature (London)*, **325**, 274–276.
Depicker, A., L. Herman, A. Jacobs, J. Schell and M. Van Montagu (1985). Frequencies of simultaneous transformation with different T-DNAs and their relevance to the *Agrobacterium*–plant cell interaction. *Mol. Gen. Genet.*, **201**, 477–484.
Deshayes, A., L. Herrera-Estrella and M. Caboche (1985). Liposome-mediated transformation of tobacco mesophyll protoplasts by an *E. coli* plasmid. *EMBO J.*, **4**, 2731–2737.
Ditta, G., S. Stanfield, D. Corbin and D. R. Helsinki (1980). Broad host range DNA cloning system for gram negative bacteria: construction of a gene bank in *Rhizobium meliloti*. *Proc. Natl. Acad. Sci. USA*, **77**, 7347–7351.
Donson, J., H. V. Gunn, C. J. Woolston, M. S. Pinner, M. I. Boulton, P. M. Mullineaux and J. W. Davies (1988). *Agrobacterium*-mediated infectivity of cloned digitaria streak virus DNA. *Virology*, **162**, 248–250.
Douglas, C. J., W. Halperin, M. Gordon and E. W. Nester (1986). Specific attachment of *Agrobacterium tumefaciens* to bamboo cells in suspension cultures. *J. Bacteriol.*, **161**, 764–766.
Draper, J., M. R. Davey, J. P.Freeman, E. C. Cocking and B. G. Cox (1982). Ti plasmid homologous sequences present in tissues from *Agrobacterium* plasmid transformed petunia tissues. *Plant Cell Physiol.*, **23**, 451–458.
Draper, J., R. Scott, P. Armitage and R. Walden (1988). Plant genetic transformation and gene expression: a laboratory manual. Blackwell, Oxford.
Dylan, T., L. Ielpi, S. Stanfield, L. Kashyap, C. Douglas, M. Yanofsky, E. Nester, D. R. Helsinki and G. Ditta (1986). *Rhizobium meliloti* genes required for nodule development are related to chromosomal virulence genes in *Agrobacterium tumefaciens*. *Proc. Natl. Acad. Sci. USA*, **83**, 4403–4407.
Elmer, J. S., G. Sunter, W. E. Gardiner, L. Brand, C. K. Browning, D. M. Bisaro and S. G. Rogers (1988). *Agrobacterium*-mediated inoculation of plants with tomato golden mosaic virus. *Plant Mol. Biol.*, **10**, 225–234.
Evans, D. A. and W. R. Sharp (1983). Single gene mutations in tomato plants regenerated from tissue culture. *Science (Washington D.C.)*, **221**, 949–951.
Fraley, R. T., S. L. Dellaporta and D. Papahadjopoulos (1982). Liposome-mediated delivery of TMV RNA into tobacco protoplasts: a sensitive assay for monitoring lipsome–protoplasts interactions. *Proc. Natl. Acad. Sci. USA*, **79**, 1859–1863.
Fraley, R. T., S. G. Rogers, R. B. Horsch, D. A. Eicholtz and J. S. Flick (1985). The SEV system: a new disarmed Ti plasmid vector for plant transformation. *Bio/Technology*, **3**, 629–635.
Fraley, R. T., S. G. Rogers and R. B. Horsch (1986). Genetic transformation in higher plants. *CRC Crit. Rev. Plant Sci.*, **4**, 1–46.
French, R., M. Janda and P. Ahlquist (1986). Bacterial gene inserted in an engineered RNA virus: efficient expression in monocotyledonous plant cells. *Science (Washington D.C.)*, **231**, 1294–1297.
Fromm, M., L. P. Taylor and V. Walbot (1985). Expression of genes transferred into monocot and dicot plant cells by electroporation. *Proc. Natl. Acad. Sci. USA*, **82**, 5824–5828.
Fromm, M., L. P. Taylor and V. Walbot (1986). Stable transformation of maize after gene transfer by electroporation. *Nature (London)*, **319**, 791–793.
Fromm, M., J. Callis, L. P. Taylor and V. Walbot (1987). Electroporation of DNA and RNA into plant protoplasts. *Methods Enzymol.*, **153**, 351–366.
Fromm, M. and V. Walbot (1987). Transient expression of DNA in plant cells. In *Plant DNA Infectious Agents (Plant Gene Research)*, ed. T. Hohn and J. Schell, pp. 304–311, Springer, Wien.
Fry, J., A. Barnason and R. B. Horsch (1987). Transformation of *Brassica napus with Agrobacterium tumefaciens* based vectors. *Plant Cell Rep.*, **6**, 321–325.
Garfinkel, D. J., R. B. Simpson, L. W. Ream, F. F. White, M. P. Gordon and E. W. Nebster (1981). Genetic analysis of crown gall: fine structure map by site-directed muatagenesis. *Cell*, **27**, 143–153.
Gelvin, S. B., S. J. Karcher and V. De Rita (1983). Methylation of T-DNA in *A. tumefaciens* and in several crown gall tumours. *Nucleic Acids Res.*, **11**, 159–174.
Gheysen, G., M. Van Montagu and P. Zambryski (1987). Integration of *A. tumefaciens* T-DNA involves rearrangement of target plant DNA sequences. *Proc. Natl. Acad. Sci. USA*, **84**, 6169–6173.
Gietl, C., Z. Koukolíková-Nicola and B. Hohn. Mobilization of T-DNA from *Agrobacterium* to plant cells involves a protein that binds single-stranded DNA. *Proc. Natl. Acad. Sci. USA*, **84**, 9006–9010.
Gluzman, Y. (1981). SV40-transformed simian cells support the replication of early SV40 mutants. *Cell*, **23**, 175–182.
Graves, A. C. F. and S. L. Goldman (1986). The transformation of *Zea mays* seedlings with *A. tumefaciens*: detection of T-DNA specific enzyme activities. *Plant Mol. Biol.*, **7**, 43–50.
Graves, A. C. F. and S. L. Goldman (1987). *A. tumefaciens* mediated transformation of the monocot genus *Galdiolus*: detection and expression of T-DNA-encoded genes. *J. Bacteriol.*, **169**, 1745–1746.
Grimsley, N., B. Hohn, T. Hohn and R. Walden (1986). 'Agroinfection', an alternative route for viral infection of plants by using the Ti plasmid. *Proc. Natl. Acad. Sci. USA*, **83**, 3282–3286.
Grimsley, N. and D. Bisaro (1987). Agroinfection. In *Plant DNA Infectious Agents (Plant Gene Research)*, ed. T. Hohn and J. Schell, pp. 88–109. Springer, Wien.
Grimsley, N., T. Hohn, J. W. Davies and B. Hohn (1987). *Agrobacterium*-mediated delivery of infectious maize streak virus into maize plants. *Nature (London)*, **325**, 177–179.
Gronenborn, B., R. C. Gardner, S. Schaefer and R. J. Shepherd (1981). Propagation of foreign DNA in plants using a cauliflower mosaic virus as vector. *Nature (London)*, **294**, 773–776.
Gronenborn, B. (1987). The molecular biology of cauliflower mosaic virus and its application as plant gene vector. In *Plant DNA Infectious Agents (Plant Gene Research)*, ed. T. Hohn and J. Schell, pp. 1–30. Springer, Wien.
Hain, R., H. H. Steinbiss and J. Schell (1984). Fusion of *Agrobacterium* and *E. Coli* spheroplasts with *Nicotiana tabacum* protoplasts: direct gene transfer from microorganism to higher plant. *Plant Cell Rep.*, **23**, 60–64.
Hain, R., P. Stabel, A. P. Czernilofsky, H. H. Steinbiss, L. Herrera- Estrella and J. Schell (1985). Uptake, integration, expression and genetic transmission of a selectable chimeric gene by plant protoplasts. *Mol. Gen. Genet.*, **199**, 161–168.
Halverson, L. J. and G. Stacey (1986). Signal exchange in plant–microbe interactions. *Microbiol. Rev.*, **50**, 193–225.
Harrison, B. D. (1985) Advances in gemini research. *Annu. Rev. Phytopathol.* **23**, 55–82.
Hauptmann, R. M., P. Ozias-Atkins, V. Vasil, Z. Tabaeizadeh, S. G. Rogers, R. B. Horsch, I. K. Vasil and R. T. Fraley (1987). Transient expression of electroporated DNA in monocotyledonous and dicotyledonous species. *Plant Cell Rep.*, **6**, 265–270.

Hayes, R. J., H. Macdonald, R. H. A. Coutts and K. W. Buck (1988a). Characterization of a subgenomic DNA isolated from *Triticum aestivum* infected with wheat dwarf virus. *J. Gen. Virol.*, **69**, 1339–1344.
Hayes, R. J., R. H. A. Coutts and K. W. Buck (1988b). Agroinfection with *Nicotiana* spp. with cloned DNA of tomato golden mosaic virus. *J. Gen. Virol.*, **69**, 1487–1496.
Hayes, R. J., I. T. D. Petty, R. H. A. Coutts and K. W. Buck (1988c). Gene amplification in plants by a replicating gemini virus vector. *Nature (London)*, **334**, 179 182.
Hemenway, C., R. Fang, W. K. Kaniewski, N. Chua and N. E. Tumer (1988). Analysis of the mechanism of protection in transgenic plants expressing the potato virus X coat protein or its antisense RNA. *EMBO J.*, **7**, 1273–1280.
Hepburn, A. G., L. E. Clark, L. Pearson and J. White (1983). The role of cytosine methylation in the control of nopaline synthase gene expression in a plant tumour. *J. Mol. Appl. Genet.*, **2**, 315–329.
Hernalsteens, J. P., L. Thia-Toong, J. Schell and M. Van Montagu (1984). An *Agrobacterium*-transformed cell culture from the monocot *Asparagus officinalis. EMBO J.*, **3**, 3039–3041.
Herrera-Estrella, L., A. Depicker, M. Van Montagu and J. Schell (1983). Expression of chimeric genes transferred into plant cells using a Ti plasmid-derived vector. *Nature (London)*, **303**, 209–213.
Hinnen, A., J. Hicks and G. Fink (1978). Transformation of yeast. *Proc. Natl. Acad. Sci. USA.*, **75**, 1929–1933.
Hoekema, A., P. J. J. Hooykaas and R. A. Schilperoort (1984). Transfer of the octopine T-DNA segment to plant cells mediated by different types of *Agrobacterium* tumour- or root-inducing plasmids: generality of the virulence systems. *J. Bacteriol.*, **158**, 383–385.
Hoekema, A., M. J. J. van Haaren, A. J. Fellinger, P. J. J. Hooykaas and R. A. Schilperoort (1985). Nononcogenic plant vectors for use in the *Agrobacterium* binary system. *Plant Mol. Biol.*, **5**, 85–89.
Hooykaas, P. J. J. (1988). Tumorigenicity of *Agrobacterium* in plants. In *Genetics of Bacterial Diversity*, eds. Hopwood and Chater, in press. Academic Press, London.
Hooykaas, P. J. J. and R. A. Schilperoort (1987). Detection of monocot transformation *via Agrobacterium tumefaciens. Methods Enzymol.*, **153**, 305–313.
Hooykaas-Van Slogteren, G. M. S., P. J. J. Hooykaas and R. A. Schilperoot (1984). Expression of Ti plasmid genes in monocotyledonous plants infected with *A. Tumefaciens. Nature (London)*, **331**, 763–764.
Horsch, R. B., and H. J.Klee (1986). Rapid assay of foreign gene expression in leaf discs transformed by *Agrobacterium*: role of T-DNA borders in the transfer process. *Proc. Natl. Acad. Sci. USA*, **83**, 4428–4432.
Horsch, R. B., J. E. Fry, N. Hoffman, D. Eichholtz, S. G. Rogers and R. T. Fraley (1985). A simple and general method for transferring genes into plants. *Science, (Washington D.C.)*, **227**, 1229–1231.
Howarth, A. J., R. C. Gardner, J. Messing and R. J. Shepherd (1981). Nucleotide sequences of naturally occurring deletion mutants of cauliflower mosaic virus. *Virology*, **112**, 1229–1231.
Howell, S. H., L. L. Walker and R. M. Walden (1981). Rescue of *in vitro* generated mutants of the cloned cauliflower mosaic virus genome in infected plants. *Nature (London)*, **293**, 483–486.
Hull, R. and J. W. Davies (1983). Genetic engineering with plant virus and their potential as vectors. *Adv. Virus Res.*, **28**, 1–33.
Jensen, J. S., K. A. Marcker, L. Otten and J. Schell (1986). Nodule specific expression of a chimeric soybean leghemoglobin gene in transgenic *Lotus corniculatus. Nature (London)*, **321**, 669–674.
Jorgensen, R., C. Snyder and J. D. G. Jones (1987). T-DNA is organized predominantly in inverted repeat structures in plants transformed with *A. tumefaciens* C58 derivatives. *Mol. Gen. Genet.*, **207**, 471–477.
Klee, H. J., M. F. Yanofsky and E. W. Nester (1985). Vectors for transformation of higher plants. *Bio/Technology*, **3**, 637–642.
Klee, H. J., R. Horsch and S. Rogers (1987). *Agrobacterium*-mediated plant transformation and its further applications to plant biology. *Annu. Rev. Plant Physiol.*, **38**, 467–486.
Klein T. M., E. D. Wolf, R. Wu and J. C. Stanford (1987). High velocity microprojectiles for delivering nucleic acids into living cells. *Nature (London)*, **327**, 70–72.
Klein, T. M., M. Fromm, A. Weissinger, D. Thomas, S. Schaaf, M. Sletten and J. C. Stamford (1988). Transfer of foreign genes into intact maize cells using high velocity microprojectiles. *Proc. Natl. Acad. Sci. USA*, in press.
Koukolíková-Nicola, Z., L. Albright and B. Hohn (1987). The mechanism of T-DNA transfer from *Agrobacterium tumefaciens* to the plant cell. In *Plant DNA Infectious Agents (Plant Gene Research)*, ed. T. Hohn and J. Schell, pp. 110–148. Springer, Wien.
Krens, F. A., L. Molendijk, G. J. Wullems and R. A. Schilperoort (1982). *In vitro* transformation of plant protoplasts with Ti plasmid DNA. *Nature (London)*, **296**, 72–74.
Kerns, F. A., R. M. W. Mans, T. M. S. Van Slogteren, J. H. C. Hoge, G. J. Wullems and R. A. Schilperoot (1985). Structure and expression of DNA transferred to tobacco *via* transformation of protoplasts with Ti plasmid: cotransfer of T-DNA and non-T-DNA sequences. *Plant Mol. Biol.*, **5**, 223–234.
Langbridge, W. H. R., B. J. Li and A. Z. Szalay (1987). Uptake of DNA and RNA into cells mediated by electroporation. *Methods Enzymol.*, **153**, 336–351.
Lebeurier, G. (1986). Cauliflower mosaic virus potential vector gene: genome organization and replication. *Symbiosis*, **2**, 19–30.
Lichtenstein, C. P. and J. Draper (1985). Genetic engineering in plants. In *DNA Cloning*, ed. D. M. Glover, vol. 2, pp. 67–119. IRL Press, Oxford.
Lichtenstein, C. P. and S. L. Fuller (1987). Vectors for the genetic engineering of plants. In *Genetic Engineering*, ed. P. W. J. Rigby, vol. 6, pp. 103–183. Academic Press, London.
Lindsey, K. and M. G. K. Jones (1987). Transient gene expression in electroporated protoplasts and intact cells of sugar beet. *Plant Mol. Biol.*, **10**, 43–52.
Lurquin, P. F. and C. I. Kado (1977). *Escherichia coli* plasmid pBR313 insertion into plant protoplasts and into their nuclei. *Mol. Gen. Genet.*, **154**, 113–121.
Matzke, A. J. M. and M. A. Matzke (1986). A novel set of Ti plasmid-derived vectors for the production of transgenic plants, *Plant Mol. Biol.*, **7**, 357–365.
McClintock, B. (1951). Mutable loci in maize. *Year Book – Carnegie Inst. Washington*, **50**, 174–181.
McCormick, S., J. Niedermeyer, J. Fry, A. Barnason, R. Horsch and R. Fraley (1986). Leaf disc transformation of cultivated tomato using *Agrobacterium tumefaciens. Plant Cell Rep.*, **5**, 81–85.
McKnight, T. D., M. T. Lillis and R. B. Simpson (1987). Segregation of genes transferred to one plant cell from two separate *Agrobacterium* strains. *Plant Mol. Biol.*, **8**, 439–445.
Melchers, L. S. and P. J. J. Hooykaas (1987). Virulence of *Agrobacterium. Oxford Surveys of Plant and Cell Biology*, **4**, 167–220.

Memelink, J., S. de Pater, J. Harry, C. Hoge and R. A. Schilperoort (1987). T-DNA hormone biosynthetic genes: phytohormones and gene expression in plants. *Dev. Genet.*, **8**, 321–337.

Menczel, L., L. S. Polsby, K. E. Steinback and P. Maliga (1986). Fusion-mediated transfer of triazine resistant chloroplasts: characterization of *Nicotiana tabacum* hybrid plants. *Mol. Gen. Genet.*, **205**, 201–205.

Meyer, P., E. Walgenbach, K. Bussmann, G. Hombrecher and and H. Saedler (1985). Synchronized tobacco protoplasts are efficiently transformed by DNA. *Mol. Gen. Genet.*, **201**, 513–518.

Mikki, L. A., T. J. Reich and V. N. Iyer (1987). Microinjection: an experimental tool for studying and modifying plant cells. In *Plant DNA Infectious Agents (Plant Gene Research)*, ed. T. Hohn and J. Schell, pp. 249–266. Springer, Wien.

Morgan, A. J., P. N. Cox, D. A. Turner, E. Peel, M. R. Davey, K. M. A. Gartland and B. J. Mulligan (1987). Transformation of tomato using an Ri plasmid vector. *Plant Sci.*, **49**, 37–49.

Morikawa, H., A. Iida, C. Matsui, M. Ikegami and Y. Yamada (1986). Gene transfer into intact plant cells by electroinfection through cell walls and membranes. *Gene*, **41**, 121–124.

Mouras, A., M. W. Saul, S. Essad and I. Potrykus (1987). Localization by *in situ* hybridization of a low copy chimeric resistance gene introduced into plants by direct gene transfer. *Mol. Gen. Genet.*, **207**, 204–209.

Müller, A. J., R. R. Mendel, J. Schiemann, C. Simoens and D. Inze (1987). High meiotic stability of a foreign gene introduced into tobacco by *Agrobacterium*-mediated transformation. *Mol. Gen. Genet.*, **207**, 171–175.

Negrutiu, I., A. Mouras, M. Horth and M. Jacobs (1987a). Direct gene transfer to plants: present developments and some future prospects. *Plant Physiol. Biochem.*, **25**, 493–503.

Negrutiu, I., R. Shillito, I. Potrykus, G. Biasini and F. Sala (1987b). Hybrid genes in the analysis of transformation conditions. *Plant Mol. Biol.*, **8**, 363–373.

Nishiguchi, M., W. H. R. Langbridge, A. A. Szalay and M. Zaitlin (1986). Electroportation-mediated infection of tobacco leaf protoplasts with TMV RNA and CLV RNA. *Plant Cell Rep.*, **5**, 57–60.

Ohta, Y. (1986). High efficiency transformation of maize by a mixture of pollen and exogenous DNA. *Proc. Natl. Acad. Sci. USA*, **83**, 715–719.

Okada, K., I. Takebi and T. Nagata (1986a). Expression and integration of genes introduced into highly synchronized plant protoplasts. *Mol. Gen. Genet.*, **205**, 398–403.

Okada, K., T. Nagata and I. Takebi (1986b). Introduction of functional RNA into plant protoplasts. *Plant Cell Rep.*, **27**, 619–626.

Ou-Lee, T., R. Turgeon and R. Wu (1986). Expression of foreign gene linked to either a plant virus or a *Drosophilia* promoter after electroporation of protoplasts of rice, wheat and sorghum. *Proc. Natl. Acad. Sci. USA*, **83**, 6815–6819.

Owens, L. D. and D. E. Cress (1985). Genotypic variability of soybean response to *Agrobacterium* strains harbouring Ti or Ri plasmids. *Plant Physiol.*, **77**, 87–94.

Parke, D., N. Ornston and E. W. Nester (1987). Chemotaxis to plant phenolic inducers of virulence genes is constitutively expressed in the absence of the Ti plasmid in *Agrobacterium tumefaciens*. *J. Bacteriol.*, **169**, 5336–5338.

Pazkowski, J., B. Pisan, R. D. Shillito, T. Hohn, B. Hohn and I. Potrykus (1986). Genetic transformation of *Brassica campestris var. rapa* protoplasts with an engineered cauliflower mosaic virus genome. *Plant Mol. Biol.*, **6**, 303–312.

Peerbolte, R., F. Krens, R. M. W. Mans, M. Floor, J. H. C. Hoge, G. J. Wullems and R. A. Schilperoort (1985). Transformation of plant protoplasts with DNA: cotransformation of nonselected calf thymus carrier DNA and meiotic segregation of transforming DNA sequences. *Plant Mol. Biol.*, **5**, 235–246.

Peerbolte, R., M. Floor, P. Ruigrok, J. H. C. Hoge and G. J. Wullems (1987a). Stability and expression of transferred DNA in F1 tobacco transformants studied at various states of differentiation. *Planta*, **172**, 448–462.

Peerbolte, R., K. Leenhouts, G. M. S. Hooykaas-van Slogteren, J. H. C. Hoge, G. J. Wullems and R. A. Schilperoort (1987b). Clones from shooty tobacco crown gall tumour I: irregular T-DNA structures and organization, T-DNA methylation and conditional expression of opine genes. *Plant Mol. Biol.*, **7**, 285–299.

Peerbolte, R., K. Leenhouts, G. M. S. Hooykaas-van Slogteren, G. J. Wullems and R. A. Schilperoort (1987c). Clones from shooty tobacco crown gall tumour II: deletions, rearrangements and amplifications resulting in irregular T-DNA structures and organizations. *Plant Mol. Biol.*, **7**, 265–284.

Peerbolte, R., P. Ruigrok, G. J. Wullems and R. A. Schilperoort (1987d). T-DNA rearrangements due to tissue culture: somaclonal variation in crown gall tissues. *Plant Mol. Biol.*, **9**, 51–57.

Perealta, G., R. Hellmiss and W. Ream (1986). Overdrive, a T-DNA transmission enhancer on the *A. tumefaciens* Ti plasmid. *EMBO J.*, **5**, 1137–1142.

Petit, A., A. Berkaloff and J. Tempen (1986). Multiple transformations of plant cells by *Agrobacterium* may be responsible for the complex organization of T-DNA in crown gall and hairy root. *Mol. Gen. Genet.*, **202**, 388–393.

Potrykus, I., J. Pazkowski, M. W. Saul, S. Krüger-Lebus, T. Müller, R. Schocher, I. Negrutiu, P. Künzler and R. D. Shillito (1985a). Direct gene transfer to protoplasts: an efficient and generally applicable method for stable alterations in plant genomes. In *Plant Genetics, UCLA Symposia Series*, ed. M. Freeling, vol. 35, pp. 181–199. Liss, New York.

Potrykus, I., J. Pazkowski, M. W. Saul, J. Petruska and R. D. Shillito (1985b). Molecular and general genetics of a hybrid foreign gene introduced into tobacco by direct gene transfer. *Mol. Gen. Genet.*, **199**, 169–177.

Potrykus, I., M. W. Saul, J. Petruska, J. Pazkowski and R. D. Shillito (1985c). Direct gene transfer to cells of a graminaceous monocot. *Mol. Gen. Genet.*, **199**, 183–188.

Potrykus, I., J. Pazkowski, M. W. Saul, I. Negrutiu and R. D. Shillito (1987a). Direct gene transfer of plants: facts and future. *Plant Biol.*, **3**, 289–302.

Potrykus, I., J. Pazkowski, R. D. Shillito and M. W. Saul (1987b). Direct gene transfer to plants. In *Plant DNA Infectious Agents (Plant Gene Research)*, ed. T. Hohn and J. Schell, pp. 230–248. Springer, Wien.

Reich, T., V. N. Iyer and B. Miki (1985). Transformation of alfalfa protoplasts by intranuclear microinjection of Ti plasmids. In *Abstracts First International Congress Plant Molecular Biology*. p. 28, University of Georgia, Centre for Continuing Education, Athens, GA.

Rhodes, C. A., D. A. Pierce, I. J. Mettler, D. Mascarenhas and J. J. Detmer (1988). Genetically transformed maize plants from protoplasts. *Science* (*Washington D.C.*), **240**, 204–207.

Riggs, C. D. and G. W. Bates (1986). Stable transformation of tobacco by electroporation: evidence for plasmid concatenation. *Proc. Natl. Acad. Sci. USA*, **83**, 5602–5606.

Rogers, S. G., R. B. Horsch and R. T. Fraley (1986a). Gene transfer in plants: production of transformed plants using Ti plasmid vectors. *Methods Enzymol.*, **118**, 627–640.

Rogers, S. G., D. M. Bisaro, R. B. Horsch, R. T. Fraley, N. L. Hoffman, L. Brand, J. S. Elmer and A. M. Lloyd (1986b). Tomato golden mosaic virus A component replicates autonomously in transgenic plants. *Cell*, **45**, 593–600.

Rogers, S. G. and H. Klee (1987). Pathways to plant genetic manipulation employing *Agrobacterium*. In *Plant DNA Infectious Agents (Plant Gene Research)*, ed. T. Hohn and J. Schell, pp. 179–204. Springer, Wien.

Rosenberg, N., A. E. Gad, A. Altman, N. Navot and H. Czosnek (1988). Liposome-mediated introduction of the CAT gene and its expression in tobacco protoplasts. *Plant Mol. Biol.*, **10**, 185–192.

Rouze, P., A. Deshayes and M. Caboche (1983). Use of liposomes for the transfer of nucleic acids: optimization of the method for tobacco mesophyll protoplasts with tobacco mosaic virus RNA. *Plant Sci. Lett.*, **31**, 55–64.

Rubin, R. A. (1986). Genetic studies on the role of octopine T-DNA border regions in crown gall tumour formation. *Mol. Gen. Genet.*, **202**, 312–320.

Saul, M. W., J. Pazkowski, R. D. Shillito and I. Potrykus (1987). Methods for direct gene transfer to plants. *Plant Physiol. Biochem.*, **25**, 361–364.

Schäfer, W., A. Görz and G. Kahl (1987). T-DNA integration and expression in a monocot crop plant after induction of *Agrobacterium. Nature (London)*, **327**, 529–532.

Schimke, R. T. (1984). Gene amplification in cultured animal cells. *Cell*, **37**, 705–713.

Schocher, R. J., R. D. Shillito, M. W. Saul, J. Pazkowski and I. Potrykus (1986). Cotransformation of unlinked foreign genes into plants by direct gene transfer. *Bio/Technology*, **4**, 1093–1096.

Scott, R. J. and J. Draper (1987). Transformation of carrot tissue derived from proembryogenic suspension cells: a useful model system for gene expression studies in plants. *Plant Mol. Biol.*, **8**, 265–274.

Shahin, E. A., K. Sukhapinda, R. B. Simpson and R. Spivey (1986). Transformation of cultivated tomato by a binary vector in *Agrobacterium rhizogenes*: transgenic plants with normal phenotype harbour binary vector T-DNA but no Ri plasmid T-DNA *Theor. Appl. Genet.*, **72**, 770–777.

Shaw, Charles H., J. Leemans, Christine H. Shaw, M. Van Montagu and J. Schell (1983). A general method for the transfer of cloned genes to plant cells. *Gene*, **23**, 315–330.

Shaw, Charles H., A. M. Ashby, A. Brown, C. Royal, G. J. Loake and Christine H. Shaw (1988). *VirA* and *G* are the Ti plasmid functions required for chemotaxis of *Agrobacterium tumefaciens* towards acetosyringone. *Mol. Microbiol.*, **6**, 413–417.

Shewmaker, C. K., J. R. Caton and C. M. Houck and R. C. Gardner (1985). Transcription of cauliflower mosaic virus integrated into plant genomes. *Virology*, **140**, 281–288.

Shillito, R. D. and I. Potrykus (1987). Direct gene transfer to protoplasts of dicotyledonous and monocotyledonous plants by a number of methods including electroporation. *Methods Enzymol.*, **153**, 312–336.

Siegel, A. (1985). Plant virus vectors for gene transfer may be of considerable use despite a presumed high error frequency during RNA synthesis. *Plant Mol. Biol.*, **4**, 327–329.

Simoens, C., T. Alliotte, R. Mendel, A. Müller, J. Schiemann, M. Van Lijsebettens, J. Schell, M. Van Montagu and D. Inze (1986). A binary vector for transferring genomic libraries to plants. *Nucleic Acids Res.*, **14**, 8073–8090.

Simpson, R. B., A. Spielmann, L. Margassian and T. D. McKinght (1986). A disarmed binary vector from *A. tumefaciens* functions in *A. rhizogenes*. Frequent cotransformation of two distinct T-DNAs. *Plant Mol. Biol.*, **6**, 403–415.

Spielmann, A. and R. B. Simpson (1986). T-DNA structure in transgenic tobacco plants with multiple independent integration sites. *Mol. Gen. Genet.*, **205**, 34–41.

Stachel, S. E., E. Messens, M. Van Montagu and P. C. Zambryski (1985). Identification of the signal molecules produced by wounded plant cells that activate T-DNA transfer in *Agrobacterium tumefaciens. Nature (London)*, **318**, 624–629.

Stachel, S. E. and P. C. Zambryski (1986). *Agrobacterium tumefaciens* and the susceptible plant cell: a novel adaptation of extracellular recognition and DNA conjugation. *Cell*, **47**, 155–157.

Stanley, J. (1985). The molecular biology of gemini viruses. *Adv. Virus Res.*, **30**, 139–177.

Stanley, J. and R. Townsend (1986). Infectious mutants of cassava latent virus generated *in vivo* from intact recombinant DNA clones containing single copies of the genome. *Nucleic Acids Res.*, **14**, 5981–5998.

Steinbiss, H. H., P. Stabel, R. Töpfer, R. D. Hirtz and J. Schell (1985). Transformation of plant cells by microinjection of DNA. In *Proceedings of the 1984 Wye International Symposium*. University of London.

Stougaard, J., D. Abildsten and K. A. Marcker (1987). The *Agrobacterium rhizogenes* pRiT$_L$-DNA segment as a gene vector system for the transformation of plants. *Mol. Gen. Genet.*, **207**, 251–255.

Szabo, L. J. and A. R. Cashmore (1987). Targetting nuclear gene products into chloroplasts. In *Plant DNA Infectious Agents (Plant Gene Research)*, ed. T. Hohn and J. Schell, pp. 321–340. Springer, Wien.

Takamatsu, N., M. Ishikawa, T. Meshi and Y. Okada (1987). Expression of bacterial chloramphenicol acetyltransferase gene in tobacco plants mediated by TMV-RNA. *EMBO J.*, **6**, 307–311.

Tepfer, D. (1984). Transformation of several species of higher plants by *Agrobacterium rhizogenes*: sexual transmission of the transformed genotype and phenotype. *Cell*, **37**, 959–967.

Tepfer, M. and F. Casse-Delbart (1987). *Agrobacterium rhizogenes* as a vector for transforming higher plants. *Microbiol. Sci.*, **4**, 24–28.

Thomashow, M. F., J. E. Karlinsey, J. R. Marks and R. E. Hurlbert (1987). Identification of a new virulence locus in *Agrobacterium tumefaciens* that affects polysaccharide composition and plant cell attachment. *J. Bacteriol.*, **169**, 3209–3216.

Tonaka, N., M. Ikegami, T. Hohn, C. Matsui and I. Watanabe (1984). *E. coli* spheroplast-mediated transfer of cloned cauliflower mosaic virus DNA into plant protoplasts. *Mol. Gen. Genet.*, **195**, 378–380.

Töpfer, R., V. Matzeit, B. Gronenborn, J. Schell and H. Steinbiss (1987). A set of plant expression vectors for transcriptional and translational fusions. *Nucleic Acids Res.*, **15**, 5890.

Townsend, R. J., J. Stanley, S. J. Curson and M. M. Short (1986). Major polyadenylated transcripts of cassava latent virus and location of the gene-encoding coat protein. *EMBO J.*, **4**, 33–37.

Vaeck, M., A. Reynaerts, H. Höfte, S. Jansens M. De Beuckleer, C. Dean, M. Zabeau, M. Van Montagu and J. Leemans (1987). Transgenic plants protected from insect attack. *Nature (London)*, **328**, 33–37.

van Haaren, M. J. J., N. J. A. Sedee, R. A. Schilperoort and P. J. J. Hooykaas (1987a). Overdrive is a T-region transfer enhancer which stimulates T-strand production in *Agrobacterium tumefaciens. Nucleic Acids Res.*, **15**, 8983–8997.

van Haaren, M. J. J., J. T. Pronk, R. A. Schilperoort and P. J. J. Hooykaas (1987b). Functional analysis of *Agrobacterium tumefaciens* octopine Ti plasmid left and right T-region border fragments. *Plant Mol. Biol.*, **8**, 95–104.

Van Lijsbettens, M., D. Inze, J. Schell and M. Van Montagu (1986). Transformed cell clones as a tool to study T-DNA integration mediated by *Agrobacterium tumefaciens. J. Mol. Biol.*, **188**, 129–145.

van Vloten-Doting, J. Bol and B. Cornelissen (1985). Plant virus based vectors for gene transfer will be of limited use because of the high error frequency during viral RNA synthesis. *Plant Mol. Biol.*, **41**, 323–326.
Walden, R. M. and S. H. Howell (1982). Intragenomic recombination events among pairs of defective cauliflower mosaic virus genomes in plants. *J. Mol. Appl. Genet.*, **1**, 447–456.
Ward, A., E. Etassami and J. Stanley (1988). Expression of a bacterial gene in plants mediated by infectious gemini virus DNA. *EMBO J.* **7**, 1583–1587.
Watts, J. W., J. M. King and N. J. Stacey (1987). Inoculation of protoplasts with viruses by electroporation. *Virology*, **157**, 40–46.
Wienand, U. and H. Saedler (1987). Plant transposable elements: unique structures for gene tagging and gene cloning. In *Plant DNA Infections Agents (Plant Gene Research)*, ed. T. Hohn and J. Schell, pp. 205–229. Springer, Wien.
Willetts, N. and B. Wilkins (1984). Processing of plasmid DNA during bacterial conjugation. *Microbiol. Rev.*, **48**, 24–41.
Willmitzer, L., M. De Beuckeleer, M. Lemmers, M. Van Montagu and J. Schell (1980). DNA from Ti-plasmid present in nucleus and absent from plastids of crown gall plant cells. *Nature (London)*, **287**, 359–361.
Yadav, N. S., J. Vanderleyden, D. R. Bennett, W. M. Barnes and M. D. Chilton (1982). Short direct repeats flank the T-DNA on a nopaline Ti plasmid. *Proc. Natl. Acad. Sci. USA*, **79**, 6322–6326.
Yanofsky, M. F., S. G. Porter, C. Young, L. M. Albright, M. P. Gordon and E. W. Nester (1986). The *virD* operon of *A. tumefaciens* encodes a site specific endonuclease. *Cell*, **47**, 471–477.
Zambryski, P., M. Hollsters, K. Kruger, A. Depicker, J. Schell, M. Van and H. M. Goodman (1980). Tumour DNA structure in plant cells transformed by *A. tumefaciens. Science (Washington D.C.)*, **209**, 1385–1391.
Zambryski, P., H. Joos, C. Genetello, J. Leemans, J. Van Montagu and J. Schell (1983). Ti plasmid vector for the introduction of DNA into plant cells without alteration of their normal regeneration capacity. *EMBO J.*, **2**, 2143–2150.
Zimmermann, U., J. Vienken and G. Pilwat (1984). Electrodiffusion of cells. In *Investigative Microtechniques in Medicine and Biology*, ed. J. Chayen and L. Bitensky, vol. 1, pp. 89–167. Dekker, New York.

# 10
# Expression of Plant Genes in Yeast and Bacteria

NICO OVERBEEKE and C. THEO VERRIPS
*Unilever Research Laboratorium, Vlaardingen, The Netherlands*



## 10.1 INTRODUCTION

There are two important reasons for the expression of plant genes in microorganisms. Firstly, it can facilitate the detailed structural and functional analysis of individual plant genes and their products. Secondly, it can create possibilities for the economically attractive production of plant proteins with a commercial impact.

Detailed analysis of the structure and function of genes and gene products in general requires the availability of a large number of mutants containing structural variations. In addition, randomly produced mutants have contributed enormously to our understanding of plant biology, and molecular biological techniques have made it possible to introduce predetermined mutations directly. With respect to microorganisms which can be genetically engineered easily (*Escherichia coli*, *Bacillus subtilis*, *Saccharomyces cerevisiae*), this approach has proved its value and will, therefore, also be used in the study of plant genes. However, for plant cells the situation is more complex. Genetic manipulation has been realized for a number of species, both dicotyledons and monocotyledons, as a result of which major and exciting progress has been made (see Chapter 9 of this volume). Yet, the construction and analysis of series of mutants for various plants is still difficult, for the methods are laborious and the crucial step of regeneration is a major obstacle. Until recently this approach was limited to genes encoded by the nucleus; however, new methods and vectors have been developed for the transformation of cell organelles (chloroplast and mitochondria; Weisbeek, personal communication). A complicating factor for nuclear genes is the presence of so-called multigene families in the plant chromosome, a number of almost identical genes encoding functionally identical products. Introduction and analysis of one mutated gene will require the elimination or modification of all the others or the multiplication of the mutated gene, otherwise the changed phenotype may be masked.

In view of the above complications it was decided to study plant genes and their products in appropriate microorganisms. Bacteria like *E. coli*, *B. subtilis* and the yeast *S. cerevisiae* have been employed intensively for the detailed analysis of both homologous and heterologous genes and gene products (for reviews see Doi *et al.*, 1986; Holland *et al.*, 1986; Kingsman *et al.*, 1985 and 1987; Marston, 1986). Specialized systems have been developed for the introduction of (foreign) genes, stable maintenance, expression of the genes and, optionally, secretion of the gene products. Employing these systems properly in genetic studies on plant genes and gene products will not only contribute to our knowledge and understanding of plant genes but will also create facilities for the commercial production of those proteins that are synthesized in the plants only in low amounts or that are hard to purify.

This chapter surveys the contributions to our knowledge of plant genes obtained from this microbial approach and evaluates future (im)possibilities.

## 10.2 EXPRESSION OF CELL ORGANELLE GENES IN MICROORGANISMS

In plant cells the cell organelles (chloroplasts, mitochondria) have an essential function in, amongst other things, respiration, photosynthetic $CO_2$ fixation and ATP synthesis. Also, economically important factors like resistance to plant pathogens and herbicides and male sterility are localized in the cell organelles. They have their own genome which encodes for a number of proteins involved in the processes mentioned above. The synthesis of these proteins takes place inside the organelle. Apart from organelle-encoded proteins, genes in the cell nucleus are involved: these are synthesized in the cytoplasm and subsequently imported into the cell organelle (for a review, see Weisbeek *et al.*, 1989).

The cell organelle encoded genes resemble prokaryotic genes in many respects. They have 5′ sequences resembling bacterial promoters and homology with the bacterial Shine and Delgarno sequence (ribosome binding). Translation starts with f-Met tRNA, and translation is on 70S ribosomes containing 23S, 16S and 5S rRNA (for a review see Whitfeld and Bottomley, 1983). So the obvious approach to the study of cell organelle genes and products is *via* prokaryotic organisms like *E. coli* and *B. subtilis*. Although methods of transforming cell organelles have been developed recently, this is not yet a routinely reproducible procedure and therefore cloning and expression of plant cell organelle genes in microorganisms remains important.

### 10.2.1 Functioning of Promoters from Chloroplast Genes in *E. coli*

The first indications for the functioning of chloroplast promoters in *E. coli* were obtained by *in vitro* transcription/translation of chloroplast genes using cell-free extracts of *E. coli* (Erion *et al.*, 1983; Whitfeld and Bottomley, 1980). After the first demonstration of the *in vivo* functioning of a chloroplast promoter in *E. coli* by Gatenby *et al.* (1981), further investigations were carried out on a number of promoters, and for a compilation see Kung and Lin (1985). A comparison by these authors of 60 chloroplast promoters showed a consensus in the −35 region of TTGACA, and in the −10 region of TATAAT comparable to prokaryotic promoters.

Replacement of the chloroplast promoter by an appropriate *E. coli* promoter (*e.g.* lac or $\lambda$PL, for a comparison of their relative strengths, see Deuschle *et al.*, 1986) results in substantially higher expression. Using such an *E. coli* promoter, the large subunit of ribulose bisphosphate carboxylase (RuBisCo) could be produced in amounts up to 10% of cellular protein (see Section 10.2.2.1). This is not too surprising, since next to structural differences one can anticipate a complex regulation of the transcription to realize the correct interaction between nuclear genes and chloroplast genes in relation to adaptations in environmental conditions (for a review, see Bohnert *et al.*, 1982).

This implies also a limitation for the study of these chloroplast promoters in *E. coli*. Nevertheless, some interesting features have been found. Surprisingly, Valle *et al.* (1988) were not able to detect mRNA, encoding the ribulose bisphosphate carboxylase large subunit of the photosynthetic bacterium *Chromatium vinosum* in *E. coli* using the *Chromatium* promoter. When the gene was placed downstream of the *E. coli* tac promoter, RuBisCo mRNA was detected. By *in vitro* transcription analysis they showed that *E. coli* RNA polymerase is unable to initiate transcription at the RuBisCo promoter.

The promoter for the maize chloroplast ATPase $\beta$-subunit was found to be active by cloning in front of the *E. coli* lac Z gene (Bradley and Gatenby, 1985). Subsequently promoter mutants were isolated after Ba-131 deletion by screening in the heterologous host *E. coli*. No effects of deletions upstream of the −35 region were found but effects of mutations in the −35 consensus TTGACA were significant. These phenomena were very similar to those of homologous *E. coli* promoters. Analysis of this series of promoter mutants in an *in vitro* chloroplast transcription system showed results similar to those obtained *in vivo* in *E. coli*.

The chloroplast PA promoter from the algae *Chlamydomonas* is stimulated by novobiocin, a DNA gyrase inhibitor, both in *Chlamydomonas* and *E. coli*. Detailed analysis using mutations or deletion revealed that upstream '−35' regions were essential for regulation in *E. coli*, which could also be essential in *Chlamydomonas* (Thompson and Mosig, 1987).

These results nicely illustrate the usefulness of *E. coli* as a heterologous system for studying mutated chloroplast promoters, which until very recently has not been possible in the host itself.

### 10.2.2 Synthesis of Chloroplast-encoded Gene Products in *E. coli*

#### *10.2.2.1 Production and analysis of the ribulose bisphosphate carboxylase/oxygenase (RuBisCo)*

The ribulose-1,5-bisphosphate carboxylase/oxygenase (RuBisCo, EC 4.1.1.39) is a key enzyme in both photosynthesis and photorespiration, in which it catalyzes the initial reaction of the $CO_2$ fixation and $O_2$ fixation, respectively. In higher plants the active enzyme occurs in a multimeric form consisting of eight large subunits (LS, 52–55 kDa) encoded by the chloroplast and eight small subunits (SS, 12–15 kDa) encoded by the nucleus and synthesized as a precursor which is processed during or after transport into the chloroplast. This enzyme has been considered as a target for mutational changes to improve either the ratio of carboxylation to oxygenation or to increase the affinity of the enzyme for $CO_2$ (Miziorko and Lorimer, 1983). One option is to combine the SS from one plant with the LS from another. Protoplast fusion of cells from both plants can result in segregation of the homologous type of chloroplasts (for a review, see Galun and Aviv, 1983). This procedure can only be applied to rather closely related species. A more direct and general approach might be to study in a microbial organism various combinations of SS and LS, as well as mutant genes obtained by site-directed mutagenesis. The availability of detailed 3D structures of the ribulose-1,5-bisphosphate carboxylase/oxygenase of *Rhodospirillum rubrum* (Lundqvist and Schneider, 1989), tobacco (Chapman *et al.*, 1988), spinach (Andersson *et al.*, 1989) and *Synechococcus* (Newman and Gutteridge, 1990) will very much stimulate this approach.

The chloroplast promoter of the *rbc L* gene encoding LS can direct the synthesis of LS in *E. coli*, but the efficiency is rather low. When preceded by a strong, inducible *E. coli* promoter (lac, $\lambda$PL), production of maize LS increases to more than 10% of cellular protein (Gatenby and Castleton, 1982; Somerville *et al.*, 1986). This LS does not have any detectable RuBisCo activity and is present in soluble form only at a low level (Somerville *et al.*, 1986). As has been described for various proteins, *e.g.* chymosin (Kawaguchi *et al.*, 1987), the LS accumulates in electron-dense amorphous granules in *E. coli*. These granules can easily be purified and solubilized in aqueous urea (8 mol $L^{-1}$). With immunoblotting the identity of the 55 kDa protein was demonstrated to be RuBisCo LS. It was hypothesized that the presence of the SS was essential to achieve a soluble form of the LS. However, coproduction of the maize LS and wheat SS (Gatenby *et al.*, 1987) did not result in an active holoenzyme either. This indicates that other factors are required for assembling the enzyme. Using mutants of *Nicotiana tabacum* Avni et al. (1989) showed that replacement of the conserved serine 112 into phenylalanine does not inhibit incorporation of LS and the binding protein into B complexes, but prevents the proper alignment of LS dimers and, oligomerization. One option could be that a nuclear-encoded, LS-binding protein is required to mediate the assembly process, as postulated by various authors (Barraclough and Ellis, 1980; Bloom *et al.*, 1983; Roy *et al.*, 1982). Another explanation is aberrant folding of the LS in *E. coli*, a phenomenon which has been found for various heterologous products in *E. coli* (for a review, see Marston, 1986). To avoid aberrant folding, another host, for instance the yeast *Saccharomyces cerevisiae*, might be more suited for these studies. There are various examples of enzymes which are produced in *E. coli* in an inactive conformation but in *S. cerevisiae* in an active, correctly folded form (*e.g.* calf prochymosin and interferon $\gamma$, for a review, see Kingsman *et al.*, 1985).

In contrast to higher plants the RuBisCo from *Cyanobacteria* can assemble in *E. coli* to give an active holoenzyme composed of 8 SS and 8 LS (Tabita and Small, 1985; Viale *et al.*, 1985). Although higher plant and cyanobacterial RuBisCo enzymes are rather similar, there are some fundamental differences. The SS is not produced in a precursor form so no maturation is required and the solubility of LS is significantly different (Andrews and Ballment, 1983; Voordouw *et al.*, 1984).

Elegant studies by van der Vies *et al.* (1986) demonstrated heterologous assembly between LS of *Synechococcus* and SS of wheat in *E. coli*, resulting in a functional enzyme but with 10% of the homologous enzyme activity and a weak binding between SS and LS. The LS from *Synechococcus* when expressed alone was found predominantly in an octameric form but without enzyme activity, indicating an essential role for the SS in enzyme activity but not in LS octamer formation.

Newman and Gutteridge (1990) used *E. coli* to produce sufficient quantities of the RuBisCo of *Synechococcus* for crystallization studies. They found that the best crystals were obtained when the RuBisCo was first activated with cofactors $CO_2$ and $Mg^{2+}$ and the intermediate analog 2′-carboxyarabinitol-1,5-bisphosphate. Based on preliminary X-ray data and biochemical studies they proposed a model for the L-dimer and the role of $Mg^{2+}$ and cysteine (247) in the formation of this complex. *E. coli* was also used by Larimer *et al.* (1990) to express mutants of the RuBisCo in *Rhodospirillum rubrum*. RuBisCo from *Rhodospirillum* is much simpler than RuBisCo from higher plants and consists of a dimer of two identical L-chains. These studies in mutants showed that replacement of lysine 191 by cysteine, serine, alanine, glutamic acid or histidine stopped the carboxylase activity completely in spite of the formation of dimers.

#### *10.2.2.2 Production of the β-subunit of the ATPase complex*

The ATP synthase (ATPase, EC 3.6.1.3) present in the chloroplast thylakoid and *E. coli* membrane are rather similar. Both contain an intrinsic membrane component $F_0$, the proton channel and attachment site for $F_1$ or $CF_1$ (chloroplast), which catalyzes ATP synthesis and hydrolysis. Both $F_1$ and $CF_1$ are composed of five subunits α–ε. In the plants, α, β and ε are encoded in the chloroplast and in the nucleus, γ and δ are encoded (Westhof *et al.*, 1981). Extensive homology has been found between the *E. coli* and chloroplast α-, β- and ε-subunits (Krebbers *et al.*, 1982; Shinozaki *et al.*, 1983; Zurawski *et al.*, 1982). Using gene fusions between the maize gene encoding the β-subunit (*atp B*) and the *E. coli* β-galactosidase gene (*lac Z*), it was shown that the first 365 *N*-terminal amino acids are involved in binding of the chloroplast β-subunit to the bacterial $F_0$ receptor (Gatenby and Rothstein, 1986). Burkovski *et al.* (1990) succeeded in the expression of subunit III of the ATP synthase from spinach in *E. coli*. The synthesized protein was found in the membrane fraction, indicating that the protein was inserted predominantly in the bacterial cytoplasm membrane. However, formation of a functional $F_0$ complex was not observed. These experiments can be seen as a first step in the further elucidation of the structure and function of the chloroplast ATPase complex.

## 10.3 EXPRESSION OF NUCLEAR GENES IN MICROORGANISMS

The structure, organization and expression of nuclear genes are essentially different from those of the cell organelles. Relevant aspects are: (i) the presence of multigene families, *i.e.*, the number of genes in such a family may range from several *e.g.* for phaseolin (Talbot *et al.*, 1984) and thaumatin (Ledeboer *et al.*, 1984), to more than 100, *e.g.* for zein (Viotti *et al.*, 1985); (ii) the abundance of introns and the transport of the mRNA across the nuclear membrane; (iii) the protein synthesis on 80S ribosomes; (iv) the topogenesis of gene products to several subcellular compartments (cell organelles; endoplasmic reticulum, vacuole) for which specific topogenic signals are present; and (v) posttranslational modification like phosphorylation, *N*-acetylation and glycosylation. It will be obvious that the eukaryotic microorganisms *S. cerevisiae*, which can carry out most of these processes, will be the preferred host for plant nuclear gene expression. Differences are, however, the elimination of introns, which for higher eukaryotic genes cannot be carried out properly by *S. cerevisiae* (Langford *et al.*, 1983) and the composition of sugar side chains extending from the core glycosylation, which in *S. cerevisiae* is only mannose while in higher eukaryotes and in plants it is variable (for reviews, see Kornfeld and Kornfeld 1985; Stanley, 1987).

### 10.3.1 Functioning of Plant Nuclear Promoters in *S. cerevisiae*

Several reports describe the functioning of nuclear plant promoters in yeast. A genomic DNA fragment from *Phaseolus vulgaris* encoding one of the major seed storage glycoproteins gave rise to transcription of the mRNA (Cramer *et al.*, 1985). Phaseolin immunoreactive proteins could only be found when the plant 5′ and 3′ regulatory sequences were fused to a cDNA clone, avoiding the intron excision by yeast. The phaseolin transcripts in yeast start and terminate in plant sequences and show a similar S1 nuclease protection pattern as plant-derived mRNA. One of the zein genes from maize (which contains no intron) was integrated into a yeast chromosome as well as inserted in a plasmid vector. The transcript originating from both started at exactly the same nucleotide as in the plant (Coraggio *et al.*, 1986; Langridge *et al.*, 1984). However, this constitutive expression in yeast cannot reflect the complex regulation system in plants resulting in specific synthesis in the maize endosperm (Burr and Burr, 1976).

The promoter from soybean leghemoglobin Lbc3 gene was shown to be active in yeast by fusing the 5′ and 3′ untranslated regions to the chloroamphenicol acetyl transferase (CAT) gene (Østergaard Jensen *et al.*, 1986). These leghemoglobins are hemoproteins synthesized exclusively in root nodules which develop through symbiosis with *Rhizobia*. The Lb genes are most likely activated by a rhizobial signal, a candidate for which is heme. In yeast, heme does not affect the transcription but does affect translation of Lbc3 (Østergaard Jensen *et al.*, 1986). Supposing there is a similarity between the heme specific regulation mechanisms in yeast and soybean, this is an indication of posttranscriptional regulation in soybean by heme, which agrees very well with the observation of Govers *et al.*, (1985) that Lb gene expression in pea is regulated by a posttranscriptional event. An interesting finding is that the −390 to −60 fragment of the promoter of the *Arabidopsis thaliana* rbcS-1A, containing the G and two I boxes, activates transcription from a truncated CYC1 promoter in *S. cerevisiae* (Donald *et al.*, 1990). The use of yeast for detailed analysis of this specific type of plant promoters seems very appropriate.

### 10.3.2 Synthesis and Analysis of Nuclear-encoded Gene Products

A number of plant nuclear genes have been successfully expressed in both yeast and *E. coli*, resulting in the synthesis of an active product. Table 1 shows a compilation of these results.

#### *10.3.2.1 Expression level*

Plasmid vector system enabling high level (inducible) expression based on host-derived promoters have been developed both for *E. coli* (*e.g.* inducible: lac) and *S. cerevisiae* (*e.g.* constitutive: GAPDH, PGK; inducible: GAL; for a review, see Kingsman *et al.*, 1985 and 1987). In addition to promoter sequences the plant gene itself can affect the final expression level. Genes which are abundantly expressed in *S. cerevisiae* show a preferred codon usage (Bennetzen and Hall, 1982) and the conversion of major codons into minor codons shows a negative effect on the expression of the abundantly expressed *PGK* gene from yeast (Hoekema *et al.*, 1987). Plant genes contain in most cases a significant amount of yeast minor codons that may be converted into yeast-preferred codons. Although the effect of this process is still a point of debate there is now some evidence that this does in fact occur. High levels of zein mRNA show adverse effects on the physiology of the yeast, especially on the growth rate, if the yeast is grown on a mineral medium (Compagno *et al.*, 1987). Furthermore, although the experiments of Edens *et al.* (1984) and Lee *et al.* (1986, 1988) cannot be compared directly, the conversion of the complete thaumatin gene into yeast-preferred codons is likely to have a positive effect on the expression level (see Section 10.5.1). The gene coding for α-galactosidase of *Cyamopsis tetragonoloba* (guar) has a codon usage quite different from that preferred by yeasts (Overbeeke *et al.*, 1989). A synthetic gene consisting of codons preferred by yeast indeed resulted in an increase in expression. Northern analysis showed that this increase was not due to increased stability of the mRNA but due to adaptation of the codons, thereby increasing the efficiency of translation (Verbakel, 1991).

For the same reasons, in the expression of the Bowman–Birk-type proteinase inhibitor in *E. coli*, a gene was synthesized according to the coding usage preferred in *E. coli* (Flecker, 1987). A high expression level was obtained, but only after the gene was fused to the β-galactosidase gene. The active inhibitor was obtained after cyanogen bromide cleavage and refolding. The 3′ end of a plant gene can influence the yield of the plant protein in a foreign host to a large degree. This was demonstrated by Utsumi *et al.* (1991) with soybean preproglycinin in *S. cerevisiae*. The 3′ sequence has three Zaret and Sherman transcription termination sequences. Deletion of one Zaret and Sherman sequence increases the yield three-fold; however, deletion of two of these sequences results in a six-fold drop in the yield compared with the original gene.

**Table 1** Expression of Nuclear-encoded Plant Genes in Yeast and Bacteria

| *Gene product* | *Plant source* | *Host* | *Promoter* | *Presequence* | *Processing* | *Secretion* | *Activity* | *Ref.* |
|---|---|---|---|---|---|---|---|---|
| Thaumatin | *Thaumatococcus daniellii* | *E. coli* | trp, lac | +(own) | – | – | a | Edens *et al.*, 1982 |
| | | | | – | | | a | |
| α-Amylase | *Triticum aestivum* (wheat) | *E. coli* | lac | +(own) | low | – | + | Gatenby *et al.*, 1986 |
| | | | | – | | | + | |
| SS-RuBisCo | *Triticum aestivum* (wheat) | *E. coli* | lac | – | | | + | van der Vies *et al.*, 1986 |
| Glutathion-*S*-transferase | *Zea mays* (maize) | *E. coli* | rec A | – | | | + | Moore *et al.*, 1986 |
| Ricin A | *Ricinus communis* | *E. coli* | $P_{N25}$ | – | | + | + | O'Hare *et al.*, 1987 |
| Ricin A | *Ricinus communis* | *E. coli* | λpl, trp | + | + | + | – | Piatak *et al.*, 1988 |
| Ricin A | *Ricinus communis* | *E. coli* | λpR | +(prot A) | – | – | + | Kim and Weaver, 1988 |
| Ricin A | *Ricinus communis* | *E. coli* | T5 $P_{N25}$ | – | – | – | + | Shire *et al.*, 1990 |
| Ricin B | *Ricinus communis* | *E. coli* | lpp, lac | + | + | + | + | Hussain *et al.*, 1989 |
| Acyl carrier protein | Spinach | *E. coli* | trc | – | – | – | + | Beremand *et al.*, 1987 |
| 5-Enolpyruvylshikimate-3-phosphate synthase | *Petunia hybrida* | *E. coli* | λpL | – | – | – | + | Padgette *et al.*, 1987 |
| Bowman–Birk-type proteinase inhibitor | *Glycine max* (soybean) | *E. coli* | lac | +(fusion β-gal) | – | – | +[b] | Flecker, 1987 |
| Glycinin | *Glycine max* | *E. coli* | trc | ± | – | – | | Utsumi *et al.*, 1988b |
| Peroxidase C | Horseradish | *E. coli* | tac | – | | | + | Burke *et al.*, 1989 |
| Fucose specific lectin | *Aleuria aurantia* | *E. coli* | tac | – | | | + | Fukumori *et al.*, 1989 |
| HW glutenin | Wheat | *E. coli* | T7 | –(δ-plant) | | | (+)[c] | Galili, 1989 |
| Strictosidine synthase | *Rauvolfia serpentina* | *E. coli* | lac | – | – | – | + | Kutchan, 1989 |
| Acetolactate synthase | *Arabidopsis* | *E. coli* | trp p/o | +(plant) | (+)[c] | + | + | Smith *et al.*, 1990 |
| Ferredoxin:NADP$^+$ reductase | Spinach | *E. coli* | lac | – | | | + | Aliverti *et al.*, 1990 |
| Glutamine synthetase α, β, Γ | *Phaseolus vulgaris* | *E. coli* | λPl | – | | | + | Bennett and Cullimore, 1990 |
| ATP synthase III | Spinach | *E. coli* | lac | +(plant) | – | – | – | Burkovski *et al.*, 1990 |
| 3-Hydroxy-3-methylglutanyl CoA reduction *L* | Radish | *E. coli* | T7 | | | | | Ferrer *et al.*, 1990 |
| *S* | | | | | | | +[b] | |
| Cyclophilin peptidylprolyl *cis–trans* isomerase | *Lycopersicon esculentum* | *E. coli* | lac p/o | – | | | + | Gasser *et al.*, 1990 |
| Antiviral protein | *Mirabilis* | *E. coli* | λPl | Omp A | + | + | + | Habuka *et al.*, 1990 |

| | | | | | | | | |
|---|---|---|---|---|---|---|---|---|
| GAPDH | *Ricinus communis* | *E. coli* | Pho A | +(plant) | + | + | | Hekman *et al.*, 1990 |
| Stilbene synthase | *Vitis* | *E. coli* | tac | – | – | – | + | Melchior and Kindl, 1990 |
| Plastocyanin | Spinach | *E. coli* | tac | – | | | – | Nordling *et al.*, 1990 |
| | | | | +(bacterial, azurin) | + | + | + | |
| Oxygen-evolving protein (psbO) | Spinach | *E. coli* | lac p/o | +(mod plant) | + | – | + | Seidler and Michel, 1990 |
| Lipoxygenase L2 | *Oryza* | *E. coli* | T7 | +(designed) | – | | + | Shirano and Shibata, 1990 |
| 1-Aminocyclopropane-1-carboxylate | *Lycopersicon esculentum* | *E. coli* | T7 | – | | | + | van der Straeten *et al.*, 1990 |
| Thaumatin | *Thaumatococus daniellii* | *B. subtilis* | amy R2 | +(amy E) | + | + | | Illingworth *et al.*, 1988 |
| α-Galactosidase | *Cyamopsis tetragonoloba* (guar) | *B. subtilis* | SPO2 | +(bacterial) | + | + | + | Overbeeke *et al.*, 1990 |
| Thaumatin | *Thaumatococcus daniellii* | *S. cerevisiae* | GAPDH | +(own) | + | – | +[b] | Edens *et al.*, 1984 |
| | | *S. cerevisiae* | GAPDH, PGK | – | | | | Huang *et al.*, 1987 |
| | | *S. cerevisiae* | GAPDH, PGK | | | | | Lee *et al.*, 1988 |
| | | *S. cerevisiae* | GAPDH, PGK | +(yeast, invertase) | + | + | + | Blair *et al.*, 1990 |
| β-Glucanase | Barley | *S. cerevisiae* | ADH | +(mouseamylase) | | + | + | Jackson *et al.*, 1986 |
| β-Glucanase | Barley | *S. cerevisiae* | ADH1 | +(mouseamylase) | + | + | + | Thomson *et al.*, 1984 |
| α-Amylase | *Triticum aestivum* | *S. cerevisiae* | PGK | +(own) | + | + | + | Rothstein *et al.*, 1984 |
| α-Amylase | *Oryza sativa* | *S. cerevisiae* | ENO | +(plant) | + | + | + | Kumagai *et al.*, 1990 |
| α-Amylase 1,2 | Barley | *S. cerevisiae* | PGK | +(plant) | + | + | + | Søgaard and Svensson, 1990 |
| Male sterility factor | *Zea mays* | *S. cerevisiae* | COX5 | +(fungal) | + | + | + | Glab *et al.*, 1990 |
| α-Galactosidase | *Cyamopsis tetragonoloba* | *S. cerevisiae* | GAPDH, GAL7 | +(yeast, invertase) | + | + | + | Overbeeke *et al.*, 1987 |
| | | *K. lactis* | GAPDH | +(yeast, invertase) | + | + | + | Overbeeke *et al.*, 1987 |
| | | *H. polymorpha* | MOX | +(yeast, invertase) | + | + | + | Fellinger *et al.*, 1991 |
| Preproglycinin | *Glycinin max* | *S. cerevisiae* | PHO5 | +(plant) | | + | (–)[d] | Utsumi *et al.*, 1991 |
| Proglycinin | *Glycinin max* | *S. cerevisiae* | PHO5 | – | | | (–)[d] | Utsumi *et al.*, 1991 |
| Glycinin | *Glycine max* | *S. cerevisiae* | PHO5 | +(own) | + | – | (+) | Utsumi *et al.*, 1988a |
| Phaseolin | *Phaseolus vulgaris* | *S. cerevisiae* | own | +(own) | | – | | Cramer *et al.*, 1985 |
| | | | PHO5 | +(own) | + | – | | Cramer *et al.*, 1987 |
| Zein | *Zea mays* (maize) | *S. cerevisiae* | own GAL1-10/CYC-1 | | | | | Coraggio *et al.*, 1986 |
| Legumin | *Pisum sativum* (pea) | *S. cerevisiae* | PGK | +(own) | – | – | | Yarwood *et al.*, 1987 |
| α-Gliadin | *Triticum aestirum* | *S. cerevisiae* | CYC1 | +(own) | + | – | | Neill *et al.*, 1987 |
| Ricin B | *Ricinus communis* | *S. cerevisiae* | PGK | + | + | – | + | Richardson *et al.*, 1988 |
| Phytohemagglutinin | *Phaseolus vulgaris* | *S. cerevisiae* | PHO5 | +(own) | + | +(vacuoles) | | Tagu and Chrispeels, 1987, 1989 |
| Vicilin | *Pisum sativum* | *S. cerevisiae* | PGK | – | – | – | | Watson *et al.*, 1988 |

[a]Not determined. [b]After renaturation. [c]In these cases activity means functionality. [d]Insoluble complex.

Although the translation of plant mRNA in microbial systems is nearly always carried out correctly, sometimes the initiation of translation occurs at the wrong position. A striking example of this phenomenon is the expression in *E. coli* of ferredoxin: $NADP^+$ reductase from spinach. Triplet 29 (GTG, coding for valine) of the mature coding sequence is used in *E. coli* to start the translation. Changing this codon into GTT (also coding for valine) prevents incorrect translation (Aliverti *et al.*, 1990).

### *10.3.2.2 Processing and secretion*

The processing of plant-derived signal sequences does not always seem to be performed efficiently in prokaryotic hosts (*e.g.* thaumatin and α-amylase in *E. coli*; Edens *et al.*, 1982; Gatenby *et al.*, 1986), but adaptation to host-originated secretion sequences seems to be helpful. Partial processing and secretion into the periplasmic space of *E. coli* was obtained for ricin B when the gene was fused to the OmpA signal peptide gene sequence (Hussain *et al.*, 1989) and for ricin A when the gene was fused behind the PhoA signal peptide gene sequence (Piatak *et al.*, 1988). Seidler and Michel (1990) studied the expression and secretion of the nuclear-encoded 33 kDa protein of the oxygen-evolving complex of spinach in *E. coli*. To that end, they replaced one amino acid (threonine to lysine at the 2-position) of that part of the plant transit sequence which is normally cleaved off after translocation across the thylakoid membrane. This part of the transit sequence is homologous to signal sequences functioning in *E. coli*. The result was the correct processing and secretion of the 33 kDa protein into the periplasm. Similarity in the secondary structure is very important in correct processing and secretion. Nordling *et al.* (1990) tried to express the nuclear-encoded plastocyanin from spinach in *E. coli*. Whereas the signal sequence of alkaline phosphatase failed to translocate plastocyanin, the use of the signal sequence of azurin, a copper protein of *Pseudomonas aeruginosa* with a secondary structure similar to plastocyanin, resulted in correct processing and translocation. While the above-mentioned examples show that plant sequences do not always function in bacteria, the use of the *Bacillus*-derived α-amylase signal sequence in front of the α- galactosidase from guar resulted in the correct processing and secretion of plant enzyme by *Bacillus subtilis* (Overbeeke *et al.*, 1990). Using similar constructs, Illingworth *et al.* (1988) have accomplished the correct processing and secretion of thaumatin by *B. subtilis*.

In yeast the hydrolytic plant enzymes β-glucanase, α-amylase and α-galactosidase were secreted for more than 50%. As secretion sequences, use was made of the mouse α-amylase sequence, the original plant sequence and the yeast invertase sequence, respectively (Jackson *et al.*, 1986; Overbeeke *et al.*, 1987; Rothstein *et al.*, 1984). This may very well illustrate the low specificity by which signal sequences are recognized by *S. cerevisiae*, which agrees with the observation of Kaiser *et al.* (1987) that in *S. cerevisiae* many random sequences can replace the invertase signal sequence. The processing was found to be exact for thaumatin and α-galactosidase by $NH_2$-terminal sequencing of the yeast-produced protein (Edens *et al.*, 1984; Overbeeke *et al.*, 1987). Plant signal sequences are often recognized and processed precisely by the yeast secretion machinery, as has been shown by the above examples and by Kumagai *et al.* (1990) for α-amylase from rice. Although the glycosylation of this enzyme in plant and yeast is different, they only found small differences in the $K$(m) and $V$(max) of the α-amylases produced by rice or by yeasts. Similar results were also obtained by Søgaard and Svensson (1990) for the secretion of α-amylase from barley. Likewise, the correct processing of plant signal sequences in *S. cerevisiae* was shown for α-gliadin, a subclass of the storage proteins of wheat (Neil *et al.*, 1987), and for glycinin, the predominant storage protein of soybean (Utsumi *et al.*, 1988a). In both cases, however, secretion of the processed protein was not found, while in the case of glycinin a potential *N*-glycosylation site was not glycosylated. This indicates the importance of the protein sequence and structure in processes like glycosylation and secretion.

### *10.3.2.3 Posttranslational modification*

Posttranslational proteolytic processing of precursor molecules (*e.g.* at the $CO_2H$ terminus of thaumatin and legumin) seems to occur less generally than $NH_2$ signal sequence processing. The reason might be that specific proteolytic enzymes are involved which are present in the plants but absent from the yeasts. Construction of genes without such precursor sequences are the obvious solution for production of proteins matching the mature form found in the plants, although this might influence the expression level (Edens *et al.*, 1984).

Posttranslational *O*- and *N*-glycosylation is restricted to eukaryotic cells, where it occurs in the endoplasmic reticulum and Golgi system. Enzymes which are present in the plants in a glycosylated form (*e.g.* phaseolin, α-galactosidase) are also found to be glycosylated in yeast, provided they are produced in a precursor form with a signal sequence targetting the protein to the endoplasmic reticulum. The consensus sequence for *N*-glycosylation is identical for plants and yeast, namely Asn-X-Ser/Thr (Struck *et al.*, 1978). The type of glycosylation is different for plant cells and yeast cells, the latter having only mannose residues in the sugar side chain and the former having complex and varying sugar chains (for a review, see Kornfeld and Kornfeld, 1985).

#### *10.3.2.4 Protein targeting*

In some cases the expression of plant genes in microorganisms can be used as the first step to study protein targeting, with the ultimate goal being to study targeting in the plant itself. Tague and Chrispeels (1987) have expressed the gene encoding the leuko-agglutinating phytohemagglutin (PHA-L), the major seed lectin of *Phaseolus vulgaris* in *S. cerevisiae* under control of the yeast PHO5 promoter. This protein, which normally is glycosylated and targeted to the protein storage vacuoles of the bean, is also glycosylated and transported to the yeast vacuoles. In a recent study with gene fusions between the PHA-L gene and yeast invertase, they were also able to target the invertase to the yeast vacuoles (Tague and Chirspeels, 1989). It was shown that only a small fragment of about 20 amino acids contains the targetting signal and a further comparison with other vacuolar proteins from plants and yeasts showed an external loop, flanked by β-sheets and a relatively conserved tetrapeptide, LQRD. Experiments are in progress to prove the importance of this sequence in targeting proteins to vacuoles in tobacco.

#### *10.3.2.5 Biological activity*

The plant proteins produced in microorganisms have been found to be biologically active in many situations. For some proteins (*e.g.* α-amylase) exact processing might be essential, as was suggested by studying expression in *E. coli* (Gatenby *et al.*, 1986). Correct folding of the proteins may also be an essential factor. Incorrect folding has been observed in the expression of the mature form of thaumatin (Lee *et al.*, 1986), but the insoluble product could be converted into an active form by a denaturation/renaturation procedure. Another obvious solution could be the secretion of the product, as has been described for calf prochymosine (Smith *et al.*, 1985).

An interesting observation has been made for α-galactosidase, which was shown to have similar enzyme activity when produced in either *B. subtilis* (nonglycosylated) or *S. cerevisiae* preceded by the invertase signal sequence (glycosylated). Thus, glycosylation does not seem to be a prerequisite for activity of this protein (Overbeeke *et al.*, 1990), although overglycosylation can decrease the enzyme activity to only 38% (Fellinger *et al.*, 1991).

Functional expression is not restricted to plant enzymes that have microbial counterparts. As given in Table 1, many genes encoding enzymes quite unrelated to microorganisms are correctly expressed in *E. coli* or *S. cerevisiae*, notably grapevine stilbene synthase, cyclophillin/peptidylprolyl *cis–trans* isomerase of tomato, radish 3-hydroxy-3-methylglutaryl CoA reductase, strictosidine synthase of *Rauvolfia* and the fucose specific lectin from *Aleuria aurantia*.

It should be evident that production of nuclear-encoded plant proteins in microorganisms can be realized such that they are produced in significant in an active form. This facilitates the analysis of individual gene products as well as the structure–function relationship by mutagenic approaches.

#### *10.3.2.6 Analysis of the structure–function relationship*

The feasibility of the approach using microorganisms to engineer and express plant proteins has been demonstrated nicely by Kim *et al.* (1989) for the sweet, plant protein monellin. Monellin consists of two peptide chains (Frank and Zuber, 1976), and recently the 3 D structure of the heterodimer of monellin has been elucidated (Ogata *et al.*, 1987). Using the information about this 3 D structure Kim's group inserted (on gene level) an amino acid linker between the B-chain (amino acid 46) and the A-chain (amino acid 6).

The five amino acid linkers were designed in such a way that the 3 D structure of each single chain monellin resembled that of the natural monellin as much as possible. To maximize the likelihood that the newly designed proteins would expose sweet site(s) the linkers contained amino acid sequences that also occur in natural monellin.

All five modified monellins proved to be sweet; one single chain monellin is as sweet as natural monellin but even a modified monellin with linker EDYKTRGR has still 10% of the sweetness of natural monellin. Moreover, the modified monellins have better thermal and pH stabilities and refold in the correct conformation very well.

## 10.4 IDENTIFICATION OF DNA REPLICATION ORIGINS IN PLANTS BY CLONING IN YEAST

The process of DNA replication involves initiation at the origin of replication, elongation and termination, in which processes a number of enzymes are involved and which for *E. coli* has been identified in great detail (Kornberg, 1980, 1982). Two routes can lead to the identification of the replication origin: (i) electron microscopic analysis of replicating DNA molecules; and (ii) cloning of the origin by conferring replication properties to a nonreplicative vector. The former approach requires the arrest of a large number of replicating DNA molecules, while cloning requires that the organism can be transformed. For higher organisms, like plants, both routes are difficult.

The observation by Stinchcomb *et al.* (1980) that eukaryotic DNA fragments from various sources were capable of initiating DNA replication in yeast initiated various other studies to identify such sequences in plants. Autonomous replicating sequences (ARSs) were cloned in yeast from the chloroplast and localized on the genetic map for *Petunia* (Overbeeke *et al.*, 1984a), tobacco (Uchimiya *et al.*, 1983) and *Chlamydomonas* (Vallet *et al.*, 1984). ARSs were also cloned from chromosomal DNA for maize (Stinchcomb *et al.*, 1980). Eckdahl *et al.* (1989) found that plant genomes contain, with a relatively high frequency, ARS consensus sequences functioning in yeasts. A striking feature is that these sequences are flanked by sequences rich in A + T. The subsequent question, whether these ARSs also function in the plant *in vivo*, is much more difficult to answer. For the *Petunia* chloroplast an *in vitro* DNA synthesizing system was developed (Overbeeke *et al.*, 1984b), in which only one of the chloroplast ARS fragments seemed to function as a specific initiator for DNA synthesis (De Haas *et al.*, 1987). The nucleotide sequence of this fragment (De Haas *et al.*, 1986) showed extensive homology with the chloroplast replication origin of the alga *Euglena gracilis*, which could be verified by electron microscopic methods due to special growth conditions yielding high levels of replicating chloroplast DNA (Koller and Delius, 1982).

It should be evident that cloning of putative plant replication origins in yeast can be considered only as a first step; much more experimental work is necessary for the method to gain biological significance.

## 10.5 COMMERCIAL PRODUCTION OF PLANT PROTEINS IN MICROORGANISMS

The motivation for commercial production of plant proteins in microorganisms is economic. A product has to be made with the desired specifications (activity, purity) more cheaply than when obtained from the plant itself. The effort in this field of research is directed towards the most effective production process by selection of the optimal microbial host, optimization of the stability of the genetically engineered constructs to cope with large scale fermentation, optimization of the expression of the gene, optimization of secretion, *etc.* For $\alpha$-galactosidase from guar a systematic study to increase the yield of this plant enzyme produced by *S. cerevisiae* has been carried out, and even at a relatively low cell density a production of 500 mg $L^{-1}$ has been obtained, showing the feasibility of an economically attractive process for the production of plant enzymes in yeasts (Verbakel, 1991). As an illustration for a general approach and demonstration that a range of specific solutions exists for specific requirements, the production of thaumatin and the ricin A chain will be reviewed in more detail in the next sections.

Next to the production of plant proteins by microorganisms the target of expressing plant genes as such may be to create microorganisms with new and desirable properties. For example, in the production of beer, yeast cells secreting $\alpha$-amylase may have potential in light beer production, while yeast cells secreting $\beta$-glucanase might alleviate the problems of beer filtration due to glucans. Recently, the construction of yeast strains that degrade $\beta$-glucans after transformation of the barley $\beta$-glucanase gene, was reported by Thomsen *et al.* (1988). The $\beta$-glucanase released to the culture medium is sufficient to degrade up to 500 mg $L^{-1}$ of $\beta$-glucan during a seven day fermentation period at 10 °C. However, amylolytic yeasts can be constructed with genes from various origins, bacterial, mould or plant (Tubb, 1986), and it has still to be established whether enzymes of plant origin are preferred for specific applications.

### 10.5.1 Production of Thaumatin in *E. coli* and *S. cerevisiae*

Thaumatin is a sweet-tasting protein originating from the fruits of the African shrub *Thaumatococcus daniellii* Benth (for a review, see van der Wel and Ledeboer, 1989; Edens and van der Wel, 1985). It is not only 100 000 times sweeter than sugar on a molar basis, but also has flavour-enhancing effects at subsweetness levels. Several very closely related forms of thaumatin are found and are most likely due to the presence of a gene family (Ledeboer *et al.*, 1984).

When considering the commercial production of thaumatin as a sweetener and taste enhancer one has to compare the cost of the production on plantations with that of the production by microorganisms. Production on plantations can only be realized in the natural environment of the fruits as they do not develop in other climates or in greenhouses (Higginbotham, 1979). The yield is estimated to be about 180 kg per hectare. To obtain a competitive price by microbial production one must produce around 1 g $L^{-1}$ of thaumatin extracellularly or even more than 10 g $L^{-1}$ intracellularly on the basis of the cost price calculations presented by Jackson (1987).

To realize expression of thaumatin in *E. coli* the cDNA encoding the precursor form of thaumatin II was fused to either the trp or lac UV5 promoter, but only low levels of protein were detected (Edens *et al.*, 1982). Because of this poor expression and the fact that for food application host organisms having the GRAS status (generally recognized as safe) are preferred, expression of thaumatin was further developed in yeasts like *S. cerevisiae* and *Kluyveromyces lactis* (for a review, see Overbeeke, 1989). Substantial amounts of thaumatin can be produced in yeast from both the gene encoding the precursor form and the gene in which only an *N*-terminal methionine precedes the part of the gene encoding the mature protein (Edens *et al.*, 1984; Huang *et al.*, 1987; Lee *et al.*, 1986). Although experiments cannot be compared directly, a major improvement in the expression of thaumatin seems to result from the use of yeast-preferred codons. The plant gene encoding thaumatin contains a significant amount of minor codons. To convert these (minor) codons which are not preferred by yeast into codons which are, the gene was built completely from *in vitro* synthesized oligonucleotides (Lee *et al.*, 1986). Production of thaumatin from the gene encoding only the mature part results in cytoplasmic localization of an insoluble monomer, which can be converted into an active product after de- and re-naturation (Lee *et al.*, 1986). Indeed, during subsequent protein refolding studies, Lee *et al.* (1988) were able to convert the intracellularly produced inactive thaumatin into an immunoreactive, sweet conformation. However, such a method will never lead to an economical production process, even though 20% of the yeast protein consists of thaumatin. An obvious solution which might yield a directly active product is the secretion of the thaumatin. Therefore, either its own presequence, which is processed correctly in yeast (Edens *et al.*, 1984), can be used, or alternatively a yeast-derived secretion sequence with an optional glycosylation consensus in the leader that might improve secretion, as has been described by Smith *et al.* (1985) for the secretion of calf prochymosine. Hahm and Batt (1990) succeeded in the expression and secretion of preprothaumatin by *Aspergillus oryzae*. Although the yield was very low (50 ng $mL^{-1}$), the secreted thaumatin proved to be as sweet as the original plant protein. Blair *et al.* (1990) have described the production and secretion of a mutant of thaumatin I by *S. cerevisiae*. Using the invertase signal sequence, 0.4 mg $L^{-1}$ of sweet thaumatin was secreted. A further step in the direction of commercialization of microbiologically produced thaumatin has been the development of a stable multicopy integration system for *S. cerevisiae* (Lopes *et al.*, 1989). It may be feasible to realize the cost price targets mentioned above for thaumatin either in *S. cerevisiae* or in other yeast species like *Kluyveromyces marxianis* or *Hansenula polymorpha* (for a review, see Gleeson and Sudberry, 1988). Another solution might be the excretion of thaumatin by other microorganisms, as has been shown for *B. subtilis* (Illingworth *et al.*, 1988). Using *B. subtilis* DB104, lacking the extracellular neutral and alkaline proteases and the amyR2 promoter and amyE leader, the highest production of up to 1 mg $L^{-1}$ thaumatin was obtained 3.5 h after the end of exponential growth.

### 10.5.2 Production of the Ricin A Chain in *E. coli*

The ricin A chain, one of the two polypeptides in ricin which is the toxic lectin from *Ricinus communis* seeds, has been used for the construction of immunotoxins by coupling to a cell reactive antibody (Vitella and Uhr, 1985). The A chain, which is glycosylated, inactivates 60S ribosomal subunits. The A chain derived from native ricin has to be freed of the B chain to avoid a specific toxicity. Futhermore, this polypeptide has the disadvantage of being *N*-glycosylated, causing rapid clearance *in vivo* by liver cells.

Production of ricin A in *E. coli* (O'Hare *et al.*, 1987) proved to be a very viable approach. The nonglycosylated ricin A was still active, and could be produced in *E. coli* since the 70S ribosomes are not inactivated and could be easily purified of *E. coli* lysates. O'Hare *et al.* (1987) as well as Piatak *et al.* (1988) reported that the production of ricin A resulted in an active product at lower temperatures, while at higher temperatures, substantial amounts of ricin A were produced in an aggregated, insoluble, less active form. This process did occur only *in vivo*. To improve the expression of the ricin A gene in *E. coli*, Shire *et al.* (1990) synthesized the gene coding for this cytotoxic lectin and put this synthetic gene under the T7 promoter regime. This resulted in 1.5 $mgL^{-1}$ at an optical density of the *E. coli* culture of 1.0. In spite of the lack of glycosylation, the *E. coli* ricin A gene product was in all respects similar to the castor bean product. Also, the ricin B chain has been expressed in *E. coli* (Hussain *et al.*, 1989) and in *S. cerevisiae* (Richardson *et al.*, 1988). It is the intention of both groups to genetically modify the B chain using protein-engineering techniques. By this approach, the nonspecific binding of galactose, which causes the nonspecific toxicity, might be eliminated, while the potentiating capacity of the A chain by the B chain might be preserved.

## 10.6 CONCLUSIONS AND FUTURE OUTLOOK

Recent years have shown considerable progress on the expression of plant genes in microorganisms. A large number of genes have been expressed in yeast and *E. coli* facilitating the analysis of structure and function of (individual) plant gene products. For the study of the relationship between the structure and function of proteins the availability of detailed 3 D structures has proved essential. Therefore it is not surprising that this type of study on monellin, RuBisCo and thaumatin has delivered very promising results. A further continuation along this route should involve genetic engineering of mutant genes and their analysis in a microbial host.

The targeting of proteins to specific plant cell compartments is still very difficult but expression of these genes in microbial hosts can be very helpful in the elucidation of this process, as was nicely demonstrated by Tague and Chrispeels (1987, 1989). Using *S. cerevisiae* as intermediate host they could define the targeting sequence of PHA-L as a 20 amino acid stretch. Of course, analysis in the plant is still necessary to prove this result.

The study of plant promoters in microorganisms can in some cases lead to exciting results and perspectives (*e.g.* heme regulation for leghemoglobin mutations in chloroplast promoters). In most situations, however, the microbial host will not be completely representative due to the lack of the often complex regulatory factors present in plants.

The identification of DNA replication origins using cloning in microorganisms must be considered as only one aspect of these complex studies. It has shown to be a helpful first step, but has to be followed by proper experiments to gain biological significance.

Commercial production of plant proteins by microorganisms is now proceeding towards viable processes. Knowledge of the host and the factors improving production are crucial. When more optimal processes have been developed, the economics for enzyme production will become more attractive. This will mean an increased number of applications for specific plant enzymes with unique properties or with properties which are preferred to those of enzymes of microbial origin.

One may anticipate the economic attractiveness of bulk production in plants of enzymes both of plant origin and of microbial origin. The first cautious, but promising, step was recently reported by Vandekerckhove *et al.* (1989). They replaced part of the 2S albumin seed storage protein of *Arabidopsis thaliana* with the gene encoding the five-long amino acid chain pharmaceutical peptide leu-enkephalin. The peptide was flanked by two lysine residues, thereby facilitating the isolation of the peptide from the fusion product by subsequent trypsin and carboxypeptidase treatment. About 0.1 $mg\,g^{-1}$ seed was isolated and it could be calculated that 1 $mg\,g^{-1}$ seed had been produced. For genetically engineered plants the prerequisite is a substantially high enzyme production level that does not affect the plant itself in order to challenge microbial fermentation processes. That stage will most probably not be reached in the future, so that the production of plant products in microorganisms will, for some time to come, stay economically very attractive.

## ACKNOWLEDGEMENTS

Many thanks are due to Dr. A. M. Ledeboer for a critical examination of the manuscript and valuable assistance in its updating, and to Mrs. G. J. Paalvast for typing the manuscript.

## 10.7 REFERENCES

Aliverti, A., T. Jansen, G. Zanetti, S. Ronchi, R. G. Herrmann and B. Curti (1990). Expression in *E. coli* of ferredoxin: $NADP^+$ reductase from spinach. *Eur. J. Biochem.*, **191**, 551–555.

Andersson, I., S. Knight, G. Schneider, Y. Lindqvist, T. Lundqvist, C.-I. Brändén and G. Lorimer (1989). Crystal structure of the active site of ribulose bisphosphate carboxylase. *Nature (London)*, **337**, 229–234.

Andrews, T. J. and B. Ballment (1983). The function of the small subunits of ribulose bisphosphate carboxylase–oxygenase. *J. Biol. Chem.*, **258**, 7514–7518.

Avni, A., M. Edelman, I. Rachailovich, D. Aviv and R. Fhuhr (1989). A point mutation in the gene for the large subunit of ribulose-1,5-bisphosphate carboxylase/oxygenase affects holoenzyme assembly in *Nicotiana tabacum*. *EMBO J.*, **8**, 1915–1918.

Barraclough, R. and R. J. Ellis (1980). Protein synthesis in chloroplasts IX. Assembly of newly synthesized large subunits into ribulose bisphosphate carboxylase in isolated intact pea chloroplasts. *Biochem. Biophys. Acta*, **608**, 19–31.

Bennett, M. and J. Cullimore (1990). Expression of three types of plant glutamine synthetase cDNA in *E. coli*. *Eur. J. Biochem.*, **193**, 319–324.

Bennetzen, J. L. and B. D. Hall (1982). Codon selection in yeast. *J. Biol. Chem.*, **257**, 3026–3031.

Beremand, P. D., D. J. Hannapel, D. J. Guerra, D. N. Kuhn and J. B. Ohlrogge (1987). Synthesis, cloning and expression in *Escherichia coli* of a spinach acyl carrier protein-1 gene. *Arch. Biochem. Biophys.*, **256**, 90–100.

Blair, L. V., R. K. Koduri, J. H. Lee and J. L. Weickmann (1990). Recombinant manufacture of analogs of thaumatin I. *World Pat.*, 90 05 775 A1, 31 May 1990.

Bloom, M. V., P. Milos and H. Roy (1983). Light-dependent assembly of ribulose-1,5-bisphosphate carboxylase. *Proc. Natl. Acad. Sci. USA*, **80**, 1013–1017.

Bohnhert, H. J., E. J. Crouse and J. M. Schmitt (1982). Organization and expression of plastid genomes. In *Encyclopedia of Plant Physiology: Nucleic Acids and Proteins in Plants*, ed. D. Boulter and B. Parthier, vol. 14B, pp. 475–530, Springer-Verlag, Berlin.

Bradley, D. and A. A. Gatenby (1985). Mutational analysis of the maize chloroplast ATPase-β-subunit gene promoter: the isolation of promoter mutants in *E. coli* and their characterization in a chloroplast *in vitro* transcription system. *EMBO J.*, **4**, 3641–3648.

Burke, J. F., A. Smith, N. Santama, R. C. Bray, R. N. F. Thorneley, S. Dacy, J. Griffiths, G. Catlin and M. Edwards (1989). Expression of recombinant horseradish peroxidase C in *Escherichia coli*. *Biochem. Soc. Trans.*, **17**, 1077–1078.

Burkovski, A., G. Deckers-Hebestreit and K.-H. Alterdorf (1990). Expression of subunit III of the ATP synthase from spinach chloroplast in *E. coli*. *FEBS Lett.*, **271**, 227–230.

Burr, B. and F. A. Burr (1976). Protein synthesis in maize endosperm by polyribosomes attached to protein bodies. *Proc. Natl. Acad. Sci. USA*, **73**, 515–519.

Chapman, M. S., S. W. Sub, P. M. G. Curmi, D. Cascio, W. W. Smith and D. S. Eisenberg (1988). Tertiary structure of plant RuBisCo: domains and contacts. *Science (Washington, D.C.)*, **241**, 71–74.

Compagno, C., I. Coraggio, B. M. Ranzi, L. Alberghina, A. Viotti and E. Martegani (1987). Translational regulation of the expression of zein cloned in yeast under an inducible GAL promoter. *Biochem. Biophys. Res. Commun.*, **146**, 809–814.

Coraggio, I., C. Compagno, E. Martegani, B. M. Ranzi, E. Sala, L. Alberghina and A. Viotti (1986). Transcription and expression of zein sequences in yeast under natural plant or yeast promoters. *EMBO J.*, **5**, 459–465.

Cramer, J. H., K. Lea and J. L. Slightom (1985). Expression of phaseolin cDNA genes in yeast under control of natural plant DNA sequences. *Proc. Natl. Acad. Sci. USA*, **82**, 334–338.

Cramer, J. H., K. Lea, M. D. Schaber and R. A. Kramer (1987). Signal peptide specificity in posttranslational processing of the plant protein phaseolin in *Saccharomyces cerevisiae*. *Mol. Cell. Biol.*, **7**, 121–128.

De Haas, J. M., K. J. M. Boot, M. A. Haring, A. J. Kool and H. J. J. Nijkamp (1986). The *Petunia hybrida* chloroplast DNA region close to one of the inverted repeats shows sequence homology with the *Euglena gracilis* chloroplast DNA region that carries the putative replication origin. *Mol. Gen. Genet.*, **202**, 48–54.

De Haas, J. M., A. J. Kool, N. Overbeeke. W. van Burg and H. J. J. Nijkamp (1987). Characterization of DNA synthesis and chloroplast DNA replication initiation in *Petunia hybrida* chloroplast lysate system. *Curr. Genet.*, **12**, 377–386.

Deuschle, U., W. Kammerer, R. Gentz and H. Bujard (1986). Promoters of *Escherichia coli*: a hierarchy of *in vivo* strength indicates alternate structures. *EMBO J.*, **5**, 2987–2994.

Doi, R. H., S.-L. Wong and F. Kawamura (1986). Potential use of *Bacillus subtilis* for secretion and production of foreign proteins. *Trends Biotechnol.*, **4**, 232–235.

Donald, R. G. K., U. Schindler, A. Batschauer and A. R. Cashmore (1990). The plant G box promoter sequence activates transcription in *S. cerevisiae* and is bound *in vitro* by a yeast activity similar to GBF, the plant G box binding factor. *EMBO J.*, **9** 1727–1735.

Eckdahl, T. T., J. L. Bennetzen and J. N. Anderson (1989). DNA structures associated with autonomously replicating sequences from plants. *Plant Mol. Biol.*, **12**, 507–519.

Edens, L., L. Heslinga, R. Klok, A. M. Ledeboer, J. Maat, M. Y. Toonen, C. Visser and C. T. Verrips (1982). Cloning of cDNA encoding the sweet-tasting plant protein thaumatin and its expression in *E. coli*. *Gene*, **18**, 1–12.

Edens, L., I. Bom, A. M. Ledeboer, J. Maat, M. Y. Toonen, C. Visser and C. T. Verrips (1984). Synthesis and processing of the plant protein thaumatin in yeast. *Cell (Cambridge, Mass.)*, **37**, 629–633.

Edens, L. and H. van der Wel (1985). Microbial synthesis of the sweet-tasting plant protein thaumatin. *Trends Biotechnol.*, **3**, 1–14.

Erion, J. L., J. Tarnowski, S. Peacock, P. Caldwell, B. Redfield, N. Brot and H. Weissbach (1983). Synthesis of the large subunit of ribulose-1,5-bisphosphate carboxylase in an *in vitro*, partially defined *E. coli* system. *Plant Mol. Biol.*, **2**, 279–285.

Fellinger A. J., J. M. A. Verbakel, R. A. Vaale, P. E. Sudberry, I. J. Bom, N. Overbeeke and C. T. Verrips (1991). Expression of the α-galactosidase from *Cyamopsis tetragonoloba* (guar) by *Hansenula polymorpha*. *Yeast*, **7**, 463–473.

Ferrer, A., C. Aparicio, N. Nogues, A. Wettstein, T. J. Bach and A. Boronat (1990). Expression of catalytically active radish 3-hydroxy-3-methylglutaryl coenzyme A reductase in *E. coli*. *FEBS Lett.*, **266**, 67–71.

Flecker, P. (1987). Chemical synthesis, molecular cloning and expression of gene coding for a Bowman–Birk-type proteinase inhibitor. *Eur. J. Biochem.*, **166**, 151–156.

Frank, G. and H. Zuber (1976). The complete amino acid sequence of both subunits of the sweet protein monellin. *Hoppe-Seyler's Z. Physiol. Chem.*, **357**, 585–592.

Fukumori, F., N. Takeuchi, T. Hagiwara, K. Ito, N. Kochibe, A. Kobata and Y. Magata (1989). Cloning and expression of a functional, fucose specific lectin from an orange peel mushroom, *Aleuria aurantia*. *FEBS Lett.*, **250**, 153–156.

Galili, G. (1989). Heterologous expression of a wheat, high molecular weight glutenin gene in *E. coli*. *Proc. Natl. Acad. Sci. USA*, **86**, 7756–7760.

Galun, E. and D. Aviv (1983). Cytoplasmic hybridization: genetic and breeding application. In *Handbook of Plant Cell Culture*, ed. D. A. Evans, W. R. Sharp, P. V. Ammirate and Y. Yamada, vol. 1, pp. 358–392. Macmillan, New York.

Gasser, C. S., D. A. Gunning, K. A. Budelier and S. M. Brown (1990). Structure and expression of cytosolic cyclophilin/peptidylprolyl *cis–trans* isomerase of higher plants and production of active tomato cyclophilin in *E. coli*. *Proc. Natl. Acad. Sci. USA*, **87**, 9519–9523.

Gatenby, A. A., J. A. Castleton and M. W. Saul (1981). Expression in maize and wheat chloroplast genes of the large subunit of ribulose bisphosphate carboxylase. *Nature (London)*, **291**, 117–121.

Gatenby, A. A. and J. A. Castleton (1982). Amplification of maize ribulose bisphosphate carboxylase large subunit synthesis in *E. coli* by transcriptional fusion with the lambda N operon. *Mol. Gen. Genet.*, **185**, 424–429.

Gatenby, A. A. and S. J. Rothstein (1986). Synthesis of maize chloroplast ATP-synthase β-subunit fusion proteins in *Escherichia coli* and binding to the inner membrane. *Gene*, **41**, 241–247.

Gatenby, A. A., M. Boccara, D. C. Baulcombe and S. J. Rothstein (1986). Expression of a wheat α-amylase gene in *Escherichia coli*: recognition of the translational initiation site and the signal peptide. *Gene*, **45**, 11–18.

Gatenby, A. A., S. M. van der Vies and S. J. Rothstein (1987). Coexpression of both the maize large and wheat small subunit genes of ribulose bisphosphate carboxylase in *Escherichia coli*. *Eur. J. Biochem.*, **168**, 227–231.

Glab, N., R. P. Wise, D. R. Pring, C. Jacq and P. Slonimski (1990). Expression in *Saccharomyces cerevisiae* of a gene associated with cytoplasmic male sterility from maize: respiratory dysfunction and uncoupling of yeast mitochondria. *Mol. Gen. Genet.*, **223**, 24–32.

Gleeson, M. A. and P. E. Sudberry (1988). The methylotrophic yeasts. *Yeast*, **4**, 1–15.

Govers, F., T. Gloudemans, M. Moerman, A. van Kammen and T. Bisseling (1985). Expression of plants genes during the development of pea root nodules. *EMBO J.*, **4**, 861–867.

Habuka, N., K. Akiyama, H. Tsuge, M. Miyano, T. Matsumoto and M. Noma (1990). Expression and secretion of Mirabilis antiviral protein in *Escherichia coli* and its inhibition of *in vitro* eukaryotic protein synthesis. *J. Biol. Chem.*, **265**, 10 988–10 992.

Hahm, Y. T. and C. A. Batt (1990). Expression and secretion of thaumatin from *Aspergillus oryzae*. *Agric. Biol. Chem.*, **54**, 2513–2520.

Hekman, W. E., D. T. Dennis and J. A. Miernyk (1990). Secretion of *Ricinus communis* glyceraldehyde-3-phosphate dehydrogenase by *Escherichia coli*. *Mol. Microbiol.*, **4**, 1363–1369.

Higginbotham, J. D. (1979). In *Developments in Sweeteners II*, ed. C. A. M. Hough, K. J. Parker and A. J. Vlitos, pp. 119–155. Applied Science Publishers, London.

Hoekema, A., R. A. Kastelein, M. Vasser and H. A. de Boer (1987). Codon replacement in the *PGKI* gene of *Saccharomyces cerevisiae*: Experimental approach to study the role of biased codon usage in gene expression. *Mol. Cell. Biol.*, **7**, 2914–2924.

Holland, I. B., N. Mackman and J.-M. Nicaud (1986). Secretion of proteins from bacteria, *Biotechnol.*, **4**, 427–431.

Huang, S., R. G. Elliot, P. S. Liu, R. K. Koduri, J. L. Weickman, J. H. Lee, L. C. Blair, P. Ghosh-Oostidar, R. A. Bradshaw, K. M. Bryan, B. Enarson, R. L. Kendall, K. H. Kolacz and K. Saito (1987). Specificity of contranslational amino-terminal processing of proteins in yeast. *Biochemistry*, **26**, 8242–8246.

Hussain, K., C. Bowler, L. M. Roberts and J. M. Lord (1989). Expression of ricin B chain in *Escherichia coli*. *FEBS Lett.*, **244**, 383–387.

Illingworth, C., G. Larson and G. Hellekant (1988). Secretion of the sweet-tasting plant protein thaumatin by *Bacillus subtilis*. *Biotechnol. Lett.*, **10**, 587–592.

Jackson, D. (1987). Cost reductions in food processing using biotechnology. In *Biotechnology in Food Processing* ed. S. K. Harlander and T. P. Labuza. Noyes, New Jersey.

Jackson, E. A., G. M. Ballance and K. K. Thomsen (1986). Construction of a yeast vector directing the synthesis and release of barley (1→3,1→4)-β-glucanase. *Carlsberg Res. Commun.*, **51**, 445–458.

Kaiser, C. A., D. Preuss, P. Grisafi and D. Botstein (1987). Many random sequences functionally replace the secretion signal sequences of yeast invertase. *Science (Washington, D.C.)*, **235**, 312–317.

Kawaguchi, Y., S. Kosugi, K. Sasaki, T. Uozumi and T. Beppu (1987). Production of chymosin in *Escherichia coli* cells and its enzymatic properties. *Agric. Biol. Chem.*, **51**, 1871–1877.

Kim, J. and R. F. Weaver (1988). Construction of a recombinant expression plasmid encoding a staphylococcal protein A–ricin A fusion protein. *Gene*, **68**, 315–321.

Kim, S.-H., C. H. Kang, R. Kim, J. M. Cho, Y.-B. Lee and T.-K. Lee (1989). Redesigning a sweet protein: increased stability and renaturability. *Protein Eng.*, **2**, 571, 578.

Kingsman, S. M., A. J. Kingsman, M. J. Dobson, J. Mellor and N. A. Roberts (1985). Heterologous gene expression in *Saccharomyces cerevisiae*. *Biotechnol. Genet. Eng. Rev.*, **3**, 377–416.

Kingsman, S. M., A. J. Kingsman and J. Mellor (1987). The production of mammalian proteins in *Saccharomyces cerevisiae* *Trends Biotechnol.*, **5**, 53–57.

Koller, B. and H. Delius (1982). The origin of replication in chloroplast DNA of *Euglena gracilis* is located close to the region of variable size. *EMBO J.*, **1**, 995–999.

Kornberg, A. (1980). *DNA Replication*. Freeman, San Francisco.

Kornberg, A. (1982). *Supplement to DNA Replication*. Freeman, San Francisco.

Kornfeld, R. and S. Kornfeld (1985). Assembly of asparagine-linked oligosaccharides. *Annu. Rev. Biochem.*, **54**, 631–664.

Krebbers, E. T., I. M. Larrinua, L. McIntosh and L. Bogorad (1982). The maize chloroplast genes for the β- and ε-subunits of the photosynthetic coupling factor $CF_1$ are fused. *Nucleic Acids Res.*, **10**, 4985–5002.

Kumagai, M. H., M. Shah, M. Terashima, Z. Vrkljan, J. R. Whitaker and R. L. Rodriguez (1990). Expression and secretion of rice α-amylase by *S. cerevisiae*. *Gene*, **94**, 209–216.

Kung, S. D. and C. M. Lin (1985). Chloroplast promoters from the higher plants. *Nucleic Acids Res.*, **13**, 7543–7549.

Kutchan, T. M. (1989). Expression of enzymatically active, cloned strictosidine synthase from the higher plant *Rauvolfia serpentina* in *E. coli*. *FEBS Lett.*, **257**, 127–130.

Langford, C., N. Nellen, J. Niessing and D. Gallwitz (1983). Yeast is unable to excise foreign intervening sequences from hybrid gene transcripts. *Cell (Cambridge, Mass.)*, **33**, 519–524.

Langridge, P., H. Eibel, J. W. S. Brown and G. Feix (1984). Transcription from maize storage protein gene promoters in yeast. *EMBO J.*, **3**, 2467–2471.

Larimer, F. W., R. J. Mural and T. S. Soper (1990). Versatile protein engineering vectors for mutagenesis, expression and hybrid enzyme formation. *Protein Eng.*, **3**, 227–231.

Ledeboer, A. M., C. T. Verrips and B. M. M. Dekker (1984). Cloning of the natural gene for the sweet-tasting plant protein thaumatin. *Gene*, **30**, 23–32.

Lee, J.-H., J. Lai and J. Weickmann (1986). Production of food additives by genetically engineered microorganisms. In *The World Biotechnology Report 1986*, vol. 2, Food Processing, pp. 63–69. On line, New York.

Lee, J.-H, J. L. Weickmann, R. K. Koduri, P. Ghosh-Dastidar, K. Saito, L. C. Blair, T. Date, J. S. Lai, S. M. Hollenberg and R. L. Kendall (1988). Expression of synthetic thaumatin genes in yeast. *Biochemistry*, **27**, 5101–5107.

Lopes, T. S., J. Klootwijk, A. E. Veenstra, P. C. van der Aar, H. van Heenikhuizen, H. A. Raué and R. J. Planta (1989). Highcopy number integration into the ribosomal DNA of *S. cerevisiae*: a new vector for high level expression. *Gene*, **79**, 199–206.

Lundqvist, T. and G. Schneider (1989). Crystal structure of the binary complex of ribulose-1,5-bisphosphate carboxylase and its product, 3-phospho-D-glycerate. *J. Biol. Chem.*, **264**, 3643–3646.

Martson, F. A. O. (1986). The purification of eukaryotic polypeptides synthesized in *Escherichia coli*. *Biochem. J.*, **240**, 1–12.

Melchior, F. and H. Kindl (1990). Grapevine stilbene synthase cDNA only slightly differing from chalcone synthase cDNA is expressed in *E. coli* into a catalytically active enzyme. *FEBS Lett.*, **268**, 17–20.

Miziorko, H. M. and G. H. Lorimer (1983). Ribulose-1,5-bisphosphate carboxylase–oxygenase. *Annu. Rev. Biochem.*, **52**, 507–535.

Moore, R. E., M. S. Davies, K. M. O'Connell, E. I. Harding, R. C. Wiegand and D. C. Tiemeier (1986). Cloning and expression of a cDNA encoding a maize glutathione-*S*-transferase in *E. coli*. *Nucleic Acids Res.*, **14**, 7227–7235.

Neill, J. D., J. C. Litts, O. D. Anderson, F. C. Greene and J. I. Stiles (1987). Expression of a wheat α-gliadin gene in *Saccharomyces cerevisiae*. *Gene*, **55**, 303–317.

Newman, J. and S. Gutteridge (1990). The purification and preliminary X-ray diffraction studies of recombinant *Synechococcus* ribulose-1,5-bisphosphate carboxylase/oxygenase from *E. coli*. *J. Biol. Chem.*, **265**, 15 154–15 159.

Nordling, M., T. Olausson and L. G. Lundberg (1990). Expression of spinach plastocyanin in *E. coli*. *FEBS Lett.*, **276**, 98–102.

Ogata, C., M. Hatada, G. Tomlinson, W. C. Shin and S. H. Kim (1987). Crystal structure of the intensely sweet protein monellin. *Nature (London)*, **328**, 739–742.

O'Hare, M., L. M. Roberts, P. E. Thorpe, G. J. Watson, B. Prior and J. M. Lord (1987). Expression of ricin A chain in *Escherichia coli*. *FEBS Lett.*, **216**, 73–78.

Østergaard Jensen, E., K. A. Marcker and I. S. Villadsen (1986). Heme regulates the expression in *Saccharomyces cerevisiae* of chimaeric genes containing 5′-flanking soybean leghemoglobin sequences. *EMBO J.*, **5**, 843–847.

Overbeeke, N., M. A. Haring, H. J. J. Nijkamp and A. J. Kool (1984a). Cloning of *Petunia hybrida* chloroplast DNA sequences capable of autonomous replication in yeast. *Plant Mol. Biol.*, **3**, 235–241.

Overbeeke, N., J. H. de Waard and A. J. Kool (1984b). Characterization of *in vitro* DNA synthesis in an isolated chloroplast system of *Petunia hybrida*. In *Proteins Involved in DNA Replication*, ed. U. Hübscher and S. Sparari, pp. 107–112. Plenum Press, New York.

Overbeeke, N. (1989). Synthesis and processing of thaumatin in yeast. In *Yeast Biotechnology*, ed. P. J. Barr, A. J. Brake and P. Valenzuela. Butterworths, Stoneham, MA, USA.

Overbeeke, N., A. J. Fellinger and S. G. Hughes (1987). Production of guar α-galactosidase by hosts transformed with recombinant DNA methods. *Eur. Pat.*, 255 153; *World Pat.*, 87 07 641.

Overbeeke, N., G. H. M. Termorshuizen, M. L. F. Giuseppin, D. R. Underwood and C. T. Verrips (1990). Secretion of the α-galactosidase from *Cyamopsis tetragonoloba* (guar) by *Bacillus subtilis*. *Appl. Environ. Microbiol.*, **56**, 1429–1434.

Padgette, S. R., Q. K. Huynh, J. Borgmeyer. D. M. Shah, L. A. Brand, D. Biest-Re, B. F. Bishop, S. G. Rogers, R. T. Fraley and G. M. Kishore (1987). Bacterial expression and isolation of *Petunia hybrida* 5-enolpyruvylshikimate-3-phosphate synthase. *Arch. Biochem. Biophys.*, **258**, 564–573.

Piatak, M., J. A. Lane, W. Laird, M. J. Bjorn, A. Wang and M. Williams (1988). Expression of soluble and fully functional ricin A chain in *Escherichia coli* is temperature sensitive. *J. Biol. Chem.*, **263**, 4837–4843.

Richardson, P. T., L. M. Roberts, J. H. Gould and J. M. Lord (1988). The expression of functional ricin B-chain in *Saccharomyces cerevisiae*. *Biochem. Biophys. Acta*, **950**, 385–394.

Rothstein, S. G., C. M. Lazarus, W. E. Smith, D. C. Baulcombe and A. A. Gatenby (1984). Secretion of a wheat α-amylase expressed in yeast. *Nature (London)*, **308**, 662–665.

Roy, H., M. V. Bloom, P. Milos and M. Monroe (1982). Studies on the assembly of large subunits of ribulose bisphosphate carboxylase in isolated pea chloroplasts. *J. Cell. Biol.*, **94**, 20–27.

Seidler, A. and H. Michel (1990). Expression in *E. coli* of the psbO gene encoding the 33 kDa protein of the oxygen-evolving complex from spinach. *EMBO J.*, **9**, 1743–1748.

Shinozaki, K., H. Deno, A. Kato and M. Sugiura (1983). Overlap and cotranscription of the genes for the β- and ε-subunits to tobacco ATPase. *Gene*, **24**, 147–155.

Shirano, Y. and D. Shibata (1990). Low temperature cultivation of *E. coli* carrying a rice lipoxygenase L-2 cDNA produces a soluble and active enzyme at a high level. *FEBS Lett.*, **271**, 128–130.

Shire, D., B. J. P. Bourrié, C. Carillon, J.-M. Derocq, P. Dousset, X. Dumont, F. K. Jansen, M. Kaghad, R. Legoux, P. Lelong, B. Pességué and H. Vidal (1990). Biologically active A-chain of the plant toxin ricin expressed from a synthetic glue in *E. coli*. *Gene*, **93**, 183–188.

Smith, R. A., M. J. Duncan and D. T. Moir (1985). Heterologous protein secretion from yeast. *Science (Washington D.C.)*, **229**, 1219–1224.
Smith, J. K., J. V. Schloss and B. J. Mazur (1989). Functional expression of plant acetolactate synthase gene in *E. coli*. *Proc. Natl. Acad. Sci. USA*, **86**, 4179–4183.
Søgaard, M. and B. Svensson (1990). Expression of cDNAs encoding barley α-amylases 1 and 2 in yeast and characterization of the secreted proteins. *Gene*, **94**, 173–179.
Somerville, C. R., L. McIntosh, J. Fitchen and M. Gurevitz (1986). The cloning and expression in *Escherichia coli* of RuBisCo large subunit genes. *Methods Enzymol.*, **118**, 419–433.
Stanley, P. (1987). Glycosylation mutants and the functions of mammalian carbohydrates. *Trends Genet.*, **3**, 77–81.
Stinchcomb, D. T., M. Thomas, J. Kelly, E. Selker and R. W. Davis (1980). Eukaryotic DNA segments capable of autonomous replication in yeast. *Proc. Natl. Acad. Sci. USA*, **77**, 4559–4563.
Struck, D. K., W. J. Lennarz and K. Brew (1978). Primary structural requirements for the enzymatic formation of the *N*-glycosidic bond in glycoproteins. *J. Biol. Chem.*, **253**, 5786–5794.
Tabita, F. R. and C. L. Small (1985). Expression and assembly of active cyanobacterial ribulose-1,5-bisphosphate carboxylase/oxygenase in *Escherichia coli* containing stoichiometric amounts of large and small subunits. *Proc. Natl. Acad. Sci. USA*, **82**, 6100–6103.
Tague, B. W. and M. J. Chrispeels (1987). The plant vacuolar protein, phytohemagglutinin, is transported to the vacuole of transgenic yeast. *J. Cell. Biol.*, **105**, 1971–1979.
Tague, B. W. and M. J. Chrispeels (1989). Identification of the targeting domain of the vacuolar glycoprotein phytohemeagglutinin in transgenic yeast and tobacco. *J. Cell. Biochem., Supp.*, **13D**, 230.
Talbot, D. R., M. J. Adang, J. L. Slightom and T. C. Hall (1984). Size and organization of a multigene family encoding phaseolin, the major seed storage protein of *Phaseolus vulgaris*. *Mol. Gen. Genet.*, **198**, 42–49.
Thompson, R. J. and G. Mosig (1987). Stimulation of a *Chlamydomonas* chloroplast promoter by novobiocin *in situ* and in *E. coli* implies regulation by torsional stress in the chloroplast DNA. *Cell (Cambridge, Mass).*, **48**, 281–287.
Thomsen, K. K., E. A. Jackson and K. Brenner (1988). Genetic engineering of yeast: construction of strains that degrade β-glucans with the aid of a barley gene. *J. Am. Soc. Brew. Chem.*, **46**, 31–36.
Tubb, R. S. (1986). Amylolytic yeasts for commerical applications. *Trends Biotechnol.*, **4**, 98–104.
Uchimiya, U., T. Ohtani, T. Ohgawara, T. Harada, M. Sugita and M. Sugiura (1983). Molecular cloning of tobacco chromosomal and chloroplast DNA segments capable of replication in yeast. *Mol. Gen. Genet.*, **192**, 1–4.
Utsumi, S., T. Sato, C.-S. Kim and M. Kito (1988a). Processing of preproglycinin expressed from the cDNA-encoding AlaBlb subunit in *Saccharomyces cerevisiae*. *FEBS Lett.*, **233**, 273–276.
Utsumi, S., C.-S. Kim, T. Sato and M. Kito (1988b). Signal sequence of preproglycinin affects production of the expressed protein in *Escherichia coli*. *Gene*, **71**, 349–358.
Utsumi, S., J. Kanamori, C.-S. Kim. T. Sato and M. Kito (1991). Properties and distribution of soybean proglycinin expressed in *S. cerevisiae*. *J. Agric. Food Chem.*, **39**, 1179–1186.
Valle, E., H. Kobayashi and T. Akazawa (1988). Transcriptional regulation of genes for plant-type ribulose-1,5-bisphosphate carboxylase/oxygenase in the photosynthetic bacterium *Chromatium vinosum*. *Eur. J. Biochem.*, **173**, 483–489.
Vallet, J.-M., M. Rahire and J.-D. Rochaix (1984). Localization and sequence analysis of chloroplast DNA sequences of *Chlamydomonas reinhardii* that promote autonomous replication in yeast. *EMBO J.*, **3**, 415–421.
Vandekerckhove, J., J. van Damme, M. van Lijsebettens, V. Gossele, J. Leemans, J. Botterman and E. Krebbers (1989). Enkephalins produced in plant seeds using engineered 2S albumins. *J. Cell. Biochem., Suppl.*, **13D**, 243.
van der Straeten, D., L. van Wiemeersch, H. M. Goodman and M. van Montague (1990). Cloning and sequence of two different cDNAs encoding 1-aminocyclopropane-1-carboxylate synthase in tomato. *Proc. Natl. Acad. Sci. USA*, **87**, 4859–4863.
van der Vies, S. M., D. Bradley and A. A. Gatenby (1986). Assembly of cyanobacterial and higher plant ribulose bisphosphate carboxylase subunits into functional, homologous and heterologous enzyme molecules in *E. coli*. *EMBO J.*, **5**, 2439–2444.
van der Wel, H. and A. M, Ledeboer (1989). The thaumatins. In *The Biochemistry of Plants: A Comprehensive Treatise*, ed. A. Marcus, vol. 15, pp. 379–391. Academic Press, New York.
Verbakel, J. M. A. (1991). Heterologous gene expression in the yeast *Saccharomyces cerevisiae*. Ph.D. Thesis, University of Utrecht.
Viale, A. M., H. Kobayashi, T. Takabe and T. Akazawa (1985). Expression of genes for subunits of plant-type RuBisCo from *Chromatium* and production of the enzymatically active molecule in *E. coli*. *FEBS Lett.*, **192**, 283–288.
Viotti. A., C. A. Viatale and E. Sala (1985). Each zein gene class can produce polypeptides of different sizes. *EMBO J.*, **4**, 1103–1110.
Vitella, E. S. and J. W. Uhr (1985). Immunotoxins. *Annu. Rev. Immunol.*, **3**, 197–238.
Voordouw, G., S. M. van der Vies and P. J. Bouwmeister (1984). Dissociation of ribulose-1,5-bisphosphate carboxylase/oxygenase from spinach by urea. *Eur. J. Biochem.*, **141**, 313–318.
Waston, M. D., N. Lambert, A. Delauney, J. N. Yarwood, R. R. D. Croy, J. A. Gatehouse, D. J. Wright and D. Boulter (1988). Isolation and expression of a pea vicilin cDNA in the yeast *Saccharomyces cerevisiae*. *Biochem. J.*, **251**, 857–864.
Weisbeek, P., J. Hageman, D. de Boer, R. Pilon and J. Smeekens (1989). Import of proteins into the chloroplast lumens. *J. Cell. Sci.*, **11** (suppl.), 199–223.
Whitfeld, P. R. and W. Bottomley (1980). Mapping of the gene for the large subunit of ribulose bisphosphate carboxylase on spinach chloroplast DNA. *Biochem. Int.*, **1**, 172–178.
Whitfeld, P. R. and W. Bottomley (1983). Organization and structure of chloroplast genes. *Annu. Rev. Plant Physiol.*, **34**, 279–310.
Yarwood, J. N., N. Harris, A. Delauney, R. R. D. Croy, J. A. Gatehouse, M. D. Watson and D. Boulter (1987). Construction of a hybrid cDNA encoding a major legumin precursor polypeptide and its expression and localization in *Saccharomyces cerevisiae*. *FEBS Lett.*, **222**, 175–180.
Zurawski, G., W. Bottomley and P. R. Whitfeld (1982). Structures of the genes for the β- and ε-subunits of spinach chloroplast ATPase indicate a dicistronic mRNA and an overlapping translation stop/start signal. *Proc. Natl. Acad. Sci. USA*, **79**, 6260–6264.

# 11
# Protoplast Fusion

HIROMICHI MORIKAWA and YASUYUKI YAMADA
*Kyoto University, Japan*



## 11.1 INTRODUCTION

The most significant aspect of the use of protoplast fusion in plant biotechnology and cell biology is that it can produce nuclear–cytoplasmic combinations that are difficult, or impossible, to obtain by conventional sexual crossing. Because angiosperms usually have cytoplasmic genomes that are maternally inherited, new cytoplasmic–nuclear combinations can be produced sexually only by backcrossing, which is both time consuming and unidirectional. Although organelle DNA represents only $10^{-3}$ to $10^{-4}$ of the total higher plant DNA, it carries important genes that control chlorophyll content, herbicide resistances such as triazine resistance, and cytoplasmic male sterility (cms) as described in Section 11.5.1. Because organelle enzymes are generally composite proteins made up of chondriome- or plastome-encoded polypeptides and nuclear-encoded polypeptides, the establishment of new nuclear–cytoplasmic genome combinations is important both for fundamental studies of plant cell and molecular biology, and for the utilization of fusion products in plant breeding programs.

**Table 1** Experimental Conditions for Production of Somatic Hybrids by Electrofusion in Higher Plants

| Plant material | Pair formation | kV cm$^{-1}$ | DC pulse condition $\Delta t^{a}$ (μs) | $CR^{b}$ (μs) | $n^{c}$ | Medium | Selection method | Hybrid | Refs. |
|---|---|---|---|---|---|---|---|---|---|
| *N. tabacum* (mesophyll) + *N. plumbaginifolia* (suspension) | AC, 0.6 MHz 15 V/chamber | 50 V/ chamber | 50 | — | 2 | No added salts | Pick up | Plantlets | Bates and Hasenkampf (1985) |
| *N. tabacum* (suspension) NR deficient | AC, 0.9 MHz 400 V cm$^{-1}$ | 1.2–1.5 | 50 | — | 1 | No added salts | Minimum medium | Shoots | Kohn *et al.* (1985) |
| *N. plumbaginifolia* (mesophyll and suspension) NR deficient | AC, 1 MHz 150 V cm$^{-1}$ | 1.2–2.0 | — | 50 | 1 | No added salts | Pick up | Calli | Puite *et al.* (1985) |
| *N. glauca* + *N. langsdorffii* (mesophyll) | multi-protoplast layer | 1.0 | — | 100–200 | 1 | 2.5 mM $CaCl_2$ | Hormone-free medium | Fertile plants | Morikawa *et al.* (1986, 1987) |
| *N. glauca* + *N. langsdorffii* (mesophyll) | 1–10% PEG | 1.0 | 100 | — | 1 | 1 mM $MgCl_2$, 50 mM NaCl, O.75 mM spermine, 0.01% Tween 80 + 1–10% PEG | Hormone-free medium | Calli | Chapel *et al.* (1986) |
| *S. tuberosum* + *S. phureja* (shoots) | AC, 1 MHz 73–150 V cm$^{-1}$ | 1.2–2.0 | 50 | — | 1–2 | No added salts | Pick up, flow sorting | Plants | Puite *et al.* (1986, 1988) |
| *B. napus* (hypocotyl protoplasts + subprotoplasts) | AC, 1 MHz 60–80 V cm$^{-1}$ several s | 0.9–1.8 | 50 | — | ⩾1 | No added salts | Pick up | Calli (homofusion) | Spangenberg and Schweiger (1986) |
| *S. tuberosum* (suspension culture) amino acid analog-resistant cell lines | AC, 1.5 MHz >233 V cm$^{-1}$ | >3.3 | 10–1000 | — | 1 | 0.15–1 mM $CaCl_2$ | Medium containing amino acid analogs | Calli | Vries *et al.* (1987) |

| | | | | | | | | | |
|---|---|---|---|---|---|---|---|---|---|
| *O. sataiva* + *E. oryzicola* (suspension) | AC, 4 MHz 250 V $cm^{-1}$ | 3–4 | 10 | — | 3 | 0.1% MES[d] | IOA[e] treatment | Plantlets | Terada *et al.* (1987a) |
| *N. glauca* + (*mesophyll*) *N. langsdorffii* (epidermis) | AC, 1 MHz 160 V $cm^{-1}$, 60 s 240 V $cm^{-1}$, 4 s | 0.78 | 50 | — | 1 | No added salts | Hormone-free medium | | Kamata and Nagata (1987) |
| *N. plumbaginifolia* (suspension, NR deficient) + *H. muticus* (suspension, tryptophan-requiring) | AC, 1 MHz 50 V $cm^{-1}$, 60 s | 0.63 | 50 | — | 5 | No added salts | Minimum medium | Calli | Kishinami and Widholm (1987) |
| *N. tabacum* (mesophyll, transformed) *N. tabacum* (mesophyll) | AC, 0.6 MHz 100 V $cm^{-1}$ | 1 | 25 | — | 2 | No added salts | Kanamycin medium | Plants | Bates *et al.* (1987) |
| *S. tuberosum* + *S. brevidens* (mesophyll) | AC, 1 MHz 100 V $cm^{-1}$ | 1.25–1.50 | 10 | — | 1 | 1 mM $CaCl_2$ | Plant morphology | Plants | Fish *et al.* (1988a) |
| *O. Sativa* (suspension, cms + fertile) | AC, 0.5 MHz 100 V $cm^{-1}$, 60 s | 1 | 40 | — | 1 | 2.5 mM $CaCl_2$ | IOA treatment + r-irradiation | Calli | Yang *et al.* (1988) |
| *Coptis japonica* + *Euphorbia millii* (suspension) | multi-protoplast layer | 2.0 | — | 100–200 | 1 | 2.5 mM $CaCl_2$ | Pick up | Calli | Morikawa *et al.* (1988c) |
| *B. oleracea* + *B. campestris* (mesophyll) | AC, 1 MHz 200 V $cm^{-1}$, 10 s | 1.2–2.35 | 15–30 | — | 2–3 | | IOA treatment | Plants | Nishio *et al.* (1988) |

[a]Duration of the square pulse. [b]Time constant of the exponentially decaying pulse. [c]Number of pulses. [d]2-(*N*-morpholino)ethanesulphonic acid. [e]Iodoacetamide

A number of nuclear hybrids and cytoplasmic hybrids (cybrids) that are male sterile, herbicide resistant or both have been obtained from the Cruciferae and Solanaceae, and the feasibility of using them in breeding programs has been explored. In addition, extensive studies on 'cms genes' in mitochondrial DNA (mtDNA), including the intra- and inter-genomic recombination of mtDNA, have been carried out with hybrids/cybrids obtained by protoplast fusion (see Section 11.5.1).

Another advantage of protoplast fusion is that a set of genes that is responsible for specific useful traits can be transferred without knowing the molecular details of the genes that confer the traits. The successful transfer of disease resistance traits in potato and eggplant has been made by protoplast fusion. The use of these disease-resistant somatic hybrids in breeding programs has been investigated (see Section 11.5.3). By the use of protoplast fusion, Austin *et al.* (1988) discovered that the nontuber-bearing wild species *Solanum brevidens* has a gene(s) responsible for the resistance of tubers to the decay caused by the bacterium *Erwinia* spp. This discovery has opened up a new aspect of the plant pathology of *Erwinia* and is evidence that protoplast fusion can be used to explore as yet unknown 'buried' genes (see Section 11.5.3).

Recent success in the genetic transformation of plants by 'direct gene transfer', *i.e.* transformation of chromosomal genomes without the use of special DNA vectors such as Ti or Ri plasmids (see, for example, Morikawa *et al.*, 1988e), has added to the significance of protoplast fusion as a method of genetic transformation; but no direct evidence has been published of the exchange of DNA sequences between heterologous chromosomes in fusion products.

Protoplast fusion permits the establishment of hybrids/cybrids between sexually incompatible species, but the question has yet to be answered as to whether protoplast fusion can circumvent the genetic or biological barrier of sexual incompatibility. No fertile intergeneric somatic hybrid plants have been reported. Asymmetric fusion, the transfer of limited amounts of chromosomal genomes from one fusion partner to the other, may be the solution. In fact, intergeneric transfer of functional genes has already been accomplished, for example, from carrot to tobacco (Dudits *et al.*, 1987) by asymmetric fusion. But so far all asymmetrically produced intergeneric hybrid plants are again sterile (see Section 11.5.4). The possibility of chondriomes and plastomes participating in the sterility of asymmetrically produced hybrids/cybrids is a subject for future research.

Although most of the hybrids/cybrids reviewed here have been produced by chemical fusion (mainly with PEG and dextran), the number of reports of somatic hybridization by electrofusion is increasing, as are the number of examples of hybrid plants produced by this method (see Section 11.6 and Table 1). The advantages and disadvantages of these two methods of protoplast fusion are discussed in Section 11.2.

In this chapter more than 200 original reports relating to protoplast fusion, published mainly from 1985 to 1988, are reviewed. For details of the history of the fusion methods or concepts and for studies conducted earlier than 1984, the reader is referred to recent reviews (*e.g.* Chaleff, 1981; Zimmermann *et al.*, 1981; Evans *et al.*, 1983; Ammirato *et al.*, 1984; Gleba and Sytnik, 1984; Sharp *et al.*, 1984; Vasil, 1984; Pelletier *et al.*, 1985; Yamada and Morikawa, 1985; Galun and Aviv, 1986; Zachrisson and Bornman, 1986; Kumar and Cocking, 1987; Morikawa *et al.*, 1988e).

## 11.2 PROTOPLAST FUSION METHODS: CHEMICAL FUSION AND ELECTROFUSION

Chemical fusion is a method of inducing cell fusion by use of high concentration of various chemical fusogens such as PEG (polyethylene glycol), dextran and PVA (polyvinyl alcohol). On the other hand, electrofusion employs high DC electric field pulse(s) to induce fusion of cell (protoplast) pairs that are formed by an AC field induced dielectrophoresis or other methods (see Table 1). Modifications of the PEG method by the addition of dimethyl sulfoxide (DMSO) (*e.g.* Menczel *et al.*, 1986) and of a hypotonic solution in the postfusion step (*e.g.* Takahashi *et al.*, 1986) also have been reported. In addition, PEG and PVA have been successfully used for the transformation of protoplasts.

Although the number of reports on hybridization by electrofusion is increasing, most of those reports reviewed used the PEG (or dextran) method. All the protoplast fusion products reviewed here were obtained by PEG fusion unless otherwise stated.

Negrutiu *et al.* (1986) used mesophyll protoplasts from various auxotrophic mutants of *Nicotiana plumbaginifolia* to study the optimum conditions for PEG-induced fusion and they examined the various factors in the isolation, culture and fusion of protoplasts. Based on the yield ($10^{-4}$ to $10^{-2}$) of the hybrid colonies after selection on minimal medium, they concluded that the essential factors for the fusion efficiency are the physiological conditions, degree of purification of the protoplast

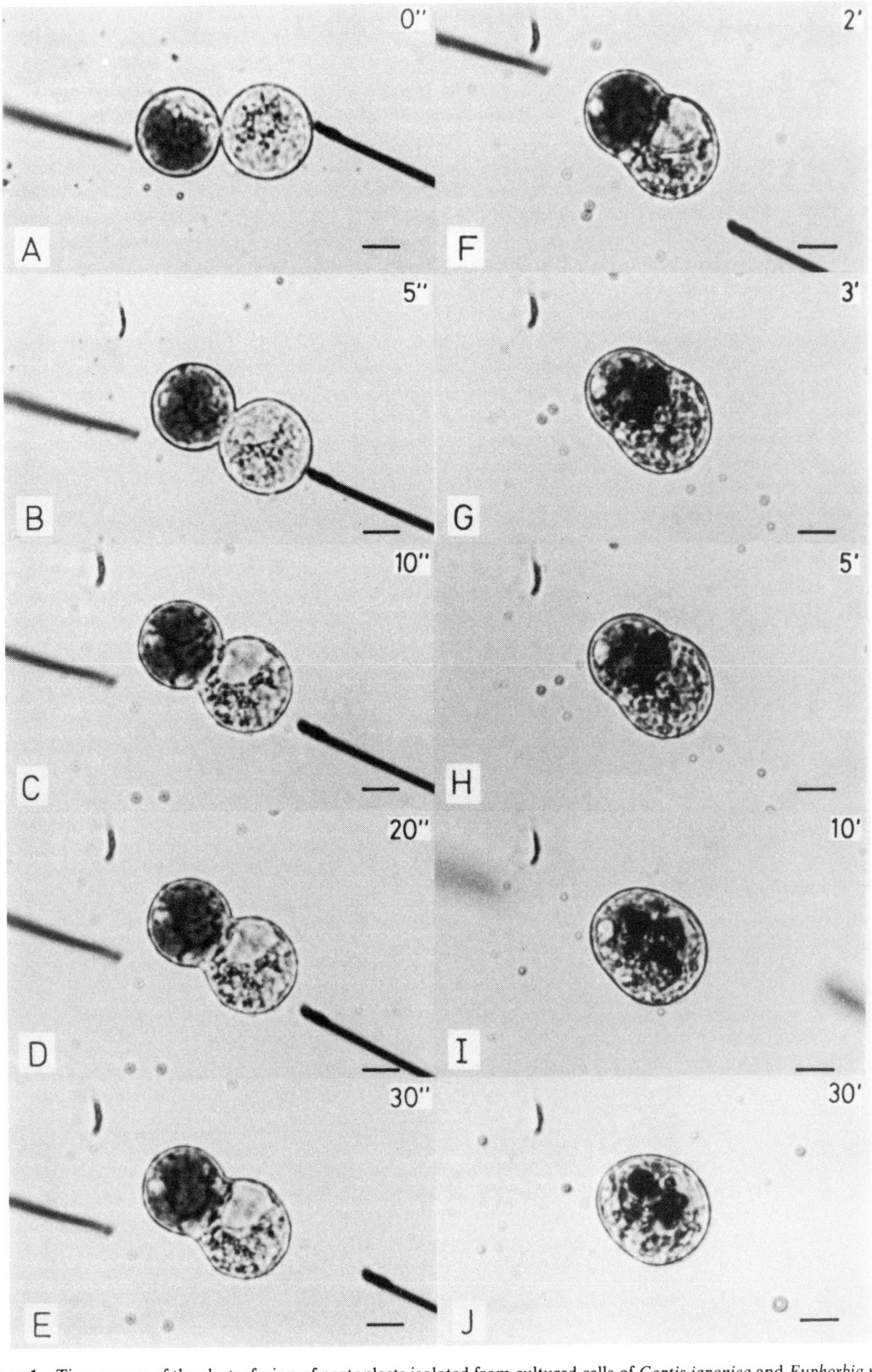


**Figure 1** Time course of the electrofusion of protoplasts isolated from cultured cells of *Coptis japonica* and *Euphorbia millii* (respective producers of large amounts of berberine and anthocyanin) by means of platinum (1 μm in diameter) microelectrodes. (A) 0 s; (B) 5 s; (C) 10 s; (D) 20 s; (E) 30 s; (F) 2 min; (G) 3 min; (H) 5 min; (I) 10 min; and (J) 30 min after the application of a single pulse (10V, 0.5 ms). Note that fusion involved two stages, cellular (A through F) and vacuolar (G through J), characterized by transient wrinkling of the membrane (E) and formation of dark red precipitates (G to J). The bar represents 20 μm. Reproduced from Morikawa *et al.* (1988b) with permission

population, preliminary incubation in salt mixtures, the population density of the protoplasts at fusion and the PEG concentration.

Dextran has been used successfully to produce hybrid plants of *Daucus* and *Brassica*. Kothari *et al.* (1986) used it to obtain intraspecific hybrids of drug-resistant strains of *Daucus carota*, and Ye *et al.* (1987) obtained intergeneric somatic hybrid calli by fusion between a double mutant of *N. plumbaginifolia* (nitrate reductase (NR) deficient and azetidine-2-carboxylate resistant) with a glyphosate-resistant *Daucus carota* strain. Toriyama *et al.* (1987b) succeeded in forming an intergeneric hybrid plantlet by fusion, in the presence of dextran, of protoplasts from a double mutant of *Sinapis turgida* (Brassicaceae; NR deficient and 5-methyltryptophan resistant) with those from the mesophyll of wild-type *B. oleracea* or *B. nigra*. Dextran fusion has also been used to resynthesize *Brassica napus* (Taguchi and Kameya, 1986) and to produce intergeneric somatic hybrids in the Brassicaceae (Toriyama *et al.*, 1987a).

Electrofusion has become an accepted plant protoplast fusion method (see *e.g.* Zimmermann *et al.*, 1981; Zachrisson and Bornman, 1986; Yamada and Morikawa, 1985; Morikawa *et al.*,1988e). The main advantages of electrofusion in comparison to chemical fusion, are: (i) it is technically easy and is quicker than PEG fusion; (ii) as a result the reproducibility of experimental procedures are better; and (iii) the number of protoplasts that can be handled in one electrofusion experiment is much higher (10 times or more) than in a chemical fusion experiment (see also Yamada and Morikawa, 1985; Negrutiu *et al.*, 1986; Zachrisson and Bornman, 1986). Electrofusion appears to be free of the cytotoxicity often associated with chemical fusogens such as PEG and PVA. Whether a high DC fusion pulse affects nuclear and organellar DNA has yet to be determined.

Terada *et al.* (1987a) reported successful intergeneric hybridization between rice and barnyard grass with electrofusion but not with PEG. Tur-Kaspa *et al.* (1986) introduced foreign genes into primary culture of animal cells by use of electric pulses (electroporation), whereas transfection of the cells by calcium phosphate or DEAE–dextran was not successful because of the cytotoxicity of these chemicals.

Chand *et al.* (1988) have reported an improved fusion procedure using a polyethylene glycol preparation, which has a low carbonyl content. In their comparison of the efficiency of PEG fusion (PF), electrofusion (EF) and PEG–electrofusion (PEF), heterokaryon formation was PF > EF > PEF, whereas the division of heterokaryons was PF ⩾ EF > PEF. They also reported that protoplast clumbing was decreased by their improved PEG fusion procedure.

An electrofusion chamber equipped with a pair of parallel plate electrodes is the most common device used, but microelectrodes of glass or metal also have been utilized (see Yamada and Morikawa, 1985; Morikawa *et al.*, 1988e). An electrofusion between protoplasts isolated from cultured cells rich in secondary metabolites that was induced with a pair of platinum microelectrodes (Morikawa *et al.*, 1988b) is shown in Figure 1.

The voltage across the membrane ($V_m$) of a protoplast exposed to an electric field ($E$) is given by equation (1) (*e.g.* Zimmermann *et al.*, 1981)

$$V_m = 1.5rE\cos\theta \qquad (1)$$

in which $r$ is the radius of the protoplast and $\theta$ the angle between the normal to the membrane surface and the field direction (*i.e.* the voltage across the membrane is at a maximum in the direction of the electric field and is equal to $1.5\,rE$). Thus, when an electric field of $1\ \mathrm{kV\ cm^{-1}}$ is applied to a protoplast or $r = 40\ \mu\mathrm{m}$, the maximum voltage across plasma membrane is 6 V. This value is 30–60 times the resting membrane potential of intact plant cells. The critical voltage for membrane breakdown (pore formation or membrane fusion) in plants is of the order of 1 V per membrane (Zimmermann *et al.*, 1981). Equation (1) shows that the $V_m$ is proportional to the radius of the cell, which indicates that the voltage necessary to induce fusion is inversely proportional to protoplast size.

## 11.3 SELECTION OF HETEROKARYONS OR SOMATIC HYBRIDS

As no specific heteroplasmic fusion methods exist, development of methods to select heterokaryons from nonfused or homoplasmically fused protoplasts is essential. Examples of recent results are given in the following subsections.

### 11.3.1 Use of Auxotrophy, Autotrophy and Resistance

One of the most common methods for heterokaryon selection is to make use of mutants that carry auxotrophy, drug resistance or autotrophy as the fusion partner in combination with inactivation of the other partner by treatments that include irradiation with X- or γ-rays and iodoacetamide (IOA) treatment. Transgenic plants that carry kanamycin or chloramphenicol resistance also have been used as fusion partners (Bates *et al.*, 1987; Brunold *et al.*, 1987) as described below.

Hormone autotrophism often is present in intraspecific and intergeneric hybrids of the Solanaceae, and somatic hybrids have been selected from certain fusion products, using this character as a marker; for example *Hyocyamus muticus* and *N. tabacum* (Potrykus *et al.*, 1984) and *N. glauca* and *N. langsdorffii* as described in Section 11.3.5. Auxin autotrophy in Ti plasmid transformed cells has been utilized as a genetic marker (*e.g.* Gleba *et al.*, 1986).

Ozias-Akins *et al.* (1986) fused protoplasts from *Pennisetum americanum* (L.) K. Schum. (pearl millet), which were resistant to (*S*)-2-aminoethyl-L-cysteine (AEC), with IOA-treated protoplasts from an embryogenic suspension culture of *Panicum maximum* Jacq. (guinea grass). Hybrid cells were selected for their AEC resistance (see Section 11.5.5.1).

Kothari *et al.* (1986) used dextran to fuse protoplasts from different *Daucus carota* L. cell strains that carry resistances to glyphosate, 5-methyltryptophan, sodium selenate and selenocystine, in three combinations. Hybrids were selected for the resistances carried by both parental strains in media that contained both inhibitors.

Saxena and King (1987) obtained paired protoplast fusion with three auxotrophic cell lines of *Datura innoxia*, Pn1, IV-1 and JM3 that have obligate growth requirement for pantothenate (Pn1), isoleucine and valine (IV-1), and threonine (JM3). The resulting intraspecific somatic hybrids were selected on minimal medium and identified by their chromosome numbers and the restoration of their ability of organogenesis.

Pirrie and Power (1986) reported the fusion on gametic protoplasts isolated form tetrads of *N. glutinosa* and mesophyll protoplasts of nitrate reductase deficient *N. tabacum*. On a medium with nitrate as the sole nitrogen, source, only the hybrids divided because *N. glutinosa* protoplasts and tetrad protoplasts are incapable of cell division. 'Gametosomatic' hybrid plants were regenerated and hybridity confirmed by their morphologies, chromosome numbers, isozyme patterns and RuBisCo polypeptide analyses. The five gametosomatic hybrid plants obtained had the expected pentaploid but functionally triploid, chromosome number of $3n = 5x = 60$. This selection method is useful for limited gene transfer of haploid chromosomes and for the efficient selection of triploid hybrids.

Kumashiro *et al.* (personal communication) have electrofused protoplasts from a transgenic tobacco resistant to either kanamycin or chloramphenicol with mesophyll protoplasts from wild-type tobacco. The maximum hybrid yield, based on selection for kanamycin or chloramphenicol resistance, was about 0.5%.

### 11.3.2 Universal Hybridizers

Loschiavo *et al.* (1983) proposed that for the selection of somatic hybrids, double mutants called 'universal hybridizers', which carry both a dominantly expressed trait such as resistance and a recessive trait such as auxotrophy, be used. When such a mutant is fused with any wild-type cell, then selected on a minimal medium containing inhibitor, only hybrids grow because the universal hybridizer is auxotrophic and the wild-type is sensitive to the inhibitor. They selected a double mutant of carrot cells that is resistant to α-amanitin and sensitive to a HAT (hypoxanthine, aminopterin, glycine and thymidine) medium. They fused this double mutant with wild-type carrot cells, after which they obtained hybrids by selection on medium with HAT and α-amanitin (Loschiavo *et al.*, 1983).

Pental *et al.* (1984) fused a similar double mutant of *N. tabacum* (NR deficient and streptomycin resistant) with protoplasts from a suspension culture of wild-type *N. rustica*. Hybrid colonies were selected for NR proficiency and streptomycin resistance, after which hybrid plants were regenerated. Hybridity was confirmed by RuBisCo polypeptide and isozyme analyses.

Pental *et al.* (1986) fused protoplasts from this double mutant with cell suspension protoplasts from wild-type *Petunia hybrida*. Intergeneric somatic hybrid colonies were selected as described above, and hybridity confirmed by isozyme, ribosomal DNA (rDNA) and chloroplast DNA (cpDNA) analyses. After 6–12 months of culture, most of the tobacco nuclear genomes had been eliminated from the hybrid cell lines. One line regenerated plants that had *N. tabacum* type chloroplasts with *P. hybrida* nuclei.

Ye *et al.* (1987) used dextran to fuse protoplasts from a universal hybridzer of *N. plumbaginifolia* (NR deficient and azetidine-2-carboxylate resistant) with those from a glyphosate-resistant *Dacus carota* strain. Intergeneric hybrid calli were selected on a medium with azetidine-2-carboxylate and nitrate as the sole nitrogen source. Hybridity was confirmed by glyphosate resistance, isozyme patterns, and chromosome number and by Southern hybridization analysis with an enolpyruvylshikimic acid-3-phosphate synthase probe.

Toriyama *et al.* (1987b) selected a double mutant of *Sinapis turgida* (Brassicaceae; NR deficient and 5-methyl tryptophan resistant) and used dextran to fuse its protoplasts with mesophyll protoplasts from wild-type *B. oleracea* or *B. nigra.* Intergeneric somatic hybrids were selected for their NR proficiency and 5-methyltryptophan resistance, and plantlets regenerated. Hybridity was confirmed by isozyme and chromosome number analyses.

Brunold *et al.* (1987) used a nonmorphogenic cell line of *N. tabacum* (NR deficient and Kanamycin resistant) that had been produced by transformation as a universal hybridizer. They fused protoplasts of this double mutant with those from streptomycin-resistant *N. tabacum* cv. Petit Havana SR1. Hybrids were selected for their NR proficiency and kanamycin resistance. Plants were regenerated and hybridity confirmed by various methods that included analysis for the neomycin phosphotransferase gene by Southern hybridization. This double mutant was also fused with the wild-type *N. glutinosa*, *N. repanda* and *N. sylvestris*, hybrid plants being obtained from the combination with *N. sylvestris* as the fusion partner.

### 11.3.3 Manual Selection

Manual selection of heterokaryons followed by microculture in the presence or absence of nurse cells has been shown to be useful for obtaining somatic hybrids of normal cells that lack genetic markers (see *e.g.* Yamada and Morikawa, 1985, and references therein). This method can also rescue new types of hybrids, such as cybrids/hybrids with mixed chloroplast types, that differ from those obtained by selection in the presence of selection pressure (minimum medium or selection medium) or with inactivation treatments (Gleba *et al.*, 1984, 1985).

Gleba *et al.* (1984) fused protoplasts from chlorophyll-deficient *N. tabacum* with mesophyll protoplasts from one of four cms analogs of tobacco bearing the cytoplasm from *N. plumbaginifolia*, *N. suaveolens*, *N. repanda* or *N. undulata*, or from wild-type *N. glauca.* The heterokaryons were selected manually, and incubated in a microdroplet culture; hybrid plants were regenerated. Hybridity was confirmed by the morphology, chromosome size and morphology, chromosome numbers, isozyme patterns and cpDNA analysis. In addition to pure green and chlorophyll- deficient plants, they obtained variegated plants of mixed chloroplast types. Variegation was inherited maternally and the variegated $F_1$ progenies also had heterozygous cpDNA compositions (Gleba *et al.*, 1985).

Sundberg and Glimelius (1986) fused mesophyll protoplasts from *B. campestris* or from *B. oleracea* with hypocotyl protoplasts from the counterpart. The heterokaryons (whose frequency was 15% 24 h after fusion treatment) were selected manually and cultured in microdroplets. Hybrid colonies were obtained (plating efficiency 14%), whose hybridity was confirmed by isozyme analysis.

Several hybrids have been successfully produced from electrofusion products, by manual selection and subsequent microculture (Table 1). Puite *et al.* (1986) electrofused protoplasts from shoots of *in vitro* cultures of *Solanum tuberosum* and *S. phureja.* One parent had been grown in the presence of herbicide to induce the bleaching of chlorophyll. The protoplasts from this parent were stained with fluorescein diacetate to facilitate visual identification of the heterokaryons. The heterokaryons were selected manually, and hybrid plants regenerated.

The authors electrofused protoplasts from suspension cultures of *Coptis japonica* and *Euphorbia millii*, which respectively produce large amounts of berberine and anthocyanin (Table 1). The resulting heterokaryons formed dark red precipitates upon vacuolar fusion (see Figure 1). These heterokaryons were selected manually and microcultured in the presence of nurse cells. Of the 2830 heterokaryons cultured, three showed cell division and one had developed into a small colony of 10–20 cells after about one year of the culture (Yamada *et al.*, 1987; Morikawa *et al.*, 1988c).

### 11.3.4 Flow Cytometry

This technique has also proven useful for sorting or concentrating the heterokaryons of normal cells that lack genetic markers. Afonso *et al.* (1985) first succeeded in producing somatic hybrid plants by this method. They isolated protoplasts from leaf or suspension cultures of *N. tabacum*, *N. sylvestris*, *N. nesophyla* and *N. stocktanii* and induced interspecific fusion. Prior to fusion treatment, the protoplasts had been labeled with fluorescein isothiocyanate (FITC) or rhodamine isothiocyanate (RITC). Heterokaryons were sorted by means of the fluorescence of these dyes and the endogenous autofluorescence of chlorophyll as markers; on sorting they were found to be enriched 33- to 391-fold (42–90% were heterokaryons). Somatic hybrid plants were regenerated and hybridity confirmed by their morphology, isozyme patterns, chromosome numbers and DNA content analyses.

Glimelius *et al.* (1986) succeeded in the sorting of heterokaryons from PEG-induced fusion products; they fused leaf protoplasts of *B. campestris* with hypocotyl protoplasts of *B. oleracea*, and then maintained the protoplast mixture at a low temperature in order to release the protoplasts from the bottom of the dish. Heterokaryons were sorted by use of the chlorophyll autofluorescence of the mesophyll protoplasts and the fluorescence of the carboxyfluorescein supplied exogenously to the hypocotyl protoplasts as markers; 80% were heterokaryons , which were subsequently cultured in microdroplets. Of the regenerated calli, 81% were identified as hybrids by isozyme analysis. Pauls and Chuong (1987) reported flow cytometric identification of *B. napus* fusion products by the use of similar markers (the autofluorescence of chlorophyll of mesophyll protoplasts and the fluorescence of the FITC supplied to hypocotyl protoplasts).

Sundberg *et al.* (1987) selected heterokaryons manually or by flow cytometry from fusion products between *B. oleracea* (one of four varieties) and *B. campestris* (one of four varieties). They obtained a number of putative hybrid calli of which 2% regenerated shoots. Isozyme analysis showed that 100% of shoots from manually selected heterokaryons and 87% from heterokaryons selected by flow cytometry were hybrids. The sum (38) of the chromosome numbers of the two fusion partners was found in 30% of the hybrids, 9% having fewer and 61% more chromosomes. Restriction enzyme analysis of the cpDNA indicated the presence of *B. oleracea* or *B. campestris* type chloroplasts in the somatic hybrids.

Puite *et al.* (1988) have compared the efficacy of manual selection and flow cytometry for the sorting of heterokaryons from electrofusion products between *S. tuberosum* and *S. phureja*. They found that flow sorting of heterokaryons gave a larger fraction of 1:1 heterofusion, whereas manual selection gave a large fraction of multiple heterofusions.

So far, flow cytometry has been used with a limited number of species, mainly those of the Solaneceae or *Brassica*, whose heterokaryons are relatively easy to culture, but in the future this technique should be useful for the sorting out of heterokaryons of low plating efficiency, such as those formed between distantly related plant species (Morikawa *et al.*, 1988c), and this should be of significance for plant biotechnology.

The development of highly efficient protoplast labeling methods is essential for manual or flow cytometry selection of heterokaryons. Fitter *et al.* (1987) has reported labeling protoplasts with monoclonal antibody to plasma–membrane antigens of *N. glutinosa* protoplasts. This antigen did not bind to *Phaseolus vulgaris* leaf protoplasts, and the antibody labeling inhibited neither callus formation nor plantlet regeneration. The effects of this antibody labeling on the ability of protoplasts to fuse was not addressed.

### 11.3.5 Density Gradient Centrifugation

Hampp and Steingraber (1985) electrofused vacuolate and evacuolated mesophyll protoplasts of oat (*Avena sativa*). Differences in buoyant density for vacuolate, evacuolated and homoplasmically fused protoplasts have been used in the separation of heterokaryons by density gradient centrifugation (Naton *et al.*, 1986).

Kamata and Nagata (1987) electrofused protoplasts from the epidermis of *N. langsdorffii* and mesophyll protoplasts of *N. glauca* that differ in specific gravity. Density gradient centrifugation with Percoll and sea water enriched heterokaryons by 70–80% in the intermediate specific gravity layer between the epidermal and mesophyll protoplast layers (see also Table 1).

## 11.4 CONFIRMATION OF HYBRIDITY

The various methods used to confirm hybridity are described in this section. The characteristics of these methods and examples from the recently published research are given in each subsection.

### 11.4.1 Isozyme Patterns

Electrophoretic analysis of the isozyme pattern (zymogram) is a rapid method used to detect hybrid-like callus colonies or plants. This method also is used to differentiate cybrids from nuclear hybrids because most enzymes are encoded on the nuclear genomes. Often, however, the zymogram changes with the physiological state of the callus or plant of a species. Moreover, not all parental isozymes are necessarily expressed, for unknown reasons, in somatic hybrids. Thus, simple isozyme pattern analysis is not usually sufficient for one to draw an unequivocal conclusion about hybridity.

Examples of the use of isozyme patterns to identify hybridity are: the phosphoglucomutase isozyme pattern used to identify interspecific somatic hybrids of *Lotus* species (forage legumes; Wright *et al.*, 1987); the isozyme patterns of alcohol dehydrogenase, 6-phosphogluconate dehydrogenase, aminopeptidase and shikimate dehydrogenase used to identify hybrids between AEC-resistant *Pennisetum americanum* (pearl millet) and *Panicum maximum* Jacq. (guinea grass; Ozias-Akins *et al.*, 1986); and the peroxidase isozyme pattern used together with mtDNA analysis to identify the cybridity of rice cybrids formed between a cms line and a fertile cultivar (Yang *et al.*, 1988).

### 11.4.2 Ribulose-1,5-bisphosphate Carboxylase (RuBisCo) Subunit Polypeptides

The gene for the small subunit polypeptides of the ribulose-1,5-bisphosphate carboxylase (RuBisCo) or Fraction I protein is encoded on the nuclear genome, whereas that for the large subunit polypeptide is encoded on the plastome. Thus, when the genes are expressed, the isoelectrofocusing patterns of those polypeptides that have been purified with the antibody can be used to identify nuclear or chloroplast hybridity. The detection limit with this method is reported to be 10%; when the gene is silent or expression is less than 10% of the dominant species, there is no detection (Pelletier *et al.*, 1985). Asahi *et al.* (1988) reported a close correlation between chloroplast hybridity determined by RuBisCo large subunit analysis and the restriction pattern of cpDNA. Examples include somatic hybrids formed between *N. glauca* and *N. langsdorffii* (Morikawa *et al.*, 1987; Chapel *et al.*, 1986) and between *N. tabacum* and *N. rustica* (Pental *et al.*, 1984).

### 11.4.3 Restriction Pattern of Nuclear DNA Coding for Ribosomal RNA (rDNA)

In several plant species combinations, the restriction enzyme patterns of rDNA differ (restriction site polymorphism), and therefore their nuclear DNA can be distinguished, after digestion with restriction enzymes, by electrophoresis and subsequent hybridization with a labeled heterologous rDNA or ribosomal RNA fragment as the probe. Examples include: EcoRI digests from the nuclear DNA of somatic hybrids between *Citrus sinensis* and *Poncirus trifoliata* made with a rice ribosomal RNA probe (Ohgawara *et al.*, 1985); XbaI and EcoRI digests from the nuclear DNA of somatic hybrids between AEC-resistant *Pennisetum americanum* (pearl millet) and *Panicum maximum* (guinea grass) made with a maize rDNA probe (Ozias-Akins *et al.*, 1986); and EcoRV and BstEII digests of the total DNA of somatic hybrids between cms *B. oleracea* and triazine-resistant *B. campestris* made with a pea rDNA probe (Robertson *et al.*, 1987).

### 11.4.4 Restriction Pattern of Mitochondrial DNA (mtDNA)

Due to restriction site polymorphism in mtDNA and in cpDNA, the electrophoretic patterns of the fragments of DNA after digestion with appropriate restriction enzymes (restriction pattern) are characteristic of the cytoplasm (mitochondrion or chloroplast) type. This can be shown after hybridization of the total cellular DNA with appropriately labeled homo- or heterologous probes or by direct staining of mtDNA (or cpDNA) wiht ethidium bromide. Use of the restriction pattern has proved useful for confirming the cytoplasm hybridity of hybrids/cybrids. Some examples are listed here.

*Nicotiana* cybrids from X-irradiated *N. tabacum* + cms *N. sylvestris*; and from X-irradiated *N. rustica* + *N. sylvestris*; and their sexual progenies (see Aviv and Galun, 1987); and male sterile tobacco plants from X-irradiated *N. debneyi* + *N. tabacum* (Asahi *et al.*, 1988).

*Daucus* hybrids from *D. carota* + *D. capillifolius* (Matthews and Widholm, 1985); and from drug-resistant strains of *D. carota* (Kothari *et al.*, 1986).

Citrus cybrids from Poorman × *Ponicirus trifoliata* + Villafranca lemon; Poorman × *Poncirus trifoliata* + Sour orange; and Sour orange + Villafranca (Vardi *et al.*, 1987; see also Section 11.5.5.4).

*Brassica* cybrids between male fertile and cms *B. napus* lines that carry the original *B. napus* and

Ogura type cytoplasms (Morgan and Maliga, 1987). Mitochondria were identified in calli (0.5–5.0 g) by hybridizing the appropriate DNA probes to the total cellular DNA.

Gramineae hybrids from *Pennisetum americanum* + *Panicum maximum* (Ozias-Akins *et al.*, 1987); and cybrids from cms rice (*Orysa sativa* L.) line 'A-58' + fertile rice cultivar 'Fujiminori' (Yang *et al.*, 1988).

### 11.4.5 Restriction Pattern of Chloroplast DNA (cpDNA)

Thanks to the restriction site polymorphism of cpDNA, the restriction pattern of cpDNA is characteristic of the chloroplast type as described in Section 11.4.4. Either the total cellular DNA is hybridized with labeled homo- or hetero-logous probes or direct observations are made by ethidium bromide staining of cpDNA. The detection limit with the former technique is 0.1–0.5% and with the latter 1–5% (Pelletier *et al.*, 1985; Clark *et al.*, 1986). Some examples are given here.

Cybrids from X-irradiated cms *N. tabacum* + *N. sylvestris*; from X-irradiated *N. tabacum* + cms *N. sylvestris*; from X-irradiated *N. tabacum* cv. Petit Havana SR1 + cms *N. tabacum*; from X-irradiated *N. rustica* + *N. sylvestris*; from alloplasmic ms *N. tabacum* + *N. tabacum*; and from alloplasmic ms *N. sylvestris* + *N. tabacum* (see Fluhr *et al.*, 1984; Galun and Aviv, 1986).

Hybrids from the *N. tabacum* mutant (albino) SR1- A15 + *N. plumbaginifolia* mutant (linocomycin-resistant) LR400, for which the recombinant cpDNA in a somatic hybrid was detected with homologous cpDNA fragments as the probe (Medgyesy *et al.*, 1985); and cybrids from the *N. tabacum* mutant (albino) SR1-A15 + *N. plumbaginifolia* mutant (triazine-resistant) TBR2, for which the presence of *N. plumbaginifolia* chloroplasts in the cybrids was confirmed by hybridization and by the chloroplast fluorescence transients characteristic of the TBR2 mutant (Menczel *et al.*, 1986).

*Brassica* cybrids from male fertile + cms *B. napus* lines. Chloroplasts were identified in calli (0.5–5.0 g) by hybridizing the appropriate DNA probes to the total cellular DNA (Morgan and Maliga, 1987).

### 11.4.6 Karyotype, DNA Content, Morphology and Others

Most of the biochemical analyses described provide only circumstantial evidence hybridity; chimeric cells cannot be distinguished from fused cells by these analyses. Karyotype analysis can be carried out to exclude chimeirsm and to confirm hybridity, provided that the sizes or shapes of the chromosomes of the fusion partners differ distinctly. There are many examples of this type of analysis, *e.g.* hybrids formed between *N. tabacum* + *N. glauca* (*e.g.* Gleba *et al.*, 1984), between *Physalis minima* + *Datura innoxia* (Gupta *et al.*, 1984), and between *S. tuberosum* + *S. phureja* (Puite *et al.*, 1986; Pijnacker *et al.*, 1987). None of those reports, however, provide evidence that the genes are really expressed on the chromosomes of the hybrids. Clarification of this point is necessary for future studies of somatic plant cell genetics.

Because mitochondrial DNA undergoes intra- and inter-generic recombination in various plant species (as described below), translocation or rearrangement of nuclear chromosomes may conceivably lead to exchanges of DNA sequences among heterologous chromosomes. In fact, Imamura *et al.* (1987) have shown that X-irradiation of one fusion partner resulted in the partial transfer of nuclear DNA from the irradiated partner to the nuclear DNA of the other.

Because chromosome aberration occurs during cell culture, the chromosome number alone is not enough to confirm hybridity. However, it is useful for deducing the ploidy of the hybrids (or the multiplicity of fusants; *e.g.* Morikawa *et al.*, 1987; Sundberg *et al.*, 1987; Terada *et al.*, 1987b; Wright *et al.*, 1987) and for confirming the transfer of limited amounts of genomes from one partner to the other by asymmetric fusion (or cybrid nature, *e.g.* Gupta *et al.*, 1984; Dudits *et al.*, 1987).

Analysis of nuclear DNA content by fluorometry or flow cytometry also can be used to confirm the hybridity or ploidy of hybrids (*e.g.* Afonso *et al.*, 1985; Puite *et al.*, 1986; Sundberg and Glimelius, 1986; Endo *et al.*, 1988). Comparisons of the plant morphologies of parents and regenerants from fusion products also has been used to select putative hybrids (*e.g.* Gleba *et al.*, 1984; Austin *et al.*, 1985; Terada *et al.*, 1987a).

Fragments of nuclear DNA from fusion partners can be used as probes to confirm hybridity. Sala *et al.* (1985) identified somatic hybrids produced by the fusion of cell culture protoplasts of carrot and rice by use of dot blot hybridization of the nuclear DNA of the fusion products with the labeled nuclear DNA of one of the partners.

## 11.5 CURRENT STATE OF THE SOMATIC HYBRIDIZATION OF PLANTS

### 11.5.1 Behavior of Organelles After Protoplast Fusion

Because important traits for plant breeding, such as male sterility and herbicide resistance, are encoded in the organellar or are regulated by interaction of the nuclear and organellar genomes (see below), gene engineering of organellar genomes (as well as nuclear genomes) is important for plant biotechnology. Studies on the behavior or fate of organelles, or their genomes, in hybrid/cybrid cells are of fundamental importance for the development of organelle gene engineering in plants. Information gained from such studies should prove useful for estimating the minimum number of foreign organelles (or their genomes) to be introduced in order to replace the organelles (or their genomes) originally present in the target cells.

Mitochondria and chloroplasts are the most widely studied plant organelles. These two organelles and their genomes generally have very different fates in the products produced by fusion. Recombination is thought to be the general rule for plant mitochondria (although this is not always the case, see below), whereas recombination of chloroplast has been reported in only one case of *Nicotiana* cells (Medgyesy *et al.*, 1985) under strong selection pressure. Belliard *et al.* (1979) first reported the appearance of novel mtDNA restriction patterns in cybrids of *Nicotiana*, an indication that recombination of mitochondrial genomes results from the mixing of cytoplasms during protoplast fusion. Similar results have since been reported for hybrids/cybrids of *Nicotiana* (see Aviv and Galun, 1987 and references therein; Asahi *et al.*, 1988), of *Petunia* (Boeshore *et al.*, 1983; Clark *et al.*, 1985, 1986), of *Brassica* (Chetrit *et al.*, 1985, Robertson *et al.*, 1987; Menczel *et al.*, 1987; Morgan and Maliga, 1987), of *Daucus* (Matthews and Widholm, 1985; Ichikawa *et al.*, 1987), of *Solanum* (Kemble *et al.*, 1986) and of the Gramineae, including intergeneric hybrids between *Pennisetum americanum* and *Panicum maximum* (Ozias-Akins *et al.*, 1987), and cybrids (Yang *et al.*, 1988). All the studies cited, except for those on the intraspecific hybrids/cybrids of *Daucus* (Kothari *et al.*, 1986) and *Orysa* (Yang *et al.*, 1988), reported that the mtDNA restriction diagrams of the hybrids/cybrids show novel bands not present in either fusion partner.

Novel restriction fragments are formed by intramolecular rearrangement or intermolecular (*i.e.* intergenomic) recombination, in fusants. Because the novel restriction fragments of mtDNA have been shown to have intergenomic recombination sites (*i.e.* DNA fragments specific to both fusion partners; see below), it is generally considered that the appearance of novel bands in the mtDNA diagram indicates the intergenomic recombination of DNA.

Kothari *et al.* (1986) reported that in one of three intraspecific hybrid combinations of *Daucus carota*, the restriction pattern of the mtDNA differed from the pattern for the parental cells, whereas in the other two combinations the restriction patterns of the mtDNA of the hybrids were parental. Yang *et al.* (1988) also reported that digestion of the mtDNA of rice cybrids gave a set of restriction fragments that differed from those of the parents and that no novel fragments were present, rather some of the unique fragments of the parents were absent from the cybrid restriction pattern. In *Brassica* cybrids (Vedel *et al.*, 1986), the number of novel bands in the restriction diagram of their mtDNA varied among the cybrids (from 0 to 8). As the molecular mechanism by which novel mtDNA fragments are formed during protoplast fusion is not yet understood, it would be premature to conclude that the absence of novel fragments indicates the absence of recombination in the mtDNA of hybrids/cybrids.

Somatic hybrid plants formed between potato (*S. tuberosum*) and tomato (*Lycopersicon esculentum*) had only potato mtDNA (Shepard *et al.*, 1983), somatic hybrids formed between tomato and *S. rickii* (O'Connell and Hanson, 1986) had none, and those between tomato and *L. pennellii* (O'Connell and Hanson, 1987) had few tomato specific restriction fragments of mtDNA. In animal somatic hybrids (*e.g.* mouse + human), only one type of parental mitochondrion has been found (see *e.g.* Kumar and Cocking, 1987). Research on the mechanism of the sorting out of tomato mtDNA in intra- and intergeneric hybrids is important for mitochondrial gene engineering and for our understanding of the regulation mechanism of mtDNA recombination.

Two aspects of the behaviour of chloroplasts and their genomes in both somatic hybrid cells and plants have been studied: (i) whether mixed chloroplast types are stably retained in hybrid/cybrid plants; and (ii) whether recombination of cpDNA takes place during protoplast fusion. Usually somatic hybrids/cybrids have one distinct chloroplast type, but those with mixed chloroplast types also have been reported (described above, and see Pelletier *et al.*, 1985; Kumar and Cocking, 1987).

Sundberg *et al.* (1987) reported that in somatic hybrid plants of *B. oleracea* and *B. campestris* only one or the other of the two parental types of chloroplast was present (see Section 11.3.4). Asahi *et al.* (1988) also reported that male sterile tobacco plants formed between *N. debneyi* and *N.*

*tabacum*, have either one or the other parental chloroplast genome and that neither a mixture of two types of cpDNA nor unique restriction fragments were detected in any of the plants they examined.

Fluhr *et al.* (1984), however, have reported cybrids with mixed chloroplast types. They produced heteroplastidic cybrids between *N. tabacum* and alloplasmic ms *N. tabacum* and between *N. tabacum* and alloplasmic *N. sylvestris* by the 'donor–recipient' method. Both types of cybrid regenerated variegated plantlets which contained mixed chloroplasts. No recombinant cpDNA types were detected. Gleba *et al.* (1984, 1985), using manual selection, obtained cybrids of *Nicotiana* that had mixed chloroplasts and showed that this heterozygocity of cpDNA was stably inherited by the progenies (see Section 11.3.3). The presence or absence of recombinant cpDNA in these cybrids was not investigated. Somatic hybrid plants formed between tomato and *L. pennellii* have been shown to inherit either or both parental chloroplast genomes, but predominantly the *L. pennellii* mitochondrial genome (O'Connell and Hanson, 1987; see also above). The presence or absence of recombinant cpDNA in this hybrid also was not addressed.

Sexual hybrid plants of *Oenothera* which are exceptional for an angiosperm plant species have biparental transmission of plastids, and there is reported to be no rearrangement of their cpDNA (Pelletier *et al.*, 1985). This is an additional indication that the presence of mixed chloroplasts in a cybrid/hybrid does not necessarily lead to the recombination of cpDNA.

Recombination of cpDNA does occur in a limited case during protoplast fusion. Medgyesy *et al.* (1985) fused protoplasts of the *N. tabacum* mutant SR1-A15 (streptomycin resistant, defective in chloroplast greening and lincomycin resistant) with those of *N. plumbaginifolia* mutant LR400 (streptomycin sensitive, normal green and lincomycin resistant). Green colonies that proliferated on a streptomycin-containing selective medium were selected, and the plants regenerated from them were identified as hybrids by their morphology. One plant expressed three *N. tabacum* and four *N. plumbaginifolia* parent-specific restriction sites for cpDNA. These chloroplasts were concluded to be neither revertants of the parents nor mixtures of both parental types but to have interspecific recombinant cpDNA.

The difference in the recombination behavior of mtDNA and cpDNA during protoplast fusion is thought to be the result of structural differences in mitochondria and chloroplasts and their genomes (Pelletier *et al.*, 1985). Microscopic observations that suggest mitochondrial fusion in plant cells have been reported (Wildman *et al.*, 1962), but fusion of chloroplasts has not been observed in higher plants. Plant mtDNA is thought to undergo recombination easily because it contains at least two direct repeats, whereas the presence of inverted repeats in cpDNA is thought to be inhibitory to its recombination.

### 11.5.2 Studies of Male Sterility Conducted with Protoplast Fusion

Cytoplasmic male sterility (cms), an important agricultural characteristic, is defined as the inhibition of male gametophyte development specified by cytoplasmically inherited factors. The finding of alloplasmic cms (produced by interspecific crosses) has led to the supposition that cms is the result of incompatibility between the nuclear genome of one species and the cytoplasmic (most plausibly the mitochondrial) genomes of another. In *Nicotiana*, for example, many cms plants have been synthesized in various interspecific nuclear–cytoplasmic combinations. The nonalloplasmic cms (produced by intraspecific crosses) that has been reported is thought to arise because of changes in the nuclear or cytoplasmic genomes within a cultivar or species (Hanson and Conde, 1985).

It has been postulated that mitochondria have a correlation with the genetic regulation of male sterility in various plant species (described above). However, because mitochondria and chloroplasts are transmitted together in sexual crossing, the role of the chloroplast in the control of male sterility cannot easily be assessed. Protoplast fusion studies have revealed that the cms trait segregates independently of the parental chloroplast type and cosegregates with the parental mitochondrial types as shown for somatic hybrid/cybrid plants of *Nicotiana* (Galun *et al.*, 1982), *Petunia* (Clark *et al.*, 1985, 1986) and *Brassica* (Chetrit *et al.*, 1985). This is unambiguous evidence that interaction between the nuclear and mitochondrial genomes is responsible for the cms trait, and has led to many studies of the cell biology of cms being conducted with cybrid/hybrid plants.

Cytoplasmic male sterility traits can be successfully transferred by protoplast fusion (*e.g.* Hanson and Conde, 1985; Pelletier *et al.*, 1985; Kumar and Cocking, 1987). Somatic fusion in which one partner is cms and the other fertile does not, however, necessarily yield only male sterile hybrids; both male fertile and male sterile hybrid plants have been reported for *Nicotiana* (Belliard *et al.*, 1979; Galun *et al.*, 1982), *Petunia* (Izhar *et al.*, 1983) and *Brassica* (Pelletier *et al.*, 1983). Stable cms progenies have been obtained from sterile plants, evidence that the cms trait can be transferred by somatic hybridization. Neither the appearance of fertile hybrids, nor the restoration of the transferred cms trait, is predicted by the fertility restorer theory (Hanson and Conde, 1985). In particular, for *Petunia* the cms and fertile fusion partners have no fertility-restorer genes (Izhar *et al.*, 1983).

These somatic hybrid plants have been used to study the molecular biology of the expression and restoration of cms traits. Boeshore *et al.* (1985) reported a 'cms-specific' mtDNA arrangement of 1.8 kb in somatic *Petunia* hybrids that was present in all 17 stable sterile somatic hybrids, but absent in the 24 stable fertile somatic hybrids tested. The intergenomic recombination of mtDNA also has been demonstrated in interspecific somatic hybrids of *Petunia* (Rothenberg *et al.*, 1985) and in intergeneric cybrids of *Brassica* (Vedel *et al.*, 1986). In those studies, novel mtDNA restriction fragments of the hybrid/cybrid plants were cloned and restriction fragments characteristic to the mtDNA of each fusion partner were identified on the same cloned novel fragments.

Restoration of transferred cms traits has been shown in three cases apart from the 'spontaneous' restoration described above. The cms of *N. sylvestris* (produced by protoplast fusion of normal *N. sylvestris* and cms *N. tabacum*) was restored by fusion with X-irradiated normal *N. tabacum* (see Galun and Aviv, 1986). This suggests that the cms trait can be restored by the transfer of cytoplasm from a fertile plant. Fusion of two alloplasmic-like cms *Nicotiana* cybrids obtained by the 'donor–recipient' method produced a fertile plant (Aviv and Galun, 1986). MtDNA analysis of its progenies revealed a novel pattern that differed from the patterns of the parental cms cybrids and from the mtDNA of the normal, male fertile *Nicotiana* species, an indication that the recombination of mtDNA from two different cms *Nicotiana* cybrid lines may lead to the restoration of male sterility. Furthermore, the cms trait of *Brassica* cybrids (Pelletier *et al.*, 1983) was restored by fusion with radish (*Raphanus sativus*), the latter being known to have the restorer gene (Chetrit *et al.*, 1985). This confirms that cybrids retain the same genetic cms trait present in the cms parent of fusion and rules out other possibilities, including that of *in vitro* induced cms.

Li *et al.* (1988) have reported the regeneration of one fertile and two cms diploid protoclones of *N. sylvestris*. Each progeny regenerated plants that possessed different mtDNA restriction patterns. Interestingly, a 40 kD mitochondrially encoded polypeptide was lacking in the cms plants.

Merely placing a nucleus in alien cytoplasm is not enough to induce cms because intergeneric cybrids in which the tobacco nucleus is placed in alien cytoplasm such as *Petunia* (Glimelius and Bonnett, 1986) or *Atropa* (Kushnir *et al.*, 1987) regenerate fully fertile plants. (The possibility that some *Petunia* or *Atropa* nuclear DNA also is transferred to the cybrid nucleus and that it affects fertility cannot be excluded.) These materials should prove useful for the study of nuclear–cytoplasm interaction in relation to the expression and restoration of cms traits.

As to details of the cms studies in *Brassica* hybrids/cybrids (including those already mentioned), Pelletier *et al.* (1983) produced intergeneric cybrids of rapeseed (*B. napus*) in two combinations: between a cms cultivar ('C') that has the Ogura-type cms cytoplasm derived from radish (*Raphanus sativus*) and a pure fertile cultivar ('Brutor'); and between the same cms cultivar and a triazine-resistant cultivar ('Tower') with triazine-resistant chloroplasts from *B. campestris*. They succeeded in regenerating cybrid plants from both combinations, some of which were sterile and some fertile. Using the second combination, Chetrit *et al.* (1985, 1986) obtained triazine-resistant cybrid plants and analyzed the genomic and plasmid-like DNA of their mitochondria. Restriction patterns of the mtDNA of the cybrid plants differed from each other and from both parents as described above. In both the parental and cybrid plants there were large quantitative variations in mitochondrial plasmid-like DNA and the conclusion was that cytoplasmic support for male sterility is located in the genomic mtDNA and not in the plasmid-like DNA.

Menczel *et al.* (1987) and Morgan and Maliga (1987) have reported similar results for cms (Ogura-type) cybrids of *B. napus* that had novel mtDNA restriction fragments and chloroplasts of fertile parent type.

Yarrow *et al.* (1986) fused protoplasts from a cytoplasmic triazine-resistant variety of *B. napus* (ctt-'Regent') and a cms variety of the same (Nap-'Regent') that has the Nap-type cms cytoplasm. They obtained cybrid regenerants which had normal morphology and which produced seeds on pollination.

Barsby *et al.* (1987) fused protoplasts from a cytoplasmic triazine-resistant variety of *Brassica napus* (ctr-'Regent' or 'Triton') and a cms variety of the same (Polima-'Regent') that has the Polima- type cms cytoplasm. Some of the protoplasts were treated with IOA or $\gamma$-rays prior to fusion. Atrazine-resistant and male sterile plants were regenerated from the fused products, the cybrid nature of these plants being confirmed by cpDNA and mtDNA analyses. Taken together, these results show that all three (Ogura, Nap and Polima) cms cytoplasms in the Cruciferae are associated with mitochondrial genomes.

A triazine-resistant, male fertile *B. napus* plant was synthesized by the fusion of protoplasts from cms *B. oleracea* ($2n = 18$, carrying Ogura-type cytoplasm) and triazine-resistant *B. campestris* ($2n = 20$) by Robertson *et al.* (1987). The hybridity of this plant was confirmed by its chromosome number ($2n = 36$–$38$) and by rDNA analyses. Its chloroplast DNA was identified as the *B. campestris* type on the basis of Southern blot analysis and the tetrazolium blue assay. The mtDNA diagram of the hybrid showed novel restriction fragments as well as fragments unique to each parent.

### 11.5.3 Successful Transfer of Disease Resistance Genes by Protoplast Fusion

Transfer of disease resistance by protoplast fusion between sexually incompatible plant species has been a dream of plant biotechnologists since this technology became feasible. Recently, the successful transfer of disease resistance has been reported in potato and tomato species. The results are very encouraging and should stimulate further development of protoplast fusion technology. It is too early to discuss the direct utilization of these fusion products in plant breeding programs but these studies have paved the way for a new aspect of plant pathology.

Austin *et al.* (1985, 1988), Helgeson *et al.* (1986) and Gibson *et al.* (1988) have successfully transferred disease resistance from wild to cultivated species of potato by protoplast fusion and Gleddie *et al.* (1985, 1986) also used protoplast fusion in their successful transfer of the pest resistance gene in eggplant.

*S. brevidens*, a non tuber-bearing wild species, has a resistance to potato leaf roll virus (PLRV) and to frost, but it is difficult to sexually cross this species with the cultivated potato (*S. tuberosum*). Austin *et al.* (1985) fused protoplasts from mesophyll protoplasts of diploid *S. brevidens* ($2n = 2x = 24$) and those from diploid *S. tuberosum* ($2n = 2x = 24$) and obtained tetraploid ($2n = 48$) somatic hybrids from which plants were regenerated. Hybridity was verified by the plant morphology and by cytological observations. Interestingly, somatic hybrid plants showed resistance to PLRV. Helgeson *et al.* (1986) produced hexaploid ($2n = 72$) somatic hybrids by protoplast fusion of a diploid *S. brevidens* carrying resistance to PLRV and a tetraploid *S. tuberosum* carrying late blight resistance. All the hybrid plants showed late blight resistance and most also were resistant to PLRV. These tetraploid and hexaploid hybrids of potato have been shown to have reasonable levels of female fertility (Ehlenfeldt and Helgeson, 1987).

From their observation that tubers of hexaploid hybrid potatoes did not decay in storage, Austin *et al.* (1988) have found that these hybrids are resistant to the soft rot (a tuber disease) caused by *Erwinia* spp. Sexual progenies of the hybrids retained their resistance to soft rot, evidence that this trait is genetically stable. Since *S. brevidens* does not form tubers, this wild species had not been known to carry the gene(s) for resistance to this bacterial soft rot. Although the genetic basis of the resistance to *Erwinia* soft rot has yet to be determined, these somatic hybrid plants should prove useful for developing molecular studies on the plant pathology of *Erwinia*. This finding indicates that somatic hybrids may serve as powerful tools to 'dig out' useful genes.

Fish *et al.* (1987) reported the chemical fusion (with PEG), and Fish *et al.* (1988a) the electrical fusion, of mesophyll protoplasts from dihaploid ($2n = 2x = 24$) *S. tuberosum* and the dihaploid wild species *S. brevidens* ($2n = 2x = 24$). Somatic hybrids were selected on the basis of phenotype from the regenerated shoots and their hybridity confirmed by isozyme pattern analysis. The chromosome number of the hybrid plants ranged from 48 (tetraploid) to 94. With electrofusion more of the hybrid plants were hexaploid (possibly products of triple fusion) than with chemical fusion. Field assessments were made of the somatic hybrid plants obtained (Fish *et al.*, 1988b). Results indicated that there was extensive variability and that the degree of somaclonal variation in somatic hybrids was greater than that found in protoclones of potato. Resistance to PLRV and potato virus Y were also shown (Gibson *et al.*, 1988).

Monohaploid ($2n = 2x = 12$) potatoes also have been used for protoplast culture and somatic hybridization (see *e.g.* Uijtewaal *et al.*, 1987, and references therein). This should provide more accurate genetic information (since they are genetically homogeneous) on the disease resistance or sensitivity of a hybrid or its parent plants.

Barsby *et al.* (1984) reported regeneration of somatic hybrid plants from fusion products between mesophyll protoplasts from a tetraploid albino potato (*Solanum tuberosum* spp. tuberosum) variant ($2n = 2x = 48$) and diploid *S. brevidens* ($2n = 2x = 24$), a non tuber-bearing species that is sexually incompatible with *S. tuberosum*. The hybrid plants were octoploid ($2n = 92$–$96$) instead of hexaploid ($2n = 72$) as had been expected and were male sterile.

Sidorov *et al.* (1987) fused $\gamma$-irradiated mesophyll protoplasts from *S. pinnatisectum* ($2n = 24$) with untreated mesophyll protoplasts of genomic chlorophyll-deficient IvP841-1 ($2n = 24$) that contained the germplasms of *S. tuberosum* and *S. phureja*. Somatic hybrid plants were selected on the basis of their morphology, and hybridity was confirmed by isozyme analysis. *S. pinnatisectum* is a wild tuber-forming species known for its high resistance to *Phytophtora* and potato viruses A and Y but the resistance of the somatic hybrids to these diseases was not addressed.

Gleddie *et al.* (1985, 1986) fused protoplasts from 6-azauracil AU-resistant cell lines of eggplant (*Solanum melongena*) with those from *S. sisymbriifoliium. S. sisymbriifoliium* is a wild species that is sexually incompatible with eggplant and is resistant to several pests that infect eggplant including root-knot nematodes and carmine spider mites. Hybrid plants were selected for their AU resistance and anthocyanin formation (*S. sisymbriifoliium* produces anthocyanin during shoot organogenesis in the presence of zeatin). Hybridity was confirmed by the isozyme patterns and by chromosome number analysis. Several of the 26 hybrid plants were resistant to nematodes and showed potential resistance to mites (Gleddie *et al.*, 1985). Although these hybrid plants are sterile, the results indicate that they may prove useful in eggplant breeding programs.

### 11.5.4 Asymmetric Fusion

When the partners of protoplast fusion are distantly related or when one of them has been irradiated with X- or $\gamma$-rays, chromosome elimination has been shown to occur in the hybrids (asymmetric fusion). The regenerated hybrid plants retain the morphological characteristics or chromosome number of one partner, and some of the biochemical traits of the other partner, as described below. In other words, asymmetric fusion makes possible the transfer of limited amounts of nuclear genes from one partner to the other. This is important because by using asymmetric fusion combined with suitable selection one can, in principle, perform asymmetric transfer of desirable genes without the transfer of nondesirable genes such as those that participate incompatibility, which is usually unavoidable in symmetric fusion. The exchange of DNA sequences between heterologous chromosome or genetic transformation of nuclear genomes by protoplast fusion is conceivable in the light of the intergenomic recombination of organellar DNA described above.

Intergeneric transfer of genes by asymmetric fusion has been reported, *e.g.* between X-irradiated wild-type *Physalis minima* and a *Datura innoxia* nuclear albino mutant (Gupta *et al.*, 1984) and between an NR-deficient *N. tabacum* mutant and $\gamma$-irradiated barley *Hordeum vulgare* (Somers *et al.*, 1986). In the first fusion, green hybrid plants that had *Datura* morphology, one or two chromosomes from *Physalis* and a unique isozyme pattern were obtained. However, no expression of transferred *Physalis* chromosomes was found in the hybrids. In the latter fusion, transfer of NR from irradiated barley to tobacco was shown by an immunological assay of the enzyme. Because tobacco plants that had barley-like NR were recovered from the culture of a mixture of barley and tobacco protoplasts without fusion treatment as well as from fusion products, the authors could not conclude that this gene transfer was mediated by protoplast fusion. Interestingly, the majority of these tobacco plants that expressed the barley-like NR were reportedly normal-appearing fertile. If these plants are somatic hybrids, this would be the first report of fertile intergeneric somatic hybrids.

Intergeneric transfer of methotrexate (a specific inhibitor of dihydrofolate reductase) resistance from carrot (*D. carota* H47) to tobacco (*N. tabacum* cv. Petit Havana SR1) has been carried out by asymmetric fusion between mesophyll protoplasts of tobacco and $\gamma$-irradiated cell culture protoplasts of carrot (Dudits *et al.*, 1987). Somatic hybrids were selected on a methotrexate medium and plants regenerated from the putative hybrid calli. Some of the regenerated somatic hybrids showed normal tobacco morphology and had chromosome numbers very close to the chromosome number of tobacco ($2n = 48$). Interestingly, as determined by the isozyme pattern and immunological characteristics of this enzyme they expressed carrot-type dihydrofolate reductase. Sexual transmission of the transferred genes also was demonstrated by pollination with tobacco. This is the first instance of the successful intergeneric transfer of functional genes by asymmetric fusion. The presence or absence of the recombination of DNA between the heterologous chromosomes that accounts for the expression of carrot enzymes in these hybrids is an intriguing and important topic for the future study. As in other cases of intergeneric hybrid/cybrid plants, these hybrid plants were male sterile and their chromosome numbers were unstable during maintenance *in vitro*, the reasons for this are not yet known.

Possible involvement of cytoplasmic genomes in the sterility and chromosome number instability of the somatic hybrids are important areas of future study. In this context, 'true' asymmetric fusion, in which both the nuclear and cytoplasmic genomes are asymmetrically delivered by protoplast fusion from one fusion parent to the other, is worth studying. This can be done, for example, by fusion between irradiated karyoplasts and unirradiated normal protoplasts. Spangenberg and Schweiger (1986) reported successful electrofusion between karyoplast and protoplasts (see Section 11.6 and Table 1).

Successful transfer of functional genes by interspecific asymmetric fusion has been reported in *Nicotiana* (Bates *et al.*, 1987). $\gamma$-Irradiated mesophyll protoplasts from kanamycin-resistant, nopaline-positive *N. plumbaginifolia* (obtained by transformation) were electrofused with unirradiated mesophyll protoplasts of wild-type *N. tabacum.* Hybrids were selected on kanamycin and regenerated. The genetic material from *N. plumbaginifolia* in the hybrid plants was characterized based on plant morphology, the esterase isozyme pattern and the presence of nopaline in leaf extracts. Irradiation reduced the amount of *N. plumbaginifolia* genetic materials in hybrids. 24 hybrid plants were obtained that had only *N. tabacum* esterases, and that produced nopaline and were kanamycin resistant. Southern blot analysis of the genomic DNA of these plants showed the presence of the neomycin phosphotransferase gene. All the regenerated plants were male sterile but female fertile. Most contained one more chromosome than normal for *N. tabacum* ($2n = 48$).

Asymmetric fusion treatment is usually performed by inactivating one fusion partner with $\gamma$- or X-rays in combination with, or without, inactivation of the other partner with IOA. A single irradiation treatment has been used, for example, by Gupta *et al.* (1984), Somers *et al.* (1986), Bates *et al.* (1987), Dudits *et al.* (1987) and Menczel *et al.* (1986, 1987). Single IOA treatment also has been used by Terada *et al.* (1987b) and Wright *et al.* (1987). The combination of irradiation and IOA treatment has been called double inactivation treatment as by Barsby *et al.* (1987), Ichikawa *et al.* (1987), Morgan and Maliga (1987) and Yang *et al.* (1988).

Recently, Imamura *et al.* (1987) produced cybrids between two auxotrophic lines, *N. tabacum* nia 115 (NR deficient) and *Hyoscyamus muticus* VIII B9 (tryptophan-requiring). One of the partners had been X-irradiated prior to fusion treatment. Based on their analysis of nuclear DNA of the hybrids with a species-specific, radioactive-labeled DNA probe, they showed that: (i) irradiation significantly reduced the amount of chromosomal DNA of the irradiated fusion partner in the somatic cybrid; (ii) irradiation with doses that completely inhibit protoplast division did not prevent the transfer of substantial amounts of chromosomal DNA to the cybrid (their data show that a high dose increased the amounts of DNA transferred); and (iii) that this method transfers functional nuclear genes from the irradiated partner to the nucleus of the nonirradiated partner. Consequently, the so-called 'cybrids' produced from protoplast fusion between irradiated and nonirradiated protoplasts must contain definite amounts of nuclear DNA.

## 11.5.5 Somatic Hybrids in Selected Genera or Families

In this section, the hybrids/cybrids in selected genera or families are summarized and details of new examples of hybrids/cybrids of these plant species are given.

### *11.5.5.1 Gramineae*

Intergeneric somatic hybrids between AEC-resistant pearl millet (*Pennisetum americanum*) and an embryogenic suspension culture of guinea grass (*Panicum maximum*) have already been described. The heterokaryons were selected on AEC-containing medium, and three somatic hybrid cell lines were identified. Hybridity was confirmed by isozyme pattern and rDNA analyses (Ozias-Akins *et al.*, 1986).

Terada *et al.* (1987a) electrofused protoplasts from suspension cultures of rice and barnyard grass (*Echinochloa oryzacola*) and obtained plantlets (see Section 3.6).

Yang *et al.* (1988) produced cybrid calli by the electrofusion of suspension-cultured cells of cms rice line 'A-58' (*Oryza sativa* L.) and the fertile rice cultivar 'Fujiminori'. The A-58 cms protoplasts were $\gamma$-irradiated and the Fujiminori protoplasts were treated with IOA prior to electrofusion. Based on peroxidase isozyme pattern analysis and the plasmid-like DNAs specific to sterile A-58 mt, seven calli were identified as cybrids with a Fujiminori nucleus and an A-58 cytoplasm (mitochondria). Interestingly, digestion of cybrid mtDNA gave a set of restriction fragments that differed from those of the parents. The same group have succeeded in regenerating plants from the cybrid calli (Yang and Yamada, unpublished). The former cultivar has high regeneration ability (Yamada *et al.*, 1986).

### *11.5.5.2 Cruciferae*

Apart from the preceding somatic hybrids described, the following three Cruciferae hybrids should be mentioned. Somatic hybrid plants were produced between cabbage (*B. oleracea*) and chinese cabbage (*B. campestris*) by the use of dextran (Taguchi and Kameya, 1986). Hybridity was confirmed by morphology, chromosome number and isozyme pattern analyses. This is the first example of production of somatic hybrids in heading-type vegetables. Also, somatic hybrid plants have been obtained between cabbage (*B. oleracea*) and komatsuna (*B. campestris*) or turnip (*B. campestris*) by electrofusion. Hybridity was confirmed by isozyme pattern analysis (Nishio *et al.*, 1988; see Table 1).

Intergeneric somatic hybrid plants, 'Brassicamoricandia' were produced between *B. oleracea* (either red cabbage or cauliflower) and *Moricandia arvensis* (Brassicaceae) by the use of dextran (Toriyama *et al.*, 1987a). Interestingly, isozyme pattern analysis showed that all the regenerants from the eight calli formed from the fusion products turned out to be somatic hybrids. The somatic hybrids showed hybrid vigor and readily produced shoots in this case.

#### *11.5.5.3 Tomato and related species*

Somatic hybrids of potato and eggplant have been described in Section 11.5.3.

Since the successful production of somatic hybrid plants of tomato (*L. esculentum*) and potato (*S. tuberosum*; Melchers *et al.*, 1978; Shepard *et al.*, 1983), various somatic hybrid plants have been reported; between tomato and *S. lycopersicoides*, between tomato and *S. rickii*, between two wild species of *Lycopersicon*, between tomato and *L. peruvianum* and between tomato and *L. pennellii* (see O'Connell and Hanson, 1986, 1987; Handley *et al.*, 1986, and references therein).

#### *11.5.5.4 Citrus*

Ohgawara *et al.* (1985) produced somatic hybrid plants between the 'Trovita' orange (*Citrus sinensis*) and the *Citrus* root-stock *Poncirus trifoliata.* Vardi *et al.* (1987) produced citrus cybrids by the 'donor–recipient' method between: (i) Poorman × *Poncirus trifoliata* with Villafranca lemon; (ii) Poorman × *Poncirus trifoliata* with Sour orange; and (iii) Sour orange with Villafranca (see Section 11.4.4). The fusion-derived plants from all three combinations had recipient morphological (nuclear-coded) features.

#### *11.5.5.5 Forage legumes*

Wright *et al.* (1987) fused protoplasts from etiolated hypocotyl of birdsfoot trefoil (*Lotus corniculatus* cv. 'Leo') that had been treated with IOA with protoplasts from suspension-cultured cells of *Lotus conimbricensis* Willd. The latter protoplasts lack plant regeneration ability. Fifteen somatic hybrid plants were obtained and their hybridity confirmed by isozyme pattern analysis of phosphoglucomutase and by the chromosome number. This is the first example of somatic hybridization between sexually incompatible *Lotus* species.

## 11.6 CURRENT STATE OF ELECTROFUSION RESEARCH

Table 1 shows recent studies done on the production of somatic hybrids of higher plants by electrofusion. Most of the hybrids belong to the Solaneceae, and include four *Nicotiana*, two *Solanum* and one *Hyocyamus* species. Terada *et al.* (1987a) have succeeded in producing hybrid Gramineae and Yang *et al.* (1988) cybrid Gramineae. The number of somatic hybrid plants produced by electrofusion continues to increase. Studies shown in Table 1 that were discussed in the preceding sections of this chapter are not repeated here (see also Morikawa *et al.*, 1988e).

Complete fertile somatic hybrid plants were first produced electrically between *N. glauca* and *N. langsdorffii* (Morikawa *et al.*, 1986, 1987). The hybrid callus isolation yield was about 0.2%. After more than eight months of culture some regenerants had developed to mature plants, flowered and set seeds. More than 10 such plants were obtained, and their hybridity was confirmed by an electrofocusing analysis of the RuBisCo polypeptides isolated from their leaves. Interestingly, the large subunits of RuBisCo polypeptides encoded on the chloroplast genomes were of the *N. langsdorffii* type in all the hybrid plants examined.

The chromosome numbers ($2n$) of the plants ranged from 60 to 66. The parental diploid chromosome numbers were 24 for *N. glauca* (GG) and 18 for *N. langsdorffii* (LL). Simple addition would give a chromosome number of 42, whereas products of triple fusion would have a chromosome number of 60 or 66: $LL + GG = 42$ chromosomes; $LL + LL + GG = 60$ chromosomes; and

LL + GG + GG = 66 chromosomes. These results offer strong evidence that these hybrid plants are products of triple fusion and that the loss or increase of chromosomes during culture gives rise to aneuploid types. Chen *et al.* (1977) have reported chromosome numbers of 54–64 for hybrid plants of the same parental combination obtained by PEG-induced fusion. They concluded that the hybrids are the products of triple fusion (see Morikawa *et al.*, 1987 for the other examples of multiple fusion with this parental combination).

Preliminary experiments on the nuclear DNA analysis of our hybrid plants indicated that some had an rDNA composition of approximately 2:1 (*N. glauca*:*N. langsdorffii* type), strongly indicating that the triple fusion involves two *N. glauca* protoplasts and one *N. langsdorffii* protoplast. Further study, including cpDNA analysis, is currently in progress.

Multiple fusion has also been observed in other plant species combinations. Puite *et al.* (1986) reported that in somatic hybrids obtained by electrofusion between diploid ($2n = 24$) *S. tuberosum* and diploid ($2n = 24$) *S. phureja*, most had chromosome numbers of about 72 (hexaploid) or 96 (octoploid) and some of about 48 (tetraploid). Fish *et al.* (1987, 1988a) reported that the chromosome numbers of chemically or electrically induced somatic hybrids between dihaploid ($2n = 2x = 24$) *S. tuberosum* and the dihaploid *S. brevidens* ($2n = 2x = 24$) were 48 (tetraploid) to 94 (octoploid). They reported that electrofusion produced more hexaploid (presumably triple fusion products) somatic hybrids than did chemical fusion. Terada *et al.* (1987b) reported that the chromosome numbers of somatic hybrids between *B. oleracea* ($2n = 18$) and *B. campestris* ($2n = 20$) varied from 38 (amphidiploid) to 56 (presumably triple fusion products). Sundberg *et al.* (1987) also reported that the chromosome numbers of somatic hybrids of the same parental combination were 38 in 30% of cases, $<38$ in 9% and $>30$ in 61%. Handley *et al.* (1986) reported that the chromosome numbers of somatic hybrids between tomato ($2n=24$) and *S. lycopersicoides* ($2n=24$) were $2n=48$ (amphidiploid) to 68 (triple fusion products?). All of these results indicate that products of multiple fusion can regenerate plants. It has yet to be determined whether multiple fusion products have a higher totipotency than binary fusion products.

Puite *et al.* (1986) were the first to produce somatic hybrid *Solanum* plants between *S. tuberosum* and *S. phureja*. They obtained 18 putative hybrid plants that were grown to maturity. The hybridity of these plants was established by determining the nuclear DNA content and by a karyotype analysis. Pijnacker *et al.* (1987) reported the elimination of specific chromosomes in these *Solanum* hybrids. Recently, Puite *et al.* (1988) obtained hybrid plants between the same parental combination after selecting heterokaryons by flow sorting as described above.

Spangenberg and Schweiger (1986) isolated protoplasts from hypocotyl of *B. napus* cv. 'Tower ATR' and fragmented it into karyoplasts and cytoplasts. These protoplasts and subprotoplasts with various combinations (karyoplast–cytoplast, karyoplast–protoplast and cytoplast–protoplast) were electrofused and the resulting homofusion products obtained were selected manually then transferred to microdroplet culture. Calli were obtained from all the combinations. This technique can be useful in studies of 'true' asymmetric fusion in which both nuclear and cytoplasm genomes are asymmetric (see Section 11.5.4).

Kishinami and Widholm (1987) electrofused protoplasts from two auxotroph mutants, NX1 (*N. plumbaginifolia* Viviani, lacking NR) and 8B9 (*Hyocyamus muticus*, tryptophan-requiring) and cultured the fusion products in a minimal medium lacking tryptophan and with nitrate as the nitrogen source. The intergeneric hybrid calli were indentified by means of the isozyme and nitrogen reductase activity analyses.

Terada *et al.* (1987a) fused protoplasts from suspension cultures of rice and barnyard grass (*Echinochloa oryzacola*). The selection of somatic hybrids was based on the inactivation of rice protoplasts by IOA and the inability of barnyard grass protoplasts to divide. A total of 166 calli were identified as hybrids by isozyme and chromosome analyses, and 44 shoots were obtained. Most shoots were abnormal, and nine grew into plantlets whose morphologies differed from that of either parent.

Pair formation prior to pulsation, a prerequisite for successful fusion, is usually induced by an AC field (dielectrophoresis) but can also be effected simply by placing a pellet of the protoplast mixture in the chamber (multiprotoplast layer), by the addition of chemical reagents such as PEG, and by manual micromanipulation with microelectrodes (Table 1). No method is known for specific heteroplasmic protoplast pair formation, except that of manual pair formation using microelectrodes connected to micromanipulators or a more sophisticated hydraulic device (see Morikawa *et al.*, 1988d,e, and references therein).

Both exponentially decaying and square pulse methods have effectively induced fusion. A rather narrow range of electrical conditions has been used for somatic hybridization (see Table 1). The field strength varies from 0.6–4kV cm$^{-1}$, the duration or time constant is 10–1000$\mu$s and a single or multiple pulses is applied. Nishio *et al.* (1988) applied two to three sequences of AC (10s) and DC (15–30 $\mu$s) pulses (Table 1).

In contrast, the chamber media used vary greatly, from a medium with no added salts (in which electric resistance would be several tens of kiloohms) to a medium containing more than 50 mM of electrolytes (the electric resistance of this medium would be less than $10^{-2}$ that of the salt-free medium). This means that the effective value of the field strength would differ for each set of conditions listed in Table 1. A systematic optimization of the DC fusion pulse in terms of field strength, width, number, interval and shape (decaying or square) of the pulse, concentration and composition (particularly of divalent cations) of the chamber medium, temperature, *etc.* has yet to be done.

Technical improvements for electrofusion, that include prior treatment of protoplasts with proteolytic enzymes, and the addition of a divalent cation ($Ca^{2+}$ or $Mg^{2+}$) or spermine, have been reported by several authors (see Zachrisson and Bornman, 1986; Morikawa *et al.*, 1988e,f). Recently, Nea and Bates (1987) and Nea *et al.* (1987) showed that lysolecithin, DMSO and $Ca^{2+}$ increased electrofusion efficiency and that brief treatment of protoplasts with Cellulysin, pronase or proteinase K markedly increased fusion efficiency.

We studied the effects of physical parameters, including the osmotic potential and density of the chamber medium and the hydrophilic/hydrophobic coating of the bottom surface of the chamber, on protoplast electrofusion. There is an optimum osmotic potential and density (or buoyant force) of the medium for electrofusion. Putting a hydrophilic coating of Gellan gum or polyacrylamide gels on the bottom surface stimulated electrofusion, but hydrophobic siliconization was entirely inhibitory (Morikawa *et al.*, 1988a).

Cell membranes are 'activated' by electric field pulses so that they easily fuse with one another or they are made permeable so as to allow entry of macromolecules which normally do not penetrate them. It is not clear, however, whether this 'activated state' of membranes is equivalent to the so-called 'electroporated' state. For example, Mehrle *et al.* (1985) and Sowers *et al.* (1986) reported that the electropermeabilization of cell membranes has a polarity which suggests that electropore formation is unlikely to be directly involved in membrane electrofusion. Also, virus particles as large as 180 × 3000 Å have been introduced into protoplasts by electroporation (Okada *et al.*, 1986; Nishiguchi *et al.*, 1987). This is evidence that the electric field activates the endocytotic activity of the plasma membrane of protoplasts rather than simple uptake through electropores. Abe and Takeda (1986) have shown that calmodulin antagonists and cytoskeleton inhibitors affected protoplast electrofusion. This indicates a possible function for calmodulin and the cytoskeleton in the electrofusion process. The molecular mechanism of membrane fusion or membrane permeabilization is also a very intriguing topic.

## 11.7 SUBJECTS FOR FUTURE STUDY IN PROTOPLAST FUSION

Possible subjects for future studies that have not been included in the preceding sections of this chapter are discussed here.

As has already been shown in this review, most of the plant materials used in protoplast fusion study are dicots; in particular, species of Solanaceae and Cruciferae. Recent successes in protoplast culture of Gramineae that include rice (Fujimura *et al.*, 1985; Yamada *et al.*, 1986; Abdullah *et al.*, 1986; Kyozuka *et al.*, 1987) have paved the way for the development of protoplast fusion technology for use with this important crop.

Hybridization of cultured plant cells rich in secondary metabolites is also an area to be more fully explored in the future (see Yamada and Morikawa, 1985; Yamada *et al.*, 1987; Endo *et al.*, 1988; Morikawa *et al.*, 1988c).

No plant hybridoma cells have been produced by protoplast fusion, which means that none of the somatic hybrid plants (or cells) obtained so far have a usefulness that is, or applications that are, comparable to those of hybridoma cells. The reasons for this have yet to be established but one may be that the number of available plant cell mutants is very limited. As shown previously, use of relevant double mutants as universal hybridizers is very recent. Another reason is that our fundamental understanding of biochemistry and the biology of plant cells at the molecular level is not yet sufficient. For example, we need to accumulate knowledge on the exact mode(s) of action of the plant hormones that regulate dedifferentiation and redifferentiation and the growth of plant cells. The possibility of somaclonal variations or genetic changes in nuclear and cytoplasmic genomes during protoplast culture and the regeneration of plants also needs to be studied in detail because these phenomena may limit the eventual usefulness of products obtained by protoplast fusion or by protoplast transformation (*e.g.* by electroporation) in breeding programs.

The outlook is that electrofusion will continue to develop as an increasingly useful fusion technique. Selective heterofusion looks to be one of the most intriguing and challenging areas in the future technological development of fusion, both by chemical and electrical methods. In addition, the mechanism off electropore formation in relation to electric field induced membrane fusion and membrane permeabilization is yet another important matter for future research.

## ACKNOWLEDGMENT

We are indebted to Drs. S. Austin, G. W. Bates, E. C. Cocking, E. Galun, J. P. Helgeson, M. K. Jones, T. Kumashiro, P. Maliga, L. P. Pijnacker, K. Puite, J. Teissie, M. J. Tempelaar and S. E. de Vries for sending preprints or reprints for use in this review. We also thank Professor T. Eriksson and Dr. T. Endo for their invaluable discussions and critical reading of the manuscripts, and Ms. M. Uchimoto for her help in arranging the references. This study was supported in part by Grants-in-aid from the Ministry of Culture, Promotion and Science of Japan.

## 11.8 REFERENCES

Abdullah, R., E. C. Cocking and J. A. Thompson (1986). Efficient plant regeneration from rice protoplasts through somatic embryogenesis. *Bio/Technology*, **4**, 1087–1090.

Abe, S. and J. Takeda (1986). Possible involvement of calmodulin and the cytoskeleton in electrofusion of plant protoplasts. *Plant Physiol.*, **81**, 1151–1155.

Afonso, C. L., K. R. Harkins, M. A. Thomas-Compton, A. E. Krejci and D. W. Galbraith (1985). Selection of somatic hybrid plants in *Nicotiana* through fluorescence-activated sorting of protoplasts. *Bio/Technology*, **3**, 811–816.

Ammirato, P. V., D. A. Evans, R. Sharp and Y. Yamada (eds.) (1984). *Handbook of Plant Cell Culture*, vol. 3, Macmillan, New York.

Asahi, T., T. Kumashiro and T. Kubo (1988). Constitution of mitochondrial and chloroplast genomes in male sterile tobacco obtained by protoplast fusion of *Nicotiana tabacum* and *N. debneyi*. *Plant Cell Physiol.*, **29**, 43–49.

Austin, S., M. A. Bear and J. P. Helgeson (1985). Transfer of resistance to potato leaf roll virus form *Solanum brevidens* into *Solanum tuberosum* by somatic fusion. *Plant Science*, **39**, 75–82.

Austin, S., E. Lojkowska, M. K. Ehlenfeldt, A. Kelman and J. P. Helgeson (1988). Fertile interspecific somatic hybrids of *Solanum*: a novel source of resistance to *Erwinia* soft rot. *Phytopathology*, (in press).

Aviv, D. and E. Galun (1986). Restoration of male fertile *Nicotiana* by fusion of protoplasts derived from two different cytoplasmic male-sterile cybrids. *Plant Mol. Biol.*, **7**, 411–417.

Aviv, D. and E. Galun (1987). Chondriome analysis in sexual progenies of *Nicotiana* cybrids. *Theor. Appl. Genet.*, **73**, 821–826.

Barsby, T. L., P. V. Chuong, A. A. Yarrow, S.-C. Wu, M. Coumans, R. J. Kemble, A. D. Powell, W. D. Beversdorf and K. P. Pauls (1987). The combination of Polima cms and cytoplasmic triazine resistance in *Brassica napus*. *Theor. Appl. Genet.*, **73**, 809–814.

Barsby, T. L., J. F. Shepard, R. J. Kemble and R. Wong (1984). Somatic hybridization in the genus *Solanum*: *S. tuberosum* and *S. brevidens*. *Plant Cell Rep.*, **3**, 165–167.

Bates, G. W. and C. A. Hasenkampf (1985). Culture of plant somatic hybrids following electrical fusion. *Theor. Appl. Genet.*, **70**, 227–233.

Bates, G. W., C. A. Hasenkampf, C. L. Contolini and W. C. Piastuch (1987). Asymmetric hybridization in *Nicotiana* by fusion of irradiated protoplasts. *Theor. Appl. Genet.*, **74**, 718–726.

Belliard, G., F. Vedel and G. Pelletier (1979). Mitochondrial recombination in cytoplasmic hybrids of *Nicotiana tabacum* by protoplast fusion. *Nature (London)*, **281**, 401–403.

Boeshore, M. L., M. R. Hanson and S. Izhar (1985). A variant mitochondrial DNA arrangement specific to *Petunia* stable sterile somatic hybrids. *Plant Mol. Biol.*, **4**, 125–132.

Boeshore, M. L., I. Lifshitz, M. R. Hanson and S. Izhar (1983). Novel composition of mitochondrial genomes in *Petunia* somatic hybrids derived from cytoplasmic male sterile and fertile plants. *Mol. Gen. Genet.*, **190**, 459–467.

Brunold, C., S. Kruger-Lebus, M. W. Saul, S. Wegmuller and I. Potrykus (1987). Combination of kanamycin resistance and nitrate reductase deficiency as selectable markers in one nuclear genome provides a universal somatic hybridizer in plants. *Mol. Gen. Genet.*, **208**, 469–473.

Chaleff, R. S. (1981). *Genetics of Higher Plants*. Cambridge University Press, Cambridge.

Chand, P. K., M. R. Davey, J. B. Power and E. C. Cocking (1988). An improved procedure for protoplast fusion using polyethylene glycol. *J. Plant Physiol.* (in press).

Chapel, M., M.-H. Montane, B. Ranty, J. Teissie and G. Alibert (1986). Viable somatic hybrids are obtained by direct current electrofusion of chemically aggregated plant protoplasts. *FEBS Lett.*, **196**, 79–86.

Chen, K., S. G. Wildman and H. H. Smith (1977). Chloroplast DNA distribution in parasexual hybrids as shown by polypeptide composition of fraction I protein. *Proc. Natl. Acad. Sci. USA*, **74**, 5109–5112.

Chetrit, P., C. Mathieu, F. Vedel, G. Pelletier and C. Primard (1985). Mitochondrial DNA polymorphism induced by protoplast fusion in Cruciferae. *Theor. Appl. Genet.*, **69**, 361–366.

Clark, E. M., S. Izhar and R. Hanson (1985). Independent segregation of the plastid genome and cytoplasmic male sterility in *Petunia* somatic hybrids. *Mol. Gen. Genet.*, **199**, 440–445.

Clark, E. M., L. Schnabelrauch, M. R. Hanson and K. C. Sink (1986). Differential fate of plastid and mitochondrial genomes in *Petunia* somatic hybrids. *Theor. Appl. Genet.*, **72**, 748–755.

Dudits, S., E. Maroy, T. Praznovszky, Z. Olah, J. Gyorgyey and R. Cella (1987). Transfer of resistance traits from carrot into tobacco by asymmetric somatic hybridization: regeneration of fertile plants. *Proc. Natl. Acad. Sci. USA*, **84**, 8434–8438.

Ehlenfeldt, M. K. and J. P. Helgeson (1987). Fertility of somatic hybrids from protoplast fusions of *Solanum brevidens* and *S. tuberosum*. *Theor. Appl. Genet.*, **73**, 395–402.

Endo, T., T. Komiya, M. Mino, K. Nakanishi, S. Fujita and Y. Yamada (1988). Genetic diversity among sublines originating from a single somatic hybrid cell of *Duboisia hopwoodii* + *Nicotiana tabacum*. *Theor. Appl. Genet.* (in press).

Evans, D. A., R. Sharp, P. V. Ammirato and Y. Yamada (eds.) (1983). *Handbook of Plant Cell Culture*, vol. 1, Macmillan, New York.

Fish, N., A. Karp and M. G. K. Jones (1987). Improved isolation of dihaploid *Solanum tuberosum* protoplasts and the production of somatic hybrids between dihaploid *S. tuberosum* and *S. brevidens*. *In Vitro Cellular & Development Biology*, **23**, 575–580.

Fish, N., A. Karp and M. G. K. Jones (1988a). Production of somatic hybrids by electrofusion in *Solanum*. *Theor. Appl. Genet.* (in press).

Fish, N., S. H. Steele and M. G. K. Jones (1988b). Field assessment of dihaploid *S. tuberosum* and *S. brevidens* somatic hybrids. *Theor. Appl. Genet.* (in press).

Fitter, M. S., P. M. Norman, M. G. Hahn, V. P. M. Wingate and C. J. Lamb (1987). Identification of somatic hybrids in plant protoplast fusions with monoclonal antibodies to plasma-membrane antigens. *Planta*, **170**, 49–54.

Fluhr, R., D. Aviv, E. Galun and M. Edelman (1984). Generation of heteroplasmic *Nicotiana* cybrids by protoplast fusion: analysis for plastid recombination types. *Theor. Appl. Genet.*, **67**, 491–497.

Fujimura, T., M. Sakurai, H. Akagi, T. Negishi and A. Hirose (1985). Regeneration of rice plants from protoplasts *Plant Tissue Cult. Lett.*, **2**, 74–75.

Galun, E. and D. Aviv (1986). Organelle transfer. In *Methods in Enzymology*, ed. A. Weiss and H. Weissback, vol. 118, pp. 595–611. Academic Press, New York.

Galun, E., P. Arzee-Gonen, R. Fluhr, M. Edelman and D. Aviv (1982). Cytoplasmic hybridization in Nicotiana: mitochondrial DNA analysis in progenies resulting from fusion between protoplasts having different organelle constitutions. *Mol. Gen. Genet.*, **186**, 50–56.

Gibson, R. G., M. G. K. Jones and N. Fish (1988). Resistance to potato leaf roll virus and potato virus Y in somatic hybrids between dihaploid *S. tuberosum* and *S. brevidens*. *Theor. Appl. Genet.* (in press).

Gleba, Y. Y., I. F. Kanevsky, M. V. Skarzhynskaya, I. K. Komarnitsky and N. N. Cherep (1986). Hybrids between tobacco crown gall cells and normal somatic cells of *Atropa belladonna Plant Cell Rep.*, **5**, 394–397.

Gleba, Y. Y., N. N. Kolesnik, I. V. Meshkene, N. N. Cherep and A. S. Parokonny (1984). Transmission genetics of the somatic hybridization process in *Nicotiana*. *Theor. Appl. Genet.*, **69**, 121–128.

Gleba, Y. Y., I. K. Komarnitsky, N. N. Kolesnik, I. Meshkene and G. I. Martyn (1985). Transmission genetics of the somatic hybridization process in *Nicotiana* II. Plastome heterozygotes. *Mol. Gen. Genet.*, **198**, 476–481.

Gleba, Y. Y. and K. M. Sytnik (1984). *Protoplast Fusion*. Springer-Verlag, Berlin.

Gleddie, S., G. Fassuliotis, W. A. Keller and G. Setterfield (1985). Somatic hybridization as a potential method of transferring nematode and mite resistance into eggplant. *Z. Pflanzenzuchtg*, **94**, 352–355.

Gleddie, S., W. A. Keller and G. Setterfield (1986). Production and characterization of somatic hybrids between *Solanum melongena* L. and *S. sisymbriifolium* Lam. *Theor. Appl. Genet.*, **71**, 613–621.

Glimelius, K. and H. T. Bonnett (1986). *Nicotiana* cybrids with *Putunia* chloroplasts. *Theor. Appl. Genet.*, **72**, 794–798.

Glimelius, K., M. Djupsjobacka and H. Fellner-Feldegg (1986). Selection and enrichment of plant protoplast heterokaryons of *Brassicaceae* by flow sorting. *Plant Sci.*, **45**, 133–141.

Gupta, P. P., O. Schieder and M. Gupta (1984). Intergeneric nuclear gene transfer between somatically and sexually incompatible plants through asymmetric protoplast fusion. *Mol. Gen. Genet.*, **197**, 30–35.

Hampp, R. and M. Steingraber (1985). Electric field induced fusion of evacuolated mesophyll protoplasts of oat. *Naturwissenschaften*, **72**, 91–92.

Handley, L. W., R. L. Nickels, M. W. Cameron, P. P. Moore and K. C. Sink (1986). Somatic hybrid plants between *Lycopersicon esculentum* and *Solanum lycopersicoides*. *Theor. Appl. Genet.*, **71**, 691–697.

Hanson, M. R. and M. F. Conde (1985). Functioning and variation of cytoplasmic genomes: lessons from cytoplasmic–nuclear interactions affecting male fertility in plants. In *International Review of Cytology*, ed. G. P. Chapman, vol. 94, pp. 213–245. Academic Press, New York.

Helgeson, J. P., G. J. Hunt, G. T. Haberlach and S. Austin (1986). Somatic hybrids between *Solanum brevidens* and *Solanum tuberosum*: expression of a late blight resistance gene and potato leaf roll resistance. *Plant Cell Rep.*, **3**, 212–214.

Ichikawa, H., L. Tanno-Suenaga and J. Imamura (1987). Selection of *Daucus* cybrids based on metabolic complementation between X-irradiated *D. capillifolius* and iodoacetamide-treated *D. carota* by somatic cell fusion. *Theor. Appl. Genet.*, **74**, 746–752.

Imamura, J., M. W. Saul and I. Potrykus (1987). X-ray irradiation promoted asymmetric somatic hybridization and molecular analysis of the products. *Theor. Appl. Genet.*, **74**, 445–450.

Izhar, S., M. Schlicter and D. Swartzberg (1983). Sorting out of cytoplasmic elements in somatic hybrids of *Petunia* and the prevalence of the heteroplasmon through several meiotic cycles. *Mol. Gen. Genet.*, **190**, 468–474.

Kamata, Y. and T. Nagata (1987). Enrichment of heterokaryocytes between mesophyll and epidermis protoplasts by density gradient centrifugation after electric fusion. *Theor. Appl. Genet.*, **75**, 26–29.

Kemble, R. J., T. L. Barsby, R. S. C. Wong and J. F. Shepard (1986). Mitochondrial DNA rearrangements in somatic hybrids of *Solanum tuberosum* and *Solanum brevidens*. *Theor. Appl. Genet.*, **72**, 787–793.

Kishinami, I. and J. M. Widholm (1987). Auxotrophic complementation in intergeneric hybrid cells obtained by electrical and dextran-induced protoplast fusion. *Plant Cell Physiol.*, **28**, 211–218.

Kohn, H., R. Schieder and O. Schieder (1985). Somatic hybrids in tobacco mediated by electrofusion. *Plant Sci.*, **38**, 121–128.

Kothari, S. L., D. C. Monte and J. M. Widholm (1986). Selection of *Daucus carota* somatic hybrids using drug resistance markers and characterization of their mitochondrial genomes. *Theor. Appl. Genet.*, **72**, 494–502.

Kumar, A. and E. C. Cocking (1987). Protoplast fusion: a novel approach to organelle genetics in higher plants. *Am. J. Bot.*, **74**, 1289–1303.

Kushnir, S. G., L. R. Shlumukov, N. J. Pogrebnyak, S. Berger and Y. Gleba (1987). Functional cybrid plants possessing a *Nicotiana* genome and an *Atropa* plastome. *Mol. Gen. Genet.*, **209**, 159–163.

Kyozuka, J., Y. Hayashi and K. Shimamoto (1987). High frequency plant regeneration from rice protoplasts by novel nurse culture methods. *Mol. Gen. Genet.*, **206**, 408–413.

Li, X. Q., P. Chetrit, C. Mathieu, F. Vedel, R. De Paepe, R. Remy and F. Ambard-Bretteville (1988). Regeneration of cytoplasmic male sterile protoclones of *Nicotiana sylvestris* with mitochondrial variations. *Curr. Genet.*, **13**, 261–266.

Loschiavo, F., G. Giovinazzo and M. Terzi (1983). 8-Azaguanine-resistant carrot cell mutants and their use as universal hybridizers. *Mol. Gen. Genet.*, **192**, 326–329.

Matthews, B. F. and J. M. Widholm (1985). Organelle DNA compositions and isoenzyme expression in an interspecific somatic hybrid of *Daucus*. *Mol. Gen. Genet.*, **198**, 371–376.

Medgyesy, P., E. Fejes and P. Maliga (1985). Interspecific chloroplast recombination in a *Nicotiana* somatic hybrid. *Proc. Natl. Acad. Sci. USA*, **82**, 6960–6964.

Mehrle, W., U. Zimmermann and R. Hampp (1985). Evidence for asymmetrical uptake of fluorescent dyes through electropermeabilized membranes of *Avena* mesophyll protoplasts. *FEBS Lett.*, **185**, 89.

Melchers, G., M. D. Sacristan and A. A. Holder (1978). Somatic hybrid plants of potato and tomato regenerated from fused protoplasts. *Carlsverg Res. Commun.*, **43**, 203–218.

Menczel, L., A. Morgan, S. Brown and P. Maliga (1987). Fusion-mediated combination of Ogura-type cytoplasmic male sterility with *Brassica napus* plastids using X-irradiated CMS protoplasts. *Plant Cell Rep.*, **6**, 98–101.

Menczel, L., L. S. Polsby, K. E. Steinback and P. Maliga (1986). Fusion- mediated transfer of triazine-resistant chloroplasts: characterization of *Nicotiana tabacum* cybrid plants. *Mol. Gen. Genet.*, **205**, 201–205.

Morgan, A. and P. Maliga (1987). Rapid chloroplast segregation and recombination of mitochondrial DNA in *Brassica* cybrids. *Mol. Gen. Genet.*, **209**, 240–246.

Morikawa, H., M. Asada and Y. Yamada (1988a). Effects of physical parameters upon protoplast electrofusion. *Plant Cell Physiol.*, **29**, 659–664.

Morikawa, H., Y. Hayashi, Y. Hirabayashi, M. Asada and Y. Yamada (1988b). Cellular and vacuolar fusion of protoplasts electrofused using platinum microelectrodes. *Plant Cell Physiol.*, **29**, 189–193.

Morikawa, H., Y. Hayashi, N. Ohnishi, Y. Yamamoto, F. Sato and Y. Yamada (1988c). Culture of electrically induced heterokaryons of plant cells rich in secondary metabolites. *Plant Cell Physiol.*, **29**, 1201–1206.

Morikawa, H., Y. Hayashi and Y. Yamada (1988d). Manual pair formation and electrofusion of protoplasts using platinum microelectrodes. *Plant Tissue Cult. Lett.*, **5**, 37–39.

Morikawa, H., A. Iida and Y. Yamada (1988e). Electric gene transfer into plant protoplasts and cells. In *Advances in Biotechnological Processes*, ed. A. Mizrahi, vol. 9, pp. 175–202. Liss, New York.

Morikawa, H., T. Kumashiro, K. Kusakari, A. Iida, A. Hirai and Y. Yamada (1987). Interspecific hybrid plant formation by electrofusion in *Nicotiana*. *Theor. Appl. Genet.*, **75**, 1–4.

Morikawa, H., J. Mitsuhashi and Y. Yamada (1988f). Interkingdom electrofusion between tobacco mesophyll protoplasts and cultured butterfly cells. *Plant Tissue Cult. Lett.*, **5**, 90–92.

Morikawa, H., K. Sugino, Y. Hayashi, J. Takeda, M. Senda, A. Hirai and Y. Yamada (1986). Interspecific plant hybridization by electrofusion in *Nicotiana*. *Bio/Technology*, **4**, 57–60.

Naton, B., W. Mehrle, R. Hampp and U. Zimmermann (1986). Mass electrofusion and mass selection of functional hybrids from vacuolate x evacuolate protoplasts. *Plant Cell Rep.*, **5**, 419–422.

Nea, L. J. and G. W. Bates (1987). Factors affecting protoplast electrofusion efficiency. *Plant Cell Rep.*, **6**, 337–340.

Nea, L. J., G. W. Bates and P. J. Gilmer (1987). Facilitation of electrofusion of plant protoplasts by membrane-active agents. *Biochim. Biophys. Acta*, **897**, 293–301.

Negrutiu, I., D. De Brouwer, J. M. Watts, V. I. Sidorov, R. Dirks and M. Jacobs (1986). Fusion of plant protoplasts: a study using auxotrophic mutants of *Nicotiana plumbaginifolia*, Viviani. *Theor. Appl. Genet.*, **72**, 279–286.

Nishiguchi, M., T. Sato and F. Motoyoshi (1987). An improved method for electroporation in plant protoplasts: infection of tobacco protoplasts by tobacco mosaic virus particles. *Plant Cell Rep.*, **6**, 90–93.

Nishio, T., K. Watanabe, T. Sato and M. Hirai (1988). Production of somatic hybrids between *Brassica oleracea* and *Brassica campestris* by electric cell fusion. *Bulletin of the National Research Institute of Vegetables, Ornamental Plants & Tea*, **A-1**, in press (in Japanese).

O'Connell, M. A. and M. R. Hanson (1986). Regeneration of somatic hybrid plants formed between *Lycopersicon esculentum* and *Solanum rickii*. *Theor. Appl. Genet.*, **72**, 59–65.

O'Connell, M. A. and M. R. Hanson (1987). Regeneration of somatic hybrid plants formed between *Lycopersicon esculentum* and *L. pennellii*. *Theor. Appl. Genet.*, **75**, 83–89.

Ohgawara, T., S. Kobayashi, E. Ohgawara, H. Uchimiya and S. Ishii (1985). Somatic hybrid plants obtained by protoplast fusion between *Citrus sinensis* and *Poncirus trifoliata*. *Theor. Appl. Genet.*, **71**, 1–4.

Okada, K., T. Nagata and I. Takebe (1986). Introduction of functional RNA into plant protoplasts by electroporation. *Plant Cell Physiol.*, **27**, 619–626.

Ozias-Akins, P., R. J. Ferl and I. K. Vasil (1986). Somatic hybridization in the gramineae: *Pennisetum americanum* (L.) K. Schum. (pearl millet) + *Panicum maximum* Jacq. (guinea grass). *Mol. Gen. Genet.*, **203**, 365–370.

Ozias-Akins, P., D. R. Pring and I. K. Vasil (1987). Rearrangements in the mitochondrial genome of somatic hybrid cell lines of *Pennisetum americanum* (L.) K. Schum. + *Panicum maximum* Jacq. *Theor. Appl. Genet.*, **74**, 15–20.

Pauls, K. P. and P. V. Chuong (1987). Flow cytometric identification of *Brassica napus* protoplast fusion products. *Can. J. Bot.*, **65**, 834–838.

Pelletier, G., C. Primard, F. Vedel, P. Chetrit, R. Remy, Rousselle and M. Renard (1983). Intergeneric cytoplasmic hybridization in Cruciferae by protoplast fusion. *Mol. Gen. Genet.*, **191**, 244–250.

Pelletier, G., F. Vedel and G. Belliard (1985). Cybrids in genetics and breeding. *Hereditas Suppl.*, **3**, 49–56.

Pental, D., J. D. Hamill and E. C. Cocking (1984). Somatic hybridization using a double mutant of *Nicotiana tabacum*. *Heredity*, **53**, 79–83.

Pental, D., J. D. Hamill, A. Pirrie and E. C. Cocking (1986). Somatic hybridization of *Nicotiana tabacum* and *Petunia hybrida*. *Mol. Gen. Genet.*, **202**, 342–347.

Pijnacker, L. P., M. A. Ferwerda, K. J. Puite and S. Roest (1987). Elimination of *Solanum phureja* nucleolar chromosomes in *S. tuberosum* + *S. phureja* somatic hybrids. *Theor. Appl. Genet.*, **73**, 878–882.

Pirrie, A. and J. B. Power (1986). The production of fertile, triploid somatic hybrid plants [*Nicotiana glutinosa* (*n*) + *N. tabacum* (2*n*)] *via* gametic:somatic protoplast fusion. *Theor. Appl. Genet.*, **72**, 48–52.

Potrykus, I., J. Jia, G. B. Lazar and M. Saul (1984). *Hyocyamus muticus* + *Nicotiana tabacum* fusion hybrids selected *via* auxotroph complementation. *Plant Cell Rep.*, **3**, 68–71.

Puite, K. J., W. T. Broeke and J. Schaart (1988). Inhibition of cell wall synthesis improves flow cytometric sorting of potato heterofusions resulting in hybrid plants. *Plant Sci.* (in press).
Puite, K. J., S. Roest and L. P. Pijnacker (1986). Somatic hybrid potato plants after electrofusion of diploid *Solanum tuberosum* and *Solanum phureja*. *Plant Cell Rep.*, **5**, 262–265.
Puite, K. J., P. Van Wikselaar and H. Verhoeven (1985). Electrofusion, a simple and reproducible technique in somatic hybridization of *Nicotiana plumbaginifolia* mutants. *Plant Cell Rep.*, **4**, 274–276.
Robertson, D., J. D. Palmer, E. D. Earle and M. A. Mutschler (1987). Analysis of organelle genomes in a somatic hybrid derived from cytoplasmic male-sterile *Brassica oleracea* and atrazine-resistant *B. campestris*. *Theor. Appl. Genet.*, **74**, 303–309.
Rothenberg, M., M. L. Boeshore, M. R. Hanson and S. Izhar (1985). Intergenomic recombination of mitochondrial genomes in a somatic hybrid plant. *Curr. Genet.*, **9**, 615–618.
Sala, C., M. G. Biasini, C. Morandi, E. Nielsen, E. Parisi and F. Sala (1985). Selection and nuclear DNA analysis of cell hybrids between *Daucus carota* and *Oryza sativa*. *J. Plant Physiol.*, **118**, 409–419.
Saxena, P. K. and J. King (1987). Complementation by intraspecific protoplast fusion of auxotrophic cell lines of *Datura innoxia* P. Mill. *Plant Cell Tissue Organ Cult.*, **9**, 61–71.
Sharp, R., D. A. Evans, P. V. Ammirato and Y. Yamada (ed.) (1984). *Handbook of Plant Cell Culture*, vol. 2. Macmillan, New York.
Shepard, J. F., D. Bidney, T. Barsby and R. Kemble (1983). Genetic transfer in plants through interspecific protoplast fusion. *Science (N.Y.)*, **219**, 683–688.
Sidorov, V. A., M. K. Zubko, A. A. Kuchko, I. K. Komarnitsky and Y. Y. Gleba (1987). Somatic hybridization in potato: use of r-irradiated protoplasts of *Solanum pinnatisectum* in genetic reconstruction. *Theor. Appl. Genet.*, **74**, 364–368.
Somers, D. A., K. R. Narayanan, A. Kleinhofs, S. Cooper-Bl and E. C. Cocking (1986). Immunological evidence for transfer of the barley nitrate reductase structural gene to *Nicotiana tabacum* by protoplast fusion. *Mol. Gen. Genet.*, **204**, 296–301.
Sowers, A. E. and M. R. Liber (1986). Electropore diameters, lifetimes, numbers, and locations in individual erythrocyte ghosts. *FEBS Lett.*, **205**, 179–184.
Spangenberg, G. and H.-G. Schweiger (1986). Controlled electrofusion of different types of protoplasts and subprotoplasts including cell reconstitution in *Brassica napus* L. *Eur. J. Cell Biol.*, **41**, 51–56.
Sundberg, E. and K. Glimelius (1986). A method for production of interspecific hybrids within *Brassiceae via* somatic hybridization, using resynthesis of *Brassica napus* as a model. *Plant Sci.* **43**, 155–162.
Sundberg, E., M. Landgren and K. Glimelius (1987). Fertility and chromosome stability in *Brassica napus* resythesized by protoplast fusion. *Theor. Appl. Genet.*, **75**, 96–104.
Taguchi, T. and T. Kameya (1986). Production of somatic hybrid plants between cabbage and chinese cabbage through protoplast fusion. *Jpn. J. Breed.*, **36**, 185–189.
Takahashi, S., Y. Fujita and Y. Yamada (1986). Effective method for protoplast fusion. *Abstracts of IVth International Congress of Plant Tissue and Cell Culture*, (Minnesota, 1986) p.317.
Terada, R., J. Kyozuka, S. Nishibayashi and K. Shimamoto (1987a). Plantlet regeneration from somatic hybrids of rice (*Oryza sativa* L.) and barnyard grass (*Echinochloa oryzicola* Vasing). *Mol. Gen. Genet.*, **210**, 39–43.
Terada, R., Y. Yamashita, S. Nishibayashi and K. Shimamoto (1987b). Somatic hybrids between *Brassica oleracea* and *B. campestris*: selection by the use of iodoacetamide inactivation and regeneration ability. *Theor. Appl. Genet.*, **73**, 379–384.
Toriyama, K., K. Hinata and T. Kameya (1987a). Production of somatic hybrid plants, 'Brassicomoricandia', through protoplast fusion between *Moricandia arvensis* and *Brassica oleracea*. *Plant Sci.*, **48**, 123–128.
Toriyama, K., T. Kameya and K. Hinata (1987b). Selection of a universal hybridizer in *Sinapis turgida* Del. and regeneration of plantlets from somatic hybrids with *Brassica* species. *Planta*, **170**, 308–313.
Tur-Kaspa, R., L. Teicher, B. J. Levine, A. I. Skoultchi and D. A. Shafritz (1986). Use of electroporation to introduce biologically active foreign genes into primary rat hepatocytes. *Mol. Cell. Biol.*, **6**, 716–718.
Uijtewaal, B. A., L. C. J. M. Suurs and E. Jacobsen (1987). Protoplast fusion of monohaploid ($2n = x = 12$) potato clones: identification of somatic hybrids using malate dehydrogenase as biochemical marker. *Plant Sci.*, **51**, 277–284.
Vardi, A., A. Breiman and E. Galun (1987). *Citrus* cybrids: production by donor–recipient protoplast-fusion and verification by mitochondrial DNA restriction profiles. *Theor. Appl. Genet.*, **75**, 51–58.
Vasil, I. K. (ed.) (1984). *Cell Culture and Somatic Cell Genetics of Plants*. vol. 1, Academic Press, New York.
Vedel, F., P. Chetrit, C. Mathieu, G. Pelletier and C. Primard (1986). Several different mitochondrial DNA regions are involved in intergenomic recombination in *Brassica napus* cybrid plants. *Curr. Genet.*, **11**, 17–24.
Vries, S. E. de, E. Jacobsen, M. G. K. Jones, A. E. H. M. Loonen, M. J. Tempelaar, J. Wijbrandi and W. J. Feenstra (1987). Somatic hybridization of amino acid analog resistant cell lines of potato (*Solanum tuberosum* L.) by electrofusion. *Theor. Appl. Genet.*, **73**, 451–458.
Wildman, S. G., T. Hongladarom and S. I. Honda (1962). Chloroplasts and mitochondria in living plant cells: cinephotomicrographic studies. *Science (N.Y.)*, **138**, 434–435.
Wright, R. L., D. A. Somers and R. L. McGraw (1987). Somatic hybridization between birdsfoot trefoil (*Lotus corniculatus* L.) and *L conimbricensis* willd. *Theor. Appl. Genet.*, **75**, 151–156.
Yamada, Y. and H. Morikawa (1985). Protoplast fusion of secondary metabolite-producing cells. In Neumann, K.-H., Barz, W., Reinhard E (eds.): *Primary and Secondary Metabolism of Plant Cell Cultures*, Springer-Verlag, Berlin, pp. 255–271.
Yamada, Y., H. Morikawa, F. Sato and Y. Yamamoto (1987). Secondary plant products from cultured hybrid cells. *Proc. Jpn. Acad., Ser. B*, **63**, 208–210.
Yamada, Y., Z.-Q. Yang and D.-T. Tang (1986). Plant regeneration from protoplast-derived callus of rice (*Oryza Sativa* L). *Plant Cell Rep.*, **5**, 85–88.
Yang, Z.-Q., T. Shikanai and Y. Yamada (1988). Asymmetric hybridization between cytoplasmic male-sterile (CMS) and fertile rice (*Oryza sativa* L.) protoplasts. *Theor. Appl. Genet.* (in press).
Yarrow, S. A., S. C. Wu, T. L. Barsby, R. J. Kemble and J. F. Shepard (1986). The introduction of CMS mitochondria to triazine tolerant *Brassica napus* L., var. 'Regent', by micromanipulation of individual heterokaryons. *Plant Cell Rep.*, **5**, 415–418.
Ye, J., R. M. Hauptmann, A. G. Smith and J. M. Widholm (1987). Selection of a *Nicotiana plumbaginifolia* universal hybridizer and its use in intergeneric somatic hybrid formation. *Mol. Gen. Genet.*, **208**, 474–480.
Zachrisson, A. and C. H. Bornman (1986). Electromanipulation of plant protoplasts. *Physiol. Plant.*, **67**, 507–516.
Zimmermann, U., P. Scheurich, G. Pilwat and R. Benz (1981). Cells with manipulated functions: new perspectives for cell biology, medicine and technology. *Angew. Chem., Int. Ed. Engl.*, **20**, 325–344.

# 12
# Genetic Engineering of Plants and Cultures

GERT OOMS
*Rothamsted Experimental Station, Harpenden, UK*



## 12.1 INTRODUCTION

Genetic engineering in plants and cultures is in no way different from other types of engineering, in that it requires technology and in that it serves a particular aim. The technology, although often loosely defined, is widely accepted to include: (i) the isolation and *in vitro* manipulation of genes; and (ii) the transfer of (manipulated) genes into plant cells; followed by (iii) culturing of the cells into transgenic plants or transgenic cell lines maintained *in vitro* as callus, suspension, roots or other organogenic tissues. Equally, the general aims pursued in genetic engineering experiments are not usually described in a unified way. They vary widely, ranging from purely scientific reasons serving many areas of plant research, to the more commercial reasons usually linked to clearly identified commercial aims and market considerations. These may, for strategic reasons, be somewhat concealed or, conversely, form the topic of promotion campaigns. The genetic engineering technology used and the immediate scientific questions pursued in academia and industry are often similar. This is simply because many of the most interesting topics in plant biology are in areas of commercial interest. Examples are flowering, reproductive biology and seed formation, growth and development which relates to maturity, yield and harvest index. Furthermore a wide range of important and interesting interactions exist between plants and their environment (soil, air, water, benign and harmful associates *etc.*). Finally, there are common interests in those research areas that are of immediate importance to any aspect of genetic engineering. These include insight into many molecular control mechanisms such as in gene expression and regulation of primary and secondary metabolism.

This review does not aim to discuss the potential impact that genetic engineering may have on society at large, and the ecological and legal considerations involved. Moreover, only a few passing comments are made on the strengths and weaknesses of genetic engineering technology relative to other genetic manipulation approaches. These include the widespread use of crosses in traditional breeding programmes, or other newly developed laboratory-based technologies like induction of somaclonal variation and somatic cell fusion which can be linked to advanced selection schemes. (See, for reviews Evans, 1989; Karp and Bright, 1985; Larkin, 1987; Lee and Philips, 1988; Scowcroft and Ryan, 1986; Withers and Alderson, 1986). Instead, the review focuses on the three principal components of genetic engineering. (i) The technology of the actual transformation process and the culturing of cells into cell lines or plants. This relates to the development of transformation protocols and to the choices in their use for transforming particular plant species, genotypes or even cell types. (ii) The isolation and manipulation of genes to be used in transformations, which relates to ways of identifying and perhaps modifying interesting specific genes. (iii) The choices of those areas of biology with best prospects for scientific or commercial application.

For several reasons initial emphasis is placed on the natural genetic engineering system employed by *Agrobacterium tumefaciens* and *A. rhizogenes* which cause crown gall and hairy-root formation in dicotyledonous plants. The emphasis in this review is on relevance for genetic engineering of plants and cell cultures in general. In complementary reviews information can be found on the general biology of *Agrobacterium* and its transformed tissues (Binns and Thomashow, 1988; Birot *et al.*, 1987; Hooykaas and Schilperoort, 1984; Nester *et al.*, 1984; White and Sinkar, 1987), on plant transformation technology in general and of *Agrobacterium* in particular (Bevan and Goldsborough, 1988; Day and Lichtenstein, Chapter 9; Melchers and Hooykaas, 1987; Zambryski, 1988; Zambryski *et al.*, 1989) and on research progress as viewed from some of the larger laboratories in the field (Klee *et al.*, 1987b; Schell, 1987). The two main reasons for this review to look at *Agrobacterium* and its transformed tissues are to use it as an illustration of the high sophistication at which a DNA transfer system can be developed (an example of the technology involved) and to highlight the far-reaching integration of genetic engineering strategies aimed at achieving particular goals (as an example of what technology can achieve). Consequently, as in other types of engineering, a close examination of what is known about the mechanism, efficiency and consequences of this natural plant genetic engineering process, which has evolved under longlasting sustained evolutionary pressure, should provide us with a great deal of insight and background knowledge which will be of value in the design of any aspect of genetic engineering of plants and cultures in the future.

An additional benefit of describing the *Agrobacterium* system first is that in the subsequent sections on choices of transformation methods and plants, on gene expression and on topics for application, many of the details of the progress in *Agrobacterium* in crowngall and hairyroot research resurface in various ways.

## 12.2 EXAMPLES OF SOPHISTICATED GENETIC ENGINEERING IN NATURE

### 12.2.1 Introduction

To date thousands and perhaps even millions of named and unnamed plants and cell cultures have already been isolated using a genetic engineering approach. Some lines are currently under evalution for marketing and others are the subject of applied and fundamental research. Most of these genetically engineered cell lines have been obtained using *Agrobacterium* transformation. The wealth of information gained in investigating these lines has helped to make the crown gall and hairyroot systems the most studied examples of genetic engineering in plants. The increasing interest in *Agrobacterium* over recent years and the volume of results obtained, is illustrated in Figure 1. The trend probably reflects the intrinsic interest in the biology of the system as well as the scope for further adaptation of the natural *Agrobacterium* transformation system for general use in an even wider variety of plant species. It should follow, however, that what is discussed here is only a sample of the progress made and many contributions have not been cited. Furthermore, the provision by

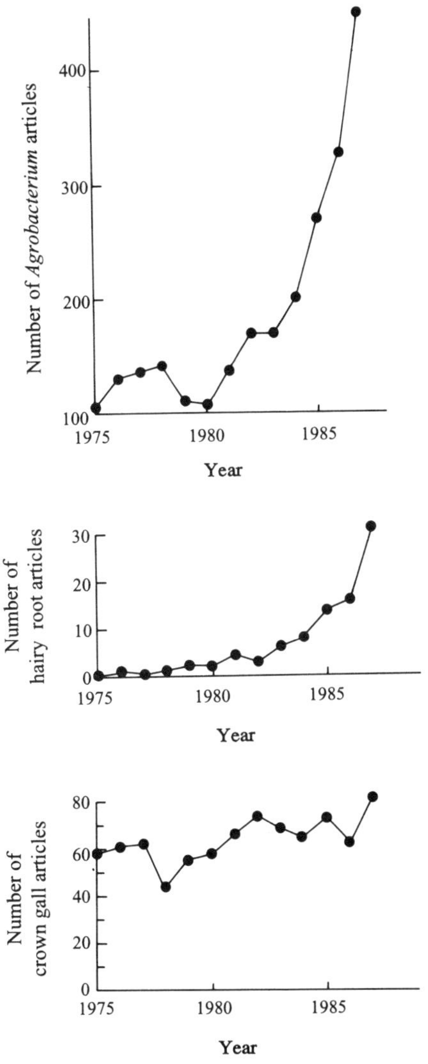


**Figure 1** The growing importance of *Agrobacterium* and hairy-roots in plant genetic engineering is illustrated by a considerable increase in published papers since 1975

*Agrobacterium* of genes and gene elements for general use in genetic engineering experiments, for example to study control of gene expression and genetic factors important in regulating growth patterns of plants and cell cultures, is likely to remain of additional relevance to plant genetic engineering in general for some time to come.

### 12.2.2 *Agrobacterium*

Agrobacteria are Gram-negative, rod-shaped bacteria that are abundant in soil and classified with bacteria of the closely related genus *Rhizobium* in the family of *Rhizobiaceae*. Smith and Townsend (1907) identified *Agrobacterium tumefaciens* as the cause of crowngall formation in plants; purified bacteria isolated from tumours in the field proved essential and sufficient to reinduce tumours on wound sites of healthy dicotyledonous plants. It is noted that this finding was preceded by less well publicized experiments by Cavora in Italy. Smith extended these basic observations and found differences between *Agrobacterium* isolates in their host range, and in the growth rate or morphology of tumours induced by them. He also observed differences between plants and plant organs in their response to infection by independent *Agrobacterium* isolates (Smith, 1917). In addition it became clear that there is an influence of environmental factors of which humidity and temperature are the most noticeable. De Cleene and De Ley (1977) have more recently compiled an extended list of responses induced by a number of *Agrobacterium* isolates on a wide range of field-grown plants.

Traditionally, *Agrobacterium* isolates were classified on the basis of differences in the morphologies of the tumours induced by them. The isolates were grouped as *A. tumefaciens* (causing crown galls), *A. rhizogenes* (causing prolific root growth or 'hairyroots'), *A. rubi* (causing cane galls which resemble crown galls but form on aerial parts of the plants) and *A. radiobacter* (the nonpathogenic agrobacteria). In some studies further distinctions were made using the more subtle differences in tumour growth of galls, on suitable test plants such as *Kalanchoe daigremontiana* (Bopp and Resende, 1966).

Classification of *Agrobacterium* isolates was subsequently refined following the discovery of a series of distinct and bacterium-dependent tumour cell specific compounds generally called 'opines', in the induced tumorous tissues (for a review, see Tempé and Goldman, 1982). Since the discovery of the first opine, lysopine (Biemann *et al.*, 1960) many additional opines have been identified. Invariably they are low molecular weight nitrogenous or sometimes phosphorus-containing compounds. Altogether there are now over a dozen opines and they have been grouped in classes like the octopine class consisting of octopine [$N^3$-(D-1-carboxyethyl)-1-arginine], lysopine, octopinic acid (or ornopine) and histopine, or the nopaline class with nopaline [$N^2$-(1,3-dicarboxypropyl)-1-arginine] and nopalinic acid (or ornaline). Generally, all the members of a particular class of opines in a tumour cell are found together and their identification thus provides an unambiguous fingerprint of the inducing *Agrobacterium* strain. In practice, however, it is common to find *Agrobacteria* that induce more than one class of opines in tumorous tissues (Petit *et al.*, 1983) and some care must be taken in unambiguous identification (Christou *et al.*, 1986).

Both of the above classifications of agrobacteria, based on tumorigenic properties and biochemistry of induced tumorous cells, are now known to be largely determined by plasmid-coded traits. Virulence in *A. tumefaciens* is coded by Ti or tumour inducing plasmids and in *A. rhizogenes* by Ri or root inducing plasmids (Van Larebeke *et al.*, 1975; Waston *et al.*, 1975; White and Nester, 1980). The Ti and Ri plasmids can move rather freely between agrobacteria, and accordingly, the tumour- or root-inducing properties and opine determinants move along with them. This implies that *Agrobacterium* classification based on chromosomally coded metabolic and physiological properties, is probably more meaningful from an evolutionary point of view, although the former classifications are the more practical and the more widely used. Recommended criteria for discrimination between agrobacteria based on stable chromosomal characters are tests for production of ketolactose from lactose, growth at 29, 35 or 37 °C, sensitivity to 2% NaCl and growth on erythritol as the C-source (for review, see Hooykaas and Schilperoort, 1984). This classifies most *Agrobacterium* strains reasonably well into three biotypes, I, II and III, and includes distinctions between *Rhizobium* strains. The classification is supported further by serological analyses (Alarcon *et al.*, 1987).

### 12.2.3 Chromosomal Virulence Genes

The existence in the *Agrobacterium* chromosome of specific genes important in tumour formation is beyond dispute. Their presence was suggested by early unsuccessful tumour induction experiments in which transfer of a Ti plasmid (as an $RP_4$::Ti cointegrate plasmid) into the distantly related *E. coli*, failed to make the bacterium tumorigenic (Holsters *et al.*, 1978). In contrast, after transfer of the Ti plasmid into *Rhizobium* the recipient became tumorigenic (Hooykaas *et al.*, 1977), whilst retaining its nodulation properties. These studies were extended by the isolation of *Agrobacterium* mutants that were avirulent or showed attenuated virulence as a result of transposon Tn5 insertions in specific chromosomal locations (Garfinkel and Nester, 1980; Matthysse, 1987; Thomashow *et al.*, 1987). The chromosomal virulence locus A (*chvA*), *chvB* (Douglas *et al.*, 1985), a single locus for determining the *A. tumefaciens* polysaccharide composition (*pscA*), and two loci affecting attachment to carrot cells, (*attC43* and *attC69*), were identified. Cloning of the first three genes and their subsequent use as probes in molecular studies demonstrated their relatedness to the chromosomal nodule development locus A (*ndvA*), *ndvB* (Dylan *et al.*, 1986) and the locus for the major acidic exopolysaccharide succinoglycan (*exoC*) important in nodule formation in *Rhizobium meliloti* (Cangelosi *et al.*, 1987; Marks *et al*; 1987), respectively. The loci also affected plant cell attachment; the *chvB* mutants no longer synthesized a low molecular weight $\beta$-1,2-glucan, the CHV protein has a role in export of the glucan (Cangelosi *et al.*, 1989; Iannino and Ugalde, 1989; O'Connell and Handelsman, 1989) and the *pscA* mutants produced little, if any, of the four species of polysaccharide synthesized by the wild-type *Agrobacterium*.

As I will be able to argue more fully in Section 12.2.10 the early observations on differences between agrobacteria in host range and tumour properties, together with the recent molecular research in chromosomal genes are impotant from a genetic engineering view point. They are among the factors that provide background for optimizing the *Agrobacterium* transformation process which is of potential use in the genetic engineering of recalcitrant plant species, particularly in cases where extension of the *Agrobacterium* host range is desired.

### 12.2.4 Ti and Ri Plasmids

In a popular review Novick (1980) suggested classifying plasmids, along with viruses, in a separate kingdom of subcellular organisms. It was argued that a plasmid needed a host cell for its 'life support systems' but that it autonomously controlled its own copy number and divergence into coexisting bacterial species and genera, and, significantly, even into competing bacteria. In other words evolutionary pressure on the bacteria and plasmids do not necessarily have a coincidal result. There is some merit in keeping this in mind when examining Ti and Ri plasmid biology, not only because the plasmids hold such a distinct position in relation to the biology of *Agrobacterium* and crown gall or hairy roots, but also when considering optimization of derived, *Agrobacterium*-based genetic engineering systems, aimed at either improving the transformation efficiency or extending the host range.

A glance at the functional organization of a typical octopine type Ti plasmid pTi Ach5, which is one of the best characterized of all Ti and Ri plasmids and is found with minor variations in the popular wild-type *A. tumefaciens* strains Ach5, 15955, B6, B6S3, A6 and B6806, shows that among the 180 or so genes on this 200 kb (120 MDa) plasmid, various genes have diverse and apparently unrelated functions (Figure 2). As expected, some genes encode distinct but functionally related products. For example, there are loci important in plasmid replication (*ori*), copy number control, plasmid stability, incompatibility (*inc*), and loci that determine plasmid transfer (the *tra* genes) *via* conjugation. One gene (*ape*) confers on *Agrobacterium* resistance or some form of exclusion to a bacteriophage, AP1 (for reviews, see Hooykaas and Schilperoort, 1984; Nester *et al.*, 1984). Directly relevant to the role Ti plasmids play in *Agrobacterium* and crown gall biology are several Ti plasmid genes that enable *Agrobacterium* to specifically take-up and metabolize opines by coding for a permease and a membrane-bound oxidase which enables catabolism of opines like octopine. A further gene for the degradation of the generated arginine has been identified (for reviews, see Klapwijk and Schilperoort, 1982; Tempé and Petit, 1982). As a result, agrobacteria containing Ti plasmids grow on media with octopine as a sole carbon and nitrogen source, whereas plasmid-free bacteria do not. This property is not entirely unique among soil borne bacteria since certain *Pseudomonas* strains also utilize opines (Tremblay *et al.*, 1987). It is illustrative of the sophistication of the Ti plasmid genetics, however, that the plasmid transfer genes and opine catabolism genes, the expression of which would normally be expected to be under independent control, are, in fact, subject to a common negative control or repressor system which is derepressable by the opine octopine (Klapwijk *et al.*, 1978). Thus, octopine enhances the spread of Ti plasmid over other bacteria, a process which is efficient under nutritionally poor conditions (Genetello *et al.*, 1977; Kerr *et al.*, 1977). Simultaneously it provides those bacteria

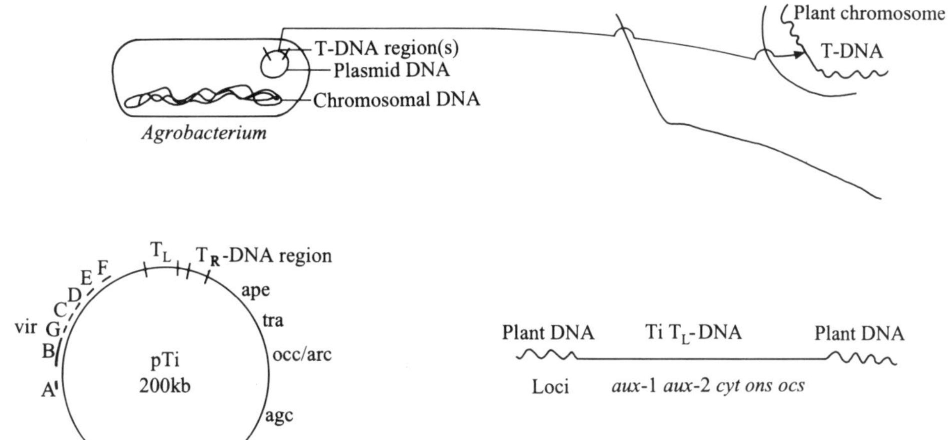


**Figure 2** Schematic representation of the key features in the natural genetic engineering process of crown gall formation by *Agrobacterium*

containing this Ti plasmid with an octopine inducible uptake and catabolism system which provides an ecological advantage under perhaps nutritionally poor, but certainly octopine enriched conditions, such as would be found in octopine-producing crown gall tissues.

### 12.2.5 *Vir* Region

Important to *Agrobacterium* and crown gall biology but also of immediate interest to plant genetic engineering in general, are the Ti plasmid functions defined by the *vir* region (Garfinkel and Nester, 1980; Ooms *et al.*, 1980) and the T-DNA region (Chilton *et al.*, 1977; Thomashow *et al.*, 1980; White *et al.*, 1982). In the *vir* region of the octopine Ti plasmid at least seven transcriptional units have been identified. Some loci comprise polycistronic operons and four (*virA*, *virB*, *virD*, *virG*) are essential for tumour formation on all plants tested. Three loci (*virC*, *virE*, *virF*) are important for some plants and thus determine host specificity (for a review, see Melchers and Hooykaas, 1987; Stachel and Zambryski, 1986; Zambryski, 1988). The two-component regulatory system of *virA* and *virG* is constitutively expressed, although *virG* together with *virB*, *virC*, *virD* and *virE* is also inducible by specific low molecular weight compounds released from wounded plant cells, such as acetosyringone and hydroxyacetosyringone (Stachel *et al.*, 1985; for a review, see Dee Cleene, 1988) or, with a similar efficiency, by a dilute cocktail of plant phenolics (Bolton *et al.*, 1986). It is noted that some of these phenols, such as 4-hydroxybenzoic acid, did not function as *vir* gene inducers in other test systems (Melchers and Hooykaas, 1987). The induction takes up to 16 h to reach a maximum, in which time two polypeptides (VIRD1 and VIRD2) coded for by the *virD* locus are induced. They are endonucleases with specific affinity for 25 bp direct imperfect repeats confining the borders flanking the T-DNA region (Ward and Barnes, 1988; Yanofski *et al.*, 1986). The VIRD1 proteins of octopine and nopaline Ti plasmids have been recognized to contain a putative DNA binding domain with the common amino acid sequence Ala-(N)3-Gly-(N)5-(Ile/Val) (Melchers and Hooykaas, 1987). Analysis of over-produced VIRD1 protein in *E. coli* showed that it had type 1 topoisomerase activity (no requirement for ATP, in some cases stimulated by $Mg^{2+}$) which catalyzes the conversion of supercoiled DNA to its relaxed form (Ghai and Das, 1989). The net result of their activity is that the endonucleases introduce nicks at the sites of the 25 bp repeats which leads to the polar synthesis of single-stranded free DNA molecules, the T-strands (Albright *et al.*, 1987; Stachel *et al.*, 1986; Timmerman *et al.*, 1988; Veluthambi *et al.*, 1988). The start or 5′ end of a T-strand can map to one of various borders including, and under certain conditions probably preferentially, to what has traditionally been named the right border (RB) repeat. High efficiency in T- strand synthesis relies on a naturally present enhancer, 'overdrive', located outside the T-DNA region but near this RB repeat (Peralta *et al.*, 1986; Toro *et al.*, 1988; Van Haaren *et al.*, 1987; Wang *et al.*, 1987). The enhancer can function with equal efficiency in *cis* up to 6714 bp away from the repeat (Van Haaren *et al.*, 1987). The generated free T-strand molecules are the likely DNA intermediates that are transferred to plant cells during transformation. In this process there is a probable further key role for strongly inducible proteins coded for by the *virE* locus. This locus specifies 7.0 and 60.5 kDa polypeptides in octopine Ti plasmids (Winans *et al.*, 1987) and a 64 kDa polypeptide in nopaline Ti plasmids (Hirooka *et al.*, 1987) corresponding to a 69 kDa polypeptide on SDS–polyacrylamide gels. These polypeptides are single-stranded DNA binding proteins, an estimated 350–700 protein molecules being required to completely protect a 20 kb T-strand molecule (Christie *et al.*, 1988; Citovsky *et al.*, 1988; Sen *et al.*, 1989). It has been suggested that the VirE protein T-strand DNA complex is the actual intermediate during DNA transfer from the bacterium to plant cells whereby the protein helps to protect the DNA against bacterial and plant nucleases and perhaps has a further involvement in DNA integration (Citovsky *et al.*, 1988; Gardner and Knauf 1986; Gietl *et al.*, 1987). This proposed (multiple) function may also explain why *Agrobacterium virE* mutants are avirulent or weakly virulent on some plants (Hirooka and Kado, 1986; Hooykaas *et al.*, 1984; Stachel and Nester, 1986). Currently it is uncertain whether circular dervatives of the T-strand or molecules with double-stranded cuts detected in *Agrobacterium* shortly following T-strand induction are important intermediates for transfer into plants cells (see, for example Koukolikova-Nicola *et al.*, 1985; Veluthambi *et al.*, 1988). Maybe they are an inevitable but noncontributory consequence of T-strand synthesis, or alternatively, different *vir* gene inducing conditions favour mechanistically different consequences. At least some of the eleven proteins probably coded for by the *virB* locus, play an active further role in mediating the transfer of T-strand molecules into plant cells, perhaps by directing and catalyzing DNA transfer (Thompson *et al.*, 1988).

One implication of the described mechanism, of relevance to genetic engineering in plants in general, is that the VirE protein may be useful in improving the precision of T-DNA integration in plant cells, either through overproduction in *Agrobacterium* or through use in DNA transformation methods with naked DNA (see also Section 12.2.8 and 12.3.3).

A second important conclusion is that it is mechanistically irrelevant what kind of genes are mobilized from *Agrobacterium* to plant cells by the T-strands but it can be expected, for example, that foreign genes containing regions of strong homology to the 25 bp repeats have a good chance of being transferred efficiently but incompletely. Furthermore, the border sequences do not have to be located on the same plasmid as the *vir* genes; *vir* gene products work successfully *in trans* on borders introduced into the *Agrobacterium* chromosome or onto separate plasmids (Hoekema *et al.*, 1983, 1984).

The demonstration that inducers of *vir* gene expression such as acetosyringone are also inducers of chemotaxis, thus causing agrobacteria to positively migrate towards wounded plant cells (Parker *et al.*, 1987; Show *et al.*, 1988; Spencer and Towers, 1988), helps to illustrate further that the success and efficiency of the *Agrobacterium* transformation process depend on the one hand on optimization of individual steps, and on the other hand on bringing together and tightly integrating a variety of seemingly unrelated biological processes. It is noted that some discrepencies exist in the relative importance of the chemotaxis inducers. This may point to genetic differences between *Agrobacterium* strains and remains to be resolved. The induction by the same *vir* gene inducers of the bacterial cytokinin biosynthesis locus *tzs* (John and Amasino, 1988) illustrates in a similar way the integration of a variety of approaches that make *Agrobacterium* efficient in tumour formation. In this latter example it is conceivable that exogenous application of cytokinins to the engineered cells by *Agrobacterium* helps to stimulate them to grow, although perhaps in certain plants only.

### 12.2.6 Ti and Ri T-DNA Genes

Relevant to plant cells transformed by *Agrobacterium* are the T-DNA genes between the border sequences. Upon introduction and integration into the plant's nuclear genome their expression may considerably affect plant cell biology. The two key features in the biology surrounding the transfer of the natural octopine T-DNA genes can briefly be summarized as follows. First, T-DNA encodes several proteins influencing cell growth, of which the most notable are enzymes enhancing endogenous auxin and cytokinin biosynthesis (for reviews, see Memelink *et al.*, 1987; Weiler and Schroder, 1987). The corresponding genetic loci were identified initially in tumour induction experiments using *Agrobacterium* mutants with insertions in the T-DNA regions (Garfinkel *et al.*, 1980, 1981; Leemans *et al.*, 1982; Ooms *et al.*, 1980, 1981). Tumours induced on specific test plants showed characteristic morphologies; for example one type of tobacco tumour showed additional root development, and another additional shoot development. In agreement with earlier experiments, demonstrating that high or low auxin-cytokinin ratios in the culture medium favoured root or shoot formation from cultured tobacco tissues (Skoog and Miller, 1975), it was found the tumours with roots were induced by agrobacteria with a mutation in the T-DNA cytokinin gene, leaving the T-DNA auxin genes intact. Conversely, tumours with shoots were induced by bacteria with mutations in the auxin genes. When none of the T-DNA genes were mutated the transformed cells grew into undifferentiated tumour tissue, whereas mutations in both types of genes made the bacteria avirulent.

The second distinct contribution by T-DNA genes to *Agrobacterium* and octopine type crown gall biology, is determined by two genes, one encoding the constitutively produced enzyme octopine synthase (Garfinkel *et al.*, 1981), which is responsible for the biosynthesis of all members of the class of octopine type opines, and the other for a permease for active secretion from transformed cells of opines such as octopine and nopaline (Messens *et al.*, 1985). Putting the hormonal and opine genes together on a single T-DNA region means that upon transfer from the bacteria, transformed plant cells proliferate into a growing cell mass and actually secrete their constantly synthesized opines. The opines are then ready for uptake and catabolism by the surrounding agrobacteria. This sequence of events is a further illustration of the tight functional integration that has evolved under natural conditions of the various *Agrobacterium* genetic engineering features which thereby synergistically serve a particular purpose.

Although the above description of the main aspects of the biology of the *Agrobacterium* system is meant to illustrate what is within reach of a plant genetic engineering approach, it is the molecular detail of the system which forms the foundation for its current impact on many areas of genetic engineering of plants and cell cultures. For example, the natural T-DNA transfer system has been adapted into a variety of versatile plant transformation vector systems. Genes affecting growth properties are widely used as markers for screening the susceptibility of plants to *Agrobacterium* transformation, and also sometimes as molecular tools for studying plant development and metabolism. Opine genes have been used as reporter genes to provide confirmation that transformed cells were indeed transformed, and various T-DNA gene elements are in use in chimaeric gene constructions and in gene expression studies. Therefore, further details of the structure and functional organization of the T-DNA regions of the best characterized Ti and Ri plasmids pTi Ach5 and pRi 1855 will be discussed for further reference and are illustrated in Figure 3.

It can be seen that both pTi Ach5 and pRi 1855 contain two distinct T-DNA regions, each bordered by direct imperfect 25 bp repeats. Because of their relative position on the traditional restriction endonuclease maps, the T-DNA regions have been called $T_L$-DNA and $T_R$-DNA. The entire sequence of the T-DNA regions has been determined except for that of Ri $T_R$-DNA (Barker *et al.*, 1983; Gielen *et al.*, 1984; Slightom *et al.*, 1986). Ti $T_L$-DNA is 13.1 kbp; Ti $T_R$-DNA 7.9 kbp and Ri $T_L$-DNA 19.4 kbp. Nine open reading frames (ORF's) longer than 380 bp were found in Ti $T_L$-DNA and eight of these corresponded with RNA transcripts detected in crown gall tissue (Gelvin *et al.*, 1982; Willmitzer *et al.*, 1982). The transcripts are low in abundance, ranging from 0.005 to less than 0.001% of the total cellular poly A-RNA. This is roughly equivalent to a maximum of five steady state RNA molecules per cell. Already in the earlier studies fluctuations were noted in abundances of some of the transcripts which related to differences in growth conditions of the cell lines implying differences in promoter activities or RNA stability.

The functions and properties of a number of Ti $T_L$-DNA expression products are known. Transcripts 1 and 2 encode a hydrolase and an oxygenase respectively, that specifically catalyze the biosynthesis of auxins such as indoleacetic acid (IAA) from tryptophan *via* the intermediate indoleacetamide (Weiler and Schroder, 1987). The two transcripts are barely detectable, if at all, on Northern blots. This is partly because of their relatively large size (Figure 3), which often results in some degradation, and partly because of their extremely low abundance. The corresponding loci have

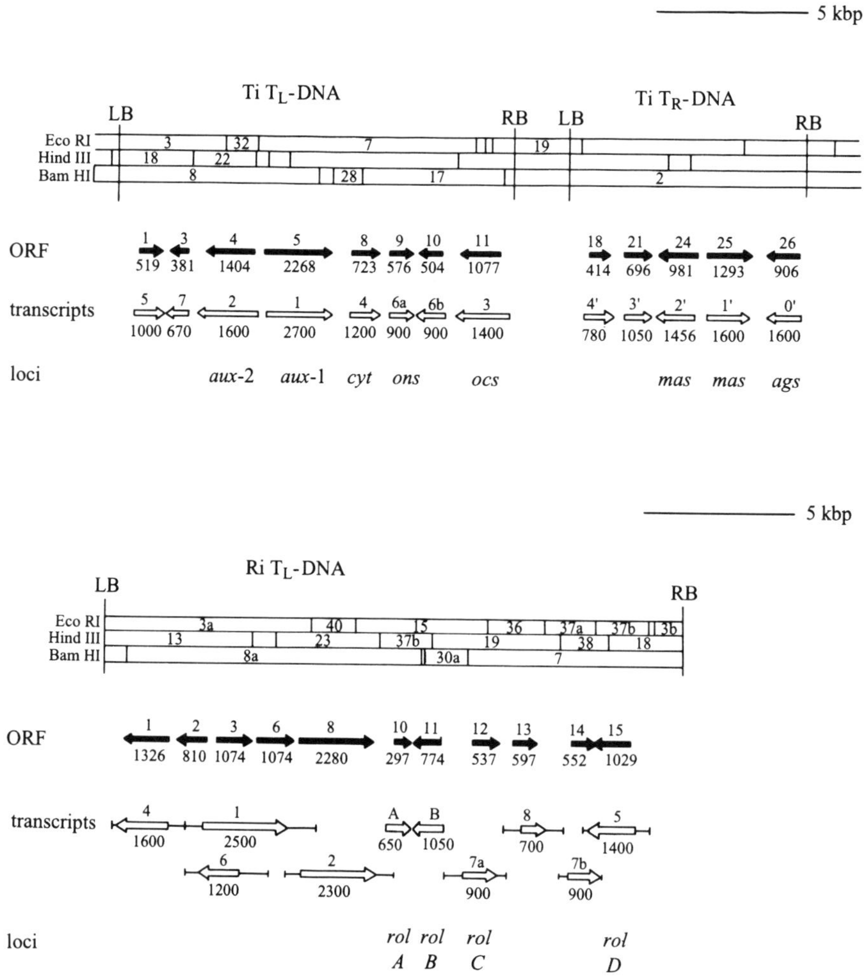


**Figure 3** Detailed physical and functional maps of the best studied T-DNA regions in an octopine-type Ti plasmid, typified by pTi Ach5, and in the Ri plasmid pRi 1855

been given the trivial names *aux*-1 and *aux*-2 (also used in this review) or the designations *iaaM* and *iaaH* emphasizing their corporate action in IAA biosynthesis, or *tms*-1 and *tms*-2 relating to tumour morphology shoots in genetic studies. Designations no longer in use are *onc* (oncogenes) or *att* (attenuated tumour growth induced by mutants). Introduction into tobacco of either the *aux*-1 or *aux*-2 gene did not noticeably alter the phenotype of regenerated transformed plants (Budar *et al.*, 1986a) although transgenic *aux*-1 plants contain high indolacetamide levels.

Transcript 3 encodes octopine synthase. It is constitutively transcribed from the *ocs* locus and forms approximately 0.5% of the cellular poly A-RNA. Based on the high Km of octopine synthase for lysine it has also been called lysopine dehydrogenase (EC.1.5.1.–), but this name is less commonly used (Otten and Schilperoort, 1978).

Transcript 4 encodes dimethylallyl pyrophosphate 5′ AMP transferase or isopentenyltransferase, which catalyzes the rate-limiting step in the biosynthesis of a broad range of biologically active and inactive cytokinins. (Letham and Palni, 1983). Following similar rationale as for the *aux* loci, the locus of the cytokinin gene has been named *cyt*, *ipt* or *tmr*. Shoots regenerated from cells transformed with this gene typically do not form roots and have reduced apical dominance.

The functions of two genes for transcripts 5 and 7 are not known and there is no obvious biological explanation why transcript 5 is differentially expressed in different organs of transgenic plants (Koncz and Schell, 1986). The two remaining Ti $T_L$-DNA genes, for transcripts 6a and 6b respectively, determine octopine and nopaline secretion (*ons*; Messens *et al.*, 1985) and a protein which perhaps modulates cytokinin or auxin activity in transformed cells (Hooykaas *et al.*, 1988).

Ti $T_R$-DNA encodes five transcripts of which two, transcripts 1′ and 2′, are involved in mannopine biosynthesis and one, transcript 0′, in subsequent agropine biosynthesis (Ellis *et al.*, 1984; Karcher *et al.*, 1984; Komro *et al.*, 1985; Salomon *et al.*, 1984). The functions of the remaining two genes is unknown but none of the $T_R$-DNA genes are essential for tumorigenesis (Komro *et al.*, 1985; Ooms *et al.*, 1982; Salomon *et al.*, 1984). It is not surprising therefore that octopine-type crown gall cells invariably contain Ti $T_L$-DNA and sometimes $T_R$-DNA.

For normal hairy-root growth induced by infection of wounded plant cells with *Agrobacterium rhizogenes* both Ri and $T_L$-DNA and $T_R$-DNA are required. Inactivation by insertional mutagenesis of any single Ri $T_L$-DNA gene does not abolish virulence although mutations at some loci have an effect on the response (White *et al.*, 1985). In this way four root loci, *rolA*, *rolB*, *rolC* and *rolD*, were identified as important. The effect of the *rolB* locus is perhaps the most dramatic since it appeared essential for root induction on Kalanchoe leaves by *A. rhizogenes* and upon introduction into transgenic tobacco it caused more pronounced changes in morphology than either the *rolA* or *rolC* loci (Spena *et al.*, 1987).

Typically, plants regenerated from hairy-roots have wrinkled leaves and more abundant plagiotropic root formation (Ackermann, 1977; Ooms *et al.*, 1985c; Tepfer, 1984). In hairy roots the Ri $T_L$-DNA coded RNAs vary in abundance with levels of up to 0.01% of the poly A-RNA in roots for the most abundant transcript 8 (Durand-Tardif *et al.*, 1985; Ooms *et al.*, 1986b; Taylor *et al.*, 1985). In plants regenerated from the hairy roots nearly all transcripts are differentially expressed in different organs. The molecular functions of those Ri $T_L$-DNA genes with an effect on cell growth are largely unknown although it is clear that their expression products directly or indirectly modify cell membranes, probably through an involvement in signal transduction of plant hormones, perhaps in modulating auxin receptor activity (Ooms *et al.*, 1986a; Shen *et al.*, 1988).

In contrast with Ri $T_L$-DNA, transposon insertions in two Ri $T_R$-DNA genes, which are structurally and functionally homologous to the *aux*-1 and *aux*-2 genes of the Ti $T_L$-DNA, do make *A. rhizogenes* avirulent (Offringa *et al.*, 1986; White *et al.*, 1985). This avirulence can be overcome by exogenous application of auxin during transformation, which indicates that auxins produced by Ri $T_R$-DNA expression products provide the initial stimulus for Ri $T_L$-DNA transformed cells to grow (Cardarelli *et al.*, 1987). This is consistent with the general observation that hairy root cultures established from such transformations invariably contain Ri $T_L$-DNA and may or may not have Ri $T_R$-DNA. In one exceptional study using two tomato genotypes (*L. esculentum* line MSK and cv Money Maker) plants transformed only with $T_R$-DNA were isolated for which there is no obvious explanation (Van der Mark *et al.*, 1988). Cell lines with $T_R$-DNA may synthesize opines related to and including mannopine and/or mannopine plus agropine, demonstrating that biosynthesis of these compounds is Ri $T_R$-DNA coded. However, no extensive sequence data, and little RNA data are available for the Ri $T_R$-DNA region.

The observation that the presence of either Ti or Ri $T_R$-DNA determined properties in crown gall or hairy root cells is optional, illustrates that the mere presence of genes in between T-DNA border sequences in *Agrobacterium* is no guarantee that the transfer-integration mechanism ensures stable introduction into plant cells or that, once introduced, the genes are faithfully expressed.

The analyses of plants and cell lines transformed with Ti or Ri T-DNA genes are still virtually the only examples of independent introductions into plants of moderately extensive blocks of genes. Indeed, examination of the spectrum of phenotypes of transformed plants and cultures both in terms of morphological variation due to cell growth-related gene expression, and in terms of opine content, has provided the most extensive data available to date on transfer of blocks of genes into plant cells.

### 12.2.7 T-DNA Transformed Cultures and Plants

Many of the pertinent features of crown galls and hairy-roots were identified well before it was recognized that these tissues arise through genetic engineering. For example, Jensen (1910), in the first decade of this century, demonstrated autonomous growth of sugar beet crown gall tissue, isolated from field-grown plants, by serial grafting onto visibly distinct redbeets. Apparently unaware of Smith's experiments in America, he only later demonstrated that his sugar beet tissues were free of the inciting *Agrobacterium*. Without current understanding of the underlying transformation mechanisms and appropriate technology he would not known, nor been able to demonstrate, that the tissues he was working with were in all probability transformed, and at some stage may well have been homogeneous and derived entirely from a single transformed cell. In other words he may have inadvertently isolated and propagated the first genetically engineered plant cell lines. Similarly, Braun and his colleagues in their early studies on the recovery of the tumourous state (Braun and Wood, 1976; Turgeon *et al.*, 1976) were not then in a position to provide molecular proof of genetic transformation for the Havanna 38 tobacco plants that they regenerated from crown galls induced by the shoot inducing nopaline type *Agrobacterium* strain T37. Examination of over 2000 T37 Havanna 38 shoots isolated from cloned tissues by grafting onto root stocks of the distinct Turkish tobacco, however, gave at least two interesting observations that are as relevant now as they were then. First, a wide range of phenotypes was found between individual transformed lines, the possible causes and implications of which are discussed below. Second, a number of the transformed lines, typically with stunted shoots, or teratomas and basal callus formation but normally without roots, could, upon grafting, develop into quite normal plants that flowered and set viable seeds. The distinct shoot morphology and the later molecular analyses of opines and T-DNA demonstrated that the plants were transformed and hence qualified as the first genetically engineered plants.

A combination of molecular, cellular, cytological and genetical approaches has, over the last decade or so, made it clear that much of the variation amongst the crown gall-derived plants can be explained by one or a combination of several possibilities. These include: (i) incomplete T-DNA transfer; (ii) modification of T-DNA following integration; (iii) modification of host-genomic DNA at or subsequent to DNA integration; (iv) position effects at the site of T-DNA integration influencing expression of introduced genes; (v) occurrence of multiple T-DNA inserts at one or more loci; (vi) introduction of T-DNA inserts of different origins, either from a single *Agrobacterium* plasmid or from different bacteria.

(i) The integration of incomplete or short T-DNA segments thereby omitting some genes from effective transfer (for examples see: Deroles and Gardner, 1988a; Fraley *et al.*, 1984; Ooms *et al.*, 1982b; Peerbolte *et al.*, 1986a, 1986b; Van Lysebettens *et al.*, 1986) was perhaps best publicized by the example of plant rGV1 (De Greve *et al.*, 1982). It was regenerated from greenish tumours covered with numerous shoots that were induced on *Nicotiana tabacum* by wound infection with shoot-inducing *Agrobacterium tumefaciens* strain pGV2100. Out of 250 shoots from nine independent tumour lines, rGV1 was the only morphologically normal and fertile plant producing octopine. It was shown to contain only about 2 kb of the Ti $T_L$-DNA and hence, rGV1 was probably the first genetically engineered plant shown to be transformed that was also morphologically and genetically normal.

(ii) Modification of the T-DNA following integration. This includes methylation leading to suppression of gene expression (Amasino *et al.*, 1984; Gelvin *et al.*, 1983; Hepburn *et al.*, 1983; Peerbolte *et al.*, 1986b; Sinkar *et al.*, 1988; Van Slogteren *et al.*, 1984). Sometimes (nearly) all T-DNA genes are suppressed and sometimes only a number, perhaps to differing extents. These studies have provided subsequent examples of T-DNA demethylation leading to resumption of T-DNA gene expression. This has been achieved by addition to culture media of derivatives of the demethylation-inducing chemical 5-azacytosine, such as 5-azacytidine, through crosses (Matzke *et al.*, 1989; see below) or, in particular examples, upon grafting of transformed shoots onto untransformed root stocks (Klaas *et al.*, 1989; Van Slogteren *et al.*, 1984). Suppression of gene expression was found also by Deroles and Gardner (1988b) in their inheritance analysis of 104 kanamycin resistant petunia 'Mitchell' plants. Forty five of the plants had progeny with lower expression of the kanamycin resistant gene than found in the parental plants. The suppression was not necessarily uniformly induced at meiosis since a number of plants showed differential results between seed capsules on the same plant. In a recent study on sequential transformations of tobacco, a plant transformed with first a kanamycin resistance plus nopaline synthase gene (T-DNA I), and then with a hygromycin resistance plus an octopine synthase gene (T-DNA II), Matzke *et al.* (1989) found that introduction of the second set of genes often gave suppression, by methylation, of the first set. In subsequent progeny plants with T-DNA I only, gene expression was restored up to normal levels, irrespective of whether the plants were obtained from self-fertilization or from backcrosses. In their study, 5-azacytidine treatment did not reverse methylation. In conclusion, suppression of the expression of introduced genes can occur frequently, but the underlying molecular causes are as yet unknown.

Another type of identified modification of T-DNA following integration involves gross structural changes. These were seen in somatic cells following subculturing (Peerbolte *et al.*, 1987), upon fusion of crown gall cells with normal cells in somatic hybrid lines (Ooms *et al.*, 1982b) and in meiotic cells in flowering, transformed tobacco plants (Yang and Simpson, 1981). In all cases significant portions of the T-DNA were lost. A further example of an extreme structural change was observed during regeneration of transformed potato plants from hairy-root cultures, when an occasional morphologically normal plant was recovered. The loss of the characteristic hairy-root phenotype was correlated with the loss of one entire chromosome (Ooms *et al.*, 1986a). These structural changes in established transformed cell lines can be regarded as examples of somoclonal variation which is not unique for T-DNA (Larkin and Scowcroft, 1981). It must be emphasized however, that in many cases one would expect a high degree of stability of foreign genes once introduced into plants by *Agrobacterium*. For example, screening of germinating seeds obtained from backcrosses using two independently isolated transgenic plants, each with single T-DNA copies carrying a kanamycin resistance gene, gave only 6/45 000 and 25/45 000 kanamycin sensitive plants which corresponded to reversion rates of 1.3 and 5.6 per $10^4$ gametes (Müller *et al.*, 1987). This contrasts with commonly observed reversion frequencies at meiosis of around $1\%$ for transgenic organisms, including plants, transformed using traditional DNA uptake protocols (Potrykus *et al.*, 1985a; see also Section 12.3.3). Therefore, the extreme example of loss of T-DNA coded traits correlated with loss of an entire chromosome, overlaps with examples of the third type of possible causes for phenotypic variability amongst plants or lines from crown galls.

(iii) Modification of genomic DNA during or following T-DNA integration. Further examples of this include the karyological analysis of 15 undifferentiated and organogenic *Nicotiana* crown gall lines which showed considerable aneuploidy, *i.e.* deviations from the normal chromosome number. There were also gross chromosomal rearrangements, in particular the occurrence of dicentric chromosomes (Mouras *et al.*, 1987). Both the induction of numerical changes and of dicentric chromosomes had been observed earlier in Ti T-DNA transformed potato plants although not as frequently (Ooms *et al.*, 1983). More subtle examples of changes in phenotype of transgenic plants are those caused by the integration event of T-DNA itself (insertional untagenesis). Possible examples of T-DNA as a mutagen, have been described by André *et al.* (1986) and solid mutants, in particular an *Arabidopsis* dwarf mutant, were isolated by Feldmann *et al.* (1989).

(iv) A fourth possible cause of phenotypic variation amongst lines or plants from crown galls is the effect of the surrounding DNA on expression of introduced genes. This is most clearly illustrated in gene expression studies whereby *Agrobacterium* mediated transfer of, for example, various light-regulated and storage protein genes gave substantial variation between independent transformants in levels of expression of the genes. In fact, only few such studies, in which *cat* was used as a reporter linked to promoter fragments of a beta conglycinin gene from soybean (Chen *et al.*, 1986), a class II patatin gene from potato (Twell and Ooms, 1987) and a pea vicillin gene (Higgins *et al.*, 1988), little (3–4 fold) variation in expression was found. Likewise in only few studies there was a correlation between copy number of introduced genes and levels of expression (Hamill *et al.*, 1987b; Stockhaus *et al.*, 1987). A specific example of only one out of many studies where a larger extent of variation (50–100 fold) was observed between independent transgenic tissues in expression of the introduced gene is a study on T-*cyt* expression (De Pater, 1987). It is noted, that in the case of T-DNA transformed cells in general and T-*cyt* transformations in particular, there may be selection involved in the choice of the material analyzed. It is likely that low levels of expression may be insufficient for transformed cells to gain a selective advantage which, unless the cells are selected in another way may exclude them from isolation. Alternatively, too much expression is likely to be detrimental. Hence, the extent of variation may be wider than that determined.

As yet, little accurate information is available on variation in expression of T-DNA as a tightly linked set of genes between independently transformed lines although comparison of results from different studies on Ri $T_L$-DNA gene expression with at least nine transcripts, suggests a perhaps surprising similarity in their relative expression patterns, even in different plant species.

The occurrence of position effects, by implication, relates to the possibility of preferences for specific integration sites of T-DNA in the plant genome, and this is discussed below (Section 12.2.8) as is the fifth possible cause of variation between transformed lines, (v) the introduction of multiple T-DNA inserts, often at a particular locus.

Additional causes of phenotypic variation between transformed plants or cell lines are not discussed further but these include: (vi) introduction of multiple T-DNAs of different origins, for example, from mixed infections with different agrobacteria, from infections with bacteria containing multiple, different plasmids or plasmids, including cointegrate plasmids, with multiple T-DNAs; and (vii) growth of heterogenous cell mixtures into lines or plants which therefore are chimaeric. The list of cases of phenotypic variation resulting from transformation can be extended even further when knock-on effects resulting from expression of introduced genes are included. As mentioned earlier (Section 12.2.2) the host plant and the particular cell type influence the newly acquired properties of transformed cells, and an interesting example of this is found in the dioecious species *Mercurialis annua* L. Male and female individuals responded differentially to *Agrobacterium* infection which correlated with different endogenous cytokinin contents between the tumours, and a feminizing effect by male tumours on the male flowers (Guerin *et al.*, 1984).

### 12.2.8 Structure and Position of Transferred DNA

Our general knowledge about the regulation of T-strand synthesis and the exact route it takes until it ends up as integrated DNA in a plant cell (Section 12.2.5) is rapidly expanding but nevertheless still limited. More, and in some ways complementary information is available on the consequences of T-DNA transfer. This provides answers to questions like where it ends up and how it is structured after integration. Normally, the DNA transferred from *Agrobacterium* becomes integrated in the nuclear genome. Transfer into chloroplast DNA has been observed (De Block *et al.*, 1985), but stable plasmid transformation proves difficult (Cornelissen *et al.*, 1987). Integration in the nuclear genome appears random with no macro preference for particular chromosomes or chromosomal regions, although perhaps there is micro preference for integration near particular sites (see below). Often multiple T-DNA copies (usually one to five but occasionally even more than twenty) are present per cell at the site of integration and normally they have a head to tail arrangement and differing unit length. Cloning and sequence analysis of border regions, either of the internal borders where left and right halves of two T-DNA copies are joined, or of the external borders forming the junctions between plant and transferred DNA, have shown that the borders are normally scrambled (Holsters *et al.*, 1983; Simpson *et al.*, 1984; Zambryski *et al.*, 1980). The RB junctions in plant cells are usually close to the RB 25 bp repeat (two thirds of 17 junctions tested in a particular study within 11 bp of the RB repeat; Jorgensen *et al.*, 1987) whereas the LB junctions are usually more heterogeneous. This is consistent with the proposed precision of the start of T-strand synthesis in *Agrobacterium* at the RB 25 bp repeat and the more variable termination of T-strand synthesis or, alternatively, with better protection of the RB border sequences by VIRD proteins.

In this context observations by Meyer *et al.* (1988) on simple integration patterns following transformation of petunia protoplasts with circular DNA (direct gene transfer) are relevent. Normally, direct gene transfer particularly using linear DNA and carrier DNA gives complex integration patterns (see below) but in their case, transformation with circular DNA of protoplasts synchronized into M phase or transformation with circular plasmids containing a specific genomic DNA fragment (TBS or transformation booster sequence) of asynchronized protoplasts both gave simple DNA integration patterns which were similar to those often seen following *Agrobacterium* transformation. Possible implications that follow for cases where *Agrobacterium* transformation gives simple integration patterns are either that *Agrobacterium* has the capacity to recognize cells in M phase specifically, or that a high proportion of cells that are transformed are in M phase. If so, it would follow that much of the observed variability in T-DNA integrations is accounted for by the exact timing and extent of duration of events leading up to DNA transfer by the bacterium and changes in cell cycle in the plant cell.

Indeed, results from earlier studies on tobacco crown gall cell growth indicated that at least some of the properties of transformed cells were determined by the method of isolation. For example it depended on whether *Agrobacterium* transformations were carried out by infecting wound sites of intact plants or by cocultivating with *in vitro* cultured protoplasts, whether there was a low or high frequency respectively of apparently irregular T-DNA structures and of phenotypic variation among cultured *Nicotiana tabacum* lines (Marton *et al.*, 1979; Ooms *et al.*, 1982b; Wullems *et al.*, 1981). This was confirmed and quantified in detail by Van Lysebettens *et al.* (1986), who suggested that, in particular, short T-DNAs are characteristic of the cocultivation method. On the other hand, studies by Van Slogteren *et al.* (1983) and Peerbolte *et al.* (1986a, 1986b) also showed considerable heterogeneity in morphology and T-DNA structure between lines, subcloned *via* protoplasts, that were originally obtained from wound infections. Subsequently, it has become particularly clear from a combined genetic and molecular analysis of transgenic plants with normal morphologies, that multiple and unlinked integration sites can be present within a genome, although again with dependence on the isolation methods used. For example, in one study in tomato, seven out of eleven lines tested had multiple integration sites, each with T-DNA copies arranged in repeats (Chyi *et al.*, 1986). The data from this study however, are not necessarily typical. Examination of nine transgenic tobacco lines produced by inoculation of leaf disks with *Agrobacterium tumefaciens* and a disarmed binary vector, containing soybean leghemoglobin Lbc3 and glycinin G2 genes, showed introduction of one to six DNA copies from the vector (Spielmann and Simpson, 1986). Thirty percent of the copies were incomplete or rearranged, and in five out of the six cases examined, these were found at multiple genetic loci, in one case at four positions. In a further example of a study with contrasting results, transformations were carried out using a binary vector and *Agrobacterium* infection of stem segments of the interspecific tomato hybrid *Lycopersicon esculentum* (cv VF36) × *L. pennelli* (A 716). Here, each of ten plants examined had an insertion at a single locus only, and the loci were rather randomly distributed over the genome (Sukhapinda *et al.*, 1987a). Confirmation of this lack of evidence for nonrandom integrations was also obtained in studies from *Nicotiana plumbaginifolia* (Horsch *et al.*, 1984) in petunia (Deroles and Gardner, 1988b; Wallroth *et al.*, 1986) in *N. tabacum* (Budar *et al.*, 1986b; Herberle-Bors *et al.*, 1988), but between these studies differences existed in, for example, the percentage of independent transformants that showed the Mendelian inheritance patterns.

It seems likely that this variation in results between studies in T-DNA structures, T-DNA copy numbers, in the number of sites for T-DNA integration and in the inheritance patterns is real, which suggests that there are relevant differences in the exact details of the transformation protocols used. For example, exact culture conditions of bacteria and plant cells may affect the enzymology of the transformation and integration processes. Conversely, the very existence of variation implies scope for optimization of the transformation process towards particular types of consequences.

It is against this background that the only study published so far on the question of possible molecular changes in the target site in the plant genome before and after T-DNA integration can be placed (Gheysen *et al.*, 1987). Cloning and sequence analysis of a unique 1.8 kb EcoRI fragment from untransformed *Nicotiana tabacum* W38 DNA, enabled a detailed comparison to be made with left and right border junction sequence data obtained earlier from a derivatory transformed *N. tabacum* line with a T-DNA insert in the same 1.8 kb EcoRI fragment. Several types of rearrangements were found of which the most dramatic was a 158 bp direct repeat of plant DNA occurring at both left and right T-DNA junctions. Both repeats had small deletions and inversions at their ends. The authors concluded that their data were not sufficient for the formulation of a precise integration mechanism but a working model was put forward. It postulated that the T-strand was transferred preferentially into the nucleus as a reasonably stable protein DNA complex (involving VirE proteins?) and that the DNA became integrated by means of plant encoded enzymes involved in recombination and local replicative and repair activities. The detection of a 15 bp plant DNA sequence with partial homology to a 16 bp T-DNA sequence located 31 bp internally to the left 25 bp T-DNA border repeat, suggested the involvement in the integration event of contact sequences, as have been proposed for Polyoma and pseudogene integration. Further comparisons by the authors with other mechanisms for DNA insertion into eukaryotic genotypes suggested that T-DNA integration in this particluar event had resulted in structures in between those commonly found in integration events for transposons or retroviruses (with efficient integration always in single units of unaltered structure and without significant consequences for the primary structure of the surrounding DNAs), and integration events for DNA introduced by microinjection or electroporation (with low efficiency of integrations of mostly tandem and often modified arrays, which is commonly associated with considerable alterations in the target DNA). In view of the above described variation in T-DNA structure, it will be of interest to see whether transformations which lead to simple T-DNA integration patterns also show 'clean' integration events which may or may not be associated with high stability (see also Section 12.2.7).

### 12.2.9 Uses of *Agrobacterium* and its Gene Elements

For a rapidly increasing number of plants, genetic engineering experiments in general, and those using *Agrobacterium* in particular, have only recently become a reality. Hence, examples of identification of specific targets for research or application are as many as the list of amenable plant species is long, and are as diverse as the range of problems and the scope for opportunities is wide. It is not surprising therefore that alongside the unravelling of the transformation mechanism, the main genetic engineering aim for *Agrobacterium* has been to adopt it for use as a widely applicable genetic engineering system. Much progress has been made towards the construction of different types of *Agrobacterium*-based, general plant cell transformation systems. This has involved the isolation of suitable *Agrobacterium* strains for plant cell transformation and the construction of compatible vector molecules, convenient for *in vitro* manipulation (see also Chapter 9, and Bevan and Goldsborough, 1988).

We now recognize two types of *Agrobacterium*-derived transformation systems based on differences in the type of vector molecules used. These are cointegrate vectors and binary vectors. Cointegrate vectors, almost by definition, consist of two plasmids which share DNA homology for cointegrate formation. One plasmid is normally an Ri or Ti plasmid or Ti plasmid derivative from which the phytohormone genes are deleted. The second plasmid is normally a mobilizable *E. coli* plasmid unable to replicate autonomously in *Agrobacterium* but with homology to the *Agrobacterium* plasmid so that upon introduction into *Agrobacterium* and under appropriate selection it will form recognizable cointegrates with the *Agrobacterium* plasmid. Availability of cloning sites in the *E. coli* plasmid should ensure general use for cloning any gene, and the correct situation of border sequences in the cointegrate plasmid will facilitate T-strand synthesis and plant cell transformation. Examples of cointegrate vectors are the disarmed *A. tumefaciens* plasmid pGV3850, a derivative of the nopaline plasmid pTi T37 from which much of the central T-DNA was deleted except for the nopaline synthase gene, and the improved plasmid pGV2260, a derivative of pTi B6S3 (Deblaere *et al.*, 1985; Zambryski *et al.*, 1983). Both pGV3850 and pGV2260 contain pBR322 DNA which makes them suitable for cointegrate formation with cloning plasmids that have homology to pBR322 DNA. In particular the combination of pGV2260 with the pBR325 derivative pGV831 has been recommended as an efficient two-component cloning system. This is partly because a consequence of the pGV3850 cointegrate formation is that genes in between T-DNA border sequences will become surrounded by repeats of pBR322-type DNA in transformed cells. This configuration is somewhat susceptible to loss by deletion from transformed plants, in particular during meiosis, and in addition, it complicates T-DNA analysis. A further example of a cointegrate vector is a derivative of a pTi B6S3, pTi B6S3-SE, in which most T-DNA, including all the phytohormone genes plus the right border, is removed whilst retaining the left border with some internal T-DNA (Fraley *et al.*, 1985). This provides a limited internal homology or LIH region of 1.6 kb for cointegrate formation with *E. coli* cloning plasmid (or intermediate vectors) containing the LIH region such as pMON120 and pMON200. The cloning plasmids also contain a right border which ensures that the cointegrate plasmids has both LB and RB DNA to make it an effective vector molecule. This particular vector system was initially named the SEV system (Split End Vector) since LB and RB sequences are present on separate plasmid prior to recombination, but the plasmids are now more commonly referred to as the Monsanto vectors.

What was once an initial step in the construction of cointegrate vectors, is the introduction into pBR322-type plasmids of some Ti or Ri plasmid derived T-DNA so that upon introduction into *Agrobacterium*, containing Ti or Ri plasmids, the *E. coli* plasmid may recombine with the *Agrobacterium* plasmid and form a cointegrate. This approach is still used, for example, in some of the studies of plant genes involved in nitrogen fixation where chimaeric gene construction combined with efficient transformation by *A. rhizogenes* and regeneration of plants from *Lotus corniculatus* hairy-roots enables detailed analysis of *cis*-acting elements and *trans*-acting factors important in nodulation (Forde *et al.*, 1989; Hansen *et al.*, 1989; Petit *et al.*, 1987; Stougaard *et al.*, 1987).

The second type of *Agrobacterium* plant transformation system is based on the concept of binary vectors first established by Hoekema *et al.* (1983). The identification that *vir* genes and T-DNA could reside on separate *Agrobacterium* plasmids and still constitute an efficient gene transfer system, stimulated the construction of a wealth of so-called binary vectors (An *et al.*, 1985, 1986a; Bevan, 1984; Hoekema *et al.*, 1985; Klee *et al.*, 1987a; Koncz and Schell, 1986; Matzke and Matzke, 1986; Muradov and Aliev, 1986; Olszewski *et al.*, 1988; Ondrej *et al.*, 1986; Rogers *et al.*, 1987; Rothstein *et al.*, 1987; Schardl *et al.*, 1987; Simoens *et al.*, 1986; Simpson *et al.*, 1986; Van den Elzen *et al.*, 1985a; Vilaine and Casse-Delbart, 1987). For effective plant transformation, the host *Agrobacterium* must provide the transacting *vir* gene products either by containing intact Ti or Ri plasmids or derivatives thereof that are void of the natural T-DNA sequences. When natural T-DNAs are present, plant cell transformation with both types of T-DNAs can take place. An example of an *Agrobacterium* plasmid depleted of all its natural T-DNAs is pAL4404, in LBA4404(pAL4404) which is a commonly used 'helper' strain (Ooms *et al.*, 1982a). It is noted that induction of *vir* gene expression by diffusable plant products sometimes does not occur, particularly in monocot seedlings (Usami *et al.*, 1987). This can be overcome by addition of acetosyringone to the culture medium or, in principle, by altering the control regions of the relevant plasmid genes.

Advantages of binary plasmids over cointegrate plasmids are the high frequency at which they can be transferred from *E. coli* to *Agrobacterium* (which is around $10^0$–$10^{-1}$ compared with around $10^{-5}$ for cointegrate formation) and the flexibility in choice of *Agrobacterium* backgrounds. This latter point is important in species that are more amenable to transformation to only certain *Agrobacterium* isolates. It must be noted, however, that very occasionally, occurrence of rearrangements in the structure of binary plasmids have been noted during or following conjugation from *E. coli* to *Agrobacterium* and that several of the binary plasmids constructed are only stable if maintained in the presence of selection pressure.

The second type of general use of the natural *Agrobacterium* system, in addition to providing the backbone for vector development, has been as a source of versatile gene elements. For example the nopaline synthase promoter, along with the cauliflower mosaic virus 16S and particularly the stronger 35S promoter (Harpster *et al.*, 1988), are probably the most commonly used plant promoters when constitutive expression of introduced genes is sought. Whether constitutive expression occurs and how to define a constitutive promoter is a different matter. For example An *et al.* (1988) showed that the *nos* promoter linked to the *cat* reporter gene introduced into transgenic tobacco plants was subject to a degree of developmental and organ specific regulation. Polyadenylation sequences 3′ to *ocs* and *nos* coding DNA are commonly used in chimaeric gene construction. The 479 bp common intergenic promoter region between the Ti $T_R$-DNA genes for transcripts 1′ and 2′ has been proposed as a suitable dual promoter for general use in genetic engineering experiments (Velten *et al.*, 1984). If it is introduced in between the coding DNA of a studied gene and that of a reporter gene, thereby initiating transcription of both genes in opposite directions, the two genes will have common 5′ regulatory DNA at the same positions of integration in the plant genome. Examination of expression of the marker gene can therefore be used to some extent to provide information about the expression of the studied gene.

The coding DNA of various T-DNA genes is also of general use in plant genetic engineering. Some genes have been used as reporter genes like the opine synthase genes *ocs, nos, ags* and *mas*. The *aux*-2 gene, for amido hydrolyze (Section 12.2.6), which catalyzes conversion of indole acetamide, α-napthalene acetamide (α-NAM) and various other substances into the corresponding auxins, has been proposed as a unique negative selectable marker (Depicker *et al.*, 1988). Tobacco cells containing the gene can grow in a medium with a low concentration (0.1–1 μM) of α-NAM whereas normal protoplast-derived tobacco cells require 1–20 μM when cultured at high density or 0.05–1 μM when cultured at low density. In media with higher α-NAM concentrations, between 30–300 μM, the extra, *aux*-2 mediated, production of auxins becomes toxic for transformed cells and these fail to grow into colonies even in mixed populations with untransformed cells. Hence, this provides a negative selection scheme of potential use, for example to study the molecular consequences of mutagenic treatments, and gene inactivation such as in certain homologous recombination experiments or in transposon trapping or perhaps in cases where the gene is linked to conditionally regulated promoters to identify genes coding for transacting regulatory proteins which interact with these promoters.

In fact, most or all Ti or Ri T-DNA genes with growth-modifying properties have potentially wide applications for studying plant growth and development (for reviews, see Kuhlemeier and Green, 1987; Ooms, 1987). For example, over- production of auxins or cytokinins in transformed cell lines or plants at particular stages of development or under various environmental conditions is feasible.

### 12.2.10 Conclusions

In conclusion, research into the *Agrobacterium* transformation system has elucidated a tightly integrated highly controlled collection of genetic engineering features. This collection provides probably the best example of sophisticated plant genetic engineering and illustrates what is within reach of the technology. It has also played a major role in the current development of various types of genetic engineering systems based on the *Agrobacterium* transformation process.

At some stage, the aim was to convert the specialized *Agrobacterium* system into versatile and flexible plant cell transformation systems. This has been successfully achieved, in general, in that current *Agrobacterium*-based vector systems are quite adequate to obtain as many transgenic cell lines and plants in several plant species as are required in most normal genetic engineering studies. Often, more lines are generated than can be examined, and thus the actual transformation process itself is no longer a limiting factor or even a factor which warrants substantial efforts towards improvement with prospects for only minor increases in efficiency. This argument, however, does not hold in more specialized cases. In some plant species transformation frequencies are low or perhaps zero. In certain types of experiments extremely high transformation frequencies are desirable, for example when seeking integration *via* homologous recombinations which may occur in only one in a thousand transformation events, in analogy to animal cell transformations. Under these circumstances, further adaptations of the *Agrobacterium* transformation system are likely to help in achieving improvements. It is possible that such improvements will be of general use in most plants but equally, they may only be of more specific use in certain cases and in only one or a few plant species. An example of this approach is research into the supervirulence of *Agrobacterium tumefaciens* strain A281 carrying plasmid pTi BO542. This bacterium induces larger and more rapidly developing tumours and has a wider host range than other *Agrobacterium* strains. The supervirulence has been correlated with increased expression of *vir* genes probably caused by the unique properties of the pTi BO542 *virG* (and *virB*) gene(s) (Jin *et al.*, 1987). Further experiments aimed at combining the most effective natural chromosomal virulence genes and plasmid-located virulence genes (or manipulated derivatives of them) from the *Agrobacterium* gene pool at large into one *Agrobacterium* strain, may contribute to optimization of *Agrobacterium*-based genetic engineering systems. In this context it is relevant to emphasize that *Agrobacterium*-based transformation technology is only one of a wide range of such technologies. Choice of the most appropriate system will often be on a case to case basis depending on plant genes, research aim, ease and availability of technology at a particular location and the anticipated consequences of the transformation system used. For example, mitotic and meiotic stability and structure and expression of integrated DNA and inheritance are set to become the important factors for consideration and can be influenced to some extent by the choice of combination of the transformation and regeneration protocols used.

## 12.3 CHOICES OF TRANSFORMATION METHODS AND PLANTS

### 12.3.1 Introduction

Ever since genetic engineering of plants and plant cell culture has become a firmly established technology, subsequent development have been and still are dominated by choices. In this and the following two sections, factors important in three types of choices are highlighted. The first type relates to the development of a wide range of transformation methods, the evaluation of the consequences of the protocols used and the choice of plants. The second briefly highlights the importance of identification, isolation and the study of expression of introduced genes, and in the third existing and future topics of application are described.

In the development of a wide range of transformation methods, the construction of genes, that enable selection of transformed cells, is important and so is the identification of plants that are amenable to a particular type of genetic engineering technology. In this section a summary is given of various types of selectable marked and the role of tobacco, in particular, in the development of a wide range of current transformation protocols. Finally, progress in the development of transformation methods in a number of agricultural and horticultural plants as well as in several other types of plants is discussed.

### 12.3.2 Selectable Markers

The recognition that the *Agrobacterium* T-DNA hormone genes gave transformed cells a selectable phenotype and concomitantly an altered phenotype in the corresponding regenerated transformed plants, prompted their replacement by engineered plant genes that were at least as efficient as selectable markers, but which had no modifying effect on plant growth and development. Initially, emphasis was placed on antibiotic resistance genes of bacterial origin. Since intact bacterial genes were not functional in plants they were equipped with 5′ and 3′ control regions of natural plant genes or at least of genes known to function in plant cells like the $T_L$-DNA *ocs* and *nos* genes. Among the most successful genes conferring antibiotic resistance to plant cells were the neomycinphosphotransferase type I and II genes which confer resistance to kanamycin and other aminoglycoside antibiotics such as neomycin and G418 (Bevan *et al.*, 1983; Fraley, 1983; Herrera-Estrella *et al.*, 1983). Tobacco and petunia cells transformed with genes comprising the coding DNA of Tn5 plus the *nos* promoter and polyadenylation sequences are usually selected by addition of around 50 $\mu g\,mL^{-1}$ kanamycin to the culture medium. Typically, however, callus growth or root and shoot growth are inhibited at different concentrations of the antibiotic and also to different extents. Other antibiotic resistance genes of bacterial origin that can be used as selectable markers include those conferring resistance to hygromycin (Van den Elzen *et al.*, 1985b; Waldron *et al.*, 1985) bleomycin (Hille *et al.*, 1986), streptomycin (Jones *et al.*, 1987), gentamycin (Hayford *et al.*, 1988) and only poorly to methotrexate (Herrera Estrella *et al.*, 1983). In contrast, conversion of the mouse dihydrofolate reductase (DHFR) into a plant gene gave a useful methotrexate-resistance plant marker (Eichholtz *et al.*, 1987). In practice however, there can be variation in the effectiveness at which the genes confer selection. This is reflected in differences in the 'selection window' *i.e.* the concentration range that can be used to select transformed cells efficiently from untransformed cells. A consequence is variation in the numbers of 'escapes' or untransformed plants amidst the regenerated transformed plants. As in the case of the kanamycin gene, this number is also dependent on the particular tissue or growth stage at which selection is applied. It is further noted that the strength of a selectable marker not only depends on the nature of the coding DNA and the toxicity of the drugs, but also on the strength of the plant promoter used and the position at which the gene is integrated following transformation, both of which affect expression and hence resistance levels (Section 12.2.7).

More recently a wider range of selectable markers for plants has been constructed based on genes that confer resistance to herbicides (see also Section 12.5.3). Examples are resistance to glyphosate and to sulfonylureas by overproduction of native target enzymes or of structurally altered enzyme derivatives (Comai *et al.*, 1985; Haughn *et al.*, 1988; Shah *et al.*, 1986). Alternatively, resistance to phosphinotricin, and bromoxymil was obtained by introduction of detoxification-degradation enzymes (De Block *et al.*, 1987; Stalker *et al.*, 1988; Wohllenben *et al.*,1988). It is evident that this development is a prime example of overlap between scientific and industrial progress. New selectable markers give more flexibility to transformation experiments, perhaps tighter selection and further scope for sequential transformations. Equally, transforming herbicide resistance into established crops is of agronomical and horticultural interest. In particular, the combination with manufacturing of patented herbicides makes it an attractive target for biotechnology and agrochemical industries.

### 12.3.3 Tobacco and the Development of Transformation Methods

Selectable marker genes have been and still are important in testing novel transformation methods. Their importance is two-fold. The first is obvious and trivial. A selectable marker linked to an efficient cell culture system allows selective amplification of transformed cells into either cultures or plants. As important, however, are posttransformation reasons. A selectable marker consisting of foreign DNA enables the establishment of definite proof that transformation has indeed occurred and the evaluation of stability and other consequences of the particular transformation protocol used. Examples of different consequences following transformation using *Agrobacterium* have already been discussed above (see Section 12.2). The coupling of transformation to an efficient cell culture system explains the main interest in *Solanaceous* plants such as tobacco and petunia in developing additional transformation systems. These plant species have superior tissue culture amenability, and with relative ease large numbers of isolated protoplasts can be grown into calli of which a high proportion can be regenerated into fertile plants.

Currently, about a dozen or so distinct transformation methods exist and most have been pioneered using these plants. The most widely used method has been discussed earlier and is based on *Agrobacterium* infection of wounded explants (leaf or stem pieces). The method has been popularized in particular by Horsch *et al.* (1985) and is sufficiently efficient for most common transformation purposes in a steadily increasing number of plant species. Not only is it used to obtain plants and cell cultures in dicotyledonous plants but successful application to transformation in the monocot asparagus has also been reported (Bytebier *et al.*, 1987). The procedures do not generally require much sophisticated science but factors recognized to be important are genotype, environment and cell type (see also Section 12.2). Some examples highlighting differences in responsiveness to *Agrobacterium* by different plant genotypes are in oilseed rape (Holbrook and Miki, 1985; Ooms *et al.*, 1985a) some forage legumes (Armstead and Webb, 1987) and medicago (White and Greenwood, 1986), in pea (Robbs *et al.*, 1987) and soybean (Byrne *et al.*,1987), in the *Cucurbitaceae* (Smarelli *et al.*, 1986; Trulson *et al.*, 1986) and in conifers (Diner and Karnosky, 1987). As outlined in the previous *Agrobacterium* section, screening of wild-type agrobacteria and genetic or experimental manipulation of relevant steps in the genetic engineering process provide a way to improve transformation efficiencies, just as *A. tumefaciens* strain A281 proved superior in transforming legumes (Hood *et al.*, 1987) and poplar (Pythoud *et al.*, 1987). Similarly, the importance of culture conditions in obtaining transformed plants is highlighted. For example in *Arabidopsis* and *Dioscorea* addition of acetosyringone helped to enhance transformation efficiency (Schafer *et al.*, 1987; Sheikholeslam and Weeks,

1987). In tomato, selection of favourable culture conditions proved useful in establishing an efficient transformation-regeneration protocol (Chyi and Phillips, 1987).

Three variations on the natural *Agrobacterium* transformation system of explant infections have been developed. One is known as cocultivation (Marton *et al.*, 1979). *Agrobacteria* are mixed with cell wall regenerating protoplasts and transformation occurs, most likely, by the same mechanism as in explant transformation but, again as noted earlier, the consequences as for the structure and expression of the introduced genes are probably influenced by the exact protocol used. A second variation is extended cocultivation (Fraley *et al.*, 1984; Muller *et al.*, 1984; Pollock *et al.*,1985). Here, agrobacteria are mixed with small rapidly growing cell clusters or calli. This superficially resembles the third variation of infecting germinating seeds and somatic embryos which has been used successfully in transforming *Arabidopsis* (Feldmann and Marks, 1987) and walnut (McGranahan *et al.*, 1988). In these cases the obtained transformed lines are not necessarily clonal or of single cell origin but this can be rectified, by sexual propagation and by repeated embryogenesis. Protoplasts also feature in the development of other transformation protocols. Treatment with chemicals such as polyethylene glycol (PEG) or calcium together with a high pH or electrical stimuli make them readily take up nucleic acids from the surrounding culture media — the direct gene transfer method (Krens *et al.*, 1982; Shillito *et al.*, 1985; see, for review, Day and Lichtenstein, Chapter 9; Potrykus *et al.*, 1985a). DNA has also been introduced by microinjection into immobilized cells, for example into nuclei of alfalfa protoplasts (Reich *et al.*, 1986) and microspore-derived embryoids in oilseed rape (Neuhaus *et al.*, 1987) by rehydration of dessicated embryos (Töpfer *et al.*, 1989) or by fusion of plant cells with either liposomes (Deshayes *et al.*, 1985) or *Agrobacterium* and *E. coli* spheroplasts (Hain *et al.*, 1984; Hasazawa *et al.*, 1985) containing the plasmid DNA of interest.

Two further established transformation methods have attracted considerable interest for different reasons. Transformation of rye by simply injecting a DNA solution into developing tillers at the right stage of development (de la Pena *et al.*, 1987), suggested an easy breakthrough for the agronomically important cereals. However, despite considerable efforts, more widespread success and even reproduction of the initial results as yet have not been forthcoming. The second transformation method of intrinsically wide applicability is based on direct penetration of the plant cell wall by incorporating the DNA into small tungsten or gold bullets which can be fired as high velocity microprojectiles into the cells, a method for which the term 'biolistic process' has been proposed (Sanford, 1988). The initial particle acceleration is provided either by gunpowder (Klein *et al.*, 1987) or by an electrical discharge which was successfully used in the isolation of transgenic soybean (McCabe *et al.*,1988). It has the promise of transformation of any type of somatic cells, or any type of organelle if it comes to that, but in principle also of any germ cell or embryonic tissue in any plant species. This latter prospect has the added advantage of potentially bypassing the induction of additional genetic changes that may occur when somatic cells are regenerated into plants (somaclonal variation) or it may be applicable to those plants where other methods fail. Finally, it has been shown that expression of foreign genes in plants can also be brought about by engineered viruses (Brisson *et al.*, 1984). In combination with the systemic spread of viruses throughout a plant, following localized infection by one or several of a range of methods, this emphasized the potential to achieve substantial production of foreign RNAs or proteins. Of course, for all these methods further evaluation of the consequences of the transformation procedures will be required to enable a more complete comparison.

### 12.3.4 Crop Plants

Alongside the development of a broader range of transformation protocols, attention has focused on transformation of plants of particular interest, most notably horticultural and agricultural crops. Broadly speaking, the initial attention in transformation was directed towards obtaining transformed cells of any kind, mainly to demonstrate principles. For example, potato transformed with Ti T-DNA from a shoot inducing *A. tumefaciens* mutant was probably the first example of a genetically engineered derivative of an established variety of a crop plant, although phenotypically it was somewhat abnormal, because of the introduced T-DNA. It was prone to tuberize (Ooms *et al.*, 1983). Similarly, transformation of protoplasts of a *Triticum monococcum* culture established the principle of transformation of wheat cells (Lörz *et al.*, 1985). Of course the latter example in particular is remote from a transformed wheat plant, let alone from a stably transformed agronomically improved derivative of a wheat cultivar or breeding line. But principles were established and scope for possibilities illustrated.

In establishing the principles, transformation by *Agrobacterium rhizogenes* has played and still does play an important role (Figure 1). In a single recent study (Mugnier, 1988) the original list of species in which hairy-root lines have been established was estimated at around 30 and was extended to approximately 70 species. Only dicotyledonous species belonging to the *Ranunculaceae* and *Papaveraceae* did not yield hairy-root lines. The study confirmed that for various purposes the ability to isolate cell lines and plants transformed with Ri T-DNA is worthwhile and can be readily achieved. The benefits of *A. rhizogenes* transformation in isolating transformed plants include the ease and speed in obtaining (clonal) transformed tissues (hairy-roots) which often are particularly amenable to regeneration into plants. These advantages have to be balanced against the possible disadvantages of Ri T-DNA associated changes in growth and development and the frequently observed abnormalities in the meiotic behaviour of the transformed plants.

In exploring scope for transformation, particularly in crop plants the Ri system has therefore attracted attention and a substantial number of Ri T-DNA transformed plants have been isolated. Examples are in carrot (Chilton *et al.*, 1982), tomato (Morgan *et al.*, 1987; Shahin *et al.*, 1986a), potato (Hanisch ten Cate, 1988; Ooms *et al.*, 1985b, 1986a), oilseed rape (Guerche *et al.*, 1987; Ooms *et al.*, 1985a; Rasmussen and Dangaard, 1988), alfalfa (*Medicago sativa L.*) (Spano *et al.*, 1987; Sukhapinda *et al.*, 1987b), cucumber (Trulson *et al.*, 1986), cauliflower (David and Tempé, 1988), flax (Zhan *et al.*, 1988) and various additional *Solanum* and soybean species (Davey *et al.*, 1987; Rech *et al.*, 1988). This list only summarizes results published in widely available refereed journals and is not exhaustive. It, together with the earlier examples and the graphs in Figure 1, illustrates, however, that in only a few years transformed cell cultures and plants have been isolated in a broad range of plant species and in rapidly increasing numbers, and the trend looks like it is set to continue.

For many species, the aim is not to just isolate transformed cells or plants but to do so with a reasonably high efficiency and to obtain transformed plants that are indistinguishable from untransformed plants with the exception of the introduced characters. This has also now been successfully achieved in several of the agronomically more important crop plants although with variation in efficiency. Examples are in tomato (De Block *et al.*, 1987; Filatti *et al.*, 1987a; Koornneef *et al.*, 1986; McCormick *et al.*, 1986; Shahin *et al.*, 1986b), potato (An *et al.*, 1986b; De Block 1988; De Block *et al.*, 1987; Ooms *et al.*, 1987; Shahin and Simpson, 1986; Sheerman and Bevan, 1988; Stiekema *et al.*, 1988; Stockhaus *et al.*, 1986; Twell and Ooms, 1987) oilseed rape, mostly in c.v. Wester (Charest *et al.*, 1988; Fry *et al.*,1987; Neuhaus *et al.*, 1987; Pua *et al.*, 1987; Radke *et al.*, 1988), cotton (Firoozabady *et al.*, 1987; Umbeck *et al.*, 1987) rye (de la Pena *et al.*, 1987), flax (Basiran *et al.*, 1987; Jordan and McHughan, 1988), lettuce (Chupeau *et al.*, 1989; Michelmore *et al.*, 1987) carrot (Scott and Draper, 1987), the grain legume *Vigna aconitofolia* (moth bean) (Eapen *et al.*, 1987), *Medicago* (alfalfa) (Chabaud *et al.*, 1988; Deak *et al.*, 1986; Shahin *et al.*, 1986b), eggplant (Guri and Sink, 1988) and celery (Catlin *et al.*, 1988) and more recently in soybean (Hinchee *et al.*, 1988) and rice (Shimamoto *et al.*, 1989; Toriayma *et al.*, 1988; Zhang *et al.*, 1988). As with Ri T-DNA transformed plants, more examples exist of successful transformation giving more or less normal plants at various stages of regeneration, but as yet these have not been reported in the widely accessible literature; forthcoming examples can be expected in sugar beet, maize and wheat.

It is clear, particularly from the studies in tomato, potato and oilseed rape, that substantial diversity exists in the details of the protocols used and sometimes in the conclusions reached. For example in tomato McCormick *et al.* (1986) were unsuccessful in obtaining transgenic plants using a binary system; a result which was later contradicted by results of Sukhapinda *et al.* (1987a). In oilseed rape, Pua *et al.* (1987) reported that, at the time, they had not been able to obtain transgenic oilseed rape plants using kanamycin as a selectable marker. This marker was successfully used however, by Fry *et al.* (1987) and Charest *et al.* (1988). These results are a further illustration of the points made in the discussion on *Agrobacterium*-based genetic engineering systems (Section 12.2) and selectable markers (Section 12.3.2), emphasizing that the whole process of transformation through to regeneration is a tightly linked multistep process, sometimes with little scope to deviate from optimized protocols without incurring a penalty in overall efficiency which can be severe in certain cases.

### 12.3.5 Other Plants

Several plants other than those important in agriculture, horticulture and floriculture have also been examined for their amenability to genetic engineering. The plant that has received most attention for various reasons is *Arabidopsis thaliana* (Meyerowitz, 1989). It is a small rapidly growing cruciferous plant with a generation time of only five weeks. It is a diploid with a small haploid nuclear genome of only $7 \times 10^7$ bp, distributed over five chromosomes. It is self-fertile, can produce more than 10 000 seeds per plant and many mutants have been isolated by examining or screening $M_2$ progeny (Estelle and Sommerville, 1986). Many of these mutations have been located on a genetic map (Koornneef *et al.*, 1983). Because of its relatively small genome with little repetitive DNA it becomes feasible to establish a genomic library in bacteria and yeast of mapped and overlapping clones. The great benefit of this system would be that it enables identification and isolation of specific cloned genes that correspond to mutated loci for which a known phenotype has been established. Corrections of a mutant phenotype by complementation following transformation, should in principle permit the identification of any such gene. Initial mutations can be introduced by treatment with a chemical mutagen (EMS) and by insertional inactivation through transformation or transposition. In the two latter procedures foreign DNA is introduced initially into *Arabidopsis* by transformation and tags the gene of interest to facilitate subsequent cloning (see also Section 12.2.7). Hence, interest in establishing transformation protocols in *Arabidopsis* is considerable. As in so many other plants initial transformations were carried out using *A. tumefaciens* and *A. rhizogenes* (Aerts *et al.*, 1979; Pavingerova *et al.*, 1983). More recently, morphologically normal *Arabidopsis* plants transformed with antibiotic resistance markers were obtained using either disarmed *Agrobacterium* and leaf discs, which gave more than a hundred hygromicin resistant transgenic plants (Lloyd *et al.*, 1986), or using *Agrobacterium* infection of germinating seeds which gave more than 200 kanamycin resistant plants (Feldmann and Marks, 1987). Even higher transformation efficiencies based on optimized transformation have subsequently been claimed by Schmidt and Willmitzer (1988) and Velvekens *et al.* (1988).

Other plants that have attracted more recent wide attention with the aim of developing genetic engineering technology for commercial reasons are the trees. Of course, it has been long known that many trees can be infected with *A. rhizogenes* and then produce additional root formation, thus achieving partial transformation, but this, in fact, is what happens naturally. The most notable successes so far in obtaining fully transformed transgenic trees are in *Populus* (Fillatti *et al.*, 1987b) and more recently in *Juglans*, walnut (McGranahan *et al.*, 1988) and *Malus*, apple (James *et al.*, 1989) where resistance against the herbicide glyphosate and twice resistance to kanamycin respectively, were introduced. Currently, the general state of tree transformation, however, resembles the earlier progress in crop plant transformation, with examples of the establishment of shoot regeneration from *A. tumefaciens*-induced tumours in the tropical timber tree *Fagraea fragrans* (Loh and Rao, 1988), gall formation in *Alnus* and *Betula* clones (Mackay *et al.*, 1988), considerable genotypic effects on the efficiency of gall formation in white spruce and other conifer species (Ellis *et al.*, 1989) and regeneration of trees of local importance such as transgenic neem (Naina *et al.*, 1989).

The third group of other plants that have attracted genetic engineering interest are those important in production of secondary metabolites which is discussed later (Section 12.5.5).

## 12.4 GENE EXPRESSION

In between the development of plant genetic engineering systems and the identification of topics for application, the technology and science centering around genes has become a paramount feature. In this chapter, methods to identify, isolate and modify genes are not dealt with in any detail, nor are control mechanisms of gene expression or factors important in protein engineering like targetting into particular cellular compartments (Della Cioppa *et al.*, 1987). The reader is referred to numerous widely available books and review articles dealing with the many aspects of gene expression. However, some of the reviews on gene expression in plants are by Hohn and Schell (1987); Kuhlemeier *et al.* (1987); Lurquin (1987); Schell (1987); and Willmitzer (1988). A review focusing on antisense technology in plants is by Van der Krol *et al.* (1988a).

From the earlier discussion on the characteristics of transformed cell lines however, it should have become clear that substantial and detailed information is rapidly becoming available on the use of gene elements in chimaeric gene constructions and on the outcome of their introduction into plants. For example, sequences very near the transcriptional start site of plant genes are often important in determining the levels and patterns of expression, such as induction by environmental stimuli (light, heat shock) or developmental stimuli (seed formation, tuber induction). Such sequences can be short, as for example a 16 bp palindromic sequences in the 5′ control region of the *ocs* gene which is an enhancer element, at least in transient expression assays (Ellis *et al.*,1987) and with some additional DNA also in stable transformation (Bruce *et al.*, 1988; Leisner and Gelvin, 1988). Of general importance to the study of gene expression is the development of various reporter genes like for chloramphenicol acetyltransferase, neomycin phosphotransferase, the more recently developed luciferase (Koncz *et al.*, 1987; Ow *et al.*, 1986), $\beta$-glucuronidase (Jefferson *et al.*, 1987) and phosphinotricinacetyltransferase (de Block *et al.*, 1987).

Finally, plant genetic engineering itself is a tool to identify and isolate the genes used for genetic engineering purposes. For example, it can be used to introduce transposable elements into plants which normally do not contain them, for purposes of gene tagging by mutagenesis. One example of transposition of the maize Ac element can be seen in tobacco (Baker *et al.*, 1986). Shotgun cloning of genes by transforming (mutant) cell lines with a library of cloned total genomic DNA followed by selection for a particular sought after change in phenotype has been evaluated (Klee *et al.*, 1987a; Olszewski *et al.*, 1988; Prosen and Simpson, 1987; Simoens *et al.*, 1986) but is generally considered a method proven to work in selected cases and being fruitful in principle but also to be laborious and 'chancy'. It should be noted however, that potential technical improvements like the development of an efficient transformation protocol for *Agrobacterium* (by electroporation) may change the outlook of such an approach. In contrast, the above mentioned *Arabidopsis* gene isolation method (Section 12.3.5) by which the phenotype of either a mutant cell line or a mutant plant is complemented by transformation with preselected DNA from a well-characterized library, is certainly within reach of current technology. It appears to provide a general method to identify and isolate any specific gene from a plant like *Arabidopsis*, provided the corresponding locus for the mutant phenotype can be identified and accurately located on a combined genetic and RFLP map.

## 12.5 TOPICS FOR APPLICATION

### 12.5.1 Introduction

A significant factor in the appeal of genetic engineering in plants or cell cultures is the specificity of the approach. From a commercial viewpoint this is particularly relevant in the identification of topics for application. For example in breeding, the established markets can be examined and the weaknesses of successful varieties identified. In principle, it is thus possible to estimate the potential gains that can be made by adding agronomical or product-related value to these varieties assuming that they will replace the parental ones. In practice, however, it is more complicated than that, and this forms the topic for discussion in Chapter 8 where the relationship between new technologies and traditional agriculture is reviewed. Within the same framework it is also becoming clear that the area of genetic engineering in plants which appears to attract most of the early attention outside the area of research for research sake, is in secondary metabolite production and in crop protection. A number of results that have been achieved in other areas of interest will be briefly summarized first. Most notable are two well-publicized ways of altering flower colour in petunia. Meyer *et al.* (1987) made use of the low substrate specificity in petunia of dihydroflavonol 4-reductase for dihydrokaempferol to introduce a novel (brick red or pink) flower colour into petunia by transformation of a mutant line with the Al maize gene encoding dihydroquercitin 4-reductase (DQR). This enzyme does reduce dihydrokaempferol. The introduced gene consisted of an A1 cDNA fragment, flanked by CaMV 35S 5′ and 3′ regulatory sequences. The corresponding DQR protein catalyzed biosynthesis of leucopelargonidin which is processed further by enzymes already in place into pelargonidin 3-glucosides responsible for the brick red flower colour. In a second study, Van der Krol *et al.* (1988b), explored the promise of antisense RNA technology in transformed plants by introducing anti-chalcone synthase transcripts in petunia flowers. They placed an inverted petunia *chs* cDNA fragment under control of the 35S CaMV promoter and host poly A sequences and selected *via Agrobacterium* transformation, transgenic petunia plants. A wide spectrum of spatial changes in flower colour were seen, consistent with suppression of the native chalcone synthase, which is an early key enzyme in anthocyanin biosynthesis.

As mentioned in the general introduction (Section 12.1), research into reproductive biology, most notably into the molecular mechanisms regulating such aspects as flower development and meiosis, in particular recombination, but also incompatibility and male sterility, are areas of interest where progress is being made and applications are possible. Plant growth and development, which relate to regulation of primary metabolism and to traditionally important breeding objectives such as field performance, tolerance to abiotic stresses, maturity, yield, harvest index and product properties is another area where genetic engineering is likely to contribute (see also Knauf, 1987; Ooms, 1987; Rogers and Klee, 1987). Early examples where availability of cloned genes has prompted manipulation of growth and development through transformation of plants followed by subsequent analyses, again are the various Ti and Ri T-DNA genes (Section 12.2) and growth inhibition in tobacco by introducing antisense small subunit rubisco transcripts (Rodermel *et al.*, 1988). A major reason for the expected slower progress towards application in some of these latter areas is that they are broad and complex. There is also the consideration that at least some of the traits are multigenic, and therefore less prominent as initial targets for a genetic engineering approach. It can be argued, however, that genetic engineering provides novel ways that are eminently suitable to find out more about the molecular mechanism underlying the complex biology, and hence to be more definite about application prospects. This reasoning is analogous to the rationale for using plant mutants in research but because of the greater intrinsic versatility of genetic engineering, it should be even more useful in this context.

### 12.5.2 Secondary Metabolites

In contrast to research into developmental biology, more immediate commercial gains can be expected from application of genetic engineering to the production of secondary products by cell cultures or even transgenic plants.

It is relevant to point out first, however, that with the exception of *Lithospermum erythrorhizon* cell suspensions which produce and to some extent secrete the antiflammatory drug shikonin (Tabata and Fujita, 1985), the synthesis of secondary metabolites by plant cell cultures obtained *via* conventional ways is commercially not yet well established despite substantial scope and promise (Hamill *et al.*, 1987a; Kurz and Constabel, 1979; Spier and Fowler, 1985). As a general rule, highest synthesis in plant cultures occurs in specially selected lines and particularly when a certain amount and type of differentiation is achieved. However, such plant cell cultures often show instability and revert to low levels of biosynthesis. Examination particularly of root cultures established following transformation using *Agrobacterium rhizogenes* in a wide range of species indicated that use of hairy-root cultures can overcome at least some of the problems and hence may help towards fulfilling earlier expressed scope and promise. Generally, the hairy-roots showed enhanced growth giving higer biomass production and simultaneously an increase in secondary metabolite accumulation per tissue weight at a high degree of stability. The wider context for hairy-root research into secondary metabolites has been reviewed in Flores *et al.* (1987) and Hamill *et al.* (1987a) and in this volume in Chapter 6, and some of the already available specific examples of alkaloid production by hairy-roots in *Calystegia sepium*, *Atropa belladonna*, *Scopolia japonica*, *Datura stramonium* and *Hygroscyamus niger* can be found in recent literature: Jaziri *et al.* (1988), Jung and Tepfer (1987), Kamada *et al.* (1986) and Mano *et al.* (1987). Further examples are of enhanced quinoline alkaloid production in *Cinchona ledgeriana* (Payne *et al.*, 1987) and of saponin in *Panax ginseng*, (Yoshikawa and Furuya, 1987).

Application of transformation using *A. rhizogenes* in this way, has the further perspective of introducing along with the Ri T-DNA genes, additional genes that may alter the secondary metabolites in a particular way, for example by making production inducible and/or introducing certain hydroxylations, methylations, glycosylations *etc.* for which specific enzymes exist, like in the anthocyanin biosynthesis pathways in flower petals. Furthermore, it is perhaps possible in certain cases to make the cells actively secrete secondary metabolites, in analogy with the opine secretion by the natural *Agrobacterium* T-DNA system, to achieve easier harvest and purification.

### 12.5.3 Herbicide Resistance

Another important area where transgenic plants already are making a contribution is in crop protection, notably in the isolation of herbicide, insect and virus resistant plants. This holds the potential for ecologically favourable reduction in herbicide and pesticide use. It is clear that agrochemical industries in particular have a vested interest in these developments, since prospects exist for a move away from the current intensive spray practices to even more selective and fewer chemical applications in combination with more extensive built-in genetic protection of the crops. The story of the isolation of glyphosate, phosphinotricin and bromoxymil resistant plants by the agrochemical company Monsanto and the biotechnology companies Calgene and Plant Genetic Systems (PSG) are illustrative of the diversity of approaches that can be applied to achieve various degrees of herbicide tolerance or resistance through genetic engineering. The target enzyme of glyphosate is the chloroplastic shikimate pathway enzyme 5-enolpyruvyl shikimate 3-phosphate (EPSP) synthase (Steinbrucken and Amrhein, 1980). The Calgene group (Comai *et al.*, 1983) isolated the corresponding wild-type *aroA* gene from the bacterium *Salmonella typhimurium*. They then selected in *E. coli* a mutant allele conferring higher resistance to glyphosate (an amino acid substitution of a proline to a serine decreased the affinity of EPSP synthesis for glyphosate) and then converted the bacterial gene into a plant gene by coupling it, in various constructs, to the Ti T-DNA promoter regions of octopine synthase (*ocs*) and mannopine synthase (*mas*) and the 3′ polyadenylation signals of the *ocs* and *tml* loci (for transcript 6a; Figure 3). These genes were then transferred by *Agrobacterium* transformation into tobacco (Comai *et al.*, 1985). Among the regenerants, plants with partial tolerance were found in which low levels of expression of the introduced gene were detected. As described in the previous section the gene was subsequently used to transform other plants like tomato and poplar.

Following a different route Monsanto constructed an EPSP synthase gene from a corresponding cDNA clone. They isolated it using a specially selected glyphosate resistant petunia culture which, as a result of 20-fold gene amplification, overproduced EPSP synthase protein and RNA, thereby making cDNA cloning feasible. A chimaeric gene was then constructed in which the cDNA was combined with a CaMV35S promoter fragment and *nos 3'* regulatory signals which, upon *Agrobacterium* transformation into petunia also gave glyphosate tolerant plants (Shah *et al.*, 1986). Subsequently both labs aimed at retargetting the enzyme to the chloroplast where it was expected that expression might give higher resistance levels and enable further evaluation of commercial use of selective herbicide tolerance in crop plants using EPSP synthase (Comai *et al.*, 1988).

Another more recent example in which herbicide resistance was achieved by overproduction in transgenic plants of a mutant target enzyme, is resistance to the sulfonylureas and imidazolinones herbicides which act on the first enzyme in the branched amino acid biosynthesis, acetohydroxy acid synthase (AHAS) which is also called acetolactate synthase (ALS) (Haughn *et al.*, 1988).

A different approach based on detoxification of herbicides rather than overproduction of target enzyme was successfully demonstrated for the herbicides bialaphos and phosphinotricin (PPT) and for bromoxynil (De Block *et al.*, 1987; Stalker *et al.*, 1988). The PPT resistance was based on introduction into various plants (tobacco, tomato, potato) of the *bar* gene from *Streptomyces hygroscopicus* equipped with the CaMV35S promoter and the 3' control region of T-DNA gene 7 (Figure 3). The corresponding gene product, phosphinotricin acetyltransferase (PAT) inactivates PPT by acetylating the free $NH_2$ group and the levels at which it is expressed are sufficient to give resistance at high doses of commercial formulations of the herbicides. Similarly, several transgenic tobacco plants and their progeny transformed with the *bxn* gene from the soil bacterium *Klebsiella azaenae*, flanked by a rubisco ss promoter and *ocs* poly A sequences, showed high resistance to commercial formulations of the herbicide bromoxymil (Bactril). Here resistance was achieved by converting bromoxynil (3,5-dibromo-4-hydroxybenzonitrile) into its primary metabolite 3,5-dibromo-4-hydroxybenzoic acid through the *bxn* encoded nitrilase.

### 12.5.4 Virus Resistance

Examples of application of genetic engineering/transformation technology in crop protection are also particularly notable in virus control. Most successful has been coat protein mediated protection; the introduction into plants of viral cDNAs, with suitable added 5' and 3' plant gene regulatory sequences, which code for viral coat proteins. In this way transgenic plants have been isolated, particularly tobacco but also 12 tomato and seven potato plants with varying degrees of resistance to a range of viruses. The list so far includes tobacco mosaic virus (TMV; Nelson *et al.*, 1987; Powell Abel *et al.*, 1987) alfalfa mosaic virus (ALMV; Loesch-Fries *et al.*, 1987; Tumer *et al.*, 1987; Van Dun *et al.*, 1988); cucumber mosaic virus (CMV; Cuotta *et al.*, 1988) and potato virus X (PVX; Hemenway *et al.*, 1988; Hoekema *et al.*, 1989). As expected from results described above (Section 12.2) there was variation in expression of the introduced genes between independent transformants. Generally, this correlated very well with the differing extents of various resistance which, depending on the particular system and experimental procedures used, was defined as delay in symptom development, reduction in lesion numbers and reduction in virus accumulation in inoculated and systemic plants. It is significant that this approach works well in different plants for different viruses such as TMV and ALMV. These viruses differ in many aspects including morphology, ways of transmission and infection, in mechanisms of replication and in genome structure and gene expression. Hence, the method potentially has very wide applicability and therefore it is anticipated that similar results will be obtained in increasing numbers of plant species and types of viruses. A second method which has been explored to give novel resistance to virus infection is the introduction into transgenic plants of antisense coat protein transcripts or antisense RNA transcribed from various regions of virus genomic cDNA. This approach does work but at best is substantially less effective than coat protein protection (for CMV in tobacco, see Cuotta *et al.*, 1988, and for PVX in tobacco, see Hemenway *et al.*, 1988). For example transgenic tobacco plants with PVX coat protein showed resistance to inoculation of 5 $\mu g\,mL^{-1}$ PVX, whereas in transgenic antisense plants only some protection was seen at 0.05 $\mu g\,mL^{-1}$ inoculum concentrations. In another study, introduction of antisense RNAs against various cucumber mosaic virus DNA segments gave no detectable protection despite expression of the antisense gene constructs (Rezaian *et al.*, 1988). One of the scientific virtues of both the coat protein and antisense approach is that the results are of considerable use in advancing insight into the molecular mechanisms underlying the occurrence of plant virus diseases. These implications are discussed in detail in the various papers cited, but one conclusion is that additional genetic engineering/transformation research into specific compounds important in virus infection, like replicases and proteins important in the migration of viruses from cell to cell, will also be fruitful in advancing the science and its potential for application.

A range of further methods exist to give virus resistance. Some of these are well publicized, like protection by an active viral satellite RNA transcribed from an introduced cDNA clone (Baulcombe *et al.*, 1986), and the suggested uses of cDNA clone coding for virus specific antibodies (Baulcombe, 1986) and of ribozyme, a catalytic RNA which can cleave specific target RNAs (Haseloff and Gerlach, 1988). In addition to this range of novel methods, of course, cloning and analyses of natural genes for virus resistance is in progress.

### 12.5.5 Insect Resistance

Analogous to protection against viruses and bacterial or fungal pathogens, naturally occurring resistance is only one of the many possible control measures against insect damage. Here, other measures include the use of pesticides. However, the same concern for environmental damage by excessive use of herbicides also exists for pesticide use which thereby stimulates examination of alternative insect pest control measures. The use of pheromones as a component in integrated pest management is one such possible way forward. Another is through genetic engineering, and indeed several examples already exist where this has been successfully pioneered. One approach was based on the observation that endotoxins produced by specific strains of *Bacillus thuringiensis* were identified as single gene products that are lethal upon ingestion by susceptible insect larvae (*Lepidoptera* and *Diptera*). Isolation and manipulation of the corresponding bacterial genes from different *B. thuringiensis* isolates gave spectacular examples of protection of transgenic tobacco plants against *Manduca sexta* (Vaeck *et al.*, 1987) and of transgenic tomato plants against *Manduca sexta*, *Heliothis virescens* and *Heliothis zea* (Barton *et al.*, 1987; Fischoff *et al.*, 1987). Similarly, plant genetic engineering experiments based on protease inhibitor studies (Gatehouse *et al.*, 1986) showed that a chimaeric gene with CaMV 5′ and *nos* 3′ control regions encoding a cowpea trypsin inhibitor gave insect resistance in tobacco against *Heliothis virescens* (Hilder *et al.*, 1987). These results indicate that several of the seed storage proteins and wound inducible proteins (or derivatives thereof), which are known proteinase inhibitors, may behave as insect resistance gene products when more widely expressed througout transgenic plants. It is emphasized, however, that the tests referred to here were done under carefully controlled conditions using selected insects or predators, and future experiments will have to show how significant the engineered plants are for practical use, in particular under standard agricultural practice.

## 12.6 CONCLUDING REMARKS

The naturally evolved plant genetic engineering system of *Agrobacterium* in which the bacterium transfers some of its own DNA into plant cells, thereby inducing tumorous growth or excessive root formation, provides the best example yet of sophisticated genetic engineering in plants. It illustrates a tightly integrated and highly regulated application of a gene-transfer system leading to biomass production, metabolite biosynthesis and secretion. Detailed examination of the transfer mechanism itself and the stability, structure and expression of the transferred DNA has provided a substantial scientific basis for design of even better *Agrobacterium*-based transfer systems that are more efficient and potentially of use in a wider range of plants. Alongside the improvement in *Agrobacterium* transformation systems, a wide range of other procedures are under development. Ultimately the relative usefulness of the various procedures will have to be balanced. Factors important in this will be the ease and efficiency of manipulation and, of considerable importance to the transgenic cell cultures and plants isolated, the stability of the introduced DNA. The immediate challenges following the establishment of transformation technology are in learning how to apply it. In particular, factors important in the control of gene expression and ways of identifying and isolating genes of interest in particular areas of plant biology are key issues. Here is where genetic engineering of cell cultures and plants becomes truly integrated into all aspects of plant biology and where it begins to affect the role plants play in society at large. Much has currently been learnt about control of plant gene expression and, in this respect, genetic engineering of plants and cultures is an increasingly important research tool. With regard to topics of application there are examples of transformed plants with differing levels of resistance to herbicides, diseases and plant pests, of transformed cell cultures with better growth properties and improved secondary metabolite production. In these areas the most immediate applications are being sought. Of obvious further potential, for example in horticulture, is the isolation of plants with sought after flower colours, growth habitat, maturity *etc.*, and in agriculture of plants that during their growth period require less intensive, or perhaps environmentally less damaging agricultural practice combined with economical yields of high quality produce. It is difficult, however, to foresee exactly how and when these prospects will come to fruition, since it is likely that various broad and well-integrated types of social, commercial, legislative and thereby invariably local and global political considerations will increasingly come into play. These considerations will be important in deciding the future of already isolated plants with agronomically improved traits and in further defining the aims genetic engineering technology will serve.

## ACKNOWLEDGEMENTS

I thank in particular Jeane Hutchins and Sally Lawson for typing the text, Liz Allsopp and Francis Teasdale for assistance in computer searches (up to Spring 1989) and Angela Karp, Paul Hooykaas and Peter Shewry for critical reading of the manuscript.

## 12.7 REFERENCES

Ackermann, C. (1977). Pflanzen aus *Agrobacterium rhizogenes* tumoren an *Nicotiana tabacum*. *Plant Sci. Lett.*, **8**, 23–30.

Aerts, M., M. Jacobs, J. P. Hernalsteens, M. Van Montagu and J. Schell (1979). Induction and *in vitro* culture of *Arabidopsis thaliana* crown galls. *Plant Sci. Lett.*, **17**, 43–50.

Alarcon, B., M. M. Lopez, M. Cambra and J. Ortiz (1987). Comparative study of *Agrobacterium* biotypes 1, 2 and 3 by electrophoresis and serological methods. *J. Appl. Bacteriol.*, **62**, 295–308.

Albright, L. M., M. F. Yanofsky, B. Leroux, D. Ma and E. W. Nester (1987). Processing of the T-DNA of *Agrobacterium tumefaciens* generates border nicks and linear, single-stranded T-DNA. *J. Bacteriol.*, **169**, 1046–1055.

Amasino, R. M., A. L. T. Powell and M. P. Gordon (1984). Change in T-DNA methylation and expression are associated with phenotypic variation and plant regeneration in a crown gall tumour line. *Mol. Gen. Genet.*, **197**, 437–446.

An, G. (1986a). Development of plant promoter expression vectors and their use for analysis of differential activity of nopaline synthase promoter in transformed tobacco cells. *Plant Physiol.*, **81**, 86–91.

An, G., B. D. Watson, S. Stachel, M. P. Gordon and E. W. Nester (1985). New cloning vehicles for transformation of higher plants. *EMBO J.*, **4**, 277–284.

An, G., B. D. Watson and C. C. Chiang (1986b). Transformation of tobacco, tomato, potato and *Arabidopsis thaliana* using a binary Ti vector system. *Plant Physiol.*, **81**, 301–305.

An, G., M. A. Costa, A. Mitra, S. B. Ha and L. Marton (1988). Organ specific and development regulation of the nopaline synthase promoter in transgenic tobacco plants. *Plant Physiol.*, **88**, 547–552.

André, D., D. Colan, J. Schell, M. van Montagu and J.-P. Hernalsteens (1986). Gene tagging in plants by a T-DNA insertion mutagen that generates APH(3')II plants gene fusions. *Mol. Gen. Genet.*, **204**, 512–518.

Armstead, I. P. and K. J. Webb (1987). Effect of age and type of tissue on genetic transformation of *Lotus corniculatus* by *Agrobacterium tumefaciens*, *Plant Cell, Tissue Organ Cult.*, **9**, 95–101

Baker, B., J. Schell, H. Lorz and N. Fedoroff (1986). Transposition of the maize controlling element 'Activator' in tobacco. *Proc. Natl. Acad. Sci. USA*, **83**, 4844–4948.

Barker, R. F., K. B. Idler, D. V. Thompson and J. D. Kemp (1983). Nucleotide sequence of the T-DNA region from the *Agrobacterium tumefaciens* octopine Ti plasmid pTi 15955. *Plant Mol. Biol.*, **2**, 335–350.

Barton, K. A., H. R. Whitely and N.-S. Yang (1987). *Bacillus-thuringiensis* delta endotoxin expressed in transgenic *Nicotiana tabacum* provides resistance to lepidopteran insects. *Plant Physiol.*, **85**, 1103–1109.

Basiran, N., P. Armitage, R. J. Scott and J. Draper (1987). Genetic transformation of flax (*Linum usitatissimum*) by *Agrobacterium tumefaciens*–regeneration of transformed shoots *via* a callus phase. *Plant Cell Rep.*, **6**, 396–399.

Baulcombe, D. C. (1986b). The use of recombinant DNA techniques in the production of virus resistant plants. *Biotechnology and crop improvement and protection. BCPC Mono.*, **34**, 13–19.

Baulcombe, D. C., G. R. Saunders, M. W. Bevan, M. A. Mayo and B. D. Harrison (1986). Expression of biologically active viral satellite RNA from the nuclear genome of transformed plants. *Nature*, **321**, 446–449.

Bevan, M. (1984). Binary *Agrobacterium* vectors for plant transformation. *Nucleic Acids Res.*, **12**, 8711–8721.

Bevan, M. W., R. B. Flavell and M. D. Chilton (1983). A chimaeric antibiotic resistance gene as a selectable marker for plant cell transformation. *Nature*, **304**, 184–187.

Bevan, M. W. and A. Goldsborough (1988). Design and use of *Agrobacterium* transformation vectors. In *Genetic Engineering Principles and Methods*, ed. J. K. Setlow, vol. 10, pp. 123–140, Premium Press, New York.

Biemann, K., C. Lioret. K. Asselimeau, E. Lederer and J. Polonski (1960). Sur la structure chimique de la lysopine, nouvel acide aminé isolé di tissue de crown gall. *Bull. Soc. Chim. Biol.*, **42**, 979–991.

Binns, A. N. and M. F. Thomashow (1988). Cell biology of *Agrobacterium* infection and transformation of plants. *Annu. Rev. Microbiol.*, **42**, 575–606.

Birot, A. M., D. Boucher, F. Casse-Delbaert, M. Durand-Tardif, L. Jouanin, J. Pantot, C. Robaglia, D. Tepfer, M. Tepfer and F. Vilaine (1987). Studies and uses of the Ri plasmids of *Agrobacterium rhizogenes. Plant Phys. Biochem* ., **25**, 1–13.
Bolton, G. W., E. W. Nester and M. P. Gordon (1986). Plant phenolic compounds induce expression of the *Agrobacterium tumefaciens* loci needed for virulence. *Science (Washington D.C.)*, **232**, 983–985.
Bopp, M. and F. Resende (1966). Crown gall tumoren bei Verscheidence Arten und Bastarden der Kalanchoideae. *Port. Acta Biol.*, **9**, 327–366.
Braun, A. C. and H. N. Wood (1976). Suppression of the neoplastic state with the acquisition of specialized functions in cells, tissues and organs of crown gall teratomas of tobacco. *Proc. Natl. Acad. Sci. USA*, **73**, 496–500.
Brisson, N., J. Paszkowski, J. R. Penswick, B. Gronenborn, I. Potrykus and T. Hohn (1984). Expression of a bacterial gene in plants using a viral vector. *Nature*, **310**, 511–514.
Bruce, W. B., Bandyopadhyay and W. Z. Gorky (1988). An enhancer-like element present in the promoter of a T-DNA gene from the TC plasmid of *Agrobacterium tumefaciens. Proc. Natl. Acad. Sci. USA*, **85**, 4310–4314.
Budar, F., F. Debock, M. Van Montagu and J.-P. Hernalsteens (1986a). Introduction and expression of the octopine T-DNA oncogenes in tobacco plants and their progeny. *Plant Sci.*, **46**, 195–206.
Budar, F., L. Thia-Toong, M. Van Montagu and J.-P. Hernalsteens (1986b). *Agrobacterium*-mediated gene transfer results mainly in transgenic plants transmitting T-DNA as a single Mendelian factor. *Genetics*, **114**, 303–314.
Byrne, M. C., R. E. McDonnell, M.S. Wright and M.G. Carnes (1987). Strain and cultivar specificity in the *Agrobacterium* soybean interaction. *Plant Cell, Tissue Organ Cult.*, **8**, 3–15.
Bytebier, B., F. Deboeck, H. De Greve, M. Van Montagu and J.P. Hernalsteens (1987). T-DNA organization in tumour cultures and transgenic plants of the monocotyledon *Asparagus officinalis. Proc. Natl. Acad. Sci. USA*, **84**, 5345–5349.
Cangelosi, G. A., L. Hung, V. Puvanesarajah, G. Stacey, D. A. Ozga, J. A. Leigh and E. W. Nester (1987). Common loci for *Agrobacterium tumefaciens* and *Rhizobium meliloti* exopolysaccharide synthesis and their roles in plant interactions. *J. Bacteriol.*, **169**, 2086–2091.
Cangelosi, G. A., G. Martinetti, J. A. Leigh, C. C. Lee, C. Theines and E. W. Nester (1989). Role of *Agrobacterium tumefaciens* ChvA protein in export of $\beta$-1,2-glycan. *J. Bacteriol.*, **171**, 1609–1615.
Cardarelli, M., L. Spano, D. Mariotti, M. L. Mauro, M. A. Van Sluys and P. Costantino (1987). The role of auxin in hairy-root induction. *Mol. Gen. Genet.*, **208**, 457–463.
Catlin, D., O. Ochoa, S. McCormick and C. F. Qhyiros (1988). Celery transformation by *Agrobacterium tumefaciens*: cytological and genetic analysis of transgenic plants. *Plant Cell Rep.*, **7**(2), 100–103.
Chabaud, M., J. E. Passiatore, F. Cannon and V. Buchanon-Wallaston (1988). Parameters affecting the frequency of kanamycin resistant alfalfa obtained by *Agrobacterium tumefaciens*-mediated transformation. *Plant Cell Rep.*, **7**, 512–516.
Charest, P. J., L. A. Holbrook, J. Gabard, V. N. Iyer and B. L. Miki (1988). *Agrobacterium*-mediated transformation of thin cell layers explants from *Brassica napus* L. *Theor. Appl. Genet.*, **75**, 488–445.
Chen, Z. L., M. A. Schuler and R. N. Beachy (1986). Functional analysis of regulatory elements in a plant embryo specific gene. *Proc. Natl. Acad. Sci. USA*, **83**, 8560–8564.
Chilton, M.-D., M. H. Drummond, D. J. Merlo, D. Saiky, A. L. Montoya, M. P. Gordon and E. W. Nester (1977). Stable incorporation of plasmid DNA into higher plant cells; the molecular basis of crown gall tumorigenesis. *Cell*, **11**, 263–271.
Chilton, M.-D., D. Tepfer, A. Petit, C. David, F. Casse Delbart and J. Tempé (1982). *Agrobacterium rhizogenes* inserts T-DNA into the genomes of the host plant root cells. *Nature*, **295**, 432–434.
Christie, P. J., J. E. Ward, S. C. Winans and E. W. Nester (1988). The *Agrobacterium tumefaciens virE* gene product is a single stranded DNA binding protein that associates with T-DNA. *J. Bacteriol.*, **170**, 2659–2667.
Christou, P., S. G. Platt and M. C. Ackerman (1986). Opine synthesis in wild-type plant tissue. *Plant Physiol.*, **82**, 218–221.
Chupeau, M.-C., C. Bellini, P. Guerche, B. Maisonneuve, G. Vastra and Y. Chupeau (1989). Transgenic plants of lettuce (*Lactuca sativa*) through electroporation of protoplasts. *Biotechnology*, **7**, 503–508.
Chyi, Y.-S., R. A. Jorgensen, D. Goldstein, S. D. Tanksley and F. Loaiza-Figueroa (1986). Locations and stability of *Agrobacterium*-mediated Ti plasmid insertions in the lycopersicon genome. *Mol. Gen. Genet.*, **204**, 64–69.
Chyi, Y. S. and G. C. Phillips (1987). High efficiency *Agrobacterium*-mediated transformation of *Lycopersicon* based on conditions favourable for regeneration. *Plant Cell Rep.*, **6**, 105–108.
Citovsky, V., G. de Vos and P. Zambryski (1988). Single-stranded DNA binding protein encoded by the *virE* locus of *Agrobacterium tumefaciens. Science* (*Washington D.C.*), **240**, 501–504.
Comai, L., L. Sen and D. Stalker (1983). An altered *aroA* gene product confers resistance to the herbicide glyphosate. *Science* (*Washington D.C.*), **221**, 370–371.
Comai, L., D. Facciotti, W. R. Hiatt, G. Thompson, R. Rose and D. Stalker (1985). Expression in plants of a mutant *aroA* gene from *Salmonella typhymurium* confers tolerance to glyphosate. *Nature*, **317**, 741–744.
Comai, L., N. Larson-Kelly, J. D. Kiser Manc, A. R. Pokalsky, C. K. Shewmaker, K. McBride, A. Jones and D. M. Stalker (1988). Chloroplast transport of a ribulose bisphosphate carboxylase small subunit 5-enolpyruvyl 3-phosphoshikimate synthase chimaeric protein requires part of the mature small subunit in addition to the transit peptide. *J. Biol. Chem.*, **263**, 15104–15109.
Cornelissen, M. J., M. De Block, M. Van Montagu, J. Leeman, P. H. Schreier and J. Schell (1987). Plastid transformation: A progress report. In *DNA Infections Agents* (*Plant Gene Research*) ed. T. Hohn and J. Schell, pp. 311–320. Springer-Verlag, New York.
Cuotta, M., K. M. O'Connell, W. Kaniewshi, R.-X. Fang, N. H. Chua and N. E. Turner (1988). Viral protection in transgenic tobacco plants expressing the cucumber mosaic virus coat protein or its antisense RNA. *Biotechnology*, **6**, 549–557.
Davey, M. R., B. J. Mulligan, K. M. A. Gartland, E. Peel, A. W. Sargent and A. J. Morgan (1987). Transformation of *Solanum* and *Nicotiana* species using an Ri plasmid vector. *J. Exp. Bot.*, **38**, 1507–1516.
David, C. and J. Tempé (1988). Genetic transformation of cauliflower (*Brassica oleracea* L. var Botrytis) by *Agrobacterium rhizogenes. Plant Cell Rep.*, **7**, 88–91.
Deak, M., G. B. Kiss, C. Koncz and D. Dudits (1986). Transformation of medicago by *Agrobacterium*-mediated gene transfer. *Plant Cell Rep.*, **5**, 97–100.
Deblaere, R., B. Bytebier, H. De Greve, F. Deboeck, J. Schell, M. Van Montagu and J. Leemans (1985). Efficient octopine Ti plasmid-derived vectors for *Agrobacterium*-mediated gene transfer to plants. *Nucleic Acids Res.*, **13**, 4777–4789.
De Block, M. (1988). Genotype-independent leaf disc transformation of potato *Solanum tuberosum* using *Agrobacterium tumefaciens. Theor. Appl. Genet.*, **76**, 767–774.

De Block, M., J. Schell and M. Van Montagu (1985). Chloroplast transformation by *Agrobacterium tumefaciens. EMBO J.*, **4**, 1367–1372.

De Block, M., J. Botterman, M. Vandewiele, J. Dockx, C. Thoen, V. Gossele, N. R. Movva, C. Thompson, M. Van Montagu and J. Leemans (1987). Engineering herbicides resistance in plants by expression of a detoxifying enzyme. *EMBO J.*, **6**, 2513–2518.

De Cleene, M. (1988). The susceptibility of plants to Agrobacterium: a discussion of the role of phenolic compounds. *FEMS Microbiol. Rev.*, **54**, 1–8.

De Cleene, M. and J. De Ley (1977). The host range of crown gall. *Bot. Rev.*, **42**, 389–466.

De Greve, H., J. Leemans, J. P. Hernalsteens, L. Thia-Toong, M. De Beuckeleer, L. Willmitzer, L. Otten, M. Van Montagu and J. Schell (1982). Regeneration of normal and fertile plants that express octopine synthase, from tobacco crown galls after deletion of tumour-controlling functions. *Nature*, **300**, 752–755.

de la Pena, A., H. Lorz and J. Schell (1987). Transgenic rye plants obtained by injecting DNA into young floral tillers. *Nature*, **325**, 274–276.

Della-Cioppa G., G. M. Kishore, R. N. Beachy and R. T. Fraley (1987). Protein trafficking in plant cells. *Plant Physiol.*, **84**, 965–968 (1987).

De Pater, S. (1987). Plant expression signals of the *Agrobacterium Tcyt* gene. Thesis (Leiden).

Depicker, A. G., A. M. Jacobs and M. Van Montagu (1988). A negative selection scheme for tobacco protoplast-derived cells expressing T-DNA gene 2. *Plant Cell Rep.*, **7**, 63–68.

Deroles, S. C. and R. C. Gardner (1988a). Analysis of the T-DNA structure in a large number of transgenic petunias generated by *Agrobacterium*-mediated transformation. *Plant Mol. Biol.*, **11**, 365–378.

Deroles, S. C. and R. C. Gardner (1988b). Expression and inheritance of kanamycin resistance in a large number of transgenic petunias generated by *Agrobacterium*-mediated transformations. *Plant Mol. Biol.*, **11**, 355–364.

Deshayes, A., L. Herrera-Estrella and M. Caboche (1985). Liposome-mediated transformation of tobacco mesophyll protoplasts by an *Esherichia coli* plasmid. *EMBO J.*, **4**., 2731–2739.

Diner, A. M. and D. F. Karnosky (1987). Differential responses of two conifers to *in vitro* inoculation with *Agrobacterium rhizogenes. Eur. J. Pathol.*, **17**, 211–216.

Douglas, C. J., R. J. Staneloni, R.A. Rubin and E.W. Nester (1985). Identification and genetic analysis of an *Agrobacterium tumefaciens* chromosomal virulence region. *J. Bacteriol.*, **161**, 850–860.

Durand-Tardif, M., R. Broglie, J. Slightom and D. Tepfer (1985). Structure and expression of Ri T-DNA from *Agrobacterium rhizogenes* in *Nicotiana tabacum. J. Mol. Biol.*, **186**, 557–564.

Dylan, T., L. Ielpi, S. Stanfield, L. Kashyap, C. Douglas, M. Yanofsky, E. W. Nester, D.R. Helinski and G. Ditta (1986). *Rhizobium meliloti* genes required for nodule development are related to chromosomal virulence genes in *Agrobacterium tumefaciens. Proc. Natl. Acad. Sci. USA*, **83**, 4403–4407.

Eapen, S., F. Kohler, M. Gerdemann and O. Schieder (1987). Cultivar dependence of transformation rates in moth bean after cocultivation of protoplasts with *Agrobacterium tumefaciens. Theor. Appl. Genet.*, **75**, 207–210.

Eichholtz, D. A., S. G. Rogers, R. B. Horsch, H. J. Klee, M. Hayford, N. L. Hoffmann, S. B. Bradford, C. Fink, J. Flick, K. M. O'Connell and R. T. Fraley (1987). Expression of mouse dihydrofolate reductase gene confers methotrexate resistance in transgenic *Petunia* plants. *Somatic Cell Mol. Genet.*, **13**, 67–76.

Ellis, J. G., M. H. Ryder and M. E. Tate (1984). *Agrobacterium tumefaciens* T-DNA encodes a pathway for agrobine biosynthesis. *Mol. Gen. Genet.*, **195**, 446–473.

Ellis, J. G., D. J. Llewellyn, J. C. Walker, E. S. Dennis and W. Peacock (1987). The *ocs* element—a 16 base pair palindrome essential for activity of the octopine synthase enhancer. *EMBO J.*, **6**, 3203–3208.

Ellis, D., D. Roberts, B. Sutton, W. Lazaroff, D. Webb and B. Flinn (1989). Transformation of white spruce and other conifer species by *Agrobacterium tumefaciens. Plant Cell Rep.*, **8**, 16–20.

Estelle, M. A. and G. R. Sommerville (1986). The mutants of *Arabidopsis. Trends Genet.*, **2**, 89–93.

Evans, D. A. (1989). Somaclonal variation genetic basis and breeding applications. *Trends Genet.*, **5**, 46–50.

Feldmann, K. A. and M. D. Marks (1987). *Agrobacterium*-mediated transformation of germinating seeds of *Arabidopsis thaliana*; a non tissue culture approach. *Mol. Gen. Genet.*, **208**, 1–9.

Feldmann, K. A., M. D. Marks, M. L. Christianson and R. S. Quatrano, (1989). A dwarf mutant of *Arabidopsis* generated by T-DNA insertion mutagenesis. *Science (Washington D.C.)*., **243**, 1351–1354.

Fillatti, J. J., J. Kiser, R. Rose and L. Comai (1987a). Efficient transfer of a glyphosate tolerance gene into tomato using a binary *Agrobacterium tumefaciens* vector. *Biotechonology*, **5**, 726–730.

Fillatti, J. J., J. Sellmer, B. McCown, B. Haissig and L. Comai (1987b). *Agrobacterium*-mediated transformation and regeneration of populus. *Mol. Gen. Genet.*, **206**, 192–199.

Firoozabady, E., D. L. De Boer, D. J. Merlo, E. L. Halk, L. N. Amerson, K. E. Rashka and E. E. Murray (1987). Transformation of cotton (*Gossypium hirsutum L*) by *Agrobacterium tumefaciens* and regeneration of transgenic plants. *Plant Mol. Biol.*, **10**, 105–116.

Fischoff, D. A., K. S. Bowdish, F. Erlak, P. G. Marrone, S. M. McCormick, J. D. Niedermeyer, D. A. Dean, K. Kusano-Kretzmer and J. Mayer (1987). Insect tolerant transgenic tomato plants. *Biotechnology*, **5**, 807–813.

Flores, H. E., M. W. Hoy and J. J. Pickard (1987). Secondary metabolites from root cultures. *Trands Biotechnol.*, **5**, 64–69.

Forde, B. G., H. M. Day, J. F. Turton, W.-J. Shen, J. V. Cullimore and J. E. Oliver (1989). Two glutamine synthetase genes from *Phaseolus vulgaris* L. display contrasting developmental and spatial patterns of expression in transgenic *Lotus corniculatus* plants. *The Plant Cell*, **1**, 391–401.

Fraley, R. T. (1983). Expression of bacterial genes in plant cells. *Proc. Natl. Acad. Sci. USA*, **80**, 4803–4807.

Fraley, R. T., R. B. Horsch, A. Matzke, M.-D. Chilton, W. S. Chilton and P. R. Sanders (1984). *In vitro* transformation of petunia cells by an improved method of cocultivation with *A. tumefaciens* strains. *Plant Mol. Biol.*, **3**, 371–378.

Fraley, R. T., S. G. Rogers, R. B. Horsch, D. A. Eichholtz, J. S. Flick, C. T. Fink, N. L. Hoffmann and P. R. Sanders (1985). The SEV system; a new disarmed Ti plasmid vector system for plant transformation. *Biotechnology*, **3**, 629–635.

Fry, J., A. Barnason and R. B. Horsch (1987). Transformation of *brassica napus* with *Agrobacterium tumefaciens* based vectors. *Plant Cell Rep.*, **6**, 321–325.

Gardner, R. C. and V. C. Knauf (1986). Transfer of *Agrobacterium* DNA to plants requires a T-DNA border but not the *virE* locus. *Science (Washington D.C.)*, **231**, 725–727.

Garfinkel, D. J. and E. W. Nester (1980). *Agrobacterium tumefaciens* mutants affected in crown gall tumorigenesis and octopine catabolism. *J. Bacteriol.*, **144**, 732–743.
Garfinkel, D. J., R. B. Simpson, L. W. Ream, F. F. White, M. P. Gordon and E. W. Nester (1981). Genetic analysis of crown gall; fine structure map of the T-DNA by site directed mutagenesis. *Cell*, **27**, 143–154.
Gatehouse, A. M. R., K. A. Fenton, I. Jepson and D. J. Pavey (1986). The effects of α-amylase inhibitors on insect storage pests; inhibition of α-amylase *in vitro* and effects on development *in vivo*. *J. Sci. Food Agric.*, **37**, 727–734.
Gelvin, S. B., M. F. Thomashow, J. C. McPherson, M. P. Gordon and E. W. Nester (1982). Sites and map positions of several plasmid DNA encoded transcripts in octopine-type crown gall tumors. *Proc. Natl. Acad. Sci. USA*, **79**, 176–180.
Gelvin, S. B., S. J. Karcher and V. J. DiRita (1983). Methylation of T-DNA in *Agrobacterium tumefaciens* and in several crown gall tumours. *Nucleic Acids Res.*, **11**, 159–174.
Genetello, C., N. Van Larebeke, M. Holsters, A. De Pincho, M. Van Montagu and J. Schell (1977). The Ti plasmids of *Agrobacterium* as conjugative plasmids. *Nature*, **265**, 561–653.
Ghai, J. and A. Das (1989). The *virD* operon of *Agrobacterium tumefaciens* Ti plasmid encodes a DNA relaxing enzyme. *Proc. Natl. Acad. Sci. USA*, **86**, 3109–3113.
Gheysen, G., M. Van Montagu and P. Zambryski (1987). Integration of *Agrobacterium tumefaciens* transfer DNA (T-DNA) involves rearrangements of target plant DNA sequences. *Proc. Natl. Acad. Sci. USA*, **84**, 6169–6173.
Gielen, J., M. De Beuckeleer, J. Seurinck, F. Deboeck, H. De Greve, M. Lemmers, M. Van Montagu and J. Schell (1984). The complete nucleotide sequence of the $T_L$-DNA of the *Agrobacterium tumefaciens* plasmid pTiAch5. *EMBO J.*, **3**, 835–846.
Gietl, C., Z. Koukolikova-Nicola and B. Hohn (1987). Mobilization of T-DNA from *Agrobacterium* to plant cells involves a protein that binds single-stranded DNA. *Proc. Natl. Acad. Sci. USA*, **84**, 9008–9010.
Guerche, P., L. Jouanin, D. Tepfer and G. Pelletier (1987). Genetic transformation of oilseed rape (*Brassica napus*) by the Ri T-DNA of *Agrobacterium rhizogenes* and analysis of inheritance of the transformed phenotype. *Mol. Gen. Genet.*, **206**, 382–386.
Guerin, B., G. Kahlem, G. Teller and B. Durand (1984). Evidence for host genome involvement in cytokinin metabolism by male and female cells of *Mercurialis annua* transformed by strain 15955 of *Agrobacterium tumefaciens*. *Plant Physiol.*, **74**, 139–145.
Guri, A. and K. C. Sink (1988). *Agrobacterium* transformation of eggplant. *J. Plant Physiol.*, **133**(1), 52–55.
Hain, R., H. H. Steinbiss and J. Schell (1984). Fusion of *Agrobacterium* and *E. coli* spheroplasts with *Nicotiana tabacum* protoplasts: Direct gene transfer from micro-organism to higher plant. *Plant Cell Rep.*, **3**, 60–64.
Hamill, J. D., A. J. Parr, M. J. C. Rhodes, R. J. Robins and N. J. Walton (1987a). New routes of plant secondary products. *Biotechnology*, **5**, 800–804.
Hamill, J. D., A. Prescott and C. Martin (1987b). Assessment of the efficiency of cotransformation of the T-DNA of disarmed binary vectors derived from *Agrobacterium tumefaciens* and the T-DNA from *Agrobacterium rhizogenes*. *Plant Mol. Biol.*, **9**, 573–584.
Hanisch ten Cate, Ch., E. Ennik, S. Roest, K. S. Ramulu, P. Dijkhuis and B. De Groot (1988). Regeneration and characterisation of plants from potato root lines transformed by *Agrobacterium rhizogenes*. *Theor. Appl. Genet.*, **75**, 452–459.
Hansen, J., J. E. Jorgenson, J. Stougard and K. A. Marcker (1989). Hairy-roots a short cut to transgenic root nodules. *Plant Cell Rep.*, **8**, 12–15.
Harpster, M. H., J. A. Townsend, J. D. G. Jones, J. Bedbrook and P. Dunsmuir (1988). Relative strengths of the 35S cauliflower mosaic virus 1′, 2′ and nopaline synthase promoters in transformed tobacco, sugar beet and oilseed rape callus tissue. *Mol. Gen. Genet.*, **212**(1), 182–190.
Haseloff, J. and W. L. Gerlach (1988). Simple RNA enzymes with new and highly specific endoribonuclease activities. *Nature*, **334**, 585–591.
Hasazawa, S., T. Nagata and K. Syomo (1985). Transformation of *Vinca* protoplasts mediated by *Agrobacterium* spheroplasts. *Mol. Gen. Genet.*, **182**, 206–210.
Haughn, G. W., J. Smith, B. Mazur and C. Sommerville (1988). Transformation with a mutant *Arabidopsis* acetolactate synthase gene renders tobacco resistant to sulfonyl urea herbicides. *Mol. Gen. Genet.*, **211**, 266–271.
Hayford, M. B., J. I. Medford, N. L. Hoffman, S. G. Rogers and H. Klee (1988). Development of a plant transformation-selection system based on expression of genes encoding gentamycin acetyl transferases. *Plant Physiol.*, **86**(4), 1216–1222.
Heberle-Bors, E., B. Charvat, D. Thompson, J. B. Schernthauer, A. Barta, A. J. M. Matzke and M. A. Matzke (1988). Genetic analysis of T-DNA insertions into the tobacco genome. *Plant Cell Rep.*, **7**, 571–574.
Hemenway, C., R.-X. Fang, W. K. Kaniewski, N.-H. Chua and N. E. Turner (1988). Analysis of the mechanism of protection in transgenic plants expression the potato virus X coat protein or its antisense RNA. *EMBO J.*, **7**, 1273–1280.
Hepburn, A. G., L. E. Clarke, L. Pearson and J. White (1983). The role of cytosine methylation in the control of nopaline synthase gene expression in a plant tumour. *J. Mol. Appl. Genet.*, **23**, 3215–3329.
Herrera-Estrella, L., M. de Block, E. Messens, J.-P. Hernalsteens, M. Van Montagu and J. Schell (1983). Chimeric genes as dominant selectable markers in plant cells. *EMBO J.*, **2**, 987–995.
Herrera-Estrella, M., Z. M. Chen, M. Van Montagu and K. Wang (1988). VirD proteins of *Agrobacterium tumefaciens* are required for the formation of a covalent DNA-protein complex at the 5′ terminus of T-strand molecules. *EMBO J.*, **7**(13), 4055–4062.
Higgins, T. J. V., E. J. Newbigin, D. Spencer, D. J. Llewellyn and S. Craig (1988). The sequence of a pea vicilin gene and its expression in transgenic tobacco plants. *Plant Mol. Biol..*, **11**, 603–696.
Hilder, V. A., A. M. R. Gatehouse, S. E. Sheerman, R. F. Barker and E. R. Boulter (1987). A novel mechanism of insect resistance engineered into tobacco. *Nature*, **330**, 150–163.
Hille, J., F. Verheggen, P. Roelvink, H. F. A. Van Kammen and P. Zabel (1986). Bleomycin resistance: a new dominant selectable marker for plant cell transformation. *Plant Mol. Biol.*, **7**, 171–176.
Hinchee, M. A. W., D. V. Connor-Ward, C. A. Newell, R. E. McDonell, S. J. Sato, C. S. Gasser, D. A. Fischhoff, D. B. Re, R. T. Fraley and R. B. Horsch (1988). Production of transgenic soybean plants using *Agrobacterium*-mediated DNA transfer. *Biotechnology*, **6**, 915–921.
Hirooka, T. and C. I. Kado (1986). Location of the right boundary of the virulence region on *Agrobacterium tumefaciens* plasmid pTi C58 and a host-specifying gene next to the boundary. *J. Bacteriol.*, **168**, 237–243.
Hirooka T., P. M. Rogowsky and C. I. Kado (1987). Characterization of the *virE* locus of *Agrobacterium tumefaciens* plasmid pTi C58. *J. Bacteriol.*, **169**, 1529–1536.

Hoekema, A., P. R. Hirsch, P. J. J. Hooykaas and R. A. Schilperoort (1983). A binary plant vector strategy based on separation of *vir* and T-region of the *Agrobacterium tumefaciens* Ti plasmid. *Nature*, **303**, 179–180.
Hoekema, A., P. W. Roelvink, P. J. J. Hooykaas and R. A. Schilperoort (1984). Delivery of T-DNA from the *Agrobacterium tumefaciens* chromosome into plant cells. *EMBO J.*, **3**, 2485–2490.
Hoekema, A., M. J. J. Van Haaren, A. J. Fellinger, P. J. J. Hooykaas and R. A. Schilperoort (1985). Nononcogenic plant vectors for use in the agrobacterium binary system. *Plant Mol. Biol.*, **5**, 85–89.
Hoekema, A., M. J. Huisman, L. Molendyk, P. J. M. Van den Elsen and B. J. C. Cornelissen (1989). The genetic engineering of two commercial potato cultivars for resistance to potato virus X. *Biotechnology*, **7**, 273–278.
Hohn, T. and J. Schell (1987). Plant gene research basic knowledge and application plant DNA infectious agents. In *Plant Gene Research: Basic Knowledge and Application: Plant DNA Infectious Agents*, ed. T. Hohn and J. Schell, pp. 1–348. Springer-Verlag, New York.
Holbrook, L. A. and B. L. Miki (1985). Brassica crown gall tumorigenesis and *in vitro* of transformed tissue. *Plant Cell Rep.*, **4**, 329–332.
Holsters, M., B. Silva, F. Van Vliet, J. P. Hernalsteens, C. Genetello, M. Van Montagu and J. Schell (1978). *In vivo* transfer of the Ti plasmid of *Agrobacterium tumefaciens* to *Escherichia coli*. *Mol. Gen. Genet.*, **163**, 335–338.
Holsters, M., R. Villaroel, J. Gielen, J. Seurinck, H. De Greve, M. Van Montagu and J. Schell (1983). An analysis of the boundaries of the octopine Ti DNA in tumors induced by *Agrobacterium tumefaciens*. *Mol. Gen. Genet.*, **190**, 35–41.
Hood, E.E., R.T. Fraley and M.-D. Chilton (1987). Virulence of *Agrobacterium tumefaciens* strain A281 on legumes. *Plant Physiol.*, **83**, 529–534.
Hooykaas, P. J. J., P. M. Klapwijk, M. P. Nuti, R. A. Schilperoort and A. Rörsch (1977). Transfer of the *Agrobacterium tumefaciens* Ti plasmid to avirulent agrobacteria and to *Rhizobium explanta*. *J. Gen. Microbiol.*, **98**, 477–484.
Hooykaas, P. J. J., M. Hofker, H. den Dulk-Ras and R. A. Schilperoort (1984). A comparison of virulence determinants in an octopine Ti plamid, aopaline Ti plasmid and an Ri plasmid by complementation analysis of *Agrobacterium tumefaciens* mutants. *Plasmid.*, **11**, 195–205.
Hooykaas, P. J. J. and R. A. Schilperoort (1984). The molecular genetics of crown gall tumorigenesis. *Adv. Genet.*, **22**, 210–283.
Hooykaas, P. J. J., H. den Dulk-Ras and R. A. Schilperoort (1988). The *Agrobacterium tumefaciens* T-DNA gene $6^b$ is an *onc* gene. *Plant Mol. Biol.*, **11**, 791–794.
Horsch, R. B., R. T. Fraley, S. G. Rogers, P. R. Sanders, A. Lloyd and N. Hoffmann (1984). Inheritance of functional foreign genes in plants. *Science (Washington D.C.)*., **203**, 496–498.
Horsch, R. B., J. E. Fry, N. L. Hoffmann, D. Eichholtz, S. G. Rogers and R. T. Fraley (1985). A simple and general method for transferring genes into plants. *Science (Washington D.C.)*., **277**, 1229–1231.
Iannino, N. I. and R. A. Ugalde (1989). Biochemical characterisation of virulent *Agrobacterium tumefaciens* ChvA mutants synthesis and excretion of $\beta$-1,2-glucan. *J. Bacteriol.*, **171**(5), 2841–2849.
James, D. J., A. J. Passey, C. J. Barbara and M. Bevan (1989). Genetic transformation of apple (*Malus pumila mill)* using a disarmed Ti-binary vector. *Plant Cell Rep.*, **7**, 658–661.
Jaziri, M., M. Legros, J. Homes and M. Vanhaelen (1988). Tropane alkaloids production by hairy-root cultures of *Datura stramonium* and *Hyoscyamus niger*. *Phytochemistry*, **27**, 419–420.
Jefferson, R. A., T. A. Kavanagh and M. W. Bevan (1987). Gus fusions beta glucuronidase as a sensitive and versatile gene fusion marker in higher plants. *EMBO J.*, **6**, 3901–3908.
Jensen. C. O. (1910). Von echten Geschwulsten bei Pflanzen. *Rapp. Conf. Int. Etude Cancer 2nd*. 243–254.
Jin, S. G., T. Komari, M. P. Gordon and E. W. Nester (1987). Genes responsible for the supervirulence phenotype of *Agrobacterium tumefaciens* A281, *J. Bacteriol.*, **169**, 4417–4425.
John, M. C. and R. M. Amasino (1988). Expression of an *Agrobacterium* Ti plasmid gene involved in cytokinin biosynthesis is regulated by virulence loci and induced by plant phenolic compounds. *J. Bacteriol.*, **170**, 790–795.
Jones, J. D. G., Z. Svab, E. C. Harper, C. D. Hurwitz and P. Maliga (1987). A dominant nuclear streptomycin resistance marker for plant cell transformation. *Mol. Gen. Genet.*, **210**, 86–91.
Jordan, M. C. and A. McHughan (1988). Glyphosate tolerant flax plants from *Agrobacterium*-mediated gene-transfer. *Plant Cell Rep.*, **7**(4), 281–284.
Jorgensen, R., C. Snyder and J. D. G. Jones (1987). T-DNA is organised predominantly in inverted repeat structures in plants transformed with *Agrobacterium tumefaciens* C58 derivatives. *Mol. Gen. Genet.*, **207**, 471–477.
Jung, G. and D. Tepfer (1987). Use of genetic transformation by the Ri T-DNA of *Agrobacterium rhizogenes* to stimulate biomass and tropane alkaloid production in *Atropa belladonna* and *Calystegia sepium* roots grown *in vitro*. *Plant Sci.*, **50**, 145–151.
Kamada, H., N. Okamura, M. Satake, H. Harada and K. Shimomura (1986). Alkaloid production by hairy-root cultures in *Atropa belladonna*. *Plant Cell Rep.*, **5**, 239–242.
Karcher, S. J., V. J. Di Rita and S. B. Gelvin (1984). Transcript analysis of $T_R$-DNA in octopine-type crown gall tumours. *Mol. Gen. Genet.*, **194**, 159–165.
Karp, A. and S. W. J. Bright (1985). On the causes and origins of somaclonal variation. In *Oxford Surveys of Plant Molecular and Cell Biology*, ed. B. J. Miflin, pp. 199–234, Oxford University Press, London.
Kerr, A., P. Manigault and J. Temple (1977). Transfer of virulence *in vivo* and *in vitro* in *Agrobacterium*. *Nature*, **265**, 560–561.
Klapwijk, P. M., T. Scheulderman and R. A. Schilperoort (1978). Coordinated regulation of octopine degradation and conjugative transfer of Ti plasmids in *Agrobacterium tumefaciens*: Evidence for a common regulatory gene and separate operons. *J. Bacteriol.*, **136**, 775–785.
Klapwijk, P. M. and R. A. Schilperoort (1982). Genetic determination of octopine degradation. In *Molecular Biology of Plant Tumors*, ed. G. Kahl and J. Schell, pp. 475–495. Academic Press, New York.
Klass, M., C. J. Manorama, O. N. Crowell and R. M. Amasino (1989). Rapid induction of genomic demethylation and T-DNA gene expression in plant cells by 5-azacytosine derivatives. *Plant Mol. Biol.*, **12**, 413–423.
Klee, H. J., M. B. Hayford and S. G. Rogers (1987a). Gene rescue in plants; a model system for 'shotgun' cloning by retransformation. *Mol. Gen. Genet.*, **210**, 282–287.
Klee, H., R. Horsch and S. Rogers (1987b). *Agrobacterium*-mediated plant transformation and its further applications to plant biology. *Annu. Rev. Plant Phys.*, **38**, 467–486.
Klein, T. M., E. D. Wolf, R. Wu and J. C. Sanford (1987). High velocity microprojectiles for delivering nucleic acids into living cells. *Nature* , **327**, 70–73.

Knauf, V. C. (1987). The application of genetic engineering to oilseed crops. *Trends Biotechnol.*, **5**, 40–46.

Komro, C. T., V. J. Di Rita, S. B. Gelvin and J. D. Kemp (1985). Site-specific mutagenesis in the $T_R$-DNA region of octopine-type Ti plasmids. *Plant Mol. Biol.*, **4**, 253–263.

Koncz, C. and J. Schell (1986). The promoter of $T_L$-DNA gene 5 controls the tissue-specific expression of chimeric genes carried by a novel type of *Agrobacterium* binary vector. *Mol. Gen. Genet.*, **204** (3), 383–396.

Koncz, C., O. Olsson, W. H. R. Langridge, J. Schell and A. A. Szalay (1987). Expression and assembly of functional bacterial luciferase in plants. *Proc. Natl. Acad. Sci. USA*, **84**, 131–135.

Koornneef, M., J. Van Eden, C. J. Hanhart, P. Stam, F. J. Braaksma and W. J. Feenstra (1983). Linkage maps of *Arabidopsis thaliana. Hered*, **74**, 265–272.

Koornneef, M., C. Hanhart, M. Jongsma, I. Toma, R. Weide, P. Zabel and J. Hille (1986). Breeding of a tomato genotype readily accessible to genetic manipulation. *Plant Sci.*, **45**, 201–208.

Koukolikova-Nicola, Z., R. D. Shillito, B. John, K.Wang, M. Van Montagu and P. Zinbryski (1985). The involvement of circular intermediates in the transfer of T-DNA from *Agrobacterium tumefaciens* to plant cells. *Nature*, **313**, 191–196.

Krens, F. A., L. Molendijk, G. J. Wullems and R. A. Schilperoort (1982). *In vitro* transformation of plant protoplasts with Ti plasmid DNA. *Nature*, **296**, 72–74.

Kuhlemeier C., P. J. Green and N. H. Chua (1987). Regulation of gene expression in higher plants. *Annu. Rev. Plant Physiol.*, **38**, 221–257.

Kuhlemeier, C. and P. Green (1987). Studying plant development: an alternative to 'spray and pray'. *Greens and Development*, **1**, 3–5.

Kurz, W. G. W. and F. Constabel (1979). Plant cell cultures, a potential source of pharmaceuticals. *Adv. Appl. Microbiol.*, **25**, 209–240.

Larkin, P. J. (1987). Somaclonal variation, history, method and meaning. *Iowa State J. Res.*, **61**, 393–434.

Larkin, P. J. and W. R. Scowcroft (1981). Somaclonal variation — a novel source of variability from cell culture for plant improvement. *Theor. Appl. Genet.*, **60**, 197–214.

Lee, M. and R. L. Phillips (1988). The chromosomal basis of somaclonal variation. *Annu. Rev. Plant Physiol. Plant. Mol. Biol.*, **39**, 413–437.

Leisner, S. M. and S. B. Gelvin (1988). Structure of the octopine synthase upstream activator sequence. *Proc. Natl. Acad. Sci. USA*, **85** (8), 2553–2557.

Leemans, J., R. Deblaere, L. Willmitzer, H. de Greve, J. P. Hernalsteens, M. Van Montagu and J. Schell (1982). Genetic identification of functions of $T_L$-DNA transcripts in octopine crown galls. *EMBO J.*, **1**, 147–152.

Letham, D. S. and L. M. S. Palni (1983). The biosynthesis and metabolism of cytokinins. *Annu. Rev. Plant Physiol.*, **34**, 163–197.

Lloyd, A. M., A. R. Barnason, S. G. Rogers, M. C. Byrne and R. T. Fraley (1986). Transformation of *Arabidopsis thaliana* with *Agrobacterium tumefaciens. Science (Washington D.C.)*, **234**, 464–466.

Loesch-Fries, E. S., D. Merlo, T. Zinner, L. Burhop, K. Hill, K. Krahn, N. Jeris, S. Nelson and E. Stalk (1967). Expression of alfalfa mosaic virus RNA4 in transgenic plants confers virus resistance. *EMBO J.*, **6**, 1845–1851.

Loh, C. S. and A. N. Rao (1988). Shoot regeneration from *Agrobacterium tumefaciens* induced tumours of a tropical timber tree *Fagrae-fragrans. Experimentia*, **44**, 72–73.

Lörz, H., B. Baker and J. Schell (1985). Gene transfer to cereal cells mediated by protoplasty transformation. *Mol. Gen. Genet.*, **199**, 178–182.

Lurquin, P. F. (1987). Foreign gene expression in plant cells. In *Progress in Nucleic Acids Research and Molecular Biology*, ed. W. E. Cohn and K. Moldave, vol. 34, pp. 173–188.

Mackay, J., A. Seguin and M. Lalonde (1988). Genetic transformation of *in vitro* clones of *Alnus* and *Betula* by *Agrobacterium tumefaciens. Plant Cell Rep.*, **7**, 229–232.

Mano, Y., S. Nabeshima, C. Matsui and H. Ohkawa (1987). Production of tropane alkaloids by hairy-root cultures of *Scopolia japonica. Agric. Biol. Chem.*, **50**, 2715–2722.

Marks, J. R., T. J. Lynch, J. E. Karlsinsey and M. F. Thomashow (1987). *Agrobacterium tumefaciens* locus *pscA* is related to the *Rhizobium melilori exoC* locus. *J. Bacteriol.*, **169**, 5835–5837.

Marton, L., G. J. Wullems, L. Molendijk and R. A. Schilperoort (1979). *In vitro* transformation of cultured cells from *Nicotiana tabacum* by *Agrobacterium tumefaciens. Nature*, **277**, 129–131.

Matthysse, A. G. (1987). Characterization of nonattaching mutants of *Agrobacterium tumefaciens. J. Bacteriol.*, **169**, 313–323.

Matzke, A. J. M. and M. A. Matzke (1986). A set of novel Ti plasmid-derived vectors for the production of transgenic plants. *Plant Mol. Biol.*, **7**, 357–366.

Matzke, M. A., M. Brimig, J. Trnovsky and A. J. M. Matzke (1989). Reversible methylation and inactivation of marker genes in sequentially transformed tobacco plants. *EMBO J.*, **8**, 643–649.

McCabe, D., W. F. Swain, B. J. Martinell and P. Christou (1988). Stable transformation of soybean (*Glycine max*) by particle acceleration. *Biotechnology*, **6**, 923–926.

McCormick, S., J. Niedermeyer, J. Fry, A. Barnason, R. Horsch and R. Fraley (1986). Leaf disk transformation of cultivated tomato (*L. esculentum*) using *Agrobacterium tumefaciens. Plant Cell Rep.*, **5**, 81–84.

McGranahan, G. H., C. A. Leslie, S. L. Uratsu, L. A. Martin and A. M. Dandekar (1988). *Agrobacterium*-mediated transformation of walnut somatic embryos and regeneration of trangenic plants. *Biotechnology*, **6** (7), 800–804.

Melchers, L. S. and P. J. J. Hooykaas (1987). Virulence of *Agrobacterium*. In *Oxford Surveys of Plant Molecular and Cell Biology*, ed. B. J. Miflin, vol. 4, pp. 167–220.

Memelink, J., S. De Pater, J. H. C. Hoge and R. A. Schilperoort (1987). T-DNA hormone biosynthetic genes: phytohormones and gene expression in plants. *Dev. Genet.*, **8**, 321–338.

Messens, E., A. Lenaerts, M. Van Montagu and R. W. Hedges (1985). Genetic basis for opine secretion from crown gall tumour cells. *Mol. Gen. Genet.*, **199**, 344–348.

Meyer, P., I. Heidmann, G. Forkmann and H. Saedler (1987). A new petunia flower colour generated by transformation of a mutant with a maize gene. *Nature*, **330**, 677–78.

Meyer, P., S. Kartzke, I. Niedenhof, I. Heidmann, K. Bussmann and H. Saedler (1988). A genomic DNA sequence from *Petunia hybrida* leads to increased transformation frequencies and simple integration patterns. *Proc. Natl. Acad. Sci. USA*, **85**, 8568–8572.

Meyerowitz, E. (1989). *Arabidopsis*, a useful weed. *Cell*, **56**, 263–269.

Michelmore, R., E. Marsh, S. Seely and B. Landry (1987). Transformation of lettuce (*Lactuca sativa*) mediated by *Agrobacterium tumefaciens*. *Plant Cell Rep.*, **6**, 439–442.
Morgan, A. J., P. N. Cox, D. A. Turner, E. Peel, M. R. Davey, K. M. A. Gartland and B. J. Mulligan (1987). Transformation of tomato using Ri Plasmid vector. *Plant Sci.*, **49**, 37–50.
Mouras, A., I. Negrutiu and Y. Dessaux (1987). Phenotypic and genetic variations in crown gall tumour cells of tobacco. *Theor. Appl. Genet.*, **74**, 253–260.
Mugnier, J. (1988). Esablishment of new axenic hairy root lines by inoculation with *Agrobacterium rhizogenes*. *Plant Cell Rep.*, **7**, 9–12.
Müller, A., T. Manzora and P. F. Lurquin (1983). Crown gall transformation to tobacco callus by cocultivation with *Agrobacterium tumefaciens*. *Biochem. Biophys Res. Commun.*, **123**, 458–462.
Müller, A. J., R. R. Mendel, J. Schiemann, C. Simoens and D. Inze (1987). High meiotic stability of a foreign gene introduced into tobacco by *Agrobacterium*-mediated transformation. *Mol. Gen. Genet.*, **207**, 171–175.
Muradov, A. Z. and D. A. Aliev (1986). Construction of intermediate vectors for the transfer and expression of foreign genes in higher plants. *Az. SSR, Baku, USSR*, **3**, 101–108.
Naina, N. S., P. K. Gupta and A. F. Mascarenhas (1989). Genetic transformation and regeneration of transgenic neem (*Azadirachta-indica*) plants using *Agrobacterium tumefaciens*. *Curr. Sci.*, **58**, 184–187.
Nelson, R. S., P. Powell Abel and R. N. Beachy (1987). Lesions and virus accumulation in inoculated transgenic tobacco plants expressing the coat protein of tobacco mosaic virus. *Virology*, **158**, 126–132.
Nester, E. W., M. P. Gordon, P. M. Amasino and M. F. Yanofsky (1984). Crown gall — a molecular and physiological analysis. *Annu. Rev. Plant Physiol.*, **35**, 387–413.
Neuhaus, G., G. Spangenberg, O. Mittelstein Scherd and H.-G. Schweiger (1987). Transgenic rape seed plants obtained by the microinjection of DNA into microspore-derived embryoids. *Theor. Appl. Genet.*, **75**, 30–36.
Novick, R. (1980). Plasmids. *Sci. Am.*, **243**, 76–90.
O'Connell, K. M. and J. Handelsman (1989). *ChvA* locus may be involved in export of neutral cyclic $\beta$-1, 2-linked D-glucan from *Agrobacterium tumefaciens*. *Mol. Plant Microbial. Int.*, **2** (1), 11–16.
Offringa, I. A., L. S. Melchers, A. J. G. Regensburg-Tuink, P. Costantino, R. A. Schiperoort and P. J. J. Hoojkaas (1986). Complementation of *Agrobacterium tumefaciens*. tumour inducing *aux* mutants by genes from the $T_R$-region of the Ri plasmid of *Agrobacterium rhizogenes*. *Proc. Natl. Acad. Sci. USA*, **83**, 6935–6939.
Olszewski, N. E., F. B. Martin and F. M. Ausubel (1988). Specialised binary vector for plant transformation; expression of the *Arabidopsis thalidna* AHAS gene in *Nicotiana tabacum*. *Nucleic Acids Res.*, **16**, 10 765–10 782.
Ondrej, M., R. Biskova and J. Vlasak (1986). Binary plant vector based on Ri plasmid and part of T-DNA of the Ti plasmid. *Inst. Exp. Bot. Czech. Acad. Sci.* **28** (4), 265–269.
Ooms, G. (1987). Controlling differentiation and endogenous growth regulator content by means of genetic transformations. In *BPGRG Monograph 16 — Advances in the chemical manipulation of plant tissue cultures*. ed. M. B. Jackson, S. H. Mantell and J. Blake, pp. 1–17. Parchments, Oxford.
Ooms, G., P. M. Klapwijk, J. A. Poulis and R. A. Schilperoort (1980). Characterization of Tn904 insertions in octopine Ti plasmid mutants of *Agrobacterium tumefaciens*. *J. Bacteriol.*, **144**, 82–91.
Ooms, G., P. J. J. Hooykaas, G. Moolenaar and R. A. Schilperoort (1981). Crown gall plant tumours of abnormal morphology, induced by *Agrobacterium tumefaciens* carrying mutated octopine Ti plasmids: analysis of T-DNA functions. *Gene*, **14**, 33–50.
Ooms, G., P. J. J. Hooykaas, R. J. M. Van Veen, P. van Beelen, T. J. G. Regensburg and R. A. Schilperoort (1982a). Octopine Ti plasmid deletion mutants of *Agrobacterium tumefaciens* with emphasis on the right side of the T-region. *Plasmid*, **7**, 15–29.
Ooms, G., A. Bakker, L. Molendijk, G. J. Wullems, M. P. Gordon, E. W. Nester and R. A. Schilperoort (1982b). T-DNA organization in homogeneous and heterogeneous octopine type crown gall tissues of *Nicotiana tabacum Cell*, **30**, 589–597.
Ooms, G., A. Karp and J. Roberts (1983). From tumour to tuber: tumour cell characteristics and chromosome numbers of crown gall-derived tetraploid potato plants (*Solanum tuberousm* cv. Maris Bard). *Theor. Appl. Genet.*, **66**, 169–172.
Ooms, G., A. Karp, M. M. Burrell. D. Twell and J. Roberts (1985a). Genetic modification of potato development using Ri T-DNA. *Theor. Appl. Genet.*, **70**, 440–446.
Ooms, G., A. Bains, M. Burrell, A. Karp. D. Twell and E. Wilcox (1985b). Genetic manipulation in cultivars of oilseed rape *Brassica napus* using *Agrobacterium*. *Theor. Appl. Genet.*, **71**, 325–329.
Ooms, G. and J. R. Lenton (1985c). Ti plasmid DNA genes to study plant development precocious tuberization and enhanced cytokinins in *Agrobacterium tumefaciens*. *Plant Mol. Biol.*, **5**, 205–212.
Ooms, G., M. E. Bossen, M. M. Burrell and A. Karp (1986a). Genetic manipulation in potato using *Agrobacterium rhizogenes*. *Potato Res.*, **29**, 367–379.
Ooms, G., D. Twell, M. E. Bossen, J. H. C. Hoge and M. M. Burrell (1986b). Development regulation of Ri Ti DNA gene expression in roots, shoots and tubers of transformed potato (*Solanum tuberosum* cv. Desiree). *Plant Mol. Biol.*, **6**, 321–330.
Ooms, G., M. M. Burrell, A. Karp, M. Bevan and J. Hille (1987). Genetic transformation in two potato cultivars with T-DNA from disarmed *Agrobacterium*. *Theor. Appl. Genet.*, **73**, 744–750.
Otten, L. A. B. M. and R. A. Schilperoort (1978). A rapid microscale method for the detection of lypsopine and nopaline dehydrogenase activities. *Biochem. Biophys. Acta*, **527**, 497–500.
Ow, D. W., K. V. Wood, M. Deluca, J. R. De Wet, D. R. Helinski and S. H. Howell (1986). Transient and stable expression of the firefly luciferase gene in plant cells and transgenic plants. *Science* (*Washington D.C*). **234**, 856–859.
Parker, D., L. N. Ornston and E. W. Nester (1987). Chemotaxis to plant phenolic inducers of virulence genes is constitutively expressed in the absence of the Ti plasmid in *Agrobacterium tumefaciens*. *J. Bacteriol.*, **169**, 5336–5338.
Pavingerova, D., M. Ondrej and J. Matonsek (1983). Analysis of progeny of *Arabidopsis thaliana* plants regenerated from crown gall tumors *Z. Pflanzenphysiol.*, **112**, 427–433.
Payne, J., M. J. C. Rhodes and R. J. Robins (1987). Quinoline alkaloid production by transformed cultures of *Cinchona ledgeriana*. *Planta Med.*, 367–372.
Peerbolte, R., K. Leenhouts. G. M. S. Hooykaas-Van Slogtern, G. J. Wullems and R. A. Schilperoort (1986a). Clones from a shooty tobacco crown gall tumor I: deletions, rearrangements and amplifications resulting in irregular T-DNA structures and organizations. *Plant Mol. Biol.*, **7**, 265–284.
Peerbolte, R., K. Leenhouts, G. M. S., Hooykaas-Van Slogteren, G. J. Wullems and R. A. Schilperoort (1986b). Clones from a

shooty tobacco crown gall tumour II. Irregular T-DNA structures and organization, T-DNA methylation and conditional expression of opine genes. *Plant Mol. Biol.*, **7**, 285–300.

Peerbolte, R., P. Ruigrok, G. Wullems and R. Schilperoort (1987). T-DNA rearrangements due to tissue culture — somaclonal variation in crown gall tissues. *Plant Mol. Biol.*, **9**, 51–57.

Peralta, E. G., R. Hellmiss and W. Ream (1986). Overdrive; a T-DNA transmission enhancer on the *Agrobacterium tumefaciens* tumour-inducing plasmid. *EMBO J.*, **5**, 1137–1142.

Petit, A., C. David, G. A. Dahl, J. G. Ellis, P. Guyon, F. Casse-Delbart and J. Tempé (1983). Further extension of the opine concept: plasmids in *Agrobacterium rhizogenes* cooperate for opine degradation. *Mol. Gen. Genet.*, **190**, 204–214.

Petit, A., J. Stougaard, A. Kuhle, K. A. Marcker and J. Tempé (1987). Transformation and regeneration of the legume *Lotus-corniculatus*: a system for molecular studies of symbiotic nitrogen fixation. *Mol. Gen. Genet.*, **207**, 245–250.

Pollock, K., D. G. Barnfield, S. J. Robinson and R. Shields (1985). Transformation of protoplast-derived cell colonies and suspension cultures by *Agrobacterium tumefaciens. Plant Cell Rep.*, **4**, 202–205.

Potrykus, I., R. D. Shillito, M. W. Saul and J. Paszkowski (1985a). Direct gene transfer: state of the art and future potential. *Plant Mol. Biol. Rep.*, **3**, 117–128.

Potrykus, I., J. Paszkowski, M. W. Saul, J. Petruska and R. D. Shillito (1985b). Molecular and general genetics of a hybrid gene introduced to tobacco by direct gene transfer. *Mol. Gen. Genet.*, **199**, 169–177.

Powell, Abel, P., R. S. Nelson, N. Hoffmann, S. G. Rogers, R. T. Fraley and R. N. Beachy (1987). Delay of disease development in transgenic plants that express the tobacco mosaic virus coat protein gene. *Science (Washington D.C.)*, **232**, 738–743.

Prosen, D. E. and R. B. Simpson (1987). Transfer of a ten member genomic library to plants using *Agrobacterium tumefaciens. Biotechnology*, **5**, 966–971.

Pua, E.-C., A. Mehra-Palta, F. Nagy and N.-H. Chua (1987). Transgenic plants of *Brassical napus* L. *Biotechnology*, **5**, 815–817.

Pythoud, F., V. P. Sinkar, E. W. Nester and M. P. Gordon (1987). Increased virulence of *Agrobacterium rhizogenes* conferred by the *vir* region of $_{P}$Ti B0542 — Application to genetic engineering of poplar. *Biotechnology*, **5**, 1323–1327.

Radke, S. E., B. M. Andrews, M. M. Moloney, M. L. Crouch, J. C. Kridl and V. C. Knauf (1988). Transformation of *Brassica napus* L. using *Agrobacterium tumefaciens*: developmentally regulated expression of a reintroduced napin gene. *Theor. Appl. Genet.*, **75**, 685–694.

Rasmussen, I. and I. Dangaard (1988). Transformation of rape *Brassica napus Agrobacterium rhizogenes. Physiol. Plant.*, **72** (2), 23.

Rech, E. L., T. J. Golds, N. Hammatt, B. J. Mulligan and M. R. Davey (1988). *Agrobacterium rhizogenes*-mediated transformation of the wild soybeans *glycine canescens* and *glycine- clandestina*: production of transgenic plants of *glycine canescens. J. Exp. Bot.*, **39**, 1275–1285.

Reich, T. J., V. N. Iyer and B. L. Miki (1986). Efficient transformation of alfalfa protoplasts by the intranuclear microinjection of Ti plasmids. *Biotechnology*, **4**, 1001–1004.

Rezaian, M. A., K. G. M. Skene and J. G. Ellis (1988). Antisense RNAs of cucumber mosaic virus in transgenic plants assessed for control of the virus. *Plant Mol. Biol.*, 463–471.

Robbs, S. L., S. G. Pueppke and M. C. Haves (1987). Genotypic variation in susceptibility of *Pisum sativum* to infection by *Agrobacterium tumefaciens. Phytopathology*, **77**, 1241–1242.

Rodermel, G. R., M. S. Abbott and L. Bogorad (1988). Nuclear-organelle interactions — nuclear antisense gene inhibits ribulosebisphosphate carboxylase enzyme levels in transformed tobacco plants. *Cell*, **55** (4), 673–681.

Rogers, S. G. and H. Klee (1987). Pathways to plant genetic manipulation employing *Agrobacterium.* In *Plant gene Research: Basic Knowledge* and *Application: Plant DNA Infectious Agents*, ed. T. Hohn and J. Schell, pp. 179–204. Springer-Verlag, New York.

Rogers, S., H. Klee, M. Byrne, R. Horsch and R. Fraley (1987). Improved vectors for plant transformation: expression cassette vectors and new selectable markers. *Methods in Enzymol.*

Rothstein, S. J., K. N. Lahners, R. J. Lotstein, N. B. Carozzi, S. M. Jayne and D. A. Rice (1987). Promoter cassettes, antibiotic resistance genes and vectors for plant transformation. *Gene*, **53**, 153–161.

Salomon, F., R. Deblaere, J. Leemans, J. P. Hernalsteen, M. Van Montagu and J. Schell (1984). Genetic identification of functions of Tr-DNA transcripts in octopine crown galls. *EMBO J.*, **3**, 141–146.

Sanford, J. C. (1988). The biolistic process. *Trends Biotechnol.*, **6** (12), 299–302.

Schafer, W., A. Gorz and G. Kahl (1987). T-DNA integration and expression in a monocot crop plant after induction of *Agrobacterium. Nature*, **327**, 529–532.

Schardl, C. L. A. D. Byrd, G. Benzion, M. A. Altschuler, D. F. Hildebrand and A. G. Hunt (1987). Design and construction of a versatile system for the expression of foreign genes in plants. *Gene*, **61**, 1–12.

Schell, J. (1987). Transgenic plants as tools to study the molecular organization plant genes. *Science (Washington D.C.)*, **237**, 1176–1183.

Schmidt, R. and L. Willmitzer (1988). High efficiency *Agrobacterium tumefaciens*-mediated transformation of *Arabdopsis thaliana* leaf and cotyledon explants. *Plant Cell Rep.*, **7** (7), 583–586.

Scott, R. J. and J. Draper (1987). Transformation of carrot tissues derived from proembryogenic suspension cells: a useful model system for gene expression studies in plants. *Plant Mol. Biol.*, **8**, 265–274.

Scowcroft, W. R. and S. A. Ryan (1986). Tissue culture and plant breeding. In *Plant Cell Culture Technology*, ed. M. M. Yeoman, pp. 67–95. Blackwell, Oxford.

Sen, P., G. J. Dazour, D. Anderson and A. Das (1986a). Co-operative binding of *Agrobacterium tumefaciens virE2* protein to single stranded DNA. *J. Bacteriol.*, **171** (5), 2573–2580.

Shah, D., R. Horsch, H. Klee, G. Kishore, J. Winter, N. Turner, C. Hironaka, P. Sanders, C. Gasser, S. Aykent, N. Siegel, S. Rogers and R. T. Fraley (1986). Engineering herbicide tolerance in transgenic plants. *Science (Washington D.C.)*, **233**, 478–481.

Shahin, E. A. and R. B. Simpson (1986). Gene transfer system for potato *Solanum tuberosum. HortScience*, **21**, 1190–1201.

Shahin, E. A., K Sukhapinda, R. B. Simpson and R. Spivey (1986a). Transformation of cultivated tomato *Lycopersicon esculentum* by a binary vector in *Agrobacterium rhizogenes*: transgenic plants with normal phenotypes harbor binary vector transferred DNA but no Ri plasmid transferred DNA. *Theor. Appl. Genet.*, **72**, 770–777.

Shahin, E. A., A. Spielmann, K. Sukhapinda, R. B., Simpson and M. Yashar (1986b). Transformation of cultivated alfalfa (*Medicago sativa*) using disarmed *Agrobacterium tumefaciens. Crop Sci.*, **26**, 1235–1239.

Shaw, C. H., A. M. Ashby, A. Brown, C. Royal, G. J. Loake and C. H. Shaw (1988). *Vir A* and *Vir G* are the Ti plasmid functions required for chemotaxis of *Agrobacterium tumefaciens* towards acetosyringone. *Mol. Microbiol.*, **2**, 413–418.

Sheerman, S. and M. W. Bevan (1988). A rapid transformation method for *Solanum tuberosum* using binary *Agrobacterium tumefaciens* vectors. *Plant Cell Rep.*, **7**, 13–16.

Sheikholeslam, S. N. and D. P. Weeks (1987). Acetosyringone promotes high efficiency transformation of *Arabidopsis thaliana* explants by *Agrobacterium tumefaciens*. *Plant Mol. Biol.*, **8**, 291–298.

Shen, W. H., A. Petit, J. Guern and J. Tempe (1988). Hairy-roots are more sensitive to auxin than normal roots. *Proc. Natl. Acad. Sci. USA*, **35**, 3417–3421.

Shillito, R. D., M. W. Saul, J. Paszkowski, M. Muller and I. Potrykus (1985). High efficiency direct gene transfer to plants. *Biotechnology*, **3**, 1099–1103.

Shimamoto K., R. Terada, T. Izawa and H. Fujimoto (1989). Fertile transgenic rice plants regenerated from transformed protoplasts. *Nature (London)*, **338**, 274–276.

Simoens, C., M. Alliotte, R. Mendel, A. Muller, J. Schiemann, M. van Lysebettens, J. Schell, M. Van Montagu and D. Inzé (1986). A binary vector for transferring genomic libraries to plants. *Nucleic Acids Res.*, **14**, 8073–8090.

Simpson, R. B., P. J. O'Hara, W. Kwok, A. L. Montoya, C. Lichtenstein, M. P. Gordon and E. W. Nester (1984). DNA from the A6 S/2 crown gall tumor contains scrambled Ti plasmid sequences near its junctions with plant DNA. *Cell*, **29**, 1005–1014.

Simpson, R. B., A. Spielmann, L. Margossian and T. D. McKnight (1986). A disarmed binary vector from *Agrobacterium tumefaciens* functions in *Agrobacterium rhizogenes* frequent cotransformation of two distinct T-DNA species. *Plant Mol. Biol.*, **6**, 403–416.

Sinkar, V. P., F. F. White, I. J. Furner, M. Abrahamsen, F. Pythoud and M. P. Gordon (1988). Reversion of aberrant plants transformed with *Agrobacterium rhizogenes* is associated with the transcriptional inactivation of the $T_L$-DNA genes. *Plant Physiol.*, **86**, 584–590.

Skoog, F. and C. O. Miller (1957). Chemical regulation of growth and organ formation in plant tissues cultured *in vitro*. *Symp. Soc. Exp. Biol.*, **11**, 118–131.

Slightom, J. L., M. Durand-Tardif, L. Jouanin and D. Tepfer (1986). Nucleotide sequence analysis of $T_L$-DNA of *Agrobacterium rhizogenes* agropine type plasmid–identification of open reading frames. *J. Biol. Chem.*, **261**, 108–121.

Smarelli, J., M. T. Watters and L. H. Diba (1986). Response of various cucurbits to infection by plasmid harbouring strains of *Agrobacterium*. *Plant Physiol.*, **82**, 622–624.

Smith, E. F. (1917). Embryomas in plants (produced by bacterial inoculations). *Bull. John Hopkins Hosp.*, **28**, 277–294.

Smith, E. F. and C. O. Townsend (1907). A plant tumour of bacterial origin. *Science (Washington D.C)*, **25**, 671–673.

Spano, L., D. Mariotti, M. Pezzotti, F. Damiani and S. Arcioni (1987). Hairy-root transformation in alfalfa *Medicago sativa* L. *Theor. Appl. Genet.*, **73**, 523–530.

Spena, A., T. Schmülling, C. Koncz and J. S. Schell (1987). Independent and synergistic activity of *rol* A, B and C loci in stimulating abnormal growth in plants. *EMBO J.*, **6**, 3891–3899.

Spencer, P. A. and G. H. N Towers (1988). Specificity of signal compounds detected by *Agrobacterium tumefaciens*. *Phytochemistry*, **27**(9), 2781–2786.

Spielmann, A. and R. B. Simpson (1986). T-DNA structure in transgenic tobacco plants with multiple independent integration sites. *Mol. Gen. Genet.*, **205**, 34–41.

Spier, R. E., M. W. Fowler (1985). Animal and plant cell cultures. In *Comprehensive Biotechnology*, ed. A. T. Bull and H. Dalton, vol. 1, pp. 301–329. Pergamon Press, Oxford.

Stachel, S. E., E. Messens, M. Van Montagu and P. Zambryski (1985). Identification of the signal molecules produced by wounded plant cells that activate T-DNA transfer in *Agrobacterium tumefaciens*. *Nature (London)*, **318**, 624–629.

Stachel, S. and P. C. Zambryski (1986). *Agrobacterium tumefaciens* and the susceptible plant cell: a novel adaptation of extracelluar recognition and DNA conjugation. *Cell*, **47**, 155–158.

Stachel, S. E., B. Timmerman and P. Zambryski (1986). Generation of single-stranded T-DNA molecules during the initial stages of T-DNA transfer from *Agrobacterium tumefaciens* to plant cells. *Nature (London)*, **322**, 706–712.

Stachel. S. and E. W. Nester (1986). The genetic and transcriptional organization of the *vir* region of the A6 Ti plasmid of *Agrobacterium tumefaciens*. *EMBO J.*, **5**, 1445–1454.

Stalker, D. M., K. E. McBride and L. D. Malyj (1988). Herbicide resistance in transgenic plants expressing a bacterial detoxification gene. *Science (Washington D.C.)*, **242**, 419–423.

Steinbrucken, H. and N. Amrhein (1980). The herbicide glyphosate is a potent inhibitor of 5-enolpyruvyl shikimic acid 3-phosphate synthase. *Biochem. Biophys. Res. Commun.*, **94**, 1207–1212.

Stiekema, W. J., F. Heidekamp, J. D. Louwerse, H. A. Verhoeven and P. Dijkhuis (1988). Introduction of foreign genes into potato cultivars Bintje and Desiree using an *Agrobacterium tumefaciens* binary vector. *Plant Cell Rep.*, **7**, 47–50.

Stockhaus. J., P. Eckes, A. Blan, J. Schell and L. Willmitzer (1987). Organ specific and dosage-dependent expression of a leaf/item specific gene from potato after tagging and transfer into potato and tobacco plants. *Nucleic Acids Res.*, **15**, 3479–3491.

Stougaard, J., D. Abildsten and K. A. Marcher (1987). The *Agrobacterium rhizogenes* pRi $T_L$-DNA segment as a gene vector system for transformation of plants. *Mol. Gen. Genet.*, **207**, 251–255.

Sukhapinda K., R. Spivey, R. B. Simpson and E. A. Shahin (1987a). Transgenic tomato (*Lycopersicon esculentum* L.) transformed with a binary vector in *Agrobacterium rhizogenes* nonchimeric origin of callus clone and low copy numbers of integrated vector T-DNA. *Mol. Gen. Genet.*, **206**, 491–497.

Sukhapinda. K., R. Spivey and E. A. Shahin (1987b). Ri plasmid as a helper for introducing vector DNA into alfalfa plants. *Plant Mol. Biol.*, **8**, 209–216.

Tabata, M. and Y. Fujita (1985). Production of shikonin by plant cell cultures. In *Biotechnology in Plant Science*, ed. M. Zaitlin, P. Day and A. Hallaender, pp. 207–218, Academic Press, New York.

Taylor, B. H., F. F. White, E. W. Nester and M. P. Gordon (1985). Transcription of *Agrobacterium rhizogenes* A4 T-DNA. *Mol. Gen. Genet.*, **201**, 546–553.

Tempé, J. and A. Goldmann (1982). Occurrence and biosynthesis of opines. In *Mol. Biol. of Plant Tumours* ed. G. Kahl, and J. Schell, pp. 428–450. Academic Press, New York.

Tempé, J. and Petit (1982). Opine utilization by *Agrobacterium*. In *Mol. Biol. of Plant Tumours* ed. G. Kahl and J. Schell, pp. 451–460. Academic Press, New York.

Tepfer, D. (1984). Transformation of several species of higher plants by *Agrobacterium rhizogenes*–sexual transmission of the transformed genotype and phenotype. *Cell*, **37**, 959–967.

Thomashow, M. F., R. Nutter, A. L. Montoya, M. P. Gordon, and E. W. Nester (1980). Integration and organization of Ti plasmid sequences in crown gall tumours. *Cell*, **19**, 729–739.

Thomashow, M. F., J. E. Korlinsey, J. R. Marles and R. F. Hurlbert (1987). Identification of a new virulence locus in *Agrobacterium tumefaciens* that affects polysaccharide composition and plant cell attachment. *J. Bacteriol.*, **169**, 3209–3216.

Thompson, D. V., L. S. Melchers, K. B. Iyer, R. A. Schilperoort and P. J. J. Hooykaas (1988). Analysis of the complete nucleotide sequence of the *Agrobacterium tumefaciens virB* operon. *Nucleic Acids Res.*, **16**, 4621–4636.

Timmerman, B., M. Van Montagu and P. Zambryski (1988). *Vir*-induced recombination in *Agrobacterium*: physical characterization of precise and inprecise T-circle formation. *J. Mol. Biol.*, **203** (2), 373–384.

Töpfer, R., B. Gronenborn, J. Schell and H. H. Steinbiss (1989). Uptake and transient expression of chimeric genes in seed-derived embryos. *Plant Cell.*, **1**, 133–139.

Toriayma K., Y. Arimoto, M. Uchimiya and K. Hinata (1988). Transgenic rice plants after direct gene transfer into protoplasts. *Biotechnology.*, **6**, 1072–1074.

Toro, N., A. Datta, M. Yanofski and E. Nester (1988). Role of overdrive sequence in T-DNA border cleavage in *Agrobacterium. Proc. Natl. Acad. Sci. USA.*, **85** (27), 8558–8562.

Tremblay, G., R. Labert, H. Lebeuf and P. Dion (1987). Isolation of bacteria from soil and crown gall tumours on the basis of their capacity for opine utilization. *Phytoprotection*, **68**, 35–42.

Trulson, A. J., R. B. Simpson and E. A. Shahin (1986). Transformation of cucumber (*Cucumis sativus* L.) plants with *Agrobacterium rhizogenes. Theor. Appl. Genet.*, **73**, 11–15.

Tumer, N. E., K. M. O'Connell, R. S. Nelson, P. R. Saunders, R. N. Beachy, R. T. Fraley and D. M. Shah (1987). Expression of alfalfa mosaic virus coat protein gene confers cross protection in transgenic tobacco and tomato plants. *EMBO J.*, **6**, 1181–1188.

Turgeon R., H. N. Wood and A. C. Braun (1976). Studies on the recovery of crown gall tumour cells. *Proc. Natl. Acad. Sci. USA*, **73**, 3562–3564.

Twell D. and G. Ooms (1987). The 5′ flanking DNA of a patatin gene directs tuber specific expression of a chimeric gene in potato. *Plant Mol. Biol.*, **9**, 365–376.

Umbeck, P., G. Johnson, K. Barton and W. Swain (1987). Genetically transformed cotton (*Gossypium hirsutum* L.) plants. *Biotechnology*, **5**, 263–266.

Usami, S., S. Morikawa, I. Takebe and Y. Machida (1987). Absence in monocotyledonous plants of the diffusible plant factors inducing T-DNA circularization and *vir* gene-expression in *Agrobacterium. Mol. Gen. Genet.*, **209**, 221–226.

Vaeck, M., A. Reynaerts, H. Höfte, S. Jansens, M. de Beuckeleer, C. Dean, M. Zabean, M. van Montagu and J. Leemans (1987). Transgenic plants protected from insect attack. *Nature.*, **328**, 33–37.

Van den Elzen, P. J. M., K. Y. Lee, J. Towsend and J. R. Bedbrook (1985a). Simple binary vectors for DNA transfer to plant cells. *Plant Mol. Biol.*, **5**, 149–154.

Van den Elzen, P. J. M., J. Townsend, K. Y. Lee and J. R. Bedbrook (1985b). A chimaeric hygromycin resistance gene as a selectable marker in plant cells. *Plant Mol. Biol..*, **5**, 299–302.

Van der Krol, A. R., J. N. M. Mol. and A. R. Stuitje (1988a). Antisense genes in plants: an overview. *Gene.*, **72**, 45–50.

Van der Krol, A. R., P. E. Lenting, J. Veenstra, I. M. Van der Meer, R. E. Koes, A. G. M. Gerats, J. N. M. Mol and A. R. Stuitje (1988b). An anti-sense chalcone synthase gene in transgenic plants inhibits flower pigmentation. *Nature.*, **333**, 866–869.

Van der Mark, F., J. Hoogendijk and J. J. M. Dons (1988). *Agrobacterium rhizogenes* transformation of tomato. In *2nd Int. Congr. of Plant Mol. Biol.*, 419.

Van Dun, C. M. P., Van Vloten-Doting and J. F. Bol (1988). Expression of alfalfa mosaic virus complementary DNA 1 and 2 in transgenic tobacco plants. *Virology.*, **163**, 572–578.

Van Haaren, M. J. J., N. J. A. Sedee, R. A. Schilperoot and P. J. J. Hooykaas (1987). Overdrive is a T-region transfer enhancer which stimulates T-strand production in *Agrobacterium tumefaciens. Nucleic Acids Res.*, **15**, 8983–8997.

Van Larebeke, N., G. Engler, M. Holsters, S. van der Elsacker, I. Zaenen, R. A. Schilperoort and J. Schell (1975). Large plasmids in *Agrobacterium tumefaciens* essential for crown gall inducing ability. *Nature*, **252**, 169–170.

Van Lysebettens, M., D. Inze, J. Schell and M. Van Montagu (1986). Transformed cell clones as a tool to study T-DNA integration mediated by *Agrobacterium tumefaciens J. Mol. Biol.*, **188**, 129–145.

Van Slogteren, G. M. S., J. H. C. Hoge, P. J. J. Hooykaas and R. A. Schilperoort (1983). Clonal analysis of heterogeneous crown gall tissues induced by wild-type and shooter mutant strains of *Agrobacterium tumefaciens* expression of T-DNA genes. *Plant Mol. Biol.*, **2**, 321–335.

Van Slogteren, G. M. S., P. J. J. Hooykaas and R. A. Schilperoort (1984). Silent T-DNA genes in plant lines transformed by *Agrobacterium tumefaciens* are activated by grafting and 5-aza-cytidine treatment. *Plant Mol. Biol.*, **3**, 333–336.

Velten J., L. Velten, R. Hain and J. Schell (1984). Isolation of a dual plant promoter fragment from the Ti plasmid of *Agrobacterium tumefaciens. EMBO J.*, **3**, 2723–2730.

Veluthambi, K., W. Ream and S. B. Gelvin (1988). Virulence genes, borders and overdrive generate single stranded T-DNA molecules from the A6 Ti plasmid of *Agrobacterium tumefaciens. J. Bacteriol.*, **170**, 1523–1532.

Velvekens, D., M. Van Montagu and M. Van Lysebettens (1988). *Agrobacterium tumefaciens*-mediated transformation of *Arabidopsis thaliana* root explants by using kanamycin selection. *Proc. Natl. Acad. Sci. USA.*, **85**, 5536–5540.

Vilaine, F. and F. Casse-Delbart (1987). A new vector derived from *Agrobacterium rhizogenes* plasmids: a micro Ri plasmid and its use to construct a mini Ri plasmid. *Gene*, **55**, 105–114.

Waldron, C., E. B. Murphy, J. L. Roberts, G. D. Gustafson, S. L. Armour and S. K. Malcolm (1985). Resistance to hygromycin B. *Plant Mol. Biol.*, **5**, 103–108.

Wallroth, M., A. G. M. Gerats, S. G. Rogers, R. T. Fraley and R. Horsch (1986). Chromosomal localization of foreign genes in *Petunia hybrida. Mol. Gen. Genet.*, **202**, 6–15.

Wang, K., C. Genetello, C. Van Montagu and P. C. Zambryski (1987). Sequence context of the T-DNA border repeat element determines its relative activity during T-DNA transfer to plant cells. *Mol. Gen. Genet.*, **210**, 338–346.

Ward, E. R. and W. M. Barnes (1988). *VirD2* protein of *Agrobacterium tumefaciens* very tightly linked to the 5′ end of T-strand DNA. *Science* (*Washington D.C.*), **242** (4880), 927–930.

Watson, B., T. C. Currier, M. P. Gordon, M. D. Chilton and E. W. Nester (1975). Plasmid required for virulence of *Agrobacterium tumefaciens. Bacteriol.*, **123**, 255–264.

Weiler. E. W. and J. Schroder (1987). Hormone genes and crown gall disease. *Trends in Biochem. Sci.*, **12**, 271–275.

White, D. W. R. and D. Greenwood (1986). Development of transformation system in white clover using modified Ti-plasmid vectors. *Agron. Soc. N. Z. Spec. Publ.*, **5**, 349–354.

White, F. F. and E. W. Nester (1980). Hairy-root: plasmid encodes virulence traits in *Agrobacterium rhizogenes. J. Bacteriol.*, **141**, 1134–1141.

White, F. F., G. Ghidossi, M. P. Gordon and E. W. Nester (1982). Tumour induction by *Agrobacterium rhizogenes* involves the transfer of plasmid DNA to the plant genome. *Proc. Natl. Acad. Sci. USA*, **79**, 3193–3197.

White, F. F., B. H. Taylor, G. A. Huffman, M. P. Gordon and E. W. Nester (1985). Molecular and genetic analysis of the transferred DNA regions of the root inducing plasmid of *Agrobacterium rhizogenes. J. Bacteriol.*, **164**, 33–44.

White, F. F. and V. P. Sinkar (1987). Molecular analysis of root induction by *Agrobacterium rhizogenes.* In *Plant Gene Research: Basic Knowledge and Application: Plant DNA Infectious Agents*, ed. T. Hohn and J. Schell, pp. 149–178. Springer-Verlag, New York.

Willmitzer, L. (1988). The use of transgenic plants to study plant gene expression. *Trends Genet.*, **4**, 13–18.

Willmitzer, L., G. Simons and J. Schell (1982). The $T_L$-DNA in octopine crown gall tumours codes for seven well defined polyadenylated transcripts. *EMBO J.*, **1**, 139–146.

Winans, S. C., P. Allanza, S. E. Stachel, K. E. McBride and E. W. Nester (1987). Characterization of the *virE* operon of the *Agrobacterium* Ti plasmid pTiA6, *Nucleic Acids Res.*, **15**, 825–837.

Withers, L. A. and P. G. Alderson (eds.) (1986). *Plant Tissue Culture and its Agricultural Applications*, pp. 1–526, Butterworths, London.

Wohllenben. W., W. Arnold, I. Broer. D. Hillemann, E. Strauch and A. Puehler (1988). Nucleotide sequence of the phosphinotrycin N-acetyltransferase gene from *Streptomyces viridochromogenes* TU494 and its expression in *Nicotiana tabacum. Gene*, **70**, 25–38.

Wullems, G. J., L. Molendijk, G. Ooms and R. A. Schilperoort (1981). Differential expression of crown gall tumour markers in transformants obtained after *in vitro Agrobacterium tumefaciens*-induced transformation of cell wall regenerating protoplasts derived from *Nicotiana tabacum. Proc. Natl. Acad. Sci. USA*, **78**, 4344–4348.

Yang, F. M. and R. B. Simpson (1981). Revertant seedlings from crown gall tumours retain a portion of the bacterial Ti plasmid DNA sequences. *Proc. Natl. Acad. Sci. USA*, **78**, 4151–4155.

Yanofski, M. F., S. G. Porter, C. Young, L. M. Albright, M. P. Gordon and E. W. Nester (1986). The *virD* operon of *Agrobacterium tumefaciens* encodes a site-specific endonuclease. *Cell*, **47**, 471–477.

Yoshikawa, T. and T. Furuya (1987). Saponin production by cultures of *Panax ginseng* transformed with *Agrobacterium rhizogenes. Plant Cell Rep.*, **6**, 449–453.

Zambryski, P. (1988). Basic processes underlying *Agrobacterium*-mediated DNA transfer to plant cells. *Ann. Rev. Genet.*, **22**, 1–30.

Zambryski P., M. Holsters, K. Kruger, A. Depicker, J. Schell, M. Van Montagu and H. M. Goodman (1980). Tumour DNA structure in plant cells transformed by *A. tumefaciens. Science (Washington D.C.)*, **209**, 1385–1391.

Zambryski, P., H. Joos, C. Genetello, J. Leemans, M. Van Montagu and J. Schell (1983). Ti plasmid vector for the introduction of DNA into plant cells without alteration of their normal regeneration capacity. *EMBO J.*, **2**, 2143–2150.

Zambryski, P., J. Tempe, and J. Schell (1989). Transfer and function of T-DNA genes from *Agrobacterium* Ti and Ri plasmids in plants. *Cell*, **56**, 193–201.

Zhan, X.-C., D. D. Jones and A. Kerr (1988). Regeneration of flax plants transformed by *Agrobacterium rhizogenes. Plant Mol. Biol.*, **11**, 551–560.

Zhang, H. M., H. Yang, E. L. Rech, T. J. Golds, A. S. Davis, B. J. Mulligan, E. C. Cocking and M. R. Davey (1988). Transgenic rice plants produced by electroporation-mediated plasmid uptake into protoplasts. *Plant Cell Rep.*, **7**, 379–384.

# 13
# Application of Unconventional Techniques in Classical Plant Production

GERHARD WENZEL
*Institute for Resistance Genetics, Grünbach, Germany*



## 13.1 INTRODUCTION

In an applied research area such as new biotechnology in plant agriculture, success can objectively be measured by the number of new cultivars protected by breeders' rights. Modern plant production is not only aiming for better yielding varieties in food production but increasingly the plant is detected as a renewable source for energy and, more importantly, for complex molecules like starch and oil. Particularly in light of such aims, increased flexibility in breeding programs is needed to serve the changing demands of a rapidly industrialized market.

Classical plant breeding has achieved so much by conventional techniques—and will continue to do so—that we suffer today more from overproduction than lack of food. This impressive success is the result of a slow but steady improvement in plant breeding, plant protection and plant cultivation, the three of which sum up to make the whole of plant production (Figure 1). In all three areas it is

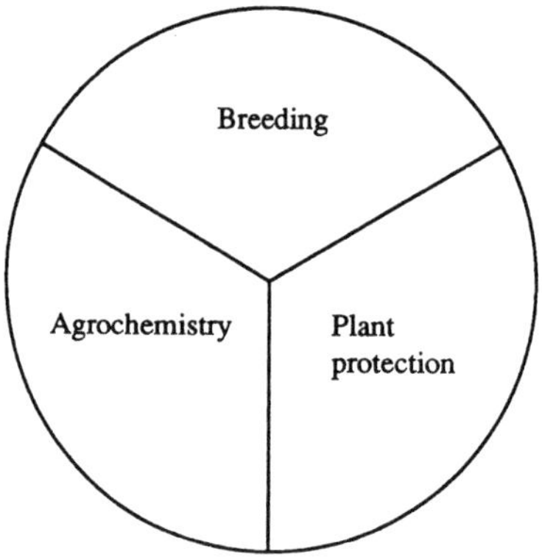


**Figure 1** The three research areas contributing in equal parts to the success of plant production

likely that in the future classical approaches will be complemented and made more efficient by biotechnology, which shall be defined here as the use of cell and tissue culture and of higher plants' molecular genetics.

Most of the new techniques work *via* the plant genome; thus the unconventional approach will have its greatest potential in that third where genetics is applied: in plant breeding. In agrochemistry, plant protection and plant breeding become a complex in which breeding supplemented with biotechnology is used to protect the plant against diseases, pests and stresses. This is also true for mineral fertilizers, where the uptake in the rhizosphere of the plant will be improved in low input varieties. The situation in plant cultivation is similar: maintenance breeding of a variety and plant cultivation are a complex as well. For biological pest control working mainly towards insect resistances, biotechnological strategies will alter not only the genome of the plant but also the genomes of pathogens' antagonists. Quality (including the character low input) and genetical resistances are the main areas where an increase in efficiency of the classical breeding approaches through unconventional steps will pay off (Table 1).

Plant breeding mainly deals with complex characteristics, the inheritance of which follows the laws of population genetics. The combination of several quantitatively inherited characteristics still demands enormous populations and at the same time quantitative selection procedures. In the case of disease resistance—a central aim in plant breeding—this applies not only to the plant but also to the pathogen populations. Due to these difficulties, breeders prefer to use in their breeding programs qualitatively inherited characteristics, normally based on a single gene. In the case of disease resistance such monogenic traits were often overcome by new virulences of the pathogen; their short durablity resulted in today's race between the action of resistance genes of the plant and new virulences of the pathogen, as presented in Figure 2a. The resistance will last longer when more alleles are combined in the plant's genome. Figure 2b represents the other extreme with a horizontal, durable

**Table 1** Summary of Breeding Goals where Unconventional Approaches may Pay Off

| | | |
|---|---|---|
| *Resistance against* | Diseases | Viruses |
| | | Bacteria |
| | | Fungi |
| | Insects | Aphids |
| | | Caterpillars |
| | Stress | Temperature |
| | | Salt |
| | | Drought |
| | Herbicides | |
| *Improvement of quality* | Primary products | Starch |
| | | Oil |
| | | Protein |
| *Improvement of energy use* | Light | Photosynthesis |
| | Nutrition (low input) | Salt |
| | | Nitrogen |

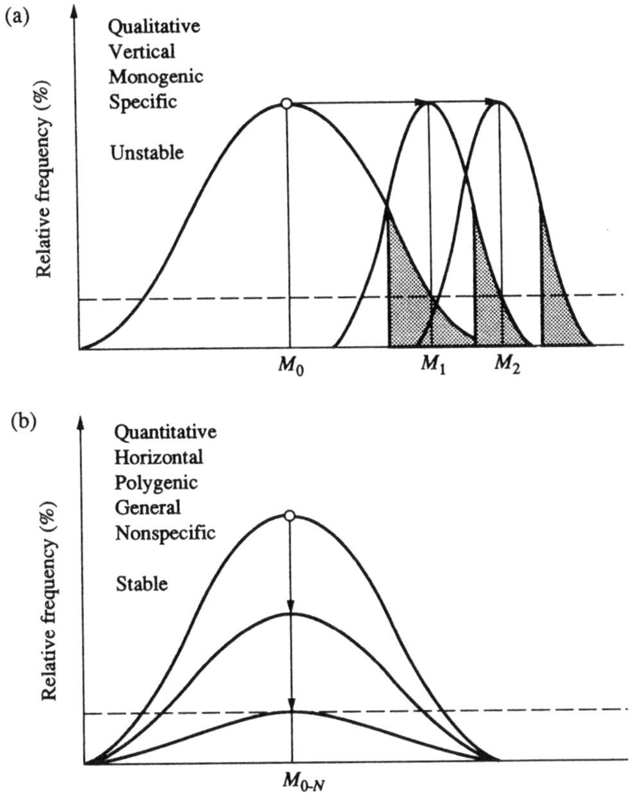


**Figure 2** Types of resistance and their effects on pathogen populations. (a) A new gene for resistance leaves a small resistant population alive, which serves as a source for a new pathogen population. Resistance and virulence are in a steady race. (b) The concept of a polygenic resistance diminishes the pathogen population without giving a great chance for the development of new virulences (after Walther, 1988)

and quantitatively inherited resistance, which is so strong that the pathogen population is depressed and new virulences which may develop have no chance to build up a new virulent population.

Because of the increasing importance of developing means for quantitative breeding programs, this aim will be emphasized when evaluating new methodologies.

## 13.2 USE OF CELL CULTURE IN PLANT BREEDING

Plant breeding has two major components: maintenance breeding for existing cultivars and breeding for new cultivars. In maintaining an existing cultivar protected through breeders' rights, biotechnological methods are already used and have become integral parts of applied breeding programs. The maintenance of vegetatively propagated crops like the potato rely on tissue culture techniques which allow for the rapid propagation of specific clones *in vitro* under more economic conditions. Far more advanced are these approaches in horticulture.

### 13.2.1 Maintaining Existing Cultivars

The first plants mass-propagated by tissue culture were ornamentals and lie at the root of the first commercial tissue culture facilities (Rice, Chapter 8). Vasil and Vasil (1980) published a table of 352 species in which adventitious buds, shoots, embryoids or whole plants were produced aseptically *in vitro* on synthetic media. Normally the explant forms a callus first, which can be induced by external treatment with phytohormones, auxins and cytokinins to regenerate shoots and roots. The process of regeneration may start either from a single cell (embryogenesis) or from a cell cluster (organogenesis). Depending on the size of the explant, the technique is called rapid propagation (up to 1 cm large pieces) or meristem culture (pieces normally beyond 0.2 mm).

#### *13.2.1.1 Rapid propagation and meristem culture*

Aside from ornamentals, rapid propagation gained economic importance in strawberries, grapes, fruit and forest trees, palms, asparagus and within the field crops, predominantly in potato (Boulay, 1987; Dore, 1987; Mullins, 1987). German potato breeders, for example, use 50 000 to 100 000 rapidly propagated plants per company per year, which are transferred from the *in vitro* culture to the greenhouse in early spring. Tubers from such plants can be harvested in May before aphids begin to fly and transfer virus diseases. This use of rapid propagation has opened new practical ways for cheaper maintenance breeding and is being used by commercial companies. In potato there is a strong tendency to automate this process, *e.g.* by the *in vitro* production of minitubers, which will make the process even cheaper. With decreasing costs acceptance of this process in practice will increase even further. Besides costs the crucial point in rapid propagation and meristem culture is the need to prevent callus formation, because undifferentiated growth increases the chance of undesired somaclonal variation; stability of the genotype is, however, an irrevocable prerequisite for the practice.

Rapid propagation techniques suffer because the total offspring from a diseased plant will be diseased as well, particularly by virus infections, which cannot be controlled chemically. Consequently, plant material for rapid propagation has to be taken first through meristem culture, where normally up to 40% of the regenerants will deliver healthy plants. By culturing the starting material at 30–35 °C, the percentage of healthy plants can be increased. However, the optimal size for the meristem should always be tested first. The larger they are, the easier the procedure, but viruses may be transferred. In a comparative study with different potato viruses it was demonstrated that only for potato virus X (PVX) was the size critical, while for the elimination of potato virus S (PVS), even 2 mm tips could give virus-free regenerants (Meyer and Foroughi-Wehr, 1987). These larger tips regenerate much more easily and never produce callus, while smaller ones tend to proliferate in an unorganized manner. Since a given genotype must remain stable, it is important to prohibit this callus formation.

Rapid propagation is easier to apply in dicots than in monocots since segments of shoots or leaves, the most handly source for rapid propagation, are only of limited use in monocots. Nevertheless regeneration of plants originating from multicellular explants — young inflorescences, immature embryos, bases of leaves, nodes, shoot primordia and root tips — has been achieved in all major cereal crops (Lörz *et al.*, 1988; Vasil, 1988). Despite this progress, rapid propagation in monocots is still so labor intensive that it is not economical.

#### *13.2.1.2 Embryo, cell and protoplast culture*

Besides rapid propagation, embryo culture is of applied interest in, for example, rescuing embryos after wide crosses (Zenkteler, 1988) or in parthenogenetic procedures for haploid induction, as is demonstrated for barley (Kasha and Kao, 1970). In all other cases the use of embryo culture is still in its experimental phase. In cereals immature embryos are most frequently used to establish *in vitro* cultures with good regeneration capacity. A similar situation is found in soybean, where normal regeneration is still difficult. Since the first report on regeneration by Christianson *et al.* (1983), who regenerated soybean plants from suspension cultures, it was possible to enhance the procedure by starting from immature embryos, either directly or *via* callus formation. Komatsuda and Ohyama (1988) found genotypes with a very high regeneration capacity of up to 58% of the original selected embryos, but there are still others that do not show any regeneration.

Plant tissues can be macerated mechanically or enzymatically into single cells with a pectinase. In a number of crops the regeneration of single cells is also possible; their use in agriculture, however, is negligible.

The propagation process can be speeded up further by digestion of cells with cellulase, resulting in protoplasts. Such protoplasts are cultured in hypertonic liquid media (*e.g.* Kao and Michayluk, 1975) until a small callus is formed. Then they can be treated as a normal callus and can regenerate functional plants (Morikawa and Yamada, Chapter 11). Such plants are cloned as in rapid propagation systems but the multiplication rate is much higher; from a given plant in one 10 cm Petri dish, protoplasts can be cultured resulting in 5000 plants. Under the influence of the culture conditions in this system, mutations easily take place (somaclonal or protoclonal variation), resulting in altered protoclones (Shepard and Totten, 1977). From a practical standpoint this reduces the applicability of such micropropagation strategies. Again it is crucial to shorten the callus phase as much as possible to prevent mutations.

Protoplast regeneration is mainly restricted to dicots and is of practical interest for agriculture at present only for tomato (Zapata *et al.*, 1977), rapeseed (Kartha *et al.*, 1974; Thomas *et al.*, 1976), potato (Binding *et al.*, 1978; Shepard and Totten, 1977) and alfalfa (Dos Santos *et al.*, 1980). In the monocots, a major breakthrough has been achieved by the regeneration of rice plants from protoplasts isolated from cell suspensions and maize (for a review, see Lörz, *et al.*, 1988; Vasil, 1988). While the maize regenerants were sterile, the rice plants regenerated from protoplasts have been tested in the field and the regenerants were found to be comparable to the protoplast donor material (Ogura *et al.*, 1987). In the other cereals only callus, incomplete plantlets or albinos have been obtained at the time of writing.

The practical interest in protoplasts as tools for rapid propagation is rather limited; much more is expected from somatic fusion of protoplasts and from their use in gene transfer. Protoplasts are able to absorb macromolecules like DNA. For dicots the transformation of protoplasts is becoming a highly reproducible technique (De Block, 1988; Klee *et al.*, 1987). However, this technique is not essential as, in dicots, others such as the *Agrobacterium*-mediated transfer technique also work. In monocots, where up till now DNA transfer has still been difficult, only the first reports on successful monocot transformation and regeneration of transformed plants are available: for maize (Rhodes *et al.*, 1988) and rice (Zhang and Wu, 1988). Substantial progress in unconventional breeding programs of cereals demand such reproducible DNA transfer techniques.

There is one additional application of cell cultures: the production of secondary compounds such as oils, flavors and drugs. As this is not an integral part of today's agriculture — although its importance is increasing — the reader is referred to recent reviews by Fowler (1984) and Mulder-Krieger *et al.* (1988).

#### *13.2.1.3 Artificial seeds*

The micropropagation system would be further advanced if the laborious transfer from *in vitro* production to *in vivo* cultivation could be avoided. Nature is using seeds with an embryo embedded in the endosperm for this purpose or vegetative organs, like the potato tuber. Since 1953 reports on *in vitro* production of minitubers have been available (for a review, see Chandra *et al.*, 1988), surely a first step in the direction of artificial seeds for practical use. In the meantime the production of embryoids and embryogenic suspension cultures also became possible for a number of dicotyledoneous and monocotyledoneous species. Consequently the idea came up to coat such embryos with an artificial nutritional medium and to produce an artificial seed. The first prerequisite for this approach is a very good embryogenic suspension, such as that available in carrots since 1958 (Reinert, 1958; Steward, 1958), in rapeseed (Lichter, 1982) and in barley (Datta and Wenzel, 1988). The quality and fidelity of induced embryos are the limiting factors for the development and scale-up of artificial seeds. According to Redenbaugh *et al.* (1987), artificial seeds will most likely to be used first for hybrid vegetable crops such as celery, *Brassica* species, carrot and tomato but also for cotton, lettuce and alfalfa. Water soluble hydrogels are suitable for making artificial seeds. Redenbaugh *et al.* (1987) coated alfalfa somatic embryos and calculated 0.026 US cent for one capsule, which is still much more expensive than for normal seeds. Thus it may only be of advantage for specific purposes, *e.g.* for hybrids or for transformed material.

#### *13.2.1.4 Living collections*

Maintenance of genetic resources is the best guarantee for solving demands of plant breeding in the future. In particular, for vegetatively propagated and/or polyploid crops, which are normally highly heterozygous, the maintenance of pathogen-tested healthy stocks is always a problem. *In vitro* cultures may overcome this difficulty. From a healthy mother clone rapid propagation is started, stopped after some time, and then the collections are transferred to modified media and stored at temperatures between 4 °C and 10 °C. In order to reduce the chance of genetic change occurring even during storage in living collections, the use of cryopreservation (preservation in the frozen state) has been proposed by Withers and Street (1977). Cryopreservation is based on the reduction and subsequent arrest of metabolic functions of plant material by imposition of the ultra low temperature of −196 °C. However, only a few biological materials in their natural state can be frozen without adverse effects on cell viability. The discovery of chemicals with cryoprotective properties was a prerequisite, in particular glycerol and DMSO. The ideal material to start with from the practical point of view would be a meristem. A list of about 40 plant species cryopreserved as cell, callus, protoplast or meristem culture is given by Kartha (1987).

Living collections are integral parts of most gene banks; cryopreservation has, however, to overcome several problems before being widely applied in agriculture.

### 13.2.2 Breeding for New Cultivars

In most crops the production of a superior new cultivar is economically more challenging than the maintenance of existing ones. As a consequence, breeding companies expect gains from unconventional strategies, particularly in cultivar production. For breeding of a new cultivar availability of a gene pool, selection of specific lines, combination and a second selection are the basic requirements. The size of the gene pool is limiting as it is equivalent to the amount of useful variability. As long as there is a chance to detect new genetic resources and to collect and evaluate them, classical breeding will theoretically be without any genetical limitation. Classical breeding is, however, very time-consuming; it takes at least 10 years from the first cross until admission with breeders' rights (and 30 years is not unusual). From the standpoint of combining quantitatively inherited traits in particular, the population size and the time needed for selection are very crucial factors. As a result the classical method is not very flexible and responds slowly to demands of the market. A new technique has a fair chance to substantially improve variety production if it is more efficient than the classical approach. Here the following items of biotechnological approaches are of interest: (i) the *in vitro* culture is free from the effects of climate and the natural environment; this makes it easier to measure slight quantitative differences in polygenically inherited (horizontal or general) characters; (ii) large numbers of individuals can be handled in a very small space; (iii) one can work with microspores and haploids, the simpler genomes of which allow uncovering of recessive traits and additive characters and reduces the population size; and (iv) *in vitro* cell cultures may serve as a tool for taking up genetic information, thus bridging genetic engineering and plant breeding.

First examples show today that procedures to create variability and to select superior genotypes can indeed be transferred partially or completely from the field into the laboratory or the Petri dish, but this technology must be accompanied by, and combined with, classical breeding procedures. Unconventional techniques can never replace or be used independently from classical breeding. Production of variability is followed by selection. While in the first area each classical method has a biotechnological equivalent (Figure 3), the shift from classical selection procedures to *in vitro* selection is smooth. Right from the start of scientific phytopathology plant organs (detached leaves or leaf pieces), for example, were kept *in vitro* and used for testing defense mechanisms against fungi or viruses. The clearcut advantage of such procedures is the large number of individuals which can be screened in a small space allowing for the use of microbiological procedures for selection within higher plants.

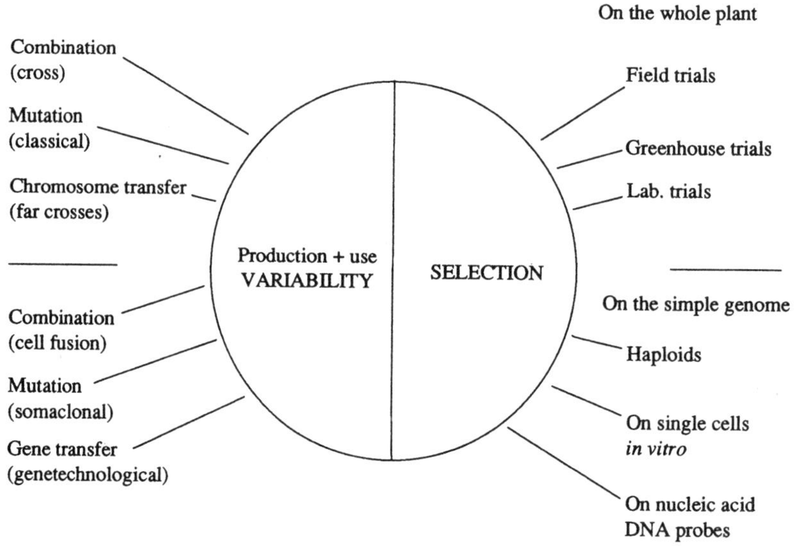


**Figure 3** Summary of the two sections of breeding new varieties: variability and selection, and the classical (upper part) and unconventional techniques available to produce a better cultivar

### *13.2.2.1 Somaclonal variation*

Variability, created normally by combination breeding through sexual recombination, may be increased artificially by mutagenesis. *In vitro* culture *per se* turned out to be mutagenic, and plants regenerated from organ cultures, calli and protoplasts often show new phenotypic variability, the somaclonal variation. Using somaclonal variation, new characters are expected from a given homogeneous cell population (homo- or hetero-zygous) without meiotic recombination. The objective is to find among the regenerated plant population spontaneous somaclonal or protoclonal (if coming from protoplasts) variation, *i.e.* mutations. Scowcroft and Larkin (1988) review variant characters for more than 20 plant species and report on their sexual or asexual transmission. They also refer to possible mechanisms, which include ploidy changes, nuclear fragmentation, somatic crossing over, gene amplification and transposable genetic elements activated under the *in vitro* culture stress. Somaclonal variation has similar limitations to classical mutation breeding. It may be superior to simple mutation breeding, however, when a selection pressure is applied during the *in vitro* phase.

Morel (1946) realized for the system *Vitis/Plasmopara* the possible advantage of growing pathogens *in vitro* together with their hosts and tried to select *in vitro* on callus cultures. The development of dual culture techniques was not very rapid. Sacristan (1985) used a dual culture of either haploid cell suspension or stem-embryogenic cultures of rapeseed together with *Phoma lingam* as a general screening system. Rape cell suspensions and *Phoma* spores were grown together in liquid media and then transferred as a mixture to agar plates. However, there was no relationship between the reaction of a plant against pathogens as tissue cultures (*in vitro*), and the reaction of the same genotype as a plant (*in vivo*) or *vice versa, i.e.* the rapeseed calli 'resistant' *in vitro* did not give rise to resistant plants.

The older experiments on field crops were done with crude culture filtrates of the pathogens; reports are available on a screening procedure in potato with culture filtrates of *Phytophthora*, and by Sacristan (1982) on screening in the *Brassica/Phoma* system (for a review, see Wenzel, 1985). Not all regenerants from selected rapeseed cultures showed increased resistance, but the proportion of resistant or tolerant plants was up to six times higher among regenerants from selected cultures than among plants from controls. The first three progeny generations were tested for their level of resistance; unexpectedly, segregation was found in the progenies of plants classified as resistant, although in all cases the proportion of resistant plants was higher for progenies obtained from resistant material, indicating genetically defined resistance. Hartman *et al.* (1984) reported on the successful selection of calli and regenerated plants of *Medicago sativa* resistant to *Fusarium oxysporum*. In addition, Hammerschlag (1988) was able to select peach cells for insensitivity to culture filtrate of *Xanthomonas campestris* and could regenerate resistant plants. In barley, comparable experiments were performed with *Helminthosporium sativum* toxin (Wenzel *et al.*, 1988). Zymograms showed one more isozyme than with the unselected sensitive callus. After regeneration, lines selected as less sensitive proved to result in less susceptible plants. Similar results were obtained after treatment of barley embryogenic calli with fusaric acid, a commercially available toxin of *Fusarium oxysporum*. It was possible to regenerate plants from surviving calli, and testing by leaf bioassay revealed that many were tolerant to the same toxin concentration employed for callus selection but the reaction was a quantitative one.

The first experiment in which a controlled selection for resistance against a pathogen was made with the responsible toxin alone was reported for maize callus (Gengenbach *et al.*, 1977). The reversion through somaclone formation from susceptible to resistant to *Helminthosporium maydis* was, however, associated with genetic changes in the cytoplasm leading to reversion from male sterility to male fertility. Thus, this result was of no practical importance. Buiatti *et al.* (1985) found a correlation between *in vivo* resistance and *in vitro* response of *Dianthus*, not only to *Fusarium*, but also to elicitors of the fungus.

The number of negative reports on *in vitro* selection is large as well, *e.g.* Vardi *et al.* (1986) found that culture filtrates of *Phytophthora citrophthora* cannot be used in *Citrus* breeding as a selective tool *in vitro*. The contrasting results demonstrate that *in vitro* screening cannot become a general procedure for a huge number of characters. It must first be checked to what extent correlations or causal dependence exist between an *in vitro* reaction and the phenotype of the whole plant. Sjödin and Glimelius (1989) obtained rather different responses to the toxin sirodesmin produced by *Phoma lingam* on protoplasts, cell aggregates and intact plants, thus, these parameters have to be identified carefully when using such an *in vitro* selection system.

The relationship between *in vitro* and *in vivo* reactions of the same genotype should be closer when the purified toxic compound responsible for the virulence instead of crude culture filtrates is used to detect sensitive individuals at an early developmental stage. Furthermore, work with single cells would increase the chance of detecting differences normally covered by cross feeding, and the number of individuals can be increased to $10^6$, which is within the range of the natural mutation rate of $10^{-6}$. Once the potato protoplast system was highly reproducible, Shepard (1981) examined the usefulness of protoplasts for creating new genotypes. He found a tremendous variability in potato clones from the tetraploid 'Russet Burbank'. Up to 30% of the regenerated protoclones were useful variants; he recommended this method for 'intracultivar improvement'. However, not all variants were genetically stable even during vegetative propagation. Thomson *et al.* (1986) regenerated isolated protoplasts of three potato varieties with the objective of finding specific improvements in resistances to *Streptomyces scabies*, potato virus Y (PVY) and potato leaf roll virus (PLRV). Wenzel *et al.* (1987) described similar findings for PVX. All resistant protoclones were very poor in other characters studied, so that despite the better resistance to PVX the clones were of no direct practical use. Probably too many mutations had taken place. In other experiments with potato protoplasts in screening for *Fusarium* resistance, attempts were made to pass the callus stage very rapidly (Wenzel *et al.*, 1987). Due to this rapid passage through the callus level, all clones looked phenotypically rather uniform after regeneration. Retesting of leaves or of a secondary callus revealed a significantly increased stability against the *Fusarium* toxins. The selected clones were propagated in the field to deliver a uniform tuber generation, which could be tested for *Fusarium* resistance in the routine way (Langerfeld, 1979). Among several clones tested, clear differences between clones with an increased or a decreased level of resistance against the fungus could be identified. The reaction was clearly of a quantitative nature, making it probable that many slight mutations had taken place *in vitro*, not all of them leading to decreased susceptibility, but often resulting in higher susceptibility compared to unselected control regenerants. Although this might sound disappointing, in view of the aim of finding genomes with quantitative resistances it looks satisfactory.

The early *Phytophthora* and these *Fusarium* experiments differ in the exposure time to the selective agent as well as in the starting material (callus or protoplasts) and the nature of the toxin (crude or semipurified). The callus was transferred several times to media with selection pressure, while the protoplasts were treated only once, shortly after their cell wall formation. The long treatment guarantees only a few escapes, but many mutations; the short one will result in many escapes, but fewer mutations. The short treatment is preferable, as it results in more normal clones, and although a second screening is necessary, the probability of maintaining the genotype of the donor material and just adding a few new characters is higher. Due to the selection agent it has not been demonstrated whether purified toxins would be better as they react more precisely. In any case, however, it would be advantageous to develop tests which give more specific answers, for example, DNA probes.

Besides disease resistance, somaclones were used for detecting specific herbicide resistances in tobacco to picloram and sulfonyl ureas (Chaleff and Mauvais, 1984), to glyphosate (Singer and McDaniel, 1985) and to chlorsulfuron (Creason and Chaleff, 1988), in tomato for paraquat (Thomas and Pratt, 1982) and in potato for 2-methyl-4-chlorophenoxy acid (Wersuhn *et al.*, 1987). Also somaclones tolerant to environmental stresses such as salts, drought and acids were found (Nabors and Dykes, 1985), but it is questionable whether for such selections cell culture systems are superior to seedling selections.

It cannot be overstressed that there is no general need to increase the genetic variability artificially *in vitro*. Sufficient variability already exists or can be found by sexual recombination in most economic plants. It is only for certain specific traits lacking in natural resources, or for certain specific crops such as trees with a long regeneration cycle, that somaclones may be preferable. Actually the many somaclones described were of rather little importance for practical agriculture; it can be concluded that breeding for characters which are available in a heritable form is on average more successful by combination than by mutation breeding. This statement counts also for *in vitro* techniques. This means that, for *in vitro* culture, along with practical aims generally there exists a need to prevent, rather than to increase, somaclonal variation.

### *13.2.2.2 Protoplast fusion*

With respect to applications, the regeneration of protoplasts had only limited importance. Protoplasts can, however, not only be regenerated, they can also be fused to result in somatic hybrids (Gleba and Sytnik, 1984; Glimelius, 1988; Morikawa and Yamada, Chapter 11). Presently the most attractive advantage under applied aspects of somatic hybridization is the chance to combine asexually two heterozygous genomes without meiotic recombination. This is very important where the need is to pool resistances or qualities that are polygenic and inherited quantitatively. Since amphidiploid, *i.e.*

polyploid, hybrids are of no immediate practical interest, it will usually be necessary to reduce the ploidy level before or after protoplast fusion to the level desired in the crop.

Fusions can be induced chemically by the polycation polyethyleneglycol (PEG) or *via* electrofusion. Fusion of crop plants is developed furthest in the *Solanaceae*. In cereals the first report on a successful fusion was for rice; Toriyama and Hinata (1988) report on obtaining somatic hybrid plants between rice cultivars.

In *Solanum* the fusion of *S. tuberosum* and *Lycopersicon esculentum* (Smillie *et al.*, 1979), resulting in plants with an increased chilling resistance, for example, was followed by fusion between *S. tuberosum* and *S. nigrum* to transfer atrazin resistance (Binding *et al.*, 1982). Austin *et al.* (1985) transferred by somatic fusion the resistance to potato leaf roll virus (PLRV) from *Solanum brevidens* into *S. tuberosum* ( × ) *S. brevidens* hybrids. They further demonstrated the presence of *Erwinia* and *Phytophthora* resistance in such hybrids (Austin *et al.*, 1988).

As long as protoplast regeneration works, the limitation of somatic fusion for applied purposes today lies no longer in the fusion itself, but in selection procedures for heterocaryotic fusion products from unfused and homocaryotically fused protoplasts. Figure 4a shows regenerating potato calli after 2*x* ( × ) 2*x* fusion; it becomes obvious that it is not practical to regenerate all plants and to leave the selection to the plant leave. Besides a number of selection systems using complementation or partial lethality (Schieder and Vasil, 1980), hybrid vigor of fusion products may be a rather universal selection system particularly under applied aspects (Schieder and Vasil, 1980). Such a system was used also in potato for hybrid detection in 2*x* ( × ) 2*x* fusion (Debnath and Wenzel, 1987; Deimling *et al.*, 1988; Waara *et al.*, 1989). After PEG-induced fusion nearly all callus-clones selected because of rapid growth possessed chromosome counts at the tetraploid level. Plants of clones from the very vigorous

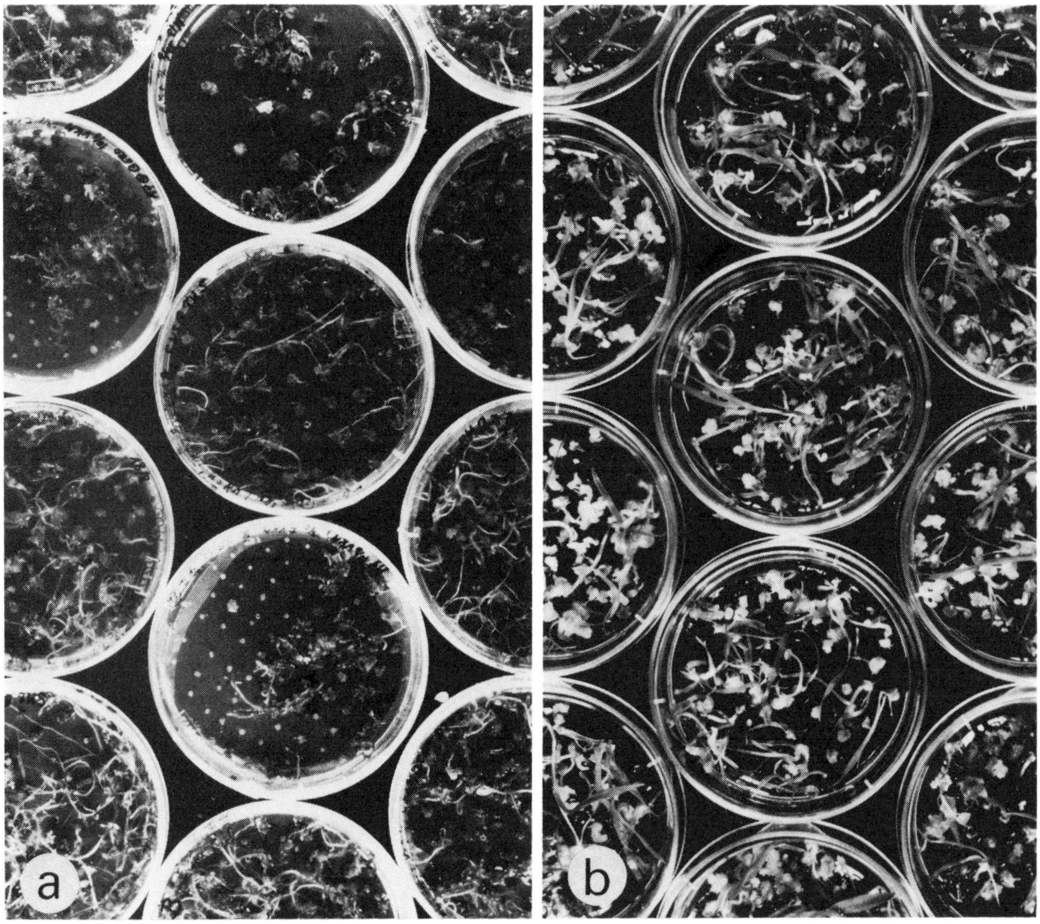


**Figure 4** Mass culture of crop plants. (a) Potato plants regenerating after fusion of two dihaploid clones (Möllers, 1990). (b) Regeneration of winter barley plants from microspores (Kuhlman, 1990)

calli showed hybrid vigor as well. Isoenzyme pattern analysis of both parents and the vigorous clones proved their hybrid nature. This procedure can, however, not be generalized, as hybrids were also found after fusion, which expressed vigor in the greenhouse, but none of the calli they descended from had shown increased growth rates.

For a wide application the fusion process is still labor intensive; there may be a chance to further automate this technique. Experiments are in the process of combining the fusion procedure for ploidy level detections with flow cytometry (Sree Ramulu and Dijkhius, 1986) or with a cell sorter (Galbraith, 1984).

For somatic hybrids, propagation procedures must be developed that will allow the continuous growth of such a heterozygous genotype without immediate segregation. Among the field crops this prerequisite is adequately fulfilled in potatoes, since breeding at the dihaploid ($2n = 2x = 24$) level and vegetative propagation are routine. The problem of ploidy level may be overcome in other crops by the donor recipient fusion technique, where one fusion partner is irradiated so as to partially or completely inactivate its nuclear genome (Aviv and Galun, 1980). During fusion only a partial genome is combined with that of the other fusion partner to produce plants with a few additional characters. The principal idea is that specific parts of a foreign genome are added to a superior genome for its further improvement. The potential use of this procedure was demonstrated in an approach to transfer a chromosome fragment from carrots bearing a gene for resistance against methotrexate into tobacco protoplasts. The regenerated plants are similar to tobacco but additionally express the methotrexate resistance (Dudits *et al.*, 1986). In contrast to sexual recombination, in such asexual procedures the cytoplasms are mixed as well and allow for combination of the cytoplasm (cybrids).

For applied agriculture, hybrid seed production is of interest. Such programs are also started in rapeseed. One problem is the transfer of the cytoplasmic male sterility (cms) allowing for economic production of hybrids. The cms character, missing in *Brassica napus*, was transferred from *Raphanus* by somatic protoplast fusion (Pelletier *et al.*, 1988). Similar experiments are reported for *Daucus carota* by Tanno-Suenaga *et al.* (1988), who used the donor recipient technique. In their experiments, from 11 cybrid plants that showed mtDNA fragment patterns different from those of the parents, 10 showed sterility. The asexual fusion allowed transfer of the cytoplasms, normally not transferred by sexual genetics, demonstrating a principal advantage of somatic fusion. Thus, in rapeseed and in potato, products from protoplast fusion are used in practical programs.

### *13.2.2.3 Use of haploids*

All higher plants are at least diploid, possessing one genome from each parent. This guarantees, in combination with the random distribution of chromosomes during meiosis, the variability necessary for further evolutionary progress. Growth of a variety demands, however, stability of the genomes. Since gametophytic cells can be regenerated into functional plants (for a review, see Nitzsche and Wenzel, 1977), the production of completely homozygous breeding lines became easily possible. In addition, microspores from an $F_1$ hybrid represent a heterogeneous cell population. As after genome-doubling all characters are expressed phenotypically, selection for desired traits is secure and simple. Figure 5 gives an example for the phenotypic variability expressed in androgenetic barley lines.

Egg cells and microspores can be regenerated into haploid plants. In barley, varieties are obtained *via* doubling haploids of egg cell origin (Ho and Jones, 1980). In most crops, the regeneration of egg cells is only possible after fertilization, but a report is available for regeneration of barley plants from egg cells (Wang and Kuang, 1981). As there normally exists many more microspores than megaspores or egg cells, a higher number of individual genotypes can be screened on the male side. Their regeneration is possible either within the anther (anther culture) or in isolated culture (isolated microspore culture or shed pollen culture). In wheat the first variety from androgenetic haploids was licensed in October 1985 (De Buyser and Henry, 1986). Commercial barley breeders have started programs using the androgenetic procedure for the rapid incorporation of resistance into winter barley lines. Of particular interest is the incorporation of resistance to the soilborne BaYMV. The use of a haploid step in the breeding program offers a rapid approach to incorporate BaYMV resistance into commercial cultivars. More than 6000 and androgenetic lines were produced from microspores from $F_1$ anther donor hybrids of crosses between susceptible lines and resistant cultivars. Since 1984, increasing numbers of homozygous doubled haploid lines carrying the BaYMV resistance were released to practical breeding companies (Foroughi-Wehr *et al.*, 1986).

It is generally accepted today that a stepwise reduction of the ploidy level of potato offers the advantage of a simpler inheritance and also a better chance to combine qualitatively inherited characters (Wenzel *et al.*, 1979). Haploidization is no problem in principle in potato (*e.g.* Powell and

**Figure 5** Visualization of the variability present in a barley microspore population. All spikes originate from doubled haploid lines of microspores from one anther (Foroughi-Wehr *et al.*, 1986)

Uhrig, 1987) and it is already widely applied. Data now available demonstrate that field resistance to PVY and PLRV is expressed on all ploidy levels over a five year period, which means that both monogenically and polygenically inherited resistances are maintained during the successive haploidization process. When a genome has passed the monohaploid level, a homozygous additive character can be transferred efficiently to the next generation *via* classical crossing. Similar results have been obtained for polygenic nematode resistance (Uhrig and Wenzel, 1987). In a vegetatively propagated crop such as potato, the $F_1$ hybrid can be propagated as a new cultivar.

In *Brassica napus*, quality characteristics are emphasized in breeding techniques with haploids (Keller, 1984). With this plant, vigor of microspores was positively correlated with high glucosinolate content in the androgenetic regenerants. More than 70% of androgentic plantlets tested had a glucosinolate content much higher than did the original material (Hoffmann *et al.*, 1982). Since the aim of breeding was for low glucosinolate content, spontaneous selection for this character worked against the program. Development of microspores with a high level of glucosinolates, these being chemically quite similar to phytohormones, may have been favored. Very low or very high glucosinolate levels are diverse qualities needed either for human consumption or for industry, respectively. A simple screening for extremes would increase breeding flexibility at an early stage.

To date, culture of isolated microspores is possible in many crops like potato (Uhrig, 1985), rapeseed (Lichter, 1982), barley (Datta and Wenzel, 1988; Sunderland and Xu, 1982), wheat (Datta *et al.*, 1988; Wei, 1982) and maize (Nitsch *et al.*, 1982). In most cases the microspores are not isolated mechanically from the anther, but are shed initially into liquid culture media. Isolated microspore cultures apparently offer a more efficient system of regenerating a random sample from the microspore population than does anther culture. This is particularly important when the desired trait is linked with low plasticity, resulting normally in rather small regeneration rates. A procedure starting from single microspores that pass through embryogenesis and develop at a high frequency into green plants is most advantageous and desirable because early selection in microspore populations becomes possible. Figure 4b shows the microspore culture of barley, where about 10% of all microspores of the anthers start to grow. Thus microspores offer the possibility of combining single cell selection procedures with the advantages of a haploid system, which are specifically: absence of cross-feeding and chimerism; direct expression of all genes independent of their recessive or dominant nature; and the possibility to regenerate directly after chromosome-doubling, homozygous lines. As this regeneration procedure additionally avoids callus formation, the normally undesired somaclonal (gametoclonal) variation will be reduced, resulting in a higher proportion of green lines with good agronomic fitness.

Ye *et al.* (1987) combined anther culture and *in vitro* selection. They cultured anthers from a barley $F_1$ hybrid, a cross between a salt tolerant and a normal sensitive cultivar, in liquid media containing up to 0.8% $Na_2SO_4$. No progeny from the $F_1$ microspores cultured in the high salt medium were as susceptible as the susceptible parent. The results indicate that in these experiments sensitive microspores were eliminated. Lines exhibiting elevated levels of tolerance to salt result from recombination rather than from gametoclonal variation. Swanson *et al.* (1988) screened microspores of rapeseed successfully for herbicide resistance. *In vitro* culture at the haploid level combined with a screening system can detect recombinations in very large populations and at the earliest developmental or breeding stage thinkable. In contrast to selections in somatic cell populations, here the desired character is delivered from one of the parents. This means that, even if one screening technique is not sufficient to detect a complex reaction resulting in a phenotypic behavior finally expressed in the field, the same technique may be sufficient to identify one step of this complex. As long as this step is linked to the whole process, and as the whole process is often transferred, such a selection system may work in segregating homozygous material.

Depending on the breeding level of the parents and on the genetical basis of the character desired, the haploidization will be performed in $F_1$ or in later generations. If, for example, a resistance gene is transferred from a primitive cultivar into a variety, numerous recombinations are advantageous; therefore, the haploid step should take place after conventional preselection. If the aim, however, is to combine complex characters of varieties, only a few recombinations are necessary; thus, start of haploidization in the $F_1$ is recommended. Repeated recombinations—as necessary in the classical procedure — will separate many alleles and, furthermore, the selection for quantitatively inherited traits is very weak. Back crosses with doubled haploids already in the $A_1$, selection and a repeated haploid step (recurrent haploid selection) is probably today the best procedure to combine quantitatively inherited characters of sexually propagated crops (Foroughi-Wehr and Wenzel, 1990).

## 13.3 USE OF MOLECULAR GENETICS IN PLANT BREEDING

Modern molecular genetics offer techniques for gene transfer and for selection. Accurate diagnosis of pathogens must precede each plant treatment and each attempt to breed for agronomic traits. Besides morphological and biological tests (phenotypic level), chromosome identification (chromosome level) and immunological procedures (gene product level) were of utmost importance in this aspect. Molecular understanding opened up the advanced possibilities of monoclonal antibodies (*e.g.* Rasmussen, 1985). However, all these techniques, though already on the protein and responsible enzyme level (gene product level), are still not dealing with the causal agent, the DNA, or differences in the DNA (DNA level). Direct DNA diagnoses allow for the immediate detection of alterations without uncovering the effects of an alteration. Before plant breeders and agriculture became aware of these possibilities human medicine gained practical interest. Today the transgenic plants produced by such tools themselves help to improve these tools further, particularly under the aspect of gene expression.

### 13.3.1 Pathogen Diagnoses

In breeding for resistances correct and quantitative diagnoses of pathogens is often difficult. Molecular techniques opened up new possibilities by DNA/DNA or RNA/DNA hybridization techniques. The first use of DNA probes for disease diagnoses was made for detection of viroids, a pathogen where immunological procedures failed because of the missing protein coat. Of particular interest for agriculture is the potato spindle tuber viroid PSTV, a quarantine disease. Van Wezenbeek *et al.* (1982) and Bernardy *et al.* (1987) developed DNA probes and a routine hybridization technique. This method was extended to detect plant viruses (Baulcombe *et al.*, 1984); for bacteria and fungi the procedures are under development. With respect to applications, it would be very beneficial to avoid the radioactive labeling of the DNA probe. Success has already been obtained for PSTV (Dibbits and Verduin, 1985). Although the sensitivity of biotin- and $^{32}P$-labeled probes is comparable ( in both cases nearly 1 pg PSTV–RNA could be detected), specificity is still better in the radioactive system, which gives less false positives (Graner, 1987).

DNA hybridization techniques to detect specific plant genes, responsible for resistances or quality characteristics, are still in the infancy stage; transfer of genes is principally solved and becoming an agricultural practice, *e.g.* in herbicide resistance. In comparison to the strategy of gene transfer it will probably be easier to look in a sexually produced population with DNA probes for desired traits. If a character is available in the crop in a heritable form, a simple and efficient selection procedure will be more economic than gene transfer. In addition the unknown problem of how a foreign gene will behave in the genome of a given variety does not count in this approach.

When gene transfer works, a breeder will need to know which transferred genes are in their materials. It will normally be more convenient to check this using DNA probes than to assay the trait in the field. DNA diagnoses by hybridization will become easier and more informative as more useful genes are known physically. Like electrophoretic procedures which are routine in variety assessment, DNA probes will be routine within the next three years.

### 13.3.2 Gene and Genome Diagnoses and Restriction Fragment Length Polymorphism

Restriction fragment length polymorphism (RFLP) represents an important tool in describing the variability in crop plants (Beckmann and Soller, 1986). This is due to their large diffusion in breeding and genetic resources, which permits saturation of the genetic maps for a given species. They are expected to be free of secondary effects on economic characters, effects which due to their pleiotropic action are frequently associated with the segregation of alleles affecting morphophysiological traits (Table 2). This concept is extended now to quantitative trait loci (QTL), which are responsible for most traits of agricultural importance. QTL have neither known protein products nor striking phenotypic effects. Soller and Beckmann (1987) explored the potential of mutagenesis as a result of integration of novel DNA sequences into the germ line as a means of cloning QTL.

Finding RFLPs and QTLs is dependent on the number of probes available. This amount of probes is steadily increasing, and for maize a RFLP kit with a high number of probes is already commercially available. There is no immediate need to always find the genome part directly responsible for a character expressed at the phenotype level. It would be sufficient for applied breeding strategies to detect characteristic RFLPs correlated to phenotypes. Particularly under the aspect that the agriculturally important characters are not coded just by one gene but are of polygenic origin, such correlations would be of utmost importance. In tomato, introgressed chromosomal segments can increase soluble solids in the fruit (Rick, 1974). Two of the introgressed segments have been identified with RFLP markers (Osborn *et al.*, 1987) and localized to the end of chromosome 10 (Tanksley and Hewitt, 1988). It should be possible to improve tomato varieties for this character by indirect selection for the linked RFLP marker. Substantial progress has also been made in soybean, where 27 markers were analyzed (Apuya *et al.*, 1988) and in rice, where a RFLP genetic map of chromosomes has been constructed. The map is comprised of 135 loci corresponding to clones selected from a Pst I genomic library. It covers 1389 cM of the rice genome (McCouch *et al.*, 1988). As a by-product, it was found that rice DNA is less *C*-methylated than tomato or maize DNA. Approximately 50% of the rice DNA is single copy. Thus, RFLP maps will be of importance in detecting variety characteristics and transferred genes. Futhermore, RFLP markers may be used to explore the origin and evolution of crop plants, as was done for the six cultivated *Brassica* species (Song *et al.*, 1988).

The RFLP maps are normally combined with isozyme markers, allowing a rather substantial genome analysis. It is still too early to judge the importance of these techniques under applied aspects, but a number of companies are investing relatively large amounts of money into this research field, indicative of the high expectations breeding companies put into tools for secure selection.

**Table 2** Advantages of RFLP Markers in Selection Programs

| | |
|---|---|
| 1. | Present everywhere |
| 2. | Mendelian inheritance |
| 3. | Codominant gene expression |
| 4. | No pleiotropic effects |
| 5. | Independent of the environment |
| 6. | Present at each developmental stage |
| 7. | Long stability of DNA probes |
| 8. | Different loci may be identified by one probe |
| 9. | Heterologous genes may be used as probes |
| 10. | Any number of probes can be produced |
| 11. | Probes are not only producible for coding sequences |
| 12. | Probes show the variablity of neighboring sequences |

### 13.3.3 Gene Transfer

The production of new variability is, in both classical and mutation approaches, the result of very accidental recombinations. Symmetric and asymmetric protoplast fusion increased predictability and today we are witnessing the first examples of a successful transfer of specific traits into higher plants by gene technology. The most obvious aim is to improve a top variety in one additional character without disturbing the rest of the genome. It should be stressed that the expression 'gene' has different meanings in agriculture and in molecular biology. Agriculture describes a gene as a segregating character, and visualizes different segregation rates on the Morgan gene map. In plant breeding, gene is a mathematical expression; nevertheless, linkage groups and crossing over rates can be symbolized on chromosomes to give a rather concrete picture. In molecular genetics a gene is known physically as a sequence of nucleotides lined up in the DNA macromolecule. At the moment, intensive research is being devoted to explaining a gene on the Morgan map as a physical base sequence. The present calculation is that 1 centimorgan equals $10^6$ base pairs. This substantiates that we are still far away from constructing the better plant by use of artificial genes. Nevertheless, the following aims are of particular practical interest: production of additional resistances against diseases; environmental stresses and herbicides; alteration of primary and secondary plant products like proteins, oils and carbohydrates; increase of photosynthetic activity; and improvement of symbiosis with soil organisms to a better uptake of nutrients.

Using gene transfer techniques, the following prerequisites have to be fulfilled: the gene has to be available as a piece of DNA; this DNA should be clonable; the gene has to be manipulable in a transfer system; integration of the DNA in the cellular genome has to be achievable; and the transformed cell has to be regenerable into a functional plant which expresses the transferred trait. In principle all prerequisites are fulfilled today. The first structural gene (RuBisCo) was isolated in 1977 by Coen *et al.* Since then the number of physically identified genes has grown rapidly. Cloning of a gene is also no longer a problem, *via* the use of plasmids or cosmids.

For gene transfer, three methods are available: microinjection by capillaries, laser microbeams or microprojectiles; vector-mediated gene transfer; and direct transfer. Based on the positive results of microinjection in animal systems, this technique was tried in plants. Although under applied aspects such a laborious method will only be of very limited importance, one procedure shall be discussed here. Neuhaus *et al.* (1987) microinjected gene constructs into multicellular structures of *Brassica napus*, which have a high competence for plant regeneration through embryogenesis. Microspore-derived embryoids of *B. napus* (Lichter, 1982) were individually microinjected with neomycin phosphotransferase II (NPT II) gene, under control of the cauliflower mosaic virus promoter 35S and the transcription terminator of a modified pDH 51 (Pietrzak *et al.*, 1986). Transformation efficiencies between 27% and 51% were determined by DNA dot blot analysis of primary regenerants. Stable integration of full length microinjected genes into high molecular weight DNA was proven by Southern blot analysis of genomic DNA isolated from regenerated plants. As the starting material was multicellular, chimeras were produced. The chimeric nature of such primary regenerants was overcome by *in vitro* segregation through secondary embryogenesis, resulting in wild type and pure transformants (Neuhaus *et al.*, 1987). Such injection procedures also work in cereals, as demonstrated by De la Pena *et al.* (1987), who obtained transgenic rye plants by injection of plasmid DNA into young floral tillers. From about 140 injected tillers, *ca.* 4000 seedlings were harvested and three plants were finally confirmed as transformed ones (Lörz *et al.*, 1987). An extension of the microinjection with capillaries was the use of high voltage microprojectiles (Klein *et al.*, 1988) or laser beams (Weber *et al.*, 1988). It is not yet clear whether these techniques will gain applied importance.

Vectors are used most commonly for gene transfer, particularly the Ti plasmid of *Agrobacterium tumefaciens*. For technical details the reader is referred to the review of Horsch *et al.* (1987). For *A. tumefaciens* based gene transfer, the infection of a plant cell by the bacterium is a prerequisite. This happens easily in nature only with dicots, with monocots not acting as hosts. Thus, this transfer technique is hampered by host range limitations, which are especially disturbing when aiming at gene transfer into cereals. The host range poses an even more severe problem in the other biological vector system, plant DNA viruses (Brisson *et al.*, 1984).

Such limitations do not exist in the direct transfer method, the basic principle of which is summarized by Potrykus *et al.* (1987). In this procedure foreign DNA is linked to a strong promoter and to marker sequences, and incubated together with protoplasts. Under the influence of PEG, heat shock and/or electroporation, the DNA is taken up and may be integrated into the nuclear DNA: It could be demonstrated that using direct gene transfer *Lolium*, *Triticum*, *Zea*, *Oryza*, *Sorghum* and *Panicum* could be transformed ( for a review, see Lörz *et al.*, 1988). The remaining problem for application of direct gene transfer in monocots, the plant regeneration from transformed single cells, has been solved in principle for maize and rice and in other cereals it looks promising. To date, transformed plants have been regenerated after direct gene transfer with electroporation from maize protoplasts transformed with the NPT II gene (Rhodes *et al.*, 1988b), and from rice protoplasts after PEG-mediated DNA uptake of the glucuronidase gene (Zhang and Wu, 1988).

Despite its restriction in host range, *Agrobacterium*-mediated transfer has gained most practical importance. In the following paragraph examples shall be described where application in agriculture is expected soon.

Most progress has been obtained in incorporating specific resistances to herbicides, and plants with a resistance to nonselective herbicides will be the first practical results of plant gene technology. One reason for the rapid progress in this field is the fact that the biochemical pathways of modern herbicides are known; they are quite often based on just one or a few enzymes, allowing for a straightforward strategy. Thus, specific resistance (tolerance) against one of the most commonly used herbicide, glyphosate, has been transferred first to petunia and subsequently to tobacco and tomato (Horsch *et al.*, 1987). The strategy used was an overexpression of 5-enolpyruvylshikimate-3-phosphate synthase (EPSPS), which is blocked at normal concentrations by the herbicide. Overproducing mutants were first screened for in cell suspension cultures of petunia by sequential step-up from 0.1 mM glyphosate to 20 mM glyphosate. In the tolerant cultures about 20 times more EPSPS was produced. The pure protein was then partially sequenced. From there the mRNA could be used to isolate a DNA, which was finally transferred *via Agrobacterium* to *Petunia* cells. Transformed callus proved to be tolerant to high glyphosate concentrations. This gene is now available in different *Solanaceae* and in *Brassica* species. The first field tests with glyphosate resistant rapeseed were made in the US and Canada. A gene from *Streptomyces*, which creates resistance to the herbicide bialaphos (phosphinothricin, BASTA), was isolated and transferred to higher plants by two groups (De Block *et al.*, 1987; Wohlleben *et al.*, 1988). Probably the potato and rapeseed are the first candidates among the crop plants, where this herbicide resistance can be applied.

Our knowledge of biochemical mechanisms of disease resistance is very limited, and a straightforward strategy, comparable to the approach in herbicide resistance, is not yet available. There is, however, an increasing understanding of reactions which take place in the infected plant, *e.g.* the production of pathogen-related proteins (PR-proteins; Rigden and Coutts, 1988). It is hoped that with such proteins a strategy can be developed to use their genes in transformation and to create resistance. Until these processes are understood well enough, experiments are performed in inducing virus resistance by gene technology. The strategy is, however, quite different from the previous one. It is known that a virus causes cross protection or praemunity, which means that a related virus cannot infect the same plant. The coat protein of the virus is responsible for this protection. Abel *et al.* (1986) transferred the gene responsible for the coat protein production of TMV *via Agrobacterium* to tobacco. It was expressed and prevented symptom formation in tobacco plants after TMV inoculation. The same strategy works now for the combinations tobacco/alfalfa mosaic virus (Loesch-Fries *et al.*, 1987), tomato/TMV (Nelson *et al.*, 1988). and tobacco/CMV (Cuozzo-*et al.*, 1988).

In the area of environmental stress resistance the strategy is to influence physiological parameters or secondary products. Seeds are quite often an important nutrient because of specific amino acid compositions. Experiments to influence the amino acid composition of the potato tuber are under investigation (Rohsal *et al.*, 1986). Unilever (R. Shields, personal communication) is interested in clarifying the pathways of fatty acids to influence the oil and lipid composition of *Brassica* (Slabas *et al.*, 1984). These approaches are of particular interest, as nature deposits in the oleosomes of seeds a wide range of very different $C_{18}$ triglycerides, thus enabling analysis of specific gene functions and transfer of such characters (detected, for example, in wild species) to crops. In cases where just a few enzymes are involved in specific product formation, gene technology has great potential. It should soon be possible to identify and transfer genetic information for one enzyme, although it will probably take quite a while before complex characters like most disease resistances can be manipulated.

Production of plants resistant to heat, frost or drought normally follows the pathways described above, using mutation and selection. Regardless of the type of stress, proteins which are very similar are often formed. These similarities give some hope for a more general strategy for different stresses. Actually only such a planned approach has a chance for long-lasting application.

A completely different strategy was followed in the case of chilling resistance by modifying leaf surface bacteria, the so-called ice bacteria, a specifically modified strain of *Pseudomonas syringae*. Ice bacteria are responsible for damage in a number of crop plants, because they produce a protein which acts at temperatures of 1.5 °C and below as a starting point for crystallization, which results in bursting cells. After cloning a *P. syringae* library in *Echerichia coli*, an active clone was found and a deletion mutant constructed *in vitro* (Hirano and Upper, 1985). This deletion mutant was then transferred by conjugation to an ice-nucleating strain of *P. syringae*. By homologous recombination the deletion mutant replaced the active gene. In 1987 the first experiments were allowed in the USA in a natural environment. By spraying stawberries at the appropriate time, colonization by ice-nucleating bacteria was reduced, with a corresponding reduction in frost injury.

## 13.4 OTHER UNCONVENTIONAL APPROACHES

Other unconventional approaches which may be of importance to plant production should be mentioned briefly: breeding for low input varieties by genetically optimizing the microflora of the soil; approaches towards nitrogen fixation; and biological pest control as connected to gene transformation. Most of these strategies are very recent and it is too early to expect cultivars to be already on the market.

### 13.4.1 Low Input

Plants always interact with soil microorganisms. The possibilities to optimize this interaction are increasingly being discussed because of the positive influence on the uptake of nutrients and their protective action against pathogens.

#### *13.4.1.1 Nutrient uptake*

Besides nitrogen-fixing root bacteria (discussed in the next paragraph), bacteria from the *Pseudomonas* group can have beneficial effects on the nutrient uptake, particularly of iron(III) (Weisbeek, 1986). Crops like potato show increases in yield after treatment of seed tubers with *Pseudomonas putida* and *P. fluorescens.* Under iron-limiting conditions some *Pseudomonas* isolates produce and excrete yellow-green fluorescent siderophores, which chelate the iron in the root environment and prevent deleterious microorganisms from growing efficiently (Weisbeek, 1986; Werner, 1987). Besides these bacteria, fungi, mycorrhizae in particular, growing endo- or exogenously, are of utmost importance; they can supply more phosphorus and micronutrients like zinc and copper to the plant (Werner, 1987). Most of the higher plants are infected with mycorrhizal fungi; the ectomycorrhizae are mostly symbionts of trees and do not only increase the uptake of ions but at the same time link the tree root systems in a forest. A genetic manipulation of such mycorrhizal fungi would have a potential impact on forest production. *Trichoderma* also belongs to the group of ectomycorrhizae (Chet, 1986); it is an antagonist of other fungi and is capable of both inhibiting and destroying its hosts. It is still an open question whether fungi or soil bacteria can be improved by genetic engineering to such an extent that they increase the uptake of nutrients or the defense against diseases so far that plant production can benefit. As natural gene transfer is common in this group, probably no additional gene technology is needed here, just a better understanding of the soil flora (Beringer, 1986; Sobieszczanski, 1986).

#### *13.4.1.2 Nitrogen fixation*

A specific subject within the topic of low input is nitrogen fixation of the higher plant by symbiosis or directly. Nature has produced a very efficient system: the symbiosis between *Leguminosae* and *Rhizobium.* Besides *Rhizobium,* there exists a number of genera which fix nitrogen, such as *Klebsiella, Azospirillum* and *Azotobacter.* The amount of air nitrogen-fixed is reasonable and surpasses the amount produced by the Haber–Bosch reaction (Werner, 1987). Not all nitrogen-fixing microorganisms rely on symbiosis, but the symbiotic ones are more efficient. The function and structure of the responsible genes have been investigated most carefully for *Klebsiella pneumoniae* (*e.g.* Dixon and Postgate, 1971) and *Rhizobium meliloti* (*e.g.* Kondorosi *et al.*, 1987). As the nitrogen fixation works very efficiently, it was hoped to be able to transfer the nitrogen-fixing (*nif*) gene from the bacteria to the plant genome. After extensive research in this area it has become clear that this will be impossible. There are, however, several reasonable steps in the direction of this goal: nitrogen-fixing by soil microorganisms and optimization of this process; optimization of the symbiosis of micro-organisms and legumes by increasing the efficiency (*e.g.* more nodules, better regulation); and extending the host range to nonlegume plants. One promising step in the direction of optimizing the interaction of plant and microorganism was the finding of soybeans with an increased number of root nodules per plant, *via* which *Rhizobium* acts (Carrol *et al.*, 1985). This success suggests that it will probably be easier to make a nonfixing plant become a host (for *Rhizobium*, for example) than to transfer by gene technology the ability to fix nitrogen.

One problem in using this system by gene technology for agriculture is the fact that 17 genes in *Klebsiella* code for nitrogen fixation. For all genes, promoters have to be adapted to the genetic conditions of the higher plant. Futhermore, these nitrogen-fixing enzymes are very sensitive to $O_2$ and need enormous amounts of ATP. As soon as enough $NH_4$ ions are available the process is stopped. Therefore, for advanced agriculture a first goal would be to block this natural regulation (Wang *et al.*, 1987). Otherwise this approach would be only sufficient for very extensive agriculture where the nitrogen could be fixed by bacteria, while for more intensive production additional nitrogen fertilizer would be needed, which would block the activity of the microorganisms.

### 13.4.2 Biological Pest Control

#### *13.4.2.1* Bacillus thuringiensis

About 75 years ago a bacillus was found in silkworms as the causal agent for the sotto disease in Japan, which is named today *Bacillus thuringiensis* ssp. *sotto*. This bacillus was also harmful to a number of other insect pests and first trials to use this disease as a biological pest control were started in 1928. This concept is based on the natural antagonists of pathogens and pests. More than a thousand insect pathogen microorganisms are known; from these, *Bacillus thuringiensis* ssp. *israelensis* (B.t.i.) is the most prominent microbial vector control agent (Krieg, 1986). Commercial preparations of B.t.i. show great insecticidal activity against mosquito larvae and against blackflies.

The bacteria produce toxic proteins in their long-lasting forms. If a larvae takes up such endotoxins they stop eating and are killed within a few days. To increase the applicability of this system four strategies are followed at present: (i) to look for other natural subspecies which might be specific to other pests (Krieg *et al.*, 1983, for example, were able to isolate a subspecies which is toxic to *Coleopterae*, particularly to the Colorado beetle); (ii) to modify the bacteria to increase their efficiency (the virulence of the toxin might be increased by amplification of the responsible genes); (iii) to transfer the responsible genes to other bacteria (experiments are in progress to transfer the genes to *Pseudomonas fluorescens* or *Rhizobium japonicum*, for example; this would combine the low input strategy with that of pest control); and (iv) to isolate the genes responsible for toxin production and to transfer these genes into crop plants (the transfer of B.t. toxin to tobacco was achieved; it could be demonstrated that larvae are killed while eating tobacco leaves, Barton *et al.*, 1987). As it is probable that the genetic information responsible for the toxin production of B.t. is located on plasmids, such gene transformation experiments can be done rather straightforwardly.

Under applied aspects one has to find in this strategy a compromise between too specific B.t. strains, which do not help agriculture too much, and too general strains, which may harm ecology.

#### *13.4.2.2 Baculo viruses*

Viruses may be used to protect against pests. Best known are the baculo viruses. They predominantly attack larvae of butterfly-like insects. The information available from biological pest control with the actual virus and the rather small genome of the virus made it attractive to influence its effectiveness by gene technology. To minimize possible dangers from such an artificially engineered virus, strategies are under investigation to delete some parts of the virus genome so that it can easily be identified. The baculo viruses are potentially of great use as vectors for the expression of foreign genes in insects. Baculo virus vectors have achieved widespread acceptance for their ability to express proteins of agricultural importance. The big advantage of baculo viruses is that they are not pathogenic to vertebrates or plants and do not employ transformed cells or transforming elements (Lucknow and Summers, 1988). There is surely a great potential in this procedure for agriculture.

#### *13.4.2.3 Other systems*

In a similar way, a gene for chitinase synthesis from the bacterium *Serratia marrescens* has been cloned in *E. coli* (Nitzsche *et al.*, 1983) and transferred to a strain of *Pseudomonas fluorescens*, which is a very efficient root colonizer. It is hoped that this strain will protect inoculated plants from fungi with chitinous cell walls and from soil-inhabiting insects, as proposed by Nitzsche (1983). Chitinase is already naturally present in plants and its activity is related to ethylene production, thus playing a role after pathogen attack (Broglie *et al.*, 1985). Consequently its manipulation might strengthen the natural defence mechanisms. Similar approaches are under development for cutinase and pectinase (Kolattukudy *et al.*, 1987). The alternative strategy is, as in the B.t.i. example, to introduce such genes directly into the plant and couple it to a root specific promoter.

## 13.5 CONCLUSIONS

For applied breeding methods, classical techniques using recombination and selection procedures are still the most efficient ones today. Only if the natural variability is too narrow or the classical way is too slow should unconventional approaches be considered. Generally it is not variability but the selection efficiency that limits progress of breeders. Until now this has also been true for most unconventional programs. It is easy to create variability, *e.g. via* somaclones; the problem is which individuals to select in the laboratory as the more resistant ones, and which one to ultimately grow in the field or release as a commercial cultivar. A selection procedure that can be done directly in the petri dish or eventually in the field should be economic and not require the screening of many thousands of protoplast-derived plants in the field, and it should also be able to detect quantitatively inherited differences under different environments. However, even the most intelligent approach, if it is too expensive, will fail in practice.

**Table 3** Subjective Prediction of which Technique Might Have which Level of Influence in Achieving Different Breeding Goals[a]

| *Goal* | *Mendelian* | In vitro *selection* | *Procedure Somatic fusion* | *Haploids* | *DNA diagnosis* | *Gene transfer* |
|---|---|---|---|---|---|---|
| Quality | + + | + − | + + | + + + | + + | + + |
| Resistance | | | | | | |
| Disease | + + + | + | + + | + + + | + + + | + + |
| Stress | + + | + − | − | + − | + + | + |
| Herbicide | − | + | + − | − | + | + + + |
| Yield | + + + | − − − | − | + | + + | − − |
| Low Input | + + | − − | − − | + | + | − − |

[a] + + + = very high probability for a contribution; − − − = very low probability

**Table 4** Summary of some Central Steps in the Application of Unconventional Techniques in Plant Breeding

| *Technique* | *Aim* | *Leading species in application* | *Ref.* | *Year* |
|---|---|---|---|---|
| Tissue culture | Rapid propagation | Potato | Wriedt | 1984 |
| | Somaclonal variation | Sugar cane | Heinz *et al.* | 1977 |
| | Living collections | Potato | Mix | 1985 |
| Cell culture | Genotype cloning | Potato | Schuchmann | 1986 |
| | Fusion | Rapeseed | Schenck and Röbbelen | 1982 |
| | Doubled haploids | Barley | Ho and Jones | 1980 |
| | Cytoplasmic male sterility | Rapeseed | Pelletier *et al.* | 1988 |
| DNA transfer | Virus resistance | Tomato | Nelson *et al.* | 1988 |
| | Insect resistance | Tomato | Barton *et al.* | 1987 |
| | Herbicide resistance | | | |
| | (BASTA) | Potato | De Block | 1988 |
| | (Roundup) | Rapeseed | Horsch *et al.* | 1987 |
| | Protein quality | Potato | Rosahl *et al.* | 1986 |
| cDNA probes | Pathogen diagnoses (PSTV) | Potato | Bernardy *et al.* | 1987 |
| RFLP marker | Fruit quality | Tomato | Tanksley and Hewitt | 1988 |

Completely logical genome constructions are not possible today, due to the huge number of structural genes (about 10 000 in the plant compared to about 2000 in bacteria) and due to the diploidy (at least several crops, particularly vegetatively propagated ones, are polyploids, which multiplies the difficulty), and even 10 000 diploid structural genes mean $4^{10\,000}$ possible gene combinations. Although our knowledge of the plant genome is very limited, the principal prerequisites for the production of a plant better adapted to man's wishes are fulfilled: regeneration of plants from organs, cells microspores and protoplasts; somatic fusion; isolation of genes in their physical form as DNA; transfer systems for such genes into the plant cell; regeneration of such cells; expression of the transferred character; and rapid multiplication of the engineered plant.

We are witnessing today the first cultivars of *in vitro* origin, namely doubled haploid cereals. Rapid propagation is already saving money for the growers and field experiments with engineered plants have been started for different resistances. Biotechnology will make its way into the field; some goals and tools of biotechnology in agriculture are listed in Table 3. It is tried as well to compare their probability with concepts of the classical approach. As can be seen, classical breeding is powerful but there are areas where only biotechnology has a fair chance. The field crop presently leading in a particular unconventional technique is listed in Table 4.

Even the most intelligent approach may fail when regulations restrict this development; it will also fail, however, when light-minded strategies create problems which are difficult to overcome. Ultimate success of biotechnology in agriculture relies on the responsible cooperation of agriculture, plant physiology, classical and molecular genetics and ecology.

## 13.6 REFERENCES

Abel, P. P., R. S. Nelson, B. De, N. Hoffmann, S. G. Rogers, R. T. Fraley and R. N. Beachy (1986). Delay of desease development in transgenic plants that express the tobacco mosaic virus coat protein. *Science (Washington D.C.)*, **232**, 738–743.

Apuya, N. R., B. L. Frazier, P. Keim, E. Jill Roth and K. G. Lark (1988). Restriction fragment length polymorphisms as genetic markers in soybean, *Glycine max* (L.) *merill. Theor. Appl. Genet.*, **75**, 889–901.

Austin, S., M. A. Baer and J. P. Helgeson (1985). Transfer or resistance to potato leaf roll virus from *Solanum brevidens* into *Solanum tuberosum* by somatic fusion. *Plant Sci. Lett.*, **39**, 75–82.

Austin, S., E. Lojkowska, M. K. Ehlenfeldt, A. Kelman and J. P. Helgeson (1988). Fertile interspecific somatic hybrids of *Solanum*: a novel source of resistance to *Erwinia* soft rot. *Phytopathology*, **78**, 1216–1220.

Aviv, D. and E. Galun (1980). Restoration of fertility in cytoplasmic male sterile (CMS) *Nicotiana sylvestris* by fusion with X-irradiated *N. tabacum* protoplasts. *Theor. Appl. Genet.*, **58**, 121–127.

Barton, K. A., H. R. Whiteley and N.-S. Yang (1987). *Bacillus thuringiensis* endotoxin expressed in transgenic *Nicotiana tabaccum* provides resistance to Lepidopteran insects. *Plant Physiol.*, **85**, 1103–1109.

Baulcombe, D. C., R. E. Bouton, R. B. Flavell and G. L. Jellis (1984). Recombinant DNA probes for detection of viruses in plants. In *British Crop Protection Conf. Pest and Diseases*, ed. British Crop Protection Council, pp. 207–213. BCPC Publications, Croydon.

Beckmann, J. S. and M. Soller (1986). Restriction fragment length polymorphisms and genetic improvement of agricultural species. *Euphytica*, **35**, 111–124.

Beringer, J. E. (1986). Plant–microbe interactions. In *Biotechnology: Potentials and Limitations*, ed. S. Silver, pp. 259–273. Springer-Verlag, Berlin.

Bernardy, M. G., G. G. Jacoli and H. W. J. Ragetti (1987). Rapid detection of potato spindle tuber viroid (PSTV) by dot blot hybridization. *Phytopathol.*, **118**, 171–180.

Binding, H., R. Nehls, O. Schieder, S. K. Sopory and G. Wenzel (1978). Regeneration of mesophyll protoplasts isolated from dihaploid clones of *Solanum tuberosum* L. *Physiol. Plant.*, **43**, 52–54.

Binding, H., S. M. Jain, J. Finger, G. Mordhorst, R. Nehls and J. Gressel (1982). Somatic hybridization of an atrazin resistant biotype of *Solanum nigrum I.* Clonal variation in morphology and in atrazin sensitivity. *Theor. Appl. Genet.*, **63**, 273–277.

Boulay, M. (1987). *In vitro* propagation of tree species. In *Plant Tissue and Cell Culture*, ed. C. E. Green *et al.*, pp. 367–382. Liss, New York.

Brisson, N., Paszkowski, J. R. Penswick, B. Gronenborn, I. Potrykus and T. Hohn (1984). Expression of a bacterial gene in plants using a viral vector. *Nature (London)*, **310**, 511–514.

Broglie, K. E., J. J. Gaynor, M. Durand-Tardif and R. Broglie (1985). Regulation of chitinase expression by ethylene. In *Biotechnology in Plant Science*, ed. M. Zaitlin *et al.*, pp. 247–258. Academic Press, San Diego.

Buiatti, M., A. Scala, P. Beltini, G. Nascari, R. Morpurgo, P. Bognai, G. Pellegrini, F. Gimelli and R. Venturo (1985). *In vitro* differential response of resistant and susceptible carnation cvs. to fungal and abiotic elicitors and to *Fusarium oxysporum* f.sp. *dianthi* culture filtrates. *Theor. Appl. Genet.*, **70**, 42–47.

Carrol, B. J., D. L. McNeil and P. M. Gresshoff (1985). A supernodulation and nitrate tolerant symbiotic (nts) soybean mutant. *Plant Physiol.*, **78**, 34–40.

Chaleff, R. S. and C. J. Mauvais (1984). Acetolactate synthase is the site of action of two sulfonylurea herbicides in higher plants. *Science (Washington D.C.)*, **224**, 1443–1445.

Chandra, A., J. H. Dodds and P. Tovar (1988). *In vitro* tuberization in potato (*Solanum tuberosum* L.). IAPTC Newsletter, **55**, 10–20.

Chet, I. (1986). *Non conventional Approaches to Plant Disease Control (Environmental and Applied Microbiology)*. Wiley, New York.

Christianson, M. L., D. A. Warnick and P. S. Carlson (1983). A morphogenetically competent soybean suspension culture. *Science* (*Washington D.C.*), **222**, 632–634.

Coen, D. M., J. R. Bedbrook, L. Bogorad and A. Rich (1977). Maize chloroplast DNA encoding the large subunit of ribulosebisphosphate carboxylase. *Proc. Natl. Acad. Sci. USA*, **74**, 5487–5491.

Creason, G. L. and R. S. Chaleff (1988). A second mutation enhances resistance of a tobacco mutant to sulfonylurea herbicides. *Theor. Appl. Genet.*, **76**, 177–182.

Cuozzo, M., K. M. O'Connell, W. Kaniewski, R.-X. Fang, N.-H. Chua and N. E. Tumer (1988). Viral protection in transgenic tobacco plants expressing the cucumber mosaic virus coat protein or its antisense RNA. *Bio/Technology*, **6**, 549–553.

Datta, S. K. and G. Wenzel (1988). Single microspore derived embryogenesis and plant formation in barley (*Hordeum vulgare* L.). *Arch. Zuechtungsforsch.*, **18**, 125–131.

Datta, S. K., M. Bolik and G. Wenzel (1991). Culture of isolated pollen of wheat (*Triticum aestivum* L.). In *Biotechnology in Agriculture and Forestry*, ed. Y. P. S. Bajaj, Vol. 13, pp. 81–86. Springer-Verlag, Berlin.

De Block, M. (1988). Genotype independent leaf disc transformation of potato (*Solanum tuberosum*) using *Agrobacterium tumefaciens*. *Theor. Appl. Genet.*, **76**, 767–774.

De Block, M., J. Bottermann, M. Vandewiele, J. Dockx, C. Thoen, V. Gossele, N. R. Movva, C. Thompson, M. Van Montagu and J. Leemans (1987). Engineering herbicide resistance in plants by expression of a detoxifying enzyme. *EMBO J.*, **9**, 2513–2518.

Debnath, S. C. and G. Wenzel (1987). Selection of somatic fusion products in potato by hybrid vigor. *Potato Res.*, **30**, 371–380.

De Buyser, J. and Y. Henry (1986). Florin: a doubled haploid wheat variety developed by anther culture method. In *IAPTC Abstracts*, ed. D. A. Somers *et al.*, p. 416. University of Minnesota, Minneapolis.

Deimling, S., J. Zitzlsperger and G. Wenzel (1988). Somatic fusion for breeding potatoes. *Z. Pflanzenuecht.*, **101**, 181–189.

De la Pēna, A., H. Lörz and J. Schell (1987). Transgenic rye plants obtained by injection of DNA into young floral tillers. *Nature* (*London*), **325**, 274–276.

Dibbits, J. G. T. and B. J. M. Verdiun (1985). Detection of potato spindle tuber viroid with cDNA using non radioactive labeling. *Acta Bot. Neerl.*, **34**, 351–352.

Dixon, R. A. and J. P. Postgate (1971). Transfer of nitrogen fixation genes by conjugation in *Klebsiella pneumoniae*. *Nature* (*London*), **234**, 47–48.

Dore, C. (1987). Application of tissue culture to vegetable crop improvement. In *Plant Tissue and Cell Culture*, ed. C. E. Green *et al.*, pp. 419–432. Liss, New York.

Dos Santos, A. V. P., D. E. Outka, E. C. Cocking and M. R. Davey (1980). Organogenesis and somatic embryogenesis in tissue derived from leaf protoplasts of *Medicago sativa*. *Z. Pflanzenphysiol.*, **99**, 261–270.

Dudits, D., E. Maroy, T. Praznovszky and J. Györgyey (1986). Gene transfer between distantly related plants by fusion with irradiated protoplasts. In *IAPTC Abstracts* ed. D. A. Somers *et al.*, p. 20. University of Minnesota, Minneapolis.

Foroughi-Wehr, B., W. Friedt, R. Schuchmann, F. Köhler and G. Wenzel (1986). *In vitro* screening for resistance. In *Somaclonal Variation in Crop Improvement*, ed. J. Semal, pp. 35–44. Nijhoff, Dordrecht.

Foroughi-Wehr, B. and G. Wenzel (1990). Recurrent selection alternating with haploid steps — a rapid breeding procedure for combining agronomic traits in inbreeders. *Theor. Appl. Genet.*, **80**, 337–382.

Fowler, M. W. (1984). Large scale cultures of cells in suspension. In *Cell Culture and Somatic Cell Genetics of Plants*, ed. I. K. Vasil, vol. 1, pp. 167–174. Academic Press, Orlando.

Galbraith, D. W. (1984). Selection of somatic hybrid cells by fluorescence-activated cell sorting. In *Cell Culture and Somatic Cell Genetics of Plants*, ed. I. K. Vasil, vol. 1, 433–447. Academic Press, Orlando.

Gengenbach, B. G., G. E. Green and C. M. Donovan (1977). Inheritance of selected pathotoxin resistance in maize plants regenerated from cell cultures. *Proc. Natl. Acad. Sci. USA*, **74**, 5113–5117.

Gleba, Y. Y. and K. M. Sytnik (1984). In *Protoplast Fusion*, ed. R. Shoeman. Springer-Verlag, Berlin.

Glimelius, K. (1988). Potentials of protoplast fusion in plant breeding programs. *Plant Cell, Tissue Organ Cult.*, **12**, 163–172.

Graner, A. (1987). Methodische Untersuchungen zum Nachweis von Potato spindle tuber viroid (PSTV). Ph.D. Thesis, Technical University, Munich.

Hammerschlag, F. A. (1988). Selection of peach cells for insensitivity to culture filtrate of *Xanthomonas campestris* Pv. *pruni* and regeneration of resistant plants. *Theor. Appl. Genet.*, **76**, 865–869.

Hartman, C. L., T. J. McCoy and T. R. Knons (1984). Selection of alfalfa (*Medicago sativa*) cell lines and regeneration of plants resistant to toxin(s) produced by *Fusarium oxysporum* f.sp. *medicagines*. *Plant Sci. Lett.*, **34**, 183–194.

Heinz, D. J., M. Krishnamurthi, L. G. Nickell and A. Maretzki (1977). Cell, tissue and organ culture in sugarcane improvement. In *Plant Cell, Tissue and Organ Culture*, ed. J. Reinert and Y. P. S. Bajaj, pp. 3–17. Springer-Verlag, Berlin.

Hirano, S. S. and C. D. Upper (1985). Ecology and physiology of *Pseudomonas syringae*. *Bio/Technology*, **3**, 1073–1078.

Ho, K. M. and G. E. Jones (1980). Mingo barley. *Can. J. Plant Sci.*, **60**, 279–280.

Hoffmann, F., E. Thomas and G. Wenzel (1982). Anther culture as a breeding tool in rape. Progeny analysis of androgenetic lines and of induced mutants from haploid cultures. *Theor. Appl. Genet.*, **61**, 225–232.

Horsch, R., R. Fraley, S. Rogers, J. Fry, H. Klee, D. Shah, S. McCormick, J. Niedermeyer and N. Hoffmann (1987). *Agrobacterium*-mediated transformation of plants. In *Plant Tissue and Cell Culture*, ed. C. E. Green *et al.*, pp. 317–329. Liss, New York.

Kao, K. N. and M. R. Michayluk (1975). Nutritional requirements for growth of *Vicia hajastana* cells and protoplasts at very low population density in liquid media. *Planta*, **126**, 105–110.

Kartha, K. K. (1987). Advances in the cryopreservation technology of plant cells and organs. In *Plant Tissue and Cell Culture*, ed. C. E. Green *et al.*, pp. 447–458. Liss, New York.

Kartha, K. K., M. R. Michayluk, K. N. Kao, O. L. Gamborg and F. Constabel, (1974). Callus formation and plant regeneration from mesophyll protoplasts of rape plants (*Brassica napus* cv. *Zephyr*). *Plant Sci. Lett*, **3**, 265–271.

Kasha, K. J. and K. N. Kao (1970). High frequency haploid production in barley (*Hordeum vulgare* L.). *Nature* (*London*), **225**, 874–875.

Keller, W. A. (1984). Anther culture of *Brassica*. In *Cell Culture and Somatic Cell Genetics of Plants*, ed. I. K. Vasil, vol. 1, pp. 302–310. Academic Press, Orlando.

Klee, H., R. Horsch and S. Rogers (1987). *Agrobacterium*-mediated plant transformation and its further applications to plant biology. *Annu. Rev. Physiol.*, **38**, 467–486.

Klein, T, M., T. Gradziel, M. E. Fromm and J. C. Sanford (1988). Factors influencing gene delivery into *Zea mays* cells by high velocity microprojectiles. *Bio/Technology*, **6**, 559–563.
Kolattukudy, P. E., J. Sebastian, W. F. Ettinger and M. S. Crawford (1987). Cutinase and pectinase in host-pathogen and plant bacterial interaction. In *Molecular Genetics of Plant–microbe Interaction*, ed. D. P. S. Verma and N. Brisson, pp. 43–50. Nijhoff, Dordrecht.
Komatsuda, T. and K. Ohyama (1988). Genotype of high competence for somatic embryogenesis and plant regeneration in soybean *Glycine max*. *Theor. Appl. Genet.*, **75**, 695–700.
Kondorosi, A., E Kondorosi, B. Horvath, M. Gottfert, C. Bachem, F. Rodriguez-Quinones, Z. Banfalvi, P. Putnoky, Z. Gyorgypal, M. John, J. Schmidt and J. Schell (1987). Common and host specific nodulation genes in *Rhizobium meliloti* and their conservation in other *Rhizobia*. In *Molecular Genetics of Plant–microbe Interaction*, ed. D. P. S. Verma and N. Brisson, pp. 217–222. Nijhoff, Dordrecht.
Krieg, A. (1986). *Bacillus thuringiensis, ein mikrobielles Insektizid. Acta Phytomedica*, **10**, Supplement to *J. Phytopath.* Parey, Berlin and Hamburg.
Krieg, A., A. M. Huger, W. Schnetter and G. A. Langenbruch (1983). Neue Ergebnisse über *Bacillus thuringiensis* var. *tenebrionis* unter besonderer Berücksichtigung seiner Wirkung auf den Kartoffekäfer (*Leptinotarsa decemlineatea*). *Anz. Schaedlingskd., Pflanzenschutz, Umweltschutz*, **57**, 145–150.
Kuhlmann, U. (1990). Mikrosporenregeneration und Untersuchungen zur Transformation bei Gerste. Ph.D. Thesis. Technical University, Munich.
Langerfeld, E. (1979). Prüfung des Resistenzverhaltens von Kartoffelsorten gegenüber *Fusarium coeruleum* (Lib.) Sacc. *Potato Res.*, **22**, 107–122.
Lichter, R. (1982). Induction of haploid plants from isolated pollen of *Brassica napus*. *Z. Pflanzenphysiol.*, **105**, 427–434.
Loesch-Fries, L. S., D. Merlo, T. Zinnen, L. Burhop, K. Hill, K. Krahn, N. Jarvis, S. Nelson and E. Halk (1987). Expression of alfalfa mosaic virus RNA 4 in transgenic plants confers virus resistance. *EMBO J.*, **6**, 1845–1851.
Lörz, H., B. Junker, J. Schell and A. De la Pena (1987). Gene transfer in cereals. In *Plant Tissue and Cell Culture*. ed. C. E. Green *et al.*, pp. 303–316. Liss, New York.
Lörz H., E. Göbel and P. Brown (1988). Advances in tissue culture and progress towards genetic transformation of cereals. *Z. Pflanzenzuecht.*, **100**, 1–25.
Lucknow, V. A. and M. D. Summers (1988). Trends in the development of baculo virus expression vectors. *Bio/Technology*, **6**, 47–52.
McCouch, S. R., G. Kochert, Z. H. Yu, Z. Y. Wang, G. S. Khush, W. R. Coffman and S. D. Tanksley (1988). Molecular mapping of rice chromosomes. *Theor. Appl. Genet.*, **76**, 815–825.
Meyer, K. and B. Foroughi-Wehr (1987). Possible ways to simplifying potato meristem culture. *Plant Res. Dev.*, **26**, 12–18.
Mix, G. (1985). Preservation of old potato varieties. In *In Vitro Techniques, Propagation and Long Term Storage*, ed. A. Schäfer-Menuhr, pp. 149–153. Nijhoff, Dordrecht.
Möllers, Ch. (1990). Protoplastenfusion als praxisgerechte Methode für die Kombinationszüchtung bei Kartoffel. Ph.D. Thesis, University of Hohenheim.
Morel, G. (1946). Essais de laboratoire sur le mildiou de la vigne. *Rev. Vitic.*, **93**, 210–213.
Mulder-Krieger T., R. Verpoorte, A. Baerheim Svendsen and J. J. C. Scheffer (1988). Production of essential oils and flavors in plant cell and tissue cultures. A review. *Plant Cell, Tissue Organ Cult.*, **13**, 85–154.
Mullins, M. G. (1987). Propagation and genetic improvement of temperate fruits: the role of tissue cultue. In *Plant Tissue and Cell Culture*. ed. C. E. Green *et al.*, pp. 395–406. Liss, New York.
Nabors, M. W. and T. A. Dykes (1985). Tissue culture of cereal cultivars with increased salt, drought and acid tolerance. In *Biotechnology in International Agricultural Research*, ed. IRRI, pp. 121–138. IRRI, Manila.
Nelson, R. S., S. M. McCormick, X. Delannay, P. Dube, J. Layton, E. J. Anderson, M. Kaniewska, R. K. Proksch, R. B. Horsch, S. G. Rogers, R. T. Fraley and R. N. Beachy (1988). Virus tolerance, plant growth and field performance of tomato plants expressing coat protein from tobacco mosaic virus. *Bio/Technology*, **6**, 403–407.
Neuhaus, G., G. Spangenberg, O. Mittelsten Scheid and H. G. Schweiger (1987). Transgenic rapeseed plants obtained by the microinjection of DNA into microspore-derived embryoids. *Theor. Appl. Genet.*, **75**, 30–36.
Nitsch, C., S. Andersen, M. Godard, M. G. Neuffer and W. F. Sheridan (1982). Production of haploid plants of *Zea mays* and *Pennisetum* through androgenesis. In *Variability in Plants Regenerated from Tissue Culture*, ed. E. D. Earle and Y. Y. Demarly, pp. 69–91. Praeger, New York.
Nitzsche, W. (1983). Chitinase as a possible resistance factor for higher plants. *Theor. Appl. Genet.*, **65** 171–172.
Nitzsche, W. and G. Wenzel (1977). *Haploids in Plant Breeding*. Parey, Berlin.
Ogura, H., J. Kyozuka, Y. Hayashi, T. Koba and K. Shimamoto (1987). Field performance and cytology of protoplast-derived rice (*Oryza sativa*): high yield and low degree of variation of four japonica cultivars. *Theor. Appl. Genet.*, **74**, 670–676.
Osborn, T. C., D. C. Alexander and J. F. Fobes (1987). Identification of restriction fragment length polymorphisms linked to genes controlling soluble content in tomato fruit. *Theor. Appl. Genet.*, **73**, 350–356.
Pelletier, G., C. Primard, M. Frerault, F. Vedel, P. Chetrit, M. Renard and R. Delourme (1988). Use of protoplasts in plant breeding: cytoplasmic aspects. *Plant Cell, Tissue Organ Cult.*, **12**, 173–180.
Pietrzak, M., R. D. Shillito, T. Hohn and I. Potrykus (1986). Expression in plants of two bacterial antibiotic resistance genes after protoplast transformation with a new plant expression vector. *Nucleic Acid Res.*, **14**, 5857–5868.
Potrykus, I., J. Paszkowski, M. W. Saul, I. Negrutiu and R. D. Shillito (1987). Direct gene transfer to plants: facts and future. In *Plant Tissue and Cell Culture.*, ed. C. E. Green *et al.*, pp. 289–302. Liss, New York.
Powell, W. and H. Uhrig (1987). Anther culture of *Solanum* genotypes. *Plant Cell, Tissue Organ Cult.*, **11**, 13–24.
Rasmussen, U. (1985). immunological screening for specific protein content in barley seeds. *Carlsberg Res. Commun.*, **50**, 83–93.
Redenbaugh, K., P. Viss, D. Slade and J. A. Fujii (1987). Scale-up: artificial seeds. In *Plant Tissue and Cell Culture*, ed. C. E. Green *et al.*, pp. 473–493. Liss, New York.
Reinert, J. (1958). Morphogenese und ihre Kontrolle an Gewebekulturen aus Carotten. *Naturwissenschaften*, **45**, 344–345.
Rhodes, C. A., D. A. Pierce, I. J. Mettler, D. Mascarenhas and J. J. Detmer (1988). Genetically transformed maize plants from protoplasts. *Science* (*Washington D.C.*), **240**, 204–207.
Rick, C. M. (1974). High soluble solids content in large-fruited tomato lines derived from a wild green-fruited species. *Hilgardia*, **42**, 493–510.

Rigden, J. and R. Coutts (1988). Pathogenesis-related proteins in plants. *Trends Genet.*, **4**, 87–89.
Rosahl, S., R. Schmidt, J. Schell and L. Willmitzer (1986). Isolation and characterization of a gene from *Solanum tuberosum* encoding patatin, the major storage protein of potato tubers. *Mol. Gen. Genet.*, **303**, 214–220.
Sacristan, M. D. (1982). Resistance response to *Phoma lingam* of plants regenerated from selected cell and embryogenic cultures of haploid *Brassica napus*. *Theor. Appl. Genet.*, **61**, 193–200.
Sacristan, M. D. (1985). Selection for disease resistance in *Brassica* cultures. *Hereditas Suppl.*, **3**, 57–63.
Schenk, H. and G. Röbbelen (1982). Somatic hybrids by fusion of protoplasts from *Brassica oleracea* and *B. compestris*. *Z. Pflanzenzuecht.*, **89**, 278–288.
Schieder, O. and I. K. Vasil (1980). Protoplast fusion and somatic hybridization. *Int. Rev. Cytol. Suppl.*, **11B**, 21–46.
Schuchmann, R. (1986). Biotechnologie—eine Bestandsaufnahme für die Kartoffelzüchtung. *Kartoffelbau*, **37**, 362–365.
Scowcroft, W. R. and P. J. Larkin (1988). Somaclonal variation, cell selection and genotype improvement. In *Comprehensive Biotechnology*, vol. 1, pp. 153–168. Pergamon Press, Oxford.
Shepard, J. F. (1981). Protoplasts as sources of disease resistance in plants. *Annu. Rev. Phytopathol.*, **19**, 145–166.
Shepard, J. F. and R. E. Totten (1977). Mesophyll cell protoplasts of potato: isolation, proliferation and plant regeneration. *Plant Physiol.*, **60**, 313–316.
Singer, S. R. and C. N McDaniel (1985). Selection of glyphosate tolerant tobacco calli and the expression of this tolerance in regenerated plants. *Plant Physiol.*, **78**, 411–416.
Sjödin, C. and K. Glimelius (1989). Differences in response to the toxin sirodesmin PL produced by *Phoma lingam* (Tode ex Fr.) Desm. on protoplasts, cell aggregates and intact plants of resistant and susceptible *Brassica* accessions. *Theor. Appl. Genet.*, **71**, 76–80.
Slabas, A. R., J. Harding, A. Hellyer, C. Sidebottom, H. Gwynne, R. Kessell and M. P. Tombs (1984). Enzymology of plant fatty acid biosynthesis. In *Structure, Function and Metabolism of Plant Lipids*, ed. P. A. Siegenthaler and W. A. Eichenberger, pp. 3–11. Elsevier, Amsterdam.
Smillie, R. M., G. Melchers and D. v. Wettstein (1979). Chilling resistance of somatic hybrids of tomato. *Carlsberg Res. Commun.*, **44**, 127–132.
Sobieszczanski, J. (1986). Importance of the rhizosphere in plant–microbe interactions. In *Biotechnology: Potentials and Limitations*, ed. S. Silver, pp. 275–282. Springer-Verlag, Berlin.
Soller, M. and J. S. Beckmann (1987). Cloning quantitative trait loci by insertional mutagenesis. *Theor. Appl. Genet.*, **74**, 39–378.
Song, K. M., T. C. Osborn and P. H. Williams (1988). *Brassica* taxonomy based on nuclear restriction fragment length polymorphisms (RFLPs). Preliminary analysis of subspecies within *B. rapa* (syn. campestris) and *B. oleracea. Theor. Appl. Genet,*, **76**, 593–600.
Sree Ramulu, K. and P. Dijkuis (1986). Flow cytometric analysis of polysomaty and *in vitro* genetic instability in potato. *Plant Cell Rep.*, **3**, 234–237.
Steward, F. M. (1958). Growth and organized development of cultured cells. Interpretations of the growth from free cell to carrot plant. *Am. J. Bot.*, **45**, 709–713.
Sunderland, N. and Z. H. Xu (1982). Shed pollen culture in *Hordeum vulgare. J. Exp. Bot.*, **33**, 1086–1095.
Swanson, E. B., M. P. Coumans, G. L. Brown, D. Patel and W. D. Beversdorf (1988). The characterization of herbicide tolerant plants in *Brassica napus* L. after *in vitro* selection of microspores and protoplasts. *Plant Cell Rep.*, **7**, 83–87.
Tanksley, S. D. and J. Hewitt (1988). Use of molecular markers in breeding for soluble solids content in tomato—a reexamination. *Theor. Appl. Genet.*, **75**, 811–823.
Tanno-Suenaga, L., H. Ichikawa and J. Imamura (1988). Transfer of CMS trait in *Daucas carota* L. by donor recipient protoplast fusion. *Theor. Appl. Genet.*, **76**, 855–860.
Thomas, B. R. and D. Pratt (1982). Isolation of parquat tolerant mutants from tomato cell cultures. *Theor. Appl. Genet.*, **63**, 169–176.
Thomas, E., F. Hoffmann, I. Potrykus and G. Wenzel (1976). Protoplast regeneration and stem embryogenesis of haploid androgenetic rape. *Mol. Gen. Genet.*, **145**, 245–247.
Thomson, A. J., R. E. Gunn, G. J. Jellis, R. E. Boulton and C. N. D. Lacey (1986). The evaluation of potato somaclones. In *Somaclonal Variation and Crop Improvement*, ed. J. Semal, pp. 236–243. Nijhoff, Dordrecht.
Toriyama, K. and K. Hinata (1988). Diploid somatic hybrid plants between cultivars in rice. *Theor. Appl. Genet.*, **76**, 655–668.
Uhrig, H. (1985). Genetic selection and liquid medium conditions improve yield of androgenetic plants from diploid potatoes. *Theor. Appl. Genet.*, **71**, 455–460.
Uhrig, H. and G. Wenzel (1987). Breeding for virus and nematode resistance in potato through microscope culture. In *Biotechnology in Agriculture and Forestry*, ed. Y. P. S. Bajaj, vol. 3, pp. 346–357, Springer-Verlag, Berlin.
Van Wezenbeek, P., P. Vos, J. van Boom and A. van Kammen (1982). Molecular cloning and characterization of a complete DNA copy of potato spindle tuber viroid RNA. *Nucleic Acids Res.*, **10**, 7947–7957.
Vardi, A., E. Epstein and A. Breiman (1986). Is the *Phytophthora citrophthora* culture filtrate a reliable tool for the *in vitro* selection of resistance *Citrus* variants? *Theor. Appl. Genet.*, **72**, 569–574.
Vasil, I. K. (1988). Progress in the regeneration and genetic manipulation of cereal crops. *Bio/Technology*, **6**, 397–402.
Vasil, I. K. and V. Vasil (1980). Clonal propagation. *Int. Rev. Cytol. Suppl.*, **11A**, 145–173.
Waara, S., H. Tegelström, A. Wallin and T. Eriksson (1989). Somatic hybridization between anther-derived dihaploid clones of potato (*Solanum tuberosum* L.) and identification of hybrid plants by isozyme analysis. *Theor. Appl. Genet.*, **77**, 49–56.
Walther, H. (1988). Strategies for quantitative resistance breeding and their impact on resistance assessment techniques. *Plant Res. Dev.*, **27**, 115–127.
Wang, C. C. and B. J. Kuang (1981). Induction of haploid plants from the female gametophyte of *Hordeum vulgare. Chih Wu Hsueh Pao*, **23**, 329–330.
Wang, S.-P., B.-F. Shen and S. C. Shen (1987). Regulation of the nitrogen fixation (*nif*) genes in *Rhizobium meliloti*. In *Molecular Genetics of Plant–microbe Interaction*, ed. D. P. S. Verma and N. Brisson, pp. 264–265. Nijhoff, Dordrecht.
Weber, G., S. Monajembashi, K. O. Greulich and J. Wolfrum (1988). Genetic manipulation of plant cells and organelles with a laser microbeam. *Plant Cell, Tissue Organ Cult.*, **12**, 219–222.
Wei, Z. M. (1982). Pollen callus in *Triticum aestivum. Theor. Appl. Genet.*, **63**, 71–73.
Weisbeek, P. J. (1986). Yield increase with culture crops by treatment of seed and seed-potatoes with bacteria. In *Biomolecular Engineering in the European Community*, ed. E. Magnien, pp. 1081–1092. Nijhoff, Dordrecht.

Wenzel, G. (1985). Strategies in unconventional breeding for disease resistance. *Annu. Rev. Phytopathol.*, **23**, 149–172.

Wenzel, G., O. Schieder, T. Przewozny, S. K. Sopory and G. Melchers (1979).Comparison of single cell culture derived *Solanum tuberosum* L. plants and a model for their application in breeding programs. *Theor. Appl. Genet.*, **55**, 49–55.

Wenzel, G., S. C. Debnath, R. Schuchmann and B. Foroughi-Wehr (1987). Combined application of classical and unconventional technniques in breeding for disease resistant potatoes. In *The Production of New Potato Varieties–Technological Advances*, ed. G. J. Jellis and D. E. Richardson, pp. 277–288. Cambridge University Press, Cambridge.

Wenzel, G., H. S. Chawla, M. Bolik, A. Graner, H. Walther and B. Foroughi-Wehr (1988). Unkonventionelle Resistenzzüchtung bei der Gerste. *Vortr. Pflanzezuecht.*, **13**, 79–88.

Werner, D. (1987). *Pflanzliche und mikrobielle Symbiosen*. Thieme, Stuttgart.

Wersuhn, G., K. Kirsch and R. Gienapp (1987). Herbicide tolerant regenerants of potato. *Theor. Appl. Genet.*, **74**, 480–482.

Withers, L. and H. E. Street (1977). The freeze-preservation of plant cell cultures. In *Plant Tissue Culture and its Biotechnological Application*, ed. W. Barz *et al.*, pp. 226–244. Springer-Verlag, Heidelberg.

Wohlleben, W., W. Arnold, I. Broer, D. Hillemann, E. Strauch and A. Pühler (1988). Nucleotide sequence of the phosphinothricin *N*-acetyltransferase gene from *Streptomyces viridachromogenes* Tü494 and its expression in *Nicotiana tabacum*. *Gene*, **70**, 25–27.

Wriedt, G. (1984). Die Gewebekultur als Möglichkeit der Erhaltungszucht in einem praktischen Zuchtbertrieb aus biologischer und ökonomischer Sicht. In *Abst. Conf. Papers EAPR, Intelaken*, ed. F. A. Wininger and A. Stöckli, p. 384. EAPR, Wageningen.

Ye, J. M., K. N. Kao, B. L. Harvey and B. G. Rossnagel (1987). Screening salt tolerant barley genotypes *via* $F_1$ anther culture in salt stress media. *Theor. Appl. Genet.*, **74**, 426–429.

Zapata, F. J., P. K. Evans, J. B. Power and E. C. Cocking (1977). The effect of temperature on the division of leaf protoplasts of *Lycopersicon exculentum* and *Lycopersicon peruvianum*. *Plant Sci. Lett.*, **8**, 119–124.

Zenkteler, M. (1988). *In vitro* fertilization as a tool of research and application. in *Pflanzliche In-vitro-Systeme für Züchtung und Stoffproduktion*, ed. H. Böhm, pp. 36–45. Biologische Gesellschaft der DDR, Halle.

Zhang, W. and R. Wu (1988). Efficient regeneration of transgenic plants from rice protoplasts and correctly regulated expression of the foreign gene in the plants. *Theor. Appl. Genet.*, **76**, 835–840.

# 14

# Cell Culture and Recombinant DNA Technology in Plant Pathology

GRAHAM S. WARREN
*University of Sheffield, UK*



## 14.1 INTRODUCTION

Traditional breeding methods have been extremely successful in producing major yield increases in crop plants. One factor involved in this plant improvement has been the incorporation of effective resistance to a wide variety of diseases. Current food surpluses, for example in the EC, have fostered a view that further effort directed towards crop improvement is not a priority. However, it is widely recognized that, in the case of disease resistance, annual yield figures hide an ongoing battle to maintain an advantage over pathogens that continually overcome plant resistance. Furthermore, there is an ever present danger of sudden increases in virulence of pathogens that can then endanger crops with little inherent resistance. The upsurge in sugar beet rhizomania (due to beet necrotic yellow vein virus spread by a fungal vector) in Europe is an example of this phenomenon. A relatively large proportion of the major crops in less-developed countries are lost or at risk from plant pathogens such as rice tungro disease virus and cocoa swollen shoot virus. Conventional resistance, obtained from related plant varieties, to either the pathogen directly or to the transmission vector, can often be overcome by mutation leading to new virulent pathogen (or vector) strains. There is a need for new, durable forms of resistance that are less easily overcome in this way.

The emphasis in this brief review is on the application of novel techniques to various aspects of the control and detection of viral diseases. This is because the majority of the advances made, especially using recombinant DNA techniques, have been in this area. The main reason for this situation is the relatively high level of understanding of plant viral genomes. However, most of the strategies discussed are being, or shortly will be, applied to other plant pathogens. The more general application of novel techniques to agriculture is considered by Wenzel in Chapter 13. Much of this work is being performed commercially and therefore complete details may not be freely available.

The rate-limiting step in pursuing the potential offered by these new methods is the relatively poor level of understanding of the molecular basis of plant resistance to pathogens. It is significant that the successes so far achieved have occurred in cases in which some molecular details of the host/pathogen interaction were appreciated, for example coat protein mediated and satellite-mediated virus resistance in transgenic plants.

## 14.2 THE USE OF PLANT CELL AND TISSUE CULTURE

Virtually all of the novel plant improvement methods (*e.g.* protoplast fusion, gene manipulation) involve a cell culture phase to recover whole plants from engineered or selected cells or tissues. However, the culture process itself has been exploited in several ways to enhance disease resistance or to recover disease-free stock.

### 14.2.1 Culture-induced (Somaclonal and Gametoclonal) Variation

It is now well established that plant cell culture is essentially a mutagenic treatment. Merely taking a plant species into culture induces genetic changes that can result in enhanced agricultural characters in the regenerated plants. Another facet of this phenomenon is that plants taken through a culture phase as part of an improvement scheme may also have undesirable characteristics introduced. This topic was reviewed earlier in this series (Scowcroft and Larkin, 1985) and is discussed elsewhere in this volume by Cullis. The mechanisms underlying culture-induced (somaclonal and gametoclonal) variation are still somewhat speculative. Lee and Phillips (1988) have recently reviewed the literature in this area and have proposed a model that links the observed chromosome effects with single gene mutations that are also seen at relatively high frequency. Chromosome aberrations are suggested to result from a relaxation of the strict coupling between DNA replication (S phase of the cell cycle) and cell division, due presumably to the absence of (unknown) control systems in the culture environment. Premature anaphase, occurring before DNA replication is complete, may result in breaks in the chromosomes at the sites of late-replicating DNA (often regions of heterochromatin). Recombination at the break points, or elimination of damaged chromosomes, can then result in the range of chromosome abnormalities seen. Chromosome rearrangement will lead to a change in genetic background for certain genes, and it has been suggested that such effects may mobilize previously cryptic transposable elements. Finally, a mismatch between insertion and subsequent excision of these elements could give rise to stable mutations in individual genes. Other processes possibly involved in the generation of culture-induced variation have also been identified (Cullis, this volume; Lee and Phillips, 1988).

The frequency of somaclonal variation is high and can occur at a similar level to variation obtained using mutagens, although the types of variation may be qualitatively different. Values of 30–40% for regenerated plants showing at least one variant character, and up to 3% for variation in a specific trait, are typical (Daub, 1986).

There are many reports of enhanced disease resistance being recovered following a period of cell culture. Resistances described by Scowcroft and Larkin (1985) include those to eyespot disease, Fiji virus, downy mildew and smut in sugar cane; late blight and early blight in potato; southern corn leaf blight in maize, and blackleg in oilseed rape. It is important to realize that, except in a very few cases (*e.g.* sugar cane), resistances have only been demonstrated under laboratory conditions, or in varieties considerably downgraded in other vital characters. In the field, 'laboratory' resistance may not necessarily be fully expressed, for instance as a result of suboptimal plant nutrition, or because of simultaneous assault by other pathogens that renders the plant more susceptible to the disease in question. The number of crop plants in commercial use that have been developed primarily through somaclonal variation is extremely small, again the case of sugar cane quoted above is one example (Daub, 1986). Subsequent to the review of Scowcroft and Larkin, 'laboratory' disease resistance in somaclonal variants of the following crops has been reported: alfalfa, turnip, celery, tobacco and tomato (Daub, 1986). Recently, field trials of somaclonal variants of plants with enhanced disease resistance have been reported. These include resistance to bacterial wilt in tobacco (Daub and Jenns,

1989) and leaf spot in poplar (Ettinger *et al.*, 1986). Daub and Jenns (1989) employed a realistic selection protocol in which infertile plants, which would not have agricultural importance, were rejected. This led to frequencies of useful resistance that were low in comparison with some other reports, but which nonetheless showed that small but significant improvements in resistance can be achieved. An important factor illustrated by this study was that the level of field resistance recorded was dependent on the location of the tests.

Culture-induced variation as a source of disease resistance does have limitations. There are several reports of an inability to recover disease resistance in the absence of mutagenization (*e.g.* Murakishi and Carlson, 1982). In addition it has not generally proved possible to obtain resistance starting from susceptible varieties with little or no inherent resistance (Daub and Jenns, 1989).

### 14.2.2 Cell Selection Methods

Selection of potentially useful varieties following culture-induced variation or mutagenization has usually been carried out on the regenerated plants. This approach has the advantage of simplicity but screening the large number of regenerants requires considerable effort and resources. The full potential of genetic variation in cell cultures can only be realized through the application of selections at the cell level. Following this approach, large numbers of genotypes can be eliminated at an early stage, thus reducing the subsequent selection required at the plant level. This type of selection presupposes a correlation between cellular and whole plant characters, a correlation that is not always found. For example salt tolerant plants frequently yield salt intolerant cell cultures and *vice versa*, and chill tolerant cells often give rise to chill intolerant regenerant plants (Cresswell, 1991; Dix, 1986; Dix and Street, 1976).

Selection for disease resistance using tissue cultures has been reviewed recently by Daub (1986), Dix (1986), Loh (Chapter 3) and Sacristan (1986). Selection strategies generally fall into one of two types: those using the pathogen as the selecting agent; and those using pathogen-derived toxic metabolites (Daub, 1986). The first approach mainly has been applied to virus resistance, but has been limited by two problems. Firstly, 100% infection of cells, or especially protoplasts, has not been achieved and consequently there are many 'escapes' making selection procedures inefficient. Secondly, infected cells or protoplasts are not usually killed by the presence of viruses and growth rates are not normally markedly different, thus selection is difficult to perform. In certain cases, however, workable systems have been devised by appropriate manipulation of the experimental conditions. Murakishi and Carlson (1982) developed specific growth conditions that favoured the division of TMV-free *Nicotiana sylvestris* cells, and used a yellow strain of the virus that produced a characteristic yellowing of infected callus colonies. In this way, seven virus resistance plants were recovered from 763 selected calli. Toyoda *et al.* (1989) were able to recognize TMV-infected tobacco cells by the presence of virus crystals. Plants were regenerated from such cells and assessed for virus resistance. Out of a total of 1072 regenerants tested, 33 showed delayed symptom formation, and three showed no symptoms following inoculation with TMV.

The second approach, namely selection using pathogen-derived toxic metabolites, has been used more extensively and has yielded laboratory resistance in a wide range of host/pathogen systems. However, few field trials have been performed, and these have sometimes yielded disappointing results. The major problems that have been encountered with toxin-based selections are: (i) lack of correlation between whole plant and cell responses to the toxin; (ii) resistance may not be compatible with normal growth in the case of toxic substances that interfere with fundamental biochemical pathways; and (iii) noncorrespondence of 'laboratory' and 'field' resistances due, for example, to synergistic reactions with other pathogens or stresses (Daub, 1986). *In vitro* selection for resistance to pathogen toxins has been applied to the following crop species: alfalfa, bean, Brassicas, lettuce, maize, oats, rice, potato, tobacco and sugar cane (Daub, 1986; Loh, this volume, and references therein).

Recently, there has been much interest in the development of RFLP (restriction fragment length polymorphism) markers for agriculturally important characters (Tanksley *et al.*, 1989). Assignments are made by comparison of RFLP banding maps following segregation of characters after crossing. Using this technique, variant clones can be screened for desirable traits immediately after isolation and thus much time can be saved, because it is not necessary to wait for expression of the characters in mature tissue RFLP screening is considered to be especially useful for characters that are difficult or time consuming to measure by conventional means; many types of disease resistance fall into this category. Many RFLP markers have already been identified for disease resistance, and examples include: resistance of tomato to *Fusarium* wilt, TMV, bacterial speck and rootknot nematode; and resistance of maize to maize dwarf mosaic virus. RFLP mapping is thought to be of such importance that recently a consortium of five major European companies initiated a project to map wheat and barley, two of the EC's most important crops. In addition to the identification of markers for useful characters, the companies hope to use the information for patent protection, by allowing a positive identification of genes from any new varieties that they may develop.

### 14.2.3 The Potential of Culture-induced Variation

The use of culture-induced variation has been somewhat eclipsed by recent technical advances in plant transformation and recovery of fertile plants from transformed cells of previously intractible genotypes.

The main advantage of somaclonal (and gametoclonal) variation is that it can be exploited without a knowledge of the molecular mechanisms underlying the traits being considered. In addition, culture-induced variation requires a minimum of sophisticated equipment and a much lower level of expertise compared with gene manipulation. It may therefore be more suitable for rapid adoption in less-developed countries.

The major disadvantages in the use of culture-induced variation are: (i) efficient culture/regeneration systems are not currently available for all agricultural species and cultivars; and (ii) downgrading of characters occurs simultaneously with enhancement of the desired trait, necessitating further selection of the regenerants recovered.

## 14.3 PROTOPLAST FUSION

The practice and application of somatic hybridization by protoplast fusion is considered by Morikawa *et al.* in Chapter 11. As described by Morkikawa *et al.*, this technique has been used in attempts to enhance disease resistance. The initial aim of protoplast fusion in agriculture is to obtain desirable characters in a form that can enter conventional breeding programmes, *i.e.* to overcome sexual incompatibility barriers. It is therefore essential that fusion hybrids are fertile. A marked correlation has been observed between the relatedness of the parental species and the fertility of the derived protoplast fusion hybrids. Only in closely related species is there a reasonable chance of high fertility, and the products of intergeneric fusions are almost always sterile. This situation is particularly favourable for the application of protoplast fusion to the enhancement of disease resistance in crop species because the wild relatives of domesticated species are often superior in this respect.

The feasibility of this approach has been demonstrated by the transfer of PLRV resistance from *Solanum brevidens* to the cultivated potato *Solanum tuberosum* (Austin *et al.*, 1985). Other cases are discussed by Morikawa *et al.* (Chapter 11).

### 14.3.1 The Potential of Protoplast Fusion for the Enhancement of Disease Resistance

As in the case of the exploitation of culture-induced variation and cell selections, protoplast fusion as a general technique has been somewhat overshadowed by developments in specific gene transfer. The technique has the advantage that the molecular basis of the characters being manipulated does not have to be understood at the molecular level, and this is the case for most agricultural traits at the present time. The major disadvantage is that genetically, somatic hybridization is very imprecise and much postfusion selection is required. However, in certain specific applications, for example the import of chromosomes or genes from closely related species as already noted, and in the manipulation of organelle-coded characters, protoplast fusion has much potential. Furthermore, through the exploitation of somatic cell genetics, a better basic understanding of the traits themselves may be obtained, thus facilitating the improvement of plants by other methods.

## 14.4 THE PRODUCTION OF VIRUS-FREE PLANTS

It has long been observed that the rapidly growing meristems of plants are usually free of viruses, or at least have much lower concentrations of systemic viruses than nonmeristem cells. This situation has been exploited for the production of virus-free plants by meristem culture. Excised apical meristems are freed of the remaining virus usually by chemo- and/or thermo-theraphy, and regenerated into whole plants. The technique is at present rather inefficient because of the difficulty in the total eradication of virus from the tissues. This subject was reviewed earlier in this series by Stace-Smith (1985), and by Garg *et al.* (1988) and Quak (1977).

The mechanisms underlying the 'resistance' of meristems to viruses are not known, this situation being due mainly to the difficulty of experimentation. However, five main possibilities have been suggested: (i) exclusion of the virus from the meristem by lack of suitable vascular or plasmodesmatal connections; (ii) competition for key metabolites by the rapidly dividing meristem cells; (iii) the production of substances in meristem cells that result in breakdown of the virus; (iv) a deficiency in some key components of the machinery of virus replication; and (v) the presence of inhibitors of virus replication (Stace-Smith, 1985). Because of the difficulty of using plant meristems as experimental systems, it has been suggested that rapidly dividing plant cell cultures may be convenient tools for the investigation of meristem resistance (Warren *et al.*, 1991).

Meristem culture has been used to free a wide range of plants from viruses. For a list of such species see Garg *et al.* (1988). Stace-Smith (1985) suggested that the existing methods for the production of virus-free plants, although somewhat inefficient, are adequate for present needs and further development work is superfluous. It is possible, however, that future research leading to a deeper understanding of the factors limiting virus accumulation in meristem cells could result in the identification of new sources of resistance suitable for incorporation into transgenic plants.

## 14.5 INCREASED VIRUS RESISTANCE IN TRANSGENIC PLANTS

Viruses can, under the appropriate environmental conditions, cause serious losses in almost all of the major crop species. The threat can be especially serious for species that are vegetatively propagated, *e.g.* potato, in which systemically infecting viruses are passed on during propagation. Plants exhibit a spectrum of response to viruses from systemic susceptibility through to apparent immunity. There are, in addition, many different types of resistance mechanism, for example acquired resistance, cross protection, meristem resistance and so on (Matthews, 1991). In the great majority of cases, the molecular details of these processes are lacking. However, in three instances, coat protein mediated, transport protein mediated and satellite-mediated resistance, the resistance mechanisms are partially understood, and have been exploited to give rise to enhanced virus resistance in transgenic plants. Strategies for increasing plant resistance to viruses have been considered in several recent reviews (Baulcombe, 1989; van den Elzen *et al.*, 1989; Gadani *et al.*, 1990; Grumet, 1990; Hemenway *et al.*, 1989; Hull, 1990; Nejidat *et al.*, 1990).

### 14.5.1 Gene Vectors and Plant Transformation Systems

Most plant gene vectors are based on *Agrobacterium* Ti plasmid components usually with some additional plant virus functions. Plant transformation is most often accomplished using *Agrobacterium*-mediated delivery for species within the host range of this microorganism. For the species outside this range, which includes most of the important monocot crops (*e.g.* maize and rice), artificial delivery systems such as electroporation of protoplasts and particle bombardment have been employed. Gene vectors and plant transformation are discussed in detail by Day and Lichtenstein (Chapter 9) and Ooms (Chapter 12) in this volume.

### 14.5.2 Coat Protein Mediated Resistance

It has long been observed that infection by a virus can 'cross protect' the plant against a second infection by the same or a related virus (Sherwood, 1987). This behaviour has been exploited previously by experimental inoculation of crops with a mild strain of a particular virus, which confers resistance to a subsequent infection by severe strains (*e.g.* tomato/TMV; Rast (1972). The precise mechanisms underlying this process have not been fully worked out, but one form of cross protection is dependent on interference by coat protein from the first virus with replication or symptom formation by the second. This has been clearly demonstrated by the resistance (delay in symptom formation) of plants transgenic in coat protein to the corresponding whole virus and related strains (Beachy *et al.*, 1990). In the case of TMV at least, resistance required the expression of the heterologous coat protein product and not merely the presence of the corresponding RNA. The following mechanisms have been proposed for coat protein mediated resistance: interference of the constitutive coat protein with (i) uncoating of the incoming virus; (ii) replication of viral RNA; and (iii) cell–cell movement of viral components (Beachy *et al.*, 1990). However, there is evidence that there may be additional factors involved (Beachy *et al.*, 1990; Wilson, 1988). Other forms of 'cross protection' also exist that do not involve coat protein. For example, certain viroids, which lack coat protein, can attenuate the effects of subsequent infections (Sherwood, 1987).

Resistance to the homologous virus has been reported in plants transgenic with respect to the coat proteins of TMV (Powell-Abel *et al.*, 1986), AlMV (Tumer *et al.*, 1987), CMV (Cuozzo *et al.*, 1988), PVX, PVY (Kaniewski *et al.*, 1990), TSV (Van Dun *et al.*, 1988) and TRV (Van Dun and Bol, 1988). Heterologous resistance has also been reported, for example in plants transgenic with respect to the coat protein of: TRV (resistant to PEBV; Van Dun and Bol, 1988); SBV (resistant to TEV and PVY; Stark and Beachy, 1989); and TMV (resistant to some other tobamoviruses; Nejidat and Beachy, 1990). For further details see Beachy *et al.* (1990).

Multiple virus resistance has also been achieved by expression of the coat proteins of PVX and PVY in transgenic Russett Burbank potato (Lawson *et al.*, 1990). Russett Burbank is a long-standing potato cultivar that still constitutes about 30% of US usage. However, it lacks resistance to several virus diseases and nematode attack. In the field, double infection by PVX and PVY results in the serious rugose mosaic disease. Researchers at Monsanto (California, USA) transformed Russett Burbank with PVX and PVY coat protein genes using an *Agrobacterium*-based vector/transformation system. The results of this study gave useful insights into factors influencing the expression of coat proteins and the level of resistance achieved. Many transformed lines were obtained that showed resistance, as judged visually and by ELISA, to PVX. However, much variation in the level of resistance between lines was noted. Variation in transformants is usually seen following *Agrobacterium*-mediated transformation and is thought to result from incomplete or multiple T-DNA incorporation and genetic background effects. Interestingly, there was not a direct correlation between the level of expression of PVX coat protein and the degree of resistance exhibited, indicating that other factors were involved in the resistance response. In contrast to the situation with PVX, only one transformed line was recovered that showed resistance to PVY. Fortunately this line was also resistant to PVX. The authors speculated that the differences in the levels of PVX and PVY coat proteins in the transgenic plants may have resulted from their different modes of expression. PVX coat protein is translated from a subgenomic mRNA, whereas PVY coat protein is derived by cleavage of a precursor polyprotein. It was thought that the absence from the construct of the authentic 5′ untranslated leader sequence, possibly containing a translation enhancer, could have limited expression levels in the latter case.

In a subsequent study, PVX plus PVY coat protein transgenic Russett Burbank potato was demonstrated to possess field resistance to these viruses (Kaniewski *et al.*, 1990). Again much variation between individual transformed clones was observed. In one clone, tuber yield was unaffected, indicating the feasibility of using recombinant DNA techniques to enhance traits in already well-developed crop species without a concomitant downgrading of other characters.

A degree of virus resistance has also been achieved by insertion into the plant of a gene that is complementary to the virus coat protein gene, such that mRNA is produced that is itself complementary to the authentic coat protein gene mRNA. This antisense RNA can then sequester the latter by base pairing to it, and thus prevent its translation. Such an approach has been reported for potato/PVX (Hemenway *et al.*, 1988) and tobacco/CMV (Rezaian *et al.*, 1988) systems. However, the degree of resistance shown by these transgenic plants was small, being effective only against very low inoculum concentrations. It has been suggested that the relatively low level of interference with coat protein biosynthesis could be due to the different sites of synthesis of the coat protein mRNA (cytosol) and the antisense RNA (nucleus) such that hybridization of only a small proportion of the coat protein RNA occurs (Hemenway *et al.*, 1988).

### 14.5.3 Satellite-mediated Resistance

It has been observed that associated with certain viruses can be small additional sequences of nucleic acid (satellites) that depend on the virus for replication, and which can greatly affect the severity of symptoms shown by the host plant. Symptoms can be intensified or alleviated depending upon the particular virus/satellite/host combination (Matthews, 1991). In some cases, the presence of a satellite can effectively suppress symptoms, such that the plant is operationally resistant to the virus. The molecular mechanisms by which these effects are mediated by the satellite are poorly understood. Proposed mechanisms include competition with the helper virus for limiting amounts of replicase enzyme, although a variety of mechanisms almost certainly exist (Baulcombe, 1989). In the case of the satellite of CMV (strain y) in tobacco, the ability to alter symptoms maps to two specific domains of the satellite RNA (Devic *et al.*, 1989). Satellite-mediated attenuation of viral disease symptoms has already been used in agriculture by inoculation of crop plants with the appropriate satellite (Tien *et al.*, 1987). Developments in molecular biology have allowed an extension of this strategy, namely the production of transgenic plants that express satellite RNA, thus conferring constitutive satellite-mediated resistance. Such resistance has been reported for CMV (Harrison *et al.*, 1987) and TRSV (Gerlach *et al.*, 1987). Further characterization of the mode of interactions between satellites, helper viruses and hosts may lead to a more general application of these techniques.

### 14.5.4 Transport Protein Mediated Resistance

Symptom formation in virus-infected plants is dependent on spread of the virus from the site of infection to other cells of the plant. This movement of the virus may involve both short range (cell–cell) movement *via* the plasmodesmata and long range transport in the vascular system. In principle, blocking either of these transport processes could result in resistance in the host plant. Short range movement appears to depend on the expression of viral transport proteins (Melcher, 1990) which interact with the plasmodesmata and allow passage of the virus or subviral particles (Robards and Lucas, 1990). Tobacco plants transgenic for the antisense CMV transport gene have been reported to exhibit some resistance to the virus (Yoshida, 1990). Long range transport mechanisms for viruses are poorly understood at present and have not been exploited as a source of resistance.

### 14.5.5 Other Sources of Resistance

The transfer of plant resistance genes to susceptible hosts is an obvious strategy for the production of new virus resistant varieties. Unfortunately our knowledge of resistance mechanisms at the molecular level is very poor. In only one case, CPMV resistance in Arlington cowpea, is a resistance well characterized (Bruening *et al.*, 1987). Some of the problems associated with the characterization of TMV resistance genes in tobacco have been discussed (Dunigan *et al.*, 1987).

Several other actual and potential sources of virus resistance suitable for transgenic plants have been identified. These include nonstructural viral genes (Golemboski *et al.*, 1990), pathogenesis-related proteins (Bol and Linthorst, 1990), antiviral factors (Loebenstein and Gera, 1981; Sela, 1981), resistance elicitation (Baulcombe, 1989) and disruption of replicase function (Hull, 1990). Resistances that prevent extensive replication of the virus are preferable because they would help to reduce the incidence of resistance-breaking mutations in the viral genome (Hull, 1990).

### 14.5.6 Other Transgenic Plants Resistant to Disease

There have been very few reports of transgenic resistance to classes of plant pathogen other than viruses. This situation reflects in part the much greater complexity of the genomes of fungi and bacteria. Recently, Anzai and Yoneyama (1990) reported resistance to a bacterial pathogen by detoxification of the bacterial toxin.

## 14.6 HAZARDS WITH THE RELEASE OF TRANSGENIC PLANTS

The risks associated with the release of transgenic, virus resistant plants have been summarized recently by Hull (1990). General issues concerned with the release of genetically engineered plants have also been reviewed (Hoffman, 1990; Tolin and Vidaver, 1989).

In the case of virus coat protein transgenic plants there is a possibity that the nucleic acid of incoming heterologous viruses could become transcapsidated with the constitutive coat protein. In certain cases, the specificity of vector transmission is thought to be determined at least in part by the coat protein, and so the former could be temporarily altered by passage through a transgenic host. This could have unpredictable effects on the host range or virulence of the virus (Hull, 1990).

In the case of satellite-mediated resistance it is possible that minor mutations in the satellite gene could have large and currently unpredictable effects on the severity of symptoms produced. It is also possible that the constitutive satellite in transformed plants could be passed *via* a helper virus to other plant species in which the satellite might increase the severity of disease symptoms. Further work is required to enable the development of disarmed satellite sequences (Baulcombe, 1989).

The widespread use of viral nucleic acid sequences in plant transformation vectors, with their associated incorporation into plant chromosomes, raises the possibility of entry of other portions of viral genomes into the plant germplasm by natural recombination (Hull, 1990). Taschner *et al.* (1991) suggested that viral replication, and therefore possibly symptom severity, could be higher in transgenic plants expressing a normally limiting component of the viral replication machinery.

While the environmental hazards related to the release of transgenic, virus resistant plants are being assessed, it is vital that appropriate containment measures are taken during field trials. During recent trials of transgenic, virus resistant sugarbeet (see Dixon, 1989), containment measures included preventing the flowering of plants to avoid the escape of transgenic pollen, and avoiding resistance tests against the whole virus to guard against the possibility of the selection of strains with increased virulence.

Related concerns exist about the use of engineered resistance to other plant pathogens. There is also a danger that a particularly successful resistant crop variety could be widely adopted and eventually constitute an extremely narrow genetic base with the attendant risk from new virulent strains of pathogen (Walbot cited in Klausner, 1989).

## 14.7 NOVEL METHODS FOR THE DIAGNOSIS OF DISEASE

An essential part of the management of plant diseases in agriculture is the ability to screen effectively for pathogens. This allows the certification of disease-free stock, and the determination of the progress of pathogens already established in crops to enable decisions concerning the appropriate timing of control measures (*e.g.* spraying) to be made.

In recent years the screening of plant material for pathogens has been greatly facilitated by the development of enzyme-linked immumosorbent assays (ELISA) and recombinant DNA or RNA probes. ELISA is based on the specific interaction between a pathogen and an antibody raised against it. Because of its specificity the assay can be used on crude tissue preparations. The assay also has the following advantages: (i) it is relatively rapid; (ii) it is sensitive, being able to detect nanograms of antigen, (iii) large numbers of samples can be handled; and (iv) the reagents are very stable (years). The practice of ELISA and its application to plant pathology has been reviewed by Clark (1981). Nucleic acid probes (Miller and Martin, 1988) are complementary sequences to part of the pathogen genome. Because of their extremely specific hybridization with the pathogen nucleic acid, these probes can offer a greater resolution between individual pathogen strains than ELISA. Detection of the hybridized probe has generally been by radiodetection and consequently this technique is not as convenient as ELISA for routine use in relatively unsophisticated laboratories. The major difficulties associated with the use of isotopes in this context are the expense of the detection equipment, the short half-life of the probes, and problems of handling and disposal. More recently, alternative methods of detection, for example based on biotin–avidin–enzyme systems, have been developed that overcome these problems (Miller and Martin, 1988). Virus detection by ELISA and nucleic acid probes have been compared (Eweida *et al.*, 1990).

## 14.8 CONCLUSIONS

Many techniques involving cell culture and recombinant DNA now exist, with the potential to produce disease-free plants, crop varieties with increased disease resistance and to facilitate the identification of diseased plant material. These techniques are also being used to gain a better understanding of the molecular basis of plant/pathogen interactions, an understanding that is vital to future progress in this area. Some of these methods, *e.g.* culture-induced variation and protoplast fusion, are most suitable in rather specific situations. With so many techniques now available, it will be important when faced with the aim of developing a particular resistance to make an informed decision about the appropriate techniques to be used. Important factors governing this choice are: (i) the amount of information available about the trait in question; (ii) the facilities available; (iii) performance of the species at the various levels of cell culture; (iv) existing traditional techniques that may more easily achieve the same ends; and (v) potential hazards associated with the final product.

Up to the present time, it has only been possible to apply these novel techniques to a relatively small number of amenable crop species. The recent reports of fertile, transgenic maize (Gordon-Kamm *et al.*, 1990) and rice (Datta *et al.*, 1990) open the way for the production, by recombinant DNA techniques, of disease resistant varieties of the world's most important crops.

## 14.9 ABBREVIATIONS

| | |
|---|---|
| AIMV | alfalfa mosaic virus |
| CMV | cucumber mosaic virus |
| CPMV | cowpea mosaic virus |
| PEBV | pea early browning virus |
| PLRV | potato leafroll virus |
| PVX | potato virus X |
| PVY | potato virus Y |
| SMV | soybean mosaic virus |
| TEV | tobacco etch virus |
| TMV | tobacco mosaic virus |
| TRSV | tobacco ringspot virus |
| TRV | tobacco rattle virus |
| TSV | tobacco streak virus |

## 14.10 REFERENCES

Anzai, H. and K. Yoneyama (1990). Transgenic resistance to a bacterial disease by the detoxification of a bacterial toxin. *Nippon Noyaku Gakkaishi*, **15**, 249–253.

Austin, S., M. A. Baer and J. P. Helgeson (1985). Transfer of resistance to potato leafroll virus from *Solanum brevidens* into *Solanum tuberosum* by somatic fusion. *Plant Sci.*, **39**, 75–82.

Baulcombe, D. (1989). Strategies for virus resistance in plants. *Trends Genet.*, **5**, 56–60.

Beachy, R. N., S. Loesch-Fries and N. E. Tumer (1990). Coat protein mediated resistance against virus infection. *Annu. Rev. Phytopathol.*, **28**, 451–474.

Bol, J. F. and H. J. M. Linthorst (1990). Plant pathogenesis-related proteins induced by virus infection. *Annu. Rev. Phytopathol.*, **28**, 113–138.

Bruening, G., F. Ponz, C. Glascock, M. L. Russell, A. Rowhani and C. Chay (1987). Resistance of cowpeas to cowpea mosaic virus and to tobacco ringspot virus. In *Plant Resistance to Viruses*, Ciba Foundation Symposium 133, ed. D. Evered and S. Harnett, pp. 23–37. Wiley, Chichester.

Clark, M. F. (1981). Immunosorbent assays in plant pathology. *Annu. Rev. Phytopathol.*, **19**, 83–106.

Cresswell, R. (1991). Improvement of plants *via* plant cell culture. In *Plant Cell and Tissue Culture*, ed. A. Stafford and G. S. Warren, pp. 101–123. Open University Press, Milton Keynes.

Cuozzo, M., K. M. O'Connell, W. Kaniewski, R.-X. Fang, N. H. Chua and N. E. Tumer (1988). Viral protection in transgenic tobacco plants expressing the cucumber mosaic virus coat protein or its antisense RNA. *Biotechnology*, **6**, 549–557.

Datta, S. K., A. Peterhans, K. Datta and I. Potrykus (1990). Genetically engineered, fertile indica rice recovered from protoplasts. *Biotechnology*, **8**, 736–740.

Daub, M. E. (1986). Tissue culture and the selection of resistance to plant pathogens. *Annu. Rev. Phytopathol.*, **24**, 159–186.

Daub, M. E. and A. E. Jenns (1989). Field trials of disease resistance in tobacco somaclones. *Phytopathology*, **79**, 600–605.

Devic, M., M. Jaegle and D. Baulcombe (1989). Symptom production on tobacco and tomato is determined by two distinct domains of the satellite RNA of cucumber mosaic virus (strain Y). *J. Gen. Virol.*, **70**, 2765–2774.

Dix, P. J. (1986). Cell line selection. In *Plant Cell Culture Technology*, ed. M. M. Yeoman, pp. 143–201. Blackwell, Oxford.

Dix, P. J. and H. E. Street (1976). Selection of plant cell lines with enhanced chilling resistance. *Ann. Bot. (London)*, **40**, 903–910.

Dixon, B. (1989). Denmark OKs trial of engineered beets. *Biotechnology*, **7**, 1001.

Dunigan, D. D., D. B. Golemboski and M. Zaitlin (1987). Analysis of the N gene of *Nicotiana*. In *Plant Resistance to Viruses*, Ciba Foundation Symposium 133, ed. D. Evered and S. Harnett, pp. 120–135. Wiley Chichester.

van den Elzen, P. J. M., M. J. Huisman, D. Posthumus-Lutke Willink, E. Jongedijk, A. Hoekema and B. J. C. Cornelissen (1989). Engineering virus resistance in agricultural crops. *Plant Mol. Biol.*, **13**, 337–346.

Ettinger, T. L., P. E. Read, W. P. Hackett, M. E. Ostry and D. D. Skilling (1986). Development of resistance in *Populus* to *Septoria musiva* utilizing somaclonal variation. In *Proceedings of the 12th Meeting of the Canadian Tree Improvement Association, Part 2*, ed. F. Carron, A. G. Corriveau and I. J. B. Boyle. Canadian Forest Service, Ottowa. Cited in Haissig *et al.*, 1987.

Eweida, M., H. Xu, R. P. Singh and M. G. Abouhoudar (1990). Comparison between ELISA and biotin-labelled probes from cloned DNA of potato virus X for the detection of virus in crude tuber extracts. *Plant Pathol.*, **39**, 623–628.

Gadani, F., L. M. Mansky, R. Medici, W. A. Miller and J. H. Hill (1990). Genetic engineering of plants for virus resistance. *Arch. Virol.*, **115**, 1–21.

Garg, G. K., U. S. Singh, R. K. Khetrapal and J. Kumar (1988). Application of tissue culture in plant pathology. In *Experimental and Conceptual Plant Pathology*, ed. W. M. Hess, R. S. Singh, U. S. Singh and D. J. Weber, pp. 83–119. Gordon and Breach, New York.

Gerlach, W. L., D. Llewellyn and J. Haseloff (1987). Construction of a plant disease resistance gene from the satellite RNA of tobacco ringspot virus. *Nature (London)*, **328**, 802–805.

Golemboski, D. B., G. P. Lomonossoff and M. Zaitlin (1990). Plants transformed with a tobacco mosaic nonstructural gene sequence are resistant to the virus. *Proc. Natl. Acad. Sci. USA*, **87**, 6311–6315.

Gordon-Kamm, W. J., T. M. Spencer, M. L. Mangano, T. R. Adams, R. J. Daines, W. G. Start, J. V. O'Brien, S. A. Chambers, W. R. Adams, N. G. Willetts, T. B. Rice, C. J. Mackey, R. W. Krueger, A. P. Kausch and P. G. Lemaux (1990). Transformation of maize cells and regeneration of fertile transgenic plants. *Plant Cell*, **2**, 603–618.

Grumet, R. (1990). Genetically engineered plant virus resistance. *Hort. Sci.*, **25**, 808–813.

Haissig, B. E., N. D. Nelson and G. H. Kidd (1987). Trends in the use of tissue culture in forest improvement. *Biotechnology*, **5**, 52–59.
Harrison, B. D., M. A. Mayo and D. C. Baulcombe (1987). Virus resistance in transgenic plants that express cucumber mosaic satellite RNA. *Nature (London)*, **328**, 799–802.
Hemenway, C., R.-X. Fan, W. K. Kaniewski, N. H. Chua and N. E. Tumer (1988). Analysis of the mechanism of protection in transgenic plants expressing the potato virus X coat protein or its antisense RNA. *EMBO J.*, **7**, 1273–1280.
Hemenway, C., N. E. Tumer, P. A. Powell and R. N. Beachy (1989). Genetic engineering of plants for viral disease resistance. In *Cell Culture and Somatic Cell Genetics of Plants*, vol. 6, ed. J. Schell and I. K. Vasil, pp. 405–423. Academic Press, New York.
Hoffman, C. A. (1990). Ecological risks of genetic engineering of crop plants. *Bioscience*, **40**, 434–437.
Hull, R. (1990). Virus resistant plants: potential and risks. *Chem. Ind. (London)*, 1990, 543–546.
Kaniewski, W., C. Lawson, D. Sammons, L. Haley, J. Hart, X. Delannay and N. E. Tumer (1990). Field resistance of transgenic Russett Burbank potato to effects of infection by potato virus X and potato virus Y. *Biotechnology*, **8**, 750–754.
Klausner, A. (1989). Abating or abetting environmental stress. *Biotechnology*, **7**, 419.
Lawson, C., W. Kaniewski, L. Haley, R. Rozman, C. Newell, P. Sanders and N. E. Tumer (1990). Engineering resistance to mixed infection into a commercial potato cultivar: resistance to potato virus X and potato virus Y in transgenic Russett Burbank. *Biotechnology*, **8**, 127–134.
Lee, M. and R. L. Phillips (1988). The chromosomal basis of somaclonal variation. *Annu. Rev. Plant Physiol.*, **39**, 413–437.
Loebenstein, G. and A. Gera (1981). Inhibitor of virus replication released from tobacco mosaic virus infected protoplasts of a local lesion responding cultivar. *Virology*, **114**, 132–139.
Matthews, R. E. F. (1991). *Plant Virology*, 3rd edn. Academic Press, New York.
Melcher, U. (1990). Similarities between putative transport proteins of plant viruses. *J. Gen Virol.*, **71**, 1009–1018.
Miller, S. A. and R. R. Martin (1988). Molecular diagnosis of plant disease. *Annu. Rev. Phytopathol.*, **26**, 409–432.
Murakishi, H. H. and P. S. Carlson (1982). *In vitro* selection of *Nicotiana sylvestris* variants with limited resistance to TMV. *Plant Cell Rep.*, **1**, 94–97.
Nejidat, A. and R. N. Beachy (1990). Transgenic tobacco plants expressing a coat protein gene of tobacco mosaic virus are resistant to some other tobamoviruses. *Mol. Plant–Microbe Interact.*, **3**, 247–251.
Nejidat, A., W. G. Clark and R. N. Beachy (1990). Engineered resistance against plant viral diseases. *Physiol. Plant.*, **80**, 662–668.
Powell-Abel, P., R. S. Nelson, B. De, N. Hoffmann, S. G. Rogers, R. T. Fraley and R. N. Beachy (1986). Delay of disease development in transgenic plants that express the tobacco mosaic virus coat protein gene. *Science (Washington D. C.)*, **232**, 738–743.
Quak, F. (1977). Meristem culture and virus-free plants. In *Applied and Fundamental Aspects of Plant Cell, Tissue and Organ Culture*, ed. J. Reinert and Y. P. S. Bajaj, pp. 598–615. Springer-Verlag, Berlin.
Rast, A. T. B. (1972). MII-16, an artificial symptomless mutant of tobacco mosaic virus for seedling inoculation of tomato crops. *Neth. J. Plant Pathol.*, **78**, 110–112.
Rezaian, M. A., K. G. M. Skeene and J. G. Ellis (1988). Antisense RNAs of cucumber mosaic virus assessed for control of the virus. *Plant Mol. Biol.*, **11**, 463–471.
Robards, A. W. and W. J. Lucas (1990). Plasmodesmata. *Annu. Rev. Plant Physiol.*, **41**, 369–419.
Sacristan, M. D. (1986). Isolation and characterisation of mutant cell lines and plants: disease resistance. In *Cell Culture and Somatic Cell Genetics of Plants*, ed. I. E. Vasil, vol. 3, pp. 513–525. Academic Press, New York.
Scowcroft, W. R. and P. J. Larkin (1985). Somaclonal variation, cell selection and genotype improvement. In *Comprehensive Biotechnology*, ed. C. W. Robinson and J. A. Howell, vol. 4, pp. 153–168. Pergamon Press, Oxford.
Sela, I. (1981). Plant–virus interactions related to resistance and localization of viral infections. *Adv. Virus Res.*, **26**, 201–237.
Sherwood, J. L. (1987). Mechanisms of cross protection between plant virus strains. In *Plant Resistance to Viruses*, CIBA Foundation Symposium 133, ed. D. Evered and S. Harnett, pp. 136–150. Wiley, Chichester.
Stace-Smith, R. (1985). Virus-free clones through plant tissue culture. In *Comprehensive Biotechnology*, ed. C. W. Robinson and J. A. Howell, vol. 4, pp. 169–179. Pergamon Press, Oxford.
Stark, D. M. and R. N. Beachy (1989). Protection against polyvirus infection in transgenic plants. Evidence for broad spectrum resistance. *Biotechnology*, **7**, 1257–1262.
Tanksley, S. D., N. D. Young, A. H. Paterson and M. W. Bonierbale (1989). RFLP mapping in plant breeding: new tools for an old science. *Biotechnology*, **7**, 257–264.
Taschner, P. E. M., A. C. Van Der Kuyl, L. Neeleman and J. F. Bol (1991). Replication of an incomplete alfalfa mosaic virus genome in plants transformed with viral replicase genes. *Virology*, **181**, 445–450.
Tien, P., X. Zhang, B. Qiu and G. Wu (1987). Satellite RNA for the control of plant diseases caused by cucumber mosaic virus. *Ann. Appl. Biol.*, **111**, 143–152.
Tolin, S. A. and A. K. Vidaver (1989). Guidelines and regulations for research with genetically modified organisms: a view from academe. *Annu. Rev. Phytopathol.*, **27**, 551–581.
Toyoda, H., K. Chatani, Y. Matsuda and S. Ouchi (1989). Multiplication of tobacco mosaic virus in tobacco tissues and *in vitro* selection for viral diseases. *Plant Cell Rep.*, **8**, 433–436.
Tumer, N. E., K. M. O'Connell, R. N. Nelson, P. R. Sanders, R. N. Beachy, R. T. Fraley and D. M. Shah (1987). Expression of alfalfa mosaic coat protein gene confers cross protection in transgenic tobacco and tomato plants. *EMBO J.*, **6**, 1181–1188.
Van Dun, C. M. P., and J. F. Bol (1988). Transgenic tobacco plants accumulating tobacco rattle coat protein resist infection with tobacco rattle virus and pea early browning virus. *Virology*, **167**, 649–652.
Van Dun, C. M. P., B. Overduin, L. van Vloting-Doting and J. F. Bol (1988). Transgenic tobacco expressing tobacco streak virus or mutated alfalfa mosaic virus coat protein does not cross protect against alfalfa mosaic virus infection. *Virology*, **164**, 383–389.
Warren, G. S., P. Thomas, M.-T. Herrera, S. J. L. Hill and R. Terry (1991). The use of plant cell cultures for studying virus resistance, and enhancing the production of virus resistant and virus-free plants. In press in *J. Biotechnol.*
Wilson, T. M. A. (1988). Structural interactions between plant RNA viruses and cells. *Oxford Surveys of Plant Molecular and Cell Biology*, **5**, 89–142.
Yoshida, M. (1990). Japan roundup. *Biotechnology*, **8**, 503.

# 15

# Exploitation of Chloroplast Systems in Biotechnology: Stabilization and Regulation of Photosynthesis

CHRISTINE H. FOYER
*INRA, Versailles, France*
and
ROBERT T. FURBANK
*CSIRO, Canberra, Australia*



## 15.1 INTRODUCTION

Chloroplasts are probably the most complex of all the organelles in the eukaryotic cell. They contain an elaborate system of energy-transducing membranes that facilitate the photochemical reactions. This enables the chloroplasts to carry out biosynthetic reactions using sunlight as the only source of energy. As a result of this pivotal role in biology the chloroplasts are the most intensely studied and best-known representatives of the plastid family of subcellular organelles.

Chloroplasts contain the 'most abundant enzyme in the world', ribulose-1,5-bisphosphate carboxylase (Ellis, 1979) and also 'perhaps the most important enzyme in the world', the water-splitting enzyme of the thylakoid electron transport chain (Prince, 1986). The isolated chloroplast is attractive both from an academic and technological viewpoint since it contains the complete machinery to capture solar energy, convert it into chemical energy and use this to fix carbon dioxide into sugar phosphates in the process of photosynthesis. In addition to carbohydrate synthesis chloroplasts contain a wide range of synthetic capabilities from the capacity to convert nitrite to ammonium to amino acids and subsequently proteins, to the ability to convert of acetate to fatty acids and lipids. Similarly, they will import sulfate and produce sulfides and sulfide-containing proteins. This wide-ranging and comprehensive biosynthetic capacity, all powered by solar energy, commends the chloroplast as an excellent candidate for technological exploitation, manipulation and emulation. It is an ideal 'model' to be copied in its ability to harvest sunlight, split water, convert solar to chemical energy and for secondary product synthesis. It has proved impossible, to date, for chemists to recreate the photochemical primary function and produce a solar-powered, electrolytic battery.

In spite of the independent origin of the chloroplast its immediate potential for biomass production is restricted because of its total dependence on its subcellular environment (the cytosol) for the maintenance of function (Heber and Walker, 1979). Isolated chloroplasts are generally very delicate structures that can only survive in a fully functional state for a period of hours in darkness, and for less than one hour when illuminated. However, there are many routes to the same goal and we may learn a great deal from chloroplast structure and function which may form the basis for technological advancement. The mere fact that photosynthesis established and maintains the biosphere is sufficient reason to study the biochemistry and molecular organization of the chloroplast, but in addition such studies lead the way to new technology and the possibility of man-made photoactive systems. Photosynthesis is the most efficient system known for the conversion and storage of solar energy as chemical energy and outranks by far any artificial system yet devised. The study of the efficiency of photosynthesis may be used to design artificial systems to store solar energy. Under the most favourable conditions a minimum of 8 mol quanta are required to convert 1 mol of $CO_2$ to carbohydrate with an absolute efficiency of energy conversion for red light of 33%.

In the chloroplast, chlorophyll is noncovalently bound to hydrophobic proteins which exist as multiprotein structural complexes embedded in the lipid phase of the photosynthetic membranes. The detailed structure and precise spatial relationships between the molecules responsible for the conversion of absorbed photons into a potential difference across the membrane has provoked a breakthrough in reconstituting an artificial solar-powered energy conversion system.

## 15.2 PROPERTIES OF INTACT CHLOROPLASTS

### 15.2.1 Preparation and Biochemical Capacity of Isolated Intact Chloroplasts

Experiments studying photosynthesis and $CO_2$ fixation using isolated intact chloroplasts have been successfully performed for many years (Walker, 1980a, 1987). The isolation procedures for functionally competent intact chloroplasts are relatively simple (Table 1) and have changed little since the methods were optimized in the 1960s (Jensen and Bassham, 1966; Walker, 1971). The chloroplast is bounded by a double membrane, each envelope membrane being about 5.5 nm thick, and the membranes are separated by a gap 2–10 nm wide (Douce and Joyard, 1979). The chloroplast envelope is a very delicate structure and some expertise is required in order to prevent rupture of the envelope during isolation. 'Intact' chloroplasts retain a whole and unbroken bounding envelope membrane that is impermeable to sucrose and ferricyanide (Lilley *et al.*, 1975). Such chloroplasts are capable of high rates of $CO_2$ fixation. A classification scheme for chloroplasts (Table 2) according to the degree of integrity and biochemical capability was introduced by Hall (1972). Isolation of intact chloroplasts can be achieved by mechanical homogenization of leaf tissue (as outlined in Table 1) or following the preparation of leaf protoplasts where the plasmalemma can be ruptured to release the intact chloroplasts (Leegood and Malkin, 1986). The procedures used to isolate intact chloroplasts have been described in several reviews (Leegood and Walker, 1985; Robinson *et al.*, 1986; Walker, 1971, 1980a, 1987). Chloroplasts prepared by these means are contaminated by other cell structures such as mitochondria, peroxisomes and sometimes intact cells.

Rapid isolation of 'intact chloroplasts' is essential in order to protect the chloroplast structure from degradative enzymes and compounds released into the cell homogenate when leaves are homogenized. The contaminating cellular components can be removed from the intact chloroplast

**Table 1** Preparation of Intact Spinach Chloroplasts

(a) *Preparation of intact chloroplasts (yielding chloroplasts with % intactness between 75 and 90%)*

1. Float the leaves on cold running water and illuminate them for 20–30 min, then remove the mid-ribs and cut the leaves into strips
2. Mix 80 g of leaf with 250 ml of the semifrozen slurried grinding medium (0.33 M sorbitol, 10 mM $Na_4P_2O_7$, 5 mM $MgCl_2$, 2 mM sodium isoascorbate, freshly prepared and adjusted to pH 6.5 with HCl). Homogenize for 3 s (or three bursts of 1 s duration) in a polytron, or 3–5 s in a Waring blender
3. Squeeze the cell brei through two layers of muslin to remove coarse or unground material, then quickly filter the brei through a sandwhich of cotton wool between eight layers of muslin (previously soaked in a little grinding medium)
4. Centrifugation should be carried out immediately after homogenization. Ideally a swing-out rotor should be accelerated to between 6000–8000 *g* and returned to rest within 90 s
5. Discard the supernatant and wash the surface of the pellet by gently swirling with a few ml of grinding or resuspension medium. Resuspend the pellet in a small quantity (1 ml) of ice cold resuspension medium (0.33 M sorbitol, 2 mM EDTA, 2 mM $MgCl_2$, 50 mM Hepes–KOH buffer [pH 7.6])

(b) *An additional purification step yielding a chloroplast preparation with over 90% intactness.*

6. Put 5 ml of 35% (v/v) Percoll into a centrifuge tube. Carefully overlay it with the chloroplast suspension and centrifuge for 1 min
7. Discard the supernatant and resuspend the pellet in fresh resuspension medium.

**Table 2** Classification of Chloroplast Preparations

| *Chloroplast type* | *Other description* | *Preparation method* | *Envelope integrity* | *Rate of $CO_2$ fixation*[a] | *Permeability* |
|---|---|---|---|---|---|
| A | Whole, intact chloroplasts | Rapid grinding and centrifugation In special media[b] | Intact | 50–250 | NAD(P), sucrose, ferricyanide Do not penetrate |
| $A_2$ | Resealed chloroplasts | As in type A, but envelope ruptures and reseals causing loss of stromal enzymes and metabolites | Intact | $<50$ | As in type A |
| B | Unbroken chloroplasts (sometimes called 'intact chloroplasts' in older papers) | Usually in isotonic sugar or salt Prolonged centrifugation (often two or three steps) | Often seems intact by microscopy, but has been damaged | $<5$ | NAD(P), sucrose, ferricyanide and all other small molecules penetrate |
| C | Broken chloroplasts | Vigorous grinding, prolonged centrifugation | Disrupted | 0 | As for B |
| D | Free-lamellar chloroplasts, thylakoids | Osmotic shock if type A followed by return to isotonic medium | Removed | 0 | As for B |
| E | Chloroplast fragments, thylakoids | Resuspension of chloroplasts in hypotonic medium | Lost | 0 | — |
| F | Subchloroplast particles, gana lamellae, stroma lamellae | Sonication, detergent, treatment, French press, chaotropic agents | — | 0 | — |

[a] $\mu$mol $h^{-1}$ $mg^{-1}$ chlorophyll. [b] Intact chloroplasts prepared as in table contain largely type 'A' chloroplasts (over 90% type 'A'). The classification is a modification of that introduced by Hall (1972) and modified by Halliwell (1984).

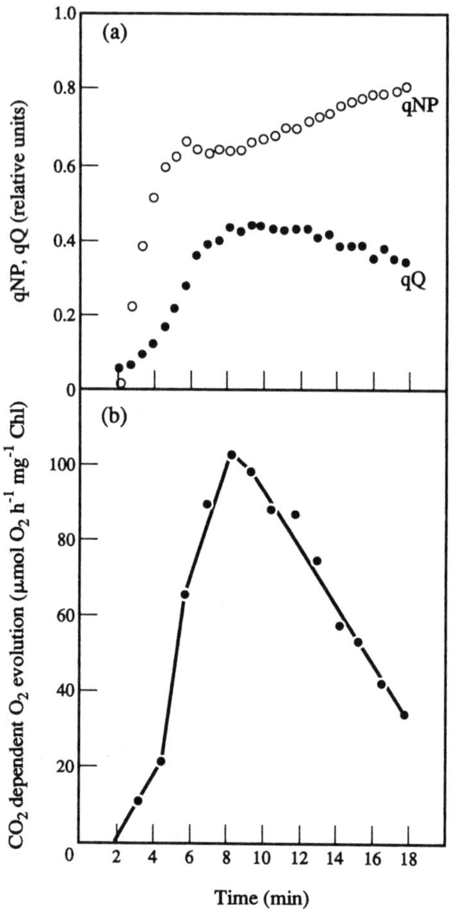


**Figure 1** The induction and decay of photosynthesis in isolated intact spinach chloroplasts as measured by $CO_2$-dependent $O_2$ evolution and chlorophyll *a* fluorescence quenching components (qQ and qNP). The chloroplast preparation was 89% intact (as measured by the ferricyanide method) and was illuminated at 1000 $\mu E\ m^{-2}\ s^{-1}$ (from time 0) in the presence of 5 mM $NaHCO_3$

fraction by centrifugation through a percoll cushion [Table 1(b)] or by differential centrifugation of leaf homogenates by isopycnic centrifugation on a silica–sol gradient (Mourioux and Douce, 1981). Such chloroplast preparations are stable in darkness for several hours. However, when isolated chloroplasts are illuminated, photosynthesis and $CO_2$ fixation can only be maintained at maximum rates for 10–15 min after which $CO_2$ fixation rapidly declines and is lost within 15–30 min (Figure 1).

### 15.2.2 Metabolic Constraints Limiting Photosynthesis

The chloroplast is not an independent entity and, in the plant cell, photosynthesis is dependent on the cytosol for a regulated supply of orthophosphate (Walker, 1980b; Walker and Sivak, 1986) and also most of its protein complement (Ellis, 1981). Triose phosphate is the end product of the Calvin cycle and, therefore, is the major end product of photosynthesis (Heldt, 1976; Robinson and Walker, 1981). Fixed carbon is exported from the chloroplast in the form of triose phosphate. Under optimal conditions one sixth of the product must be exported or converted into starch within the stroma. The export of triose phosphate is strictly regulated to prevent depletion of the levels of Calvin cycle intermediates that would severely impede the autocatalytic process. Transport occurs across a specific translocator protein that exchanges divalent anions and catalyzes the stoichiometric exchange of triose phosphate and some other divalent anions such as glycerate 3-phosphate for orthophosphate from the surrounding medium (Heldt and Rapley, 1970). One Pi (=orthophosphate) molecule must be returned to the stroma for every three molecules of $CO_2$ fixed. In the leaf, triose phosphate is converted to sucrose in the cytosol of the leaf mesophyll cell, a process that liberates Pi (Walker and Herold, 1977; Walker and Robinson, 1978). This is then returned to the chloroplast in exchange for more triose phosphate and in this way the processes of photosynthesis and sucrose synthesis are coordinated (Figure 2). The Pi optimum for photosynthesis in isolated chloroplasts is extremely sharp. The Pi concentration for any medium for isolated intact chloroplasts is, therefore, critical if optimum rates of photosynthesis are to be maintained. Too much Pi will drain the Calvin cycle of triose phosphate and inhibit the regeneration of ribulose 1,5-bisphosphate; too little Pi will cause a decrease in photosynthesis because of Pi limitation (maximal rates of photosynthesis cannot be maintained by Pi cycling *via* stromal starch synthesis). Our understanding of the regulatory mechanisms involved in the coordinate control of the photosynthetic processes and the interactions between electron transport, carbon assimilation and carbon partitioning is still far from complete (Foyer, 1986; Foyer *et al.*, 1990; Stitt, 1987). It is possible to describe specific metabolic constraints limiting photosynthesis (Leegood *et al.*, 1985). The experimental observations show that there are multiple sites of flux control in photosynthesis and that control is spread over many processes such that overcoming any single limitation, for example by the genetic improvement of the enzyme ribulose-1,5-bisphosphate carboxylase to achieve more efficient carboxylation, need not result in improved rates of carbon assimilation since other limitations then become important (Leegood *et al.*, 1985).

An understanding of the metabolic basis for the regulation of the rate of carbon assimilation and the reallocation of resources under given circumstances is a prerequisite if we hope to engineer improved photosynthetic capacity in the future. Many individual regulatory steps contribute to the control of overall flux in photosynthetic carbon assimilation (Heber *et al.*, 1988; Woodrow, 1986). The competing requirements of metabolic pathways demand that the individual reactions are always kept in balance and to determine the relative significance of environmental limitations and metabolic constraints it is important to recognize the ability of each metabolic step to control flux in a given circumstance (Kacser and Burns, 1979). To understand and quantify metabolic control systems and the responses of metabolism in changing conditions, it is essential to examine the structure and regulation of the individual metabolic events that together comprise the overall process of photosynthesis (Leegood *et al.*, 1985).

Photosynthetic carbon assimilation clearly can be limited by the supply of $CO_2$ or by the supply of ATP and NADPH to drive the Calvin cycle. Only if the chloroplast is provided with saturating light, unlimited $CO_2$ and optimal Pi will maximal rates of photosynthesis be achieved. In situations

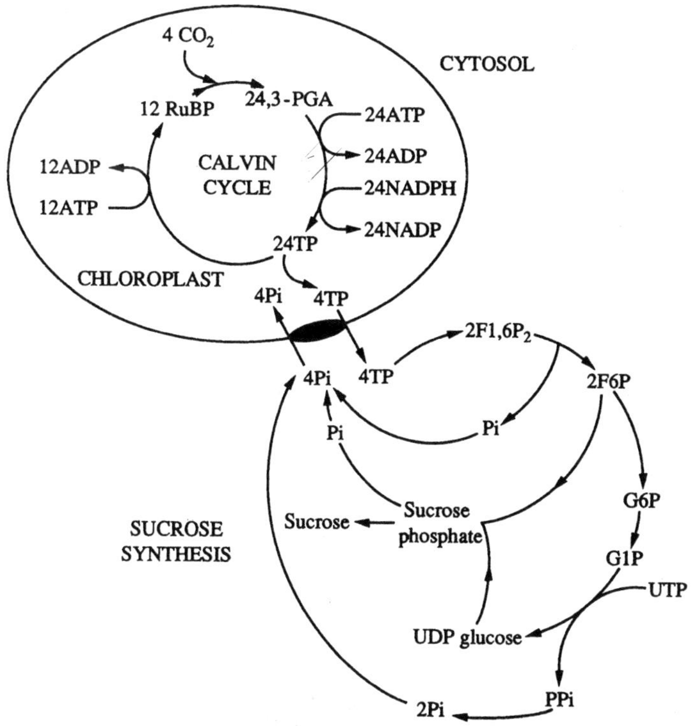


**Figure 2** The coordination of the processes of photosynthesis and sucrose synthesis through the cycling of Pi

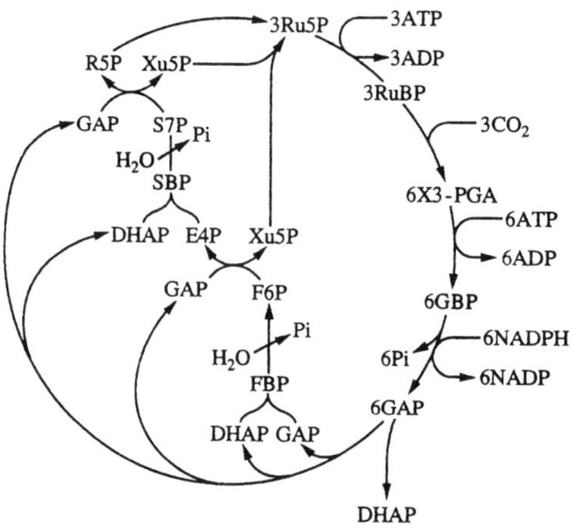


**Figure 3** The autocatalytic Calvin cycle (reductive pentose phosphate pathway)

where $CO_2$ and light energy are plentiful metabolic limitations to photosynthesis become of major importance. In the multienzyme sequence of the Calvin cycle (Figure 3) the enzyme of carbon fixation, ribulose-1,5-bisphosphate carboxylase, is a key regulatory enzyme determining the rate of flux through the pathway. The maximum activity of this enzyme is sufficient to sustain measured rates of carbon assimilation but it is most probable that in many circumstances the activity of the enzyme may be held well below the maximum catalytic activity measured *in vitro*.

### 15.2.3 $CO_2$ and Carbon Assimilation

When $CO_2$ is low the enzyme ribulose-1,5-bisphosphate carboxylase mediates the limitation on photosynthesis (Farquhar *et al.*, 1980; Von Caemmerer and Farquhar, 1981). Ribulose-1,5-bisphosphate carboxylase is therefore potentially a rate-limiting enzyme. However, it is becoming increasingly apparent that the actual amount of ribulose-1,5-bisphosphate carboxylase protein present in the chloroplasts is precisely adapted to match to the amount of light and $CO_2$ available for photosynthesis. The carboxylation rate is coordinated with the supply of ATP and NADPH required for the utilization of the glycerate 3-phosphate and the regeneration of the acceptor molecule, ribulose 1,5-bisphosphate. The synthesis of ribulose-1,5-bisphosphate carboxylase is light regulated such that in growth conditions where irradiance is varied, the amount of carboxylase protein present in the chloroplasts is found to increase with increasing light intensity (Kuhlemeier *et al.*, 1987). In addition, when chloroplasts or leaves are subjected to high $CO_2$ (1–5%) the activation state of the enzyme is down-regulated such that the initial increase in carboxylation rate facilitated by the increased availability of $CO_2$ is offset by a reduction in enzyme activation. The regulation of the catalytic activity and activation state of the enzyme is the result of several regulatory mechanisms that coordinate the supply of assimilatory power (ATP and NADPH) to the activity of the Calvin cycle (Foyer *et al.*, 1987; Furbank *et al.*, 1987a, 1987b; Portis, 1990).

In order to attain catalytic competence ribulose-1,5-bisphosphate carboxylase must first be activated with $CO_2$ and $Mg^{2+}$. The concentration of $CO_2$ required to produce maximal activity *in vitro* is an order of magnitude greater than that which occurs in $C_3$ chloroplasts *in vivo*. However, orthophosphate and several esters such as glycerate 3-phosphate, that are present in the chloroplast, cause the enzyme to reach full activity *in vitro* at ambient concentrations of $CO_2$. Calvin cycle intermediates modulate ribulose-1,5-bisphosphate carboxylase activity by stabilizing the activation state but inhibiting catalysis. Ribulose-1,5-bisphosphate carboxylase has an active site concentration of at least 4 mM and will bind reversibly chloroplast metabolites that are substrates for both thylakoid reactions (*e.g.* ADP, Pi and NADP) and stromal enzymes (*e.g.* glycerate 3-phosphate). The binding of these metabolites has a considerable inhibitory action on the reactions for which these metabolites are substrates. The presence of ribulose-1,5-bisphosphate carboxylase significantly increases the apparent Km values for Pi and ADP of coupled electron flow to methyl viologen (Furbank *et al.*, 1987a). In the presence of the specific inhibitor, carboxyarabinitol bisphosphate, the binding of ADP and Pi is reversed suggesting that these metabolites are bound to the ribulose-1,5-bisphosphate binding site. However, we have shown that ribulose-1,5-bisphosphate carboxylase can inhibit photophosphorylation in situations of limiting Pi *via* a magnesium-dependent binding of ribulose-1,5-bisphosphate carboxylase to the thylakoid membrane (Furbank *et al.*, 1986). It is probable that both of these processes contribute to higher Km (ADP) and Km (Pi) values for photophosphorylation. Similarly it is apparent that ribulose-1,5-bisphosphate carboxylase reduces the effective concentrations of stromal metabolites such as glycerate 3-phosphate (Figure 4). This suggests that ribulose-1,5-bisphosphate carboxylase has a rôle in synchronizing its catalytic activity with other stromal enzymes and with the rate of photophosphorylation. Indeed, feedback inhibition of ribulose-1,5-bisphosphate carboxylase by glycerate 3-phosphate ensures a close correspondence between the rate of energy production in the thylakoid and the rate of carboxylation.

The recent discovery of a protein called ribulose-1,5-bisphosphate carboxylase 'activase' that can activate ribulose-1,5-bisphosphate carboxylase at low $CO_2$ concentrations in the presence of ribulose 1,5-bisphosphate (Portis, 1990; Salvucci *et al.*, 1985) indicates the existence of a further means of regulating ribulose-1,5-bisphosphate carboxylase activity to match energy production. It has been suggested that the activation of ribulose-1,5-bisphosphate carboxylase could be related to $\Delta$pH. The current theory of photosynthetic control of electron flow by the requirement for assimilatory power proposes that high $\Delta$pH exerts feedback inhibition of electron transport. This situation could occur if photophosphorylation was limited by the availability of either ADP or Pi. However, simultaneous measurements of enzyme activation and 9-aminoacridine fluorescence showed that there was no direct correlation between $\Delta$pH and activation (Parry *et al.*, 1988). The activation of ribulose-1,5-bisphosphate carboxylase *via* the 'activase' system requires prior inactivation with impure ribulose-1,5-bisphosphate to stabilize inactive enzymes. In these circumstances fructose 1,6-bisphosphate or ATP-mediated reversal of the inhibition is observed. Robinson and Portis (1988) have shown that there is a strong correlation between the activation state of ribulose-1,5-bisphosphate carboxylase and the stromal ATP content in isolated intact spinach chloroplasts. Thus the extent of activation of this enzyme in the light appears to require ATP and to be proportional to the stromal ATP concentration.

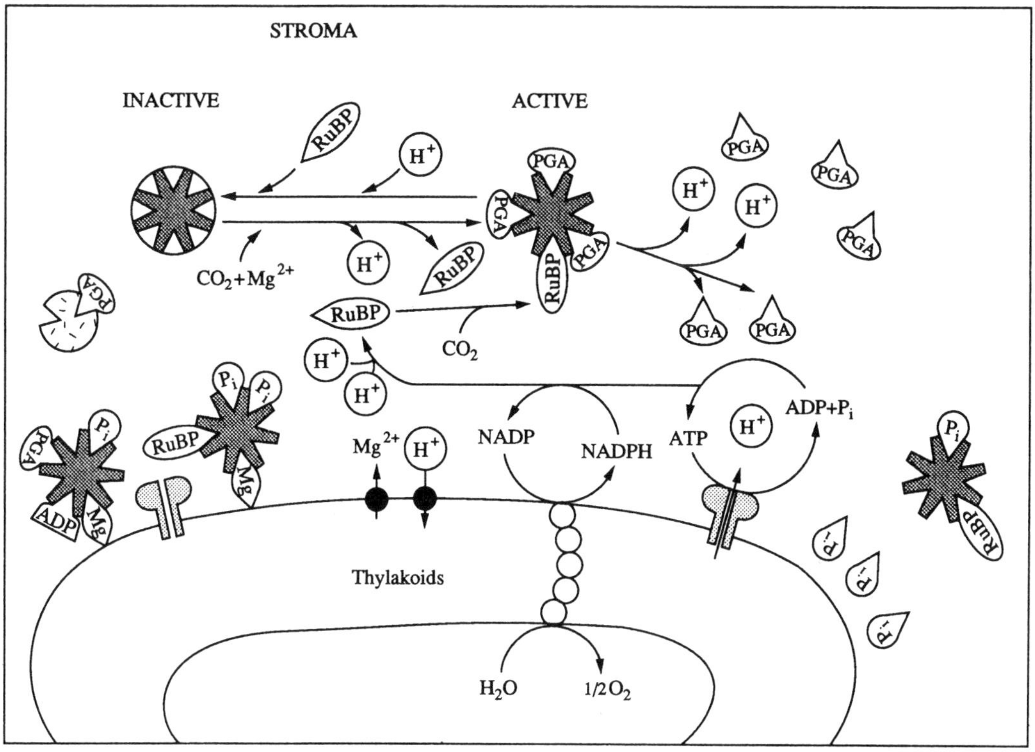


**Figure 4** A diagrammatic representation of metabolite binding to ribulose-1,5-bisphosphate carboxylase in the stroma and the binding of this enzyme to the thylakoid membranes. The wheel-shaped structures represent ribulose-1,5-bisphosphate carboxylase and the closed circles represent phosphoglycerate kinase

### 15.2.4 Light

Light drives photosynthesis by the generation of assimilatory power (ATP + NADPH). At saturating light intensities electron transport is limited by regulation through ΔpH, photosynthetic control and dissipative pathways (Foyer *et al.*, 1990). Too much light is destructive to the photosynthetic apparatus; in the short term this will result in the process of photoinhibition (Powles, 1984) where reversible loss of thylakoid function helps prevent irreversible loss of chlorophyll and photooxidative damage (Asada and Takahashi, 1987). Photosynthesizing tissues and chloroplasts are especially prone to the generation of toxic oxygen radicals because of their high internal $O_2$ concentration during photosynthesis, the presence of photosensitizing pigments which can generate singlet states of $O_2$ and an electron transport chain that can generate the superoxide free radical, $O_2^-$. In addition the Calvin cycle enzymes are highly sensitive to inhibition by $H_2O_2$ and there is a high percentage of unsaturated lipids in the thylakoid membranes. Photosynthetic cells can tolerate elevated oxygen levels because of the presence of several endogenous protective systems that effectively scavenge and remove the toxic products before cellular damage occurs. Oxygen-related cell damage can occur as a result of an increased rate of oxygen activation or when one or more of the cellular components of the defence system is destroyed so that even relatively normal levels of active oxygen species cannot be detoxified. Chloroplasts contain no catalase but are protected against free radical mediated damage *via* a cycle of $H_2O_2$ and $O_2^-$ detoxification (Figure 5) involving ascorbic acid and glutathione at the eventual expense of NADPH (Anderson *et al.*, 1983; Asada and Takahashi, 1987; Foyer and Halliwell, 1976).

At low irradiance photosynthesis is limited by the supply of ATP and NADPH. In these circumstances the efficiency of ATP production is optimized. At low irradiance the efficiency of excitation energy distribution between the photosystems is improved by the phosphorylation of specific membrane polypeptides (Allen, 1983; Horton, 1983; Staehelin and Arntzen, 1983). The thylakoid polypeptides that become phosphorylated are the light-harvesting *a/b* binding protein (LHC II) and other polypeptides of photosystem II. The regulation of excitation energy distribution between the photosystems optimizes the flow of electrons between the cyclic and noncyclic pathways and thus helps to match the supply of ATP to the requirements of the Calvin cycle (Horton and Foyer, 1983; Horton, 1985). The reversible phosphorylation of the LHC II, that is instrumental in directing the flow of excitation energy, is catalyzed by a thylakoid protein kinase that is distinct from the protein kinase that phosphorylates the 9–10 kDa and other photosystem II polypeptides. A suggested function of the reversible 9–10 kDa phosphoprotein of photosystem II is the modulation of electron flow from the water oxidation complex to cycling of electrons around PSII (Ikeuchi *et al.*, 1987).

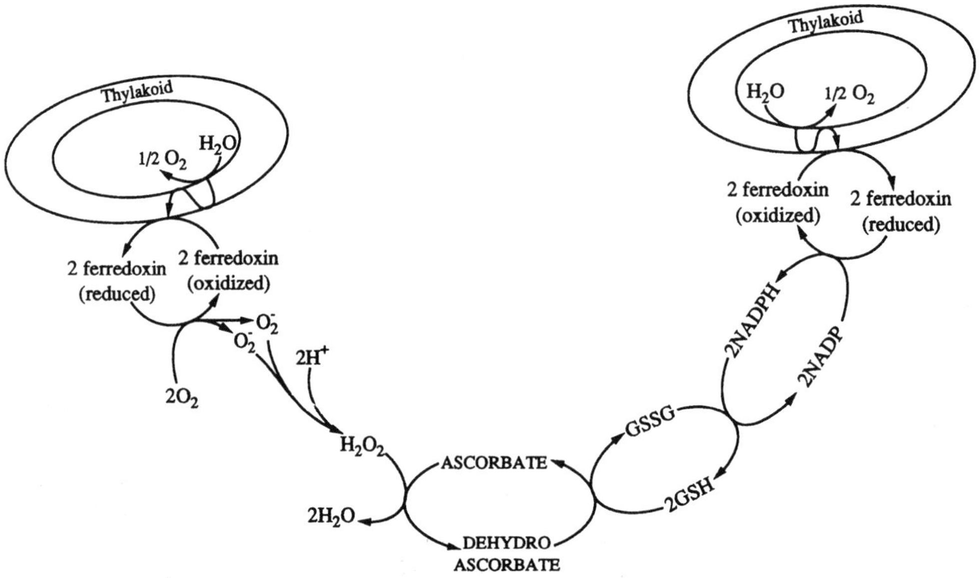


**Figure 5** The ascorbate/glutathione cycle for the detoxification of $H_2O_2$ in the chloroplasts. The first product of oxidation of ascorbic acid by ascorbate peroxidase is probably the monodehydroascorbate radical which undergoes a disproportionation to form ascorbate and dehydroascorbate. It is possible that not only dehydroascorbate but also an NAD(P)H dependent semidehydroascorbate reductase are involved in the regeneration of ascorbic acid in chloroplasts

### 15.2.5 Pi Limitation

Chloroplasts illuminated with $CO_2$ but without Pi show a progressive loss of photosynthetic capacity. However, photoinhibitory damage to photosystem II is not the major cause of the inhibition of photosynthesis observed with low Pi. There are multiple sites of flux control exerted when the chloroplast is subjected to suboptimal Pi (Furbank *et al.*, 1987b). At least three factors contribute to the inhibition of photosynthesis under phosphate limitation: (i) there appears to be a direct effect of Pi on the energy-transducing system; (ii) there is direct inhibition of the Calvin cycle decreasing the ability of the pathway to act as a sink for ATP and NADPH; and (iii) feedback inhibition of primary processes occurs either *via* $\Delta$pH or the redox state of electron carriers. However, $\Delta$pH does not appear to be a limiting factor but rather an inability to regenerate NADP as electron acceptor is suggested. There clearly appears to be a complex series of interactions between thylakoid and stromal reactions that provides an efficient synchronous regulation of photosynthetic flux.

### 15.2.6 Chloroplast Stability

Intact, isolated chloroplasts (called 'class A' chloroplasts in the terminology of Hall, 1972; Table 2) capable of high rates of $CO_2$-dependent $O_2$ evolution are inherently unstable. Commonly, chloroplasts require 1–5 min illumination to reach a steady-state rate of $CO_2$ fixation (see Walker, 1976) and this activity is often only stable for 10–15 min thereafter. Stored on ice, chloroplasts will survive with minimal loss of activity for 4–6 h. The causes for the rapid decline in activity in the light and at room temperature are the subject of controversy. It has been suggested that, during storage, Pi and Calvin cycle intermediates slowly leak out of the chloroplast, depressing the rate of photosynthesis and changing the Pi optimum (Walker, 1976). Seftor and Jensen (1986) have shown that this decline is initially caused by a decrease in the activity of ribulose-1,5-bisphosphate carboxylase with the specific loss of the enzyme–$CO_2$–$Mg^{2+}$ complex form of the enzyme. The chloroplast envelope apparently then loses its integrity but it tends to break and reseal protecting the thylakoids such that the ferricyanide-dependent $O_2$ evolution is retained but stromal metabolism is lost. We can assume that these deleterious events are light-mediated since under identical conditions in the dark the loss of ribulose-1,5-bisphosphate carboxylase activity and protein from the chloroplasts did not occur. Since chloroplasts are fragile, osmotically sensitive organelles, breakage of the chloroplast envelope during assay is a problem because stirring of the solution is generally necessary to avoid localized $O_2$ and $CO_2$ gradients and to distribute products, reactants and light throughout the preparation.

The electron transport components of isolated chloroplasts are generally more robust than those of carbon metabolism and rupture of the chloroplast envelope can leave these processes unaffected. Thus, thylakoid preparations (Class C or D chloroplasts; Table 2), although incapable of $CO_2$-dependent $O_2$ evolution or NADP reduction without added enzymes and cofactors, are generally more stable than their intact counterparts. It has been shown that thylakoids stored at 4 °C in the dark lose 50% of their electron transport (with the artificial acceptor $KFe(CN)_6$) after 5 d, while at 17 °C this half-life decreases to 24 h (Hardt and Kok, 1976).

Chloroplast ageing, at the level of electron transport, correlates with a decrease in PSII activity and a release of lipoprotein and fatty acids through the action of lipolytic enzymes (see Papageorgiou, 1980). Also, loss of activity during illumination may occur due to photoinhibition (Powles, 1984) possibly enhanced by photoperoxidation of unsaturated fatty acids, resulting in chlorophyll bleaching (see Papageorgiou, 1980; Powles, 1984). The stability of thylakoids is changed by the presence of bovine serum albumin, isotonic, nonionic osmotica, superoxide and $H_2O_2$ scavengers (superoxide dismutase and catalase), and by storage in dense suspensions at higher pH values (see Papageorgiou, 1980).

Although electron transport *per se* is relatively stable in thylakoid preparations, photophosphorylation is not. Photophosphorylation requires a high degree of thylakoid structural integrity in order to support the transthylakoid pH gradient required to produce ATP *via* the ATPase. The stability of this process is less than 12 h at or below 4 °C and may decrease to less than 30 min under illumination at room temperature. The 'uncoupling' of electron flow from photophosphorylation in thylakoids can occur for a variety of reasons. (i) Mechanical or osmotic disruption of the thylakoid membrane can result in ionic 'leakiness' and inefficiency in pumping $H^+$ across the membrane. (ii) The ATPase itself may detach from the membranes under conditions of low cation concentration or in the presence of EDTA. (iii) ATPase activity can be suppressed by oxidizing conditions or through chemical modification during preparation. (iv) Certain endogenous polyphenols, amines and other similar compounds may affect membrane permeability, causing 'uncoupling'.

An alternative chloroplast system often used for electron transport measurements is the chloroplast fragment/particle preparation. In this procedure, thylakoids are prepared and disrupted either by high speed blending or sonication and treatment with French press. The product is a suspension of small thylakoid fragments or membrane vesicles containing thylakoid proteins. The stability of these particles seems similar to that of purified thylakoids but may have advantages over thylakoids in permeability and ease of manipulation. Also, individual components of PSII and PSI may be separated in these particles (see Sane *et al.*, 1970).

### 15.2.7 Reactions Centres, Charge Separation and Water Oxidation

The photosynthetic membranes of the chloroplast contain arrays of pigment protein complexes that absorb light and cooperate to funnel the absorbed light energy to special 'traps' called reaction centres (Diner, 1986; Glazer and Melis, 1987). The reaction centre is where light energy is converted to chemical energy by the formation of an excited singlet state. Excitation transfer from an antenna pigment to a reaction centre generates an excited state with a redox potential that makes it a powerful reducing agent which immediately transfers an electron to an adjacent low potential acceptor achieving separation of charge. In this process both spin and angular momentum are conserved such that the radical pair is formed in the pure singlet state where the two unpaired electrons have anti-parallel spins. The photochemical products of the charge separation step are then stabilized by a reduction–oxidation system of electron carriers and in this way provides a chemical energy store used in the synthesis of ATP and NADPH. In order to drive electrons from a very poor electron donor, water, (an oxidizing potential of approximately +1 V is necessary to remove electrons and protons from water) with the consequent reduction of a weak electron acceptor, NADP ($\Delta E$ 1600 mV), two distinct types of reaction centre associated with separate photosystems arranged in series are used in conjunction with the thylakoid electron transport chain (Figure 6). Reaction centres are denoted by the letter P and the wavelength of absorption in the red or near infrared region that is characteristic for each reaction centre. In the thylakoid membranes the two kinds of reaction centre absorb either at 700 nm ($P_{700}$, associated with photosystem I and the reduction of NADP) or 680 nm ($P_{680}$, associated with photosystem II and water oxidation). The position of these reaction centres and the electron carriers which join the two are shown in Figure 6.

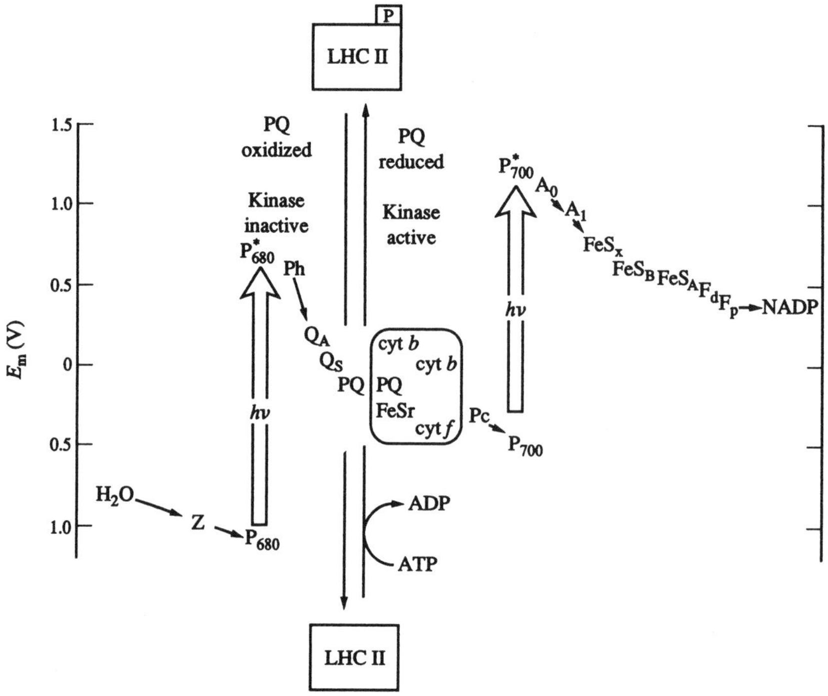


**Figure 6** The 'Z' Scheme model for the photosynthetic electron transport chain (using excited state redox potentials). The regulation of the protein kinase that phosphorylates LHC II is also represented (see Section 15.2.4). It is possible that the redox state of a component of the cytochrome-$b_6$/f protein complex is also important in the regulation of this protein kinase

Photosystem II consists of the reaction centre chlorophyll ($P_{680}$), phaeophytin, Fe, cytochrome $b_{559}$, two plastoquinone electron acceptors $Q_A$ and $Q_B$, the core antenna chlorophylls and a primary electron donor (Diner, 1986; Glazer and Melis, 1987; Govindjee *et al.*, 1985). These functional components of photosystem II are associated with five integral polypeptides with molecular weights of 47, 44, 34 (D2), 32 (D1) and 10 kDa. Similarities between the bacterial reaction centre (and component polypeptides) of *Rhodobacter sphaeroides* and *R. viridis* and the photosystem II reaction have led to the accepted proposal that the D1 and D2 polypeptides form the reaction centre of photosystem II (Barber, 1987; Michel and Deisenhofer, 1986; Trebst, 1986). $P_{680}$ and phaeophytin are bound to the D1/D2 heterodimer. The *Rhodobacter sphaeroides* reaction centre consists of four bacteriochlorophylls, two bacteriophaeophytins, one ubiqinone and one iron that are associated with three polypeptides of 20 kDa (small), 22 kDa (medium) and 28 kDa (large). the X-ray cystallographic data show that the reaction centre has a high degree of symmetry between right and left but electron flow follows a preferred route through the right hand chain of electron carriers (Deisenhofer *et al.*, 1985; Michel and Deisenhofer, 1986). A significant difference between the bacterial reaction centre and that of photosystem II is the presence of a cytochrome component in the latter. This feature may be associated with the capacity to cycle electrons around photosystem II at times when electron flow from the reaction centre is prevented. Similarities in the amino acid sequences suggest that the D1/D2 heterodimer is likely to have a very similar structure to that of the large and medium subunits of the reaction centre of the purple bacteria.

Unlike the bacterial reaction centre, photosystem II can oxidize water. The primary electron donor to $P_{680}$ is denoted as Z, and this in turn removes electrons from the water-splitting complex that facilitates the four-electron oxidation of water and liberates oxygen (Figure 7). Z must therefore turn over four times to accumulate the oxidizing equivalents to release one $O_2$ molecule. The donor side of photosystem II has been monitored by electron paramagnetic resonance (EPR) spectroscopy (the photosystem II donor components yield a spectrum at $g = 2.0046$). In darkness following illumination the signal at 2.0046 $g$ is found to consist of two kinetically distinct decay phases, a fast phase attributable to the decay of Z+ and a slow phase that is attributed to the presence of a second donor, called D. Recent evidence has shown that both Z and D are tyrosine residues within the photosystem II heterodimer (Barber, 1988; Ikeuchi and Inoue, 1987). In *Synechocytis* D+ there has been shown to be a tyrosine radical at position 160 on the D2 polypeptide and Z is almost certainly the symmetrically related tyrosine on the D1 polypeptide. However, although these tyrosine residues occupy symmetrical positions Z and D have different functions and show different redox kinetics. While Z is directly involved in the electron transfer to $P_{680}$, D is not but serves instead to achieve accumulation of the $S_1$ state in the dark (Figure 7).

The site of water oxidation that donates electrons to Z, and hence to $P_{680}$, is located in clefts on the

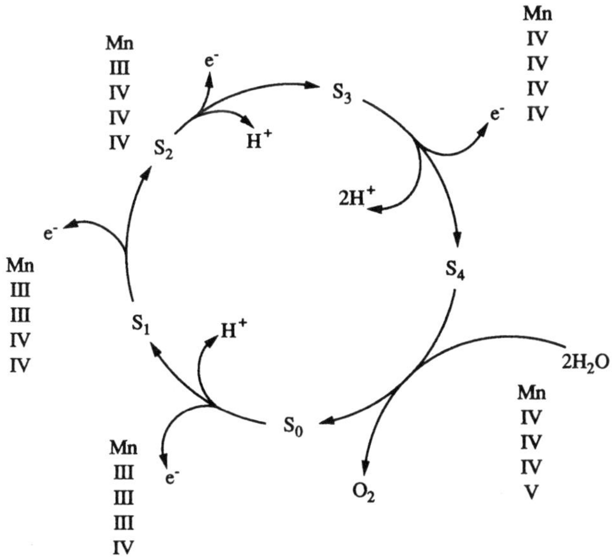


**Figure 7** The S-state model for the sequential reactions of the four-electron oxidation of water incorporating the proposed oxidation states of the four manganese atoms in the model suggested by Brudvig and Crabtree (1986) and Prince (1986). The S-states represent the oxidation states of the manganese of the water-splitting enzyme. Electrons are removed from the complex by the photosystem II donor, 'Z'. The oxidation of 'Z' by photosystem II drives the oxidation reactions of the S-state transitions

inner surface of the thylakoid membrane and has associated extrinsic polypeptides of 33 000, 23 000 and 16 000 molecular weights. There appears to be no strong binding of manganese to these polypeptides or, indeed, to any of the photosystem II polypeptides. It had been suggested that the 33 kDa polypeptide was the manganese binding site and that it lost its bound manganese upon extraction from the membrane. However, it now appears more likely that the manganese cluster is stabilized by this 33 kDa hydrophilic protein. Recent evidence suggests that the manganese cluster is closely associated with the D1/D2 heterodimer (Barber, 1987; Coleman and Govindjee, 1987). It is clear that the water-splitting enzyme contains a cluster of four manganese atoms that are required for activity. These manganese atoms are considered to act as charge accumulators with sequential reactions which were formulated into the S-state scheme shown in Figure 7. The structure of the manganese complex is unknown and much controversy surrounds the nature of the manganese cluster since there is an apparent minimum stoichiometry of two Mn atoms per PSII reaction centre and two additional atoms of Mn or other cation required for maximal activity. Models ranging from an $Mn_4O_4$ cubane/$Mn_4O_6$ (with oxidation states of the four manganese atoms of III, IV and V) (Brudvig and Crabtree, 1986) structure to binuclear clusters that have suggested oxidation states of II or III have been proposed (Prince, 1986).

The detailed X-ray structural analysis of a bacterial reaction centre has given much insight into the conversion of absorbed photons into a voltage difference across the membrane. Within picoseconds of a photon reaching the reaction centre an electron is transferred from pair of bacteriochlorophyll molecules to a nearby bacteriophaeophytin. Within 200 ps the electron is transferred from the bacteriophaeophytin to a quinone molecule which provides a large degree of stabilization of the charge separation. It is of particular interest to note that the molecules involved in these primary electron transfer steps are rather widely separated by distances of approximately 10 Å. In addition, the active components are rather rigidly embedded in a nonpolar region of the membrane and there is no evidence of connections for the electrons to flow through. In a view of these constraints Warman (1987) has suggested that charge separation must occur by quantum mechanical electron tunnelling between components that enables electrons to move across large atomic distances 'without ever being in between'.

## 15.3 IMMOBILIZATION OF CHLOROPLASTS

All chloroplast systems suffer from the same basic shortcomings in their application to biotechnology. First, for prolonged illumination, low stability is a serious problem. This cannot be completely overcome because the chloroplast gemone only provides a fraction of the genetic information required to reproduce stromal proteins and electron transport components. Most chloroplast proteins are coded for by the nuclear genome. It is therefore unlikely that a self-repairing, self-replicating chloropast system, similar to cell culture and algal culture systems already in existence, could be developed. The second problem with chloroplast systems relates to all photosynthetic reactor biotechnology: the physical difficulty of illumination, supply of substrates and recovery of products in an aqueous medium. The latter problem is of particular significance because of the fragility of chloroplasts. Carbon compounds produced in a reactor cannot be easily removed since centrifugation is not desirable. Also, in photosynthetic systems, light is the principle reactant, necessitating a reactor design where light utilization is optimum but dilution of products minimal. This is a difficult engineering problem, as mutual shading can be a serious limitation of productivity in liquid culture (see Gisby *et al.*, 1987).

One way in which these problems may be partially overcome is by discarding liquid media to some degree and using chloroplasts 'immobilized' in a gel or solid phase. This technique has been used successfully for bacteria, yeasts and algae. Immobilization of chloroplasts considerably improves longevity of both electron transport and carbon metabolism (Cerovic *et al.*, 1987; Gisby *et al.*, 1987; Karube *et al.*, 1979) and can provide the physical conditions necessary for application of chloroplasts to biotechnology. A variety of techniques and materials have been used for the immobilization of chloroplasts. The more popular alternatives are discussed in the following section, subdivided according to the material used.

### 15.3.1 Agar and Alginates

Agar has been successfully used for entrapment of isolated chloroplasts, thylakoids and cyanobacteria (Gisby *et al.*, 1987; Karube *et al.*, 1981). A number of agar grades with different gelation temperatures have used with agar concentrations ranging from 0.5 to 5%. The following procedure is generally applicable to thylakoid preparations and was described by Gisby *et al* (1987). The agar powder (OXOID #3 or DIFCO Bacto Czapek) is dissolved in boiling 0.15 M NaCl to give a final concentration of 2% (w/v). The solution is cooled to 45 °C and added to the chloroplast/thylakoid suspension in a petri dish containing 0.5% bovine serum albumin. The preparation is mixed on ice to reduce the problem of heat inactivation of chloroplasts. A film of immobilized chloroplasts is immediately formed which can be cut into segments for experimental manipulation (in an $O_2$ electrode vessel, for example) after suspension in assay buffer. These immobilized preparations are incapable of $CO_2$-dependent $O_2$ evolution and consist solely of broken chloroplasts or thylakoids. However, rates of ferricyanide reduction suggest that there is little or no damage to the electron transport chain (Gisby *et al.*, 1987; Karube *et al.*, 1981). Longevity of these chloroplasts is more than doubled by agar immobilization (Gisby *et al.*, 1987).

Immobilization of chloroplasts in alginate differs from the agar method in that a heating step is avoided. Alginate is a polymer of mannuronic acid and guluronic acid which is extracted from marine algae. The degree of water solubility of this polymer depends upon the cation used as a gelling agent. Sodium alginate is water soluble, forming a viscous but workable solution, while calcium or barium alginate is virtually insoluble. Two types of procedure are commonly used for alginate immobilization producing either alginate beads or films. In the former method (described in Gisby and Hall, 1980; Gisby *et al.*, 1987), equal volumes of chloroplast suspension and 2% sodium alginate (in 50 mM Hepes buffer, pH 7.5 and 1% bovine serum albumin) are combined and dispensed drop by drop into a solution of 0.1 M $CaCl_2$ (in 50 mM Hepes–KOH buffer, pH 7.5). The product is a suspension of beads approximately 2–3 mm in diameter which require around 1 h to harden at 0.4 °C. Smaller beads can be produced by spraying the Na alginate/chloroplast mixture into the $CaCl_2$ with a fine atomizer (Gisby *et al.*, 1987). Alginate films are made in much the same way but 1% sodium alginate is used. The alginate/chloroplast mixture is spread onto a petri dish and the $CaCl_2$ spread onto the solution to initiate gel formation (Gisby *et al.*, 1987). These films may be supported by meshes made from cheesecloth, nylon grid or aluminum mesh embedded in the alginate.

Alginate immobilization confers considerable stability to thylakoid electron transport and when coupled to $H_2$ production (Gisby and Hall, 1980) or ferricyanide reduction (Gisby *et al.*, 1987), immobilization can more than double the $t_{\frac{1}{2}}$ for activity loss compared to free chloroplasts. Alginate films appear to be preferable to beads as no loss in electron transport occurs during immobilization and the $t_{\frac{1}{2}}$ for $O_2$ evolution is extended from 24 h to 72 h (Gisby *et al.*, 1987). In contrast, alginate bead immobilization can cause a 50% loss of electron transport activity (Gisby *et al.*, 1987). These techniques have been used almost exclusively for stabilizing electron transport activity, but, providing a suitable cation is chosen for gelation, they should be useful for intact chloroplast preparations fixing $CO_2$.

### 15.3.2 Polyurethane

Immobilization in polyurethane foam is a technique used successfully for cyanobacteria and chenopodium thylakoids (Gisby *et al.*, 1987). This procedure involves a heat-dependent polymerization of hydrophilic urethane monomers and it is this heating step (40 °C or above) which may be a disadvantage for retention of full biological activity. Despite this shortcoming, minimal losses of electron transport in thylakoids have been reported following immobilization (Gisby *et al.*, 1987) and the resulting material is porous and easily cut into convenient shapes and sizes. The longevity of the urethane/chloroplast system has not been extensively studied.

### 15.3.3 Cross-linked Proteins

Immobilization of organelles and organelle fragments in cross-linked protein media is the most popular method reported in the literature and was recently reviewed by Barbotin *et al.* (1987). A variety of cross-linking methods have been used with chloroplast preparations, most using the chemical cross-linker and fixative glutaraldehyde. This method, developed by Broun *et al.* (1973) involves the cross-linking of free proteins in solution with the amino groups of protein in the chloroplast membranes. The reactions are complex but primarily appear to involve the lysyl—$NH_2$ groups of amino acids. Two 'carrier' proteins are commonly used to produce the cross-linked matrix, either albumin or gelatin (Barbotin and Thomasset, 1980; Papageorgiou. 1983). These materials can be used to produce either a foam structure or a thin film of immobilized chloroplasts.

A general procedure (Barbotin *et al.*, 1987) for glutaraldehyde/protein immobilization is as follows. For albumin films, a mixture of 5.45% bovine serum albumin, 0.47% glutaraldehyde, phosphate buffer (0.05 M, pH 7) and 0.3 mg chlorophyll (as a thylakoid suspension) is stirred in a total volume of 0.1 mL then spread over a glass surface (area = 20 $cm^2$). Following air drying, a proteinaceous membrane is produced. For gelatin films, 5% ossein gelatin is mixed with buffer, as before, but 1% glutaraldehyde is added as a cross-linker after the film has been formed. To produce albumin foams, thylakoids (2 mg chlorophyll) are mixed with 5.86% bovine serum albumin, 0.33% glutaraldehyde in phosphate buffer in a volume of 2–3 mL. The mixture is then quickly frozen to −20 °C for 2 h and thawed slowly at 4 °C. Similarly, for gelatin foams, 5% ossein gelatin is added to 0.625% glutaraldehyde in a 4 mL volume and treated as for bovine serum albumin foams.

The effects of glutaraldehyde on chloroplast photosynthesis are manifold and often destructive; however, electron transport activity can be stabilized at least three-fold by cross-linkage immobilization (Hardt and Kok, 1976). Unfortunately, in addition to this stabilizing effect on remaining activity, there is substantial initial inhibition of electron transport in glutaraldehyde-treated thylakoids (Hardt and Kok, 1976; Packer, 1976; Papageorgiou, 1980). Also, the amount of glutaraldehyde required for cross-linkage and stabilization overlaps considerably with that required to inhibit whole chain electron flow (see Packer, 1976). Papageorgiou (1980) summarized the effects of glutaraldehyde on thylakoids as follows. There is a loss of osmotic response of thylakoids, loss of photophosphorylation, inhibition of the fast rise of chlorophyll *a* fluorescence and its subsequent quenching. Glutaraldehyde appears to have no effect on the $H_2O$ splitting but does cause drastic inhibition of electron transport between PSII and PSI and at the level of plastocyanin.

Alternative cross-linking agents may be used for protein immobilization, in particular members of the aliphatic diimide group. These agents are somewhat more specific in their action, reacting only with $NH_2$ groups to yield positively charged amidines. This retains the charge conservation of the protein, which glutaraldehyde treatment can destroy. These cross-linkers inhibit photosystem II activity but allow retention of photophosphorylation and PSI activity (Packer, 1976). The use of diimide esters may be limited, however, as they appear to confer no increase in thylakoid longevity (Papageorgiou, 1980). Cross-linkers such as toluidine isocyanate have also been used to entrap thylakoid fragments in vesicles formed from protamine and gelatine (Kitajima and Butler, 1976). Although PSII activity was lost, PSI activity remained and longevity was improved. Electron microscopy has shown that cross link stabilized, intact chloroplasts retain their morphological integrity both at the level of the envelope and thylakoid membranes (Barbotin and Thomasset, 1980). However, there are no reports in the literature of substantial rates of $CO_2$ fixation by chloroplasts immobilized with this technique.

### 15.3.4 Polyacrylamide Gels

Karube *et al.* (1979) have used polyacrylamide gel as an immobilization medium for isolated, intact chloroplasts from Chinese mustard. Although $CO_2$ fixation was inhibited 35% by immobilization, these chloroplasts were much more stable under alkaline conditions and at high temperature than native chloroplasts. The lifetime of these immobilized chloroplasts was three-fold longer than free chloroplasts and sugar phosphates were continously produced and excreted from the gel matrix.

Using the method of Kurabe *et al.* (1979) chloroplasts are immobilized by adding acrylamide and bisacrylamide to 2.5 mL of chloroplast resuspension medium containing (1.0 mg Chl) isolated chloroplasts. Polymerization was initiated by the addition of 0.05 ml 10% dimethylaminoproportionate and 0.5 mL 5% potassium persulfate. This reaction was complete within 30 min at 25 °C. The optimum acrylamide concentration for chlorophyll longevity was found to be 7–8% w/v. This system appears to be potentially very useful for the production of specific sugar phosphates as only two major products are released by intact chloroplasts, triose phosphates and glycerate 3-phosphate. It has been suggested, however, that polyacrylamide may inhibit PSII activity of immobilized thylakoids, as the polymerization reaction involves free radical production (see Ochiai *et al.*, 1978).

### 15.3.5 Polyvinyl Alcohol

To overcome the inhibitory effects of acrylamide polymerization on thylakoid reactions, Ochiai *et al.* (1978) developed a procedure for immobilizing thylakoids in polyvinyl alcohol (PVA). Flakes of polyvinyl alcohol/thylakoids were prepared by suspending thylakoids in 0.4 M sucrose, 0.05 M Tris buffer (pH 7.2), 0.01 M NaCl, 80 mM Na borate and 1% bovine serum albumin then mixing this with

20% polyvinyl alcohol-105. After drying *in vacuo* in the dark, a green flaky substance was obtained. These flakes retained 80% of native PSII activity and 100% of PSI activity. After 5 weeks, less than 20% loss in activity was observed and photophosphorylation was 44% of the control even after 2 weeks dark storage. This system may also be of considerable use for immobilizing intact chloroplasts.

### 15.3.6 Polyethylene Glycol—Viscous Solutions and Gels

Viscous solutions, such as polyethylene glycol of various molecular weights have been used to stabilize isolated, intact chloroplasts by Kaetsu *et al.* (1980). The addition of polyethylene glycol 300, 400 or 600 to chloroplast assay medium considerably increased the lifetime of PSII activity and when used in conjunction with radiation polymerization (see below) PSII activity could be maintained for over 40 d (Kaetsu *et al.*, 1980). The optimum polyethylene glycol concentration was between 70 and 90% v/v. Longer chain polyethylene glycol is not suitable as it is solid below 25 °C.

### 15.3.7 Radiation Polymerization

This technique is based on the polymerization of vinyl monomers in the presence of a chloroplast preparation. Fujimura *et al.* (1981) have used this method is immobilize intact chloroplasts using a glass-forming monomer, 2-hydroxyethyl acetate. These authors added 1 mL of chloroplast suspension to 0.4 mL of 10% bovine serum albumin, 0.4 mL of 0.1% mannitol, 0.4 mL of 0.6% sodium ascorbate and 0.4 mL of monomer and cooled this mixture to −24 °C. The preparation was then irradiated with $\gamma$-rays for 1 h at $1 \times 10^6$ rad $h^{-1}$. The product was then cut into small cubes and suspended in assay buffer. $CO_2$ fixation was not reported for these chloroplasts and they did not appear to be intact after immobilization as high rates of ferricyanide reduction were observed. However, PSII activity was stabilized, resulting in a $t_{\frac{1}{2}}$ for loss of ferricyanide reduction of 10 d at 4 °C and 2–3 d at 20 °C. Free chloroplasts showed complete decay within 18 h. Thermal stability was also improved by immobilization and there was no substantiated inhibition of PSII activity by the treatment.

### 15.3.8 Adsorption Techniques

The methods for immobilization of chloroplasts described above all involved the physical incorporation of biological material into a stable matrix. It is also possible to immobilize chloroplasts by trapping them in cavities of membrane filters (Cerovic *et al.*, 1987) or by adhesion onto diethylaminoethyl(DEAE)–cellulose (Shioi and Sasa, 1979).

In the membrane entrapment method (Cerovic *et al.*, 1987), a suspension of intact chloroplasts in buffer is passed through a cellulose nitrate membrane filter with a pore size of around 8 $\mu$m (Figure 8). The intact chloroplasts are trapped in the pores of the filter, forming not only a film on the surface but a layer penetrating a third of the depth of the cellulose nitrate matrix. Chloroplasts immobilized in this way showed no loss of $CO_2$ fixation over a 30 min period of continuous illumination. The advantages of such an 'artificial leaf' are that optical and gas analysis measurements are easily made and products of $CO_2$ fixation can be washed out of the preparation without chloroplast damage.

Adsorption of chloroplasts onto DEAE–cellulose, although suited to the stabilization of electron transport, does not appear to be useful for immobilizing intact chloroplasts fixing $CO_2$. This is because DEAE addition appears to uncouple totally thylakoid electron transport (Shioi and Sasa, 1979). However, with this technique, high rates of whole chain electron transport are retained (over 80% of that in free chloroplasts) and reduction of artificial electron acceptors will proceed for over 5 h in the light in a flow through system. The procedure is simple: a chloroplast suspension in isotonic buffer is stirred into precycled DEAE–cellulose (Whatman DE23) for 3 min in an ice bath then loaded onto a chromatography column or suspended in buffer. Chloroplasts are retained by adsorption onto the cellulose beads.

### 15.3.9 Which Technique is Best?

The criteria for selecting an immobilization system must include: (i) retention of high rates of reactions and efficiency of light utilization; (ii) stabilization of activity so that biological material does not require replacement causing economic inviability; (iii) low cost and ease of production of the

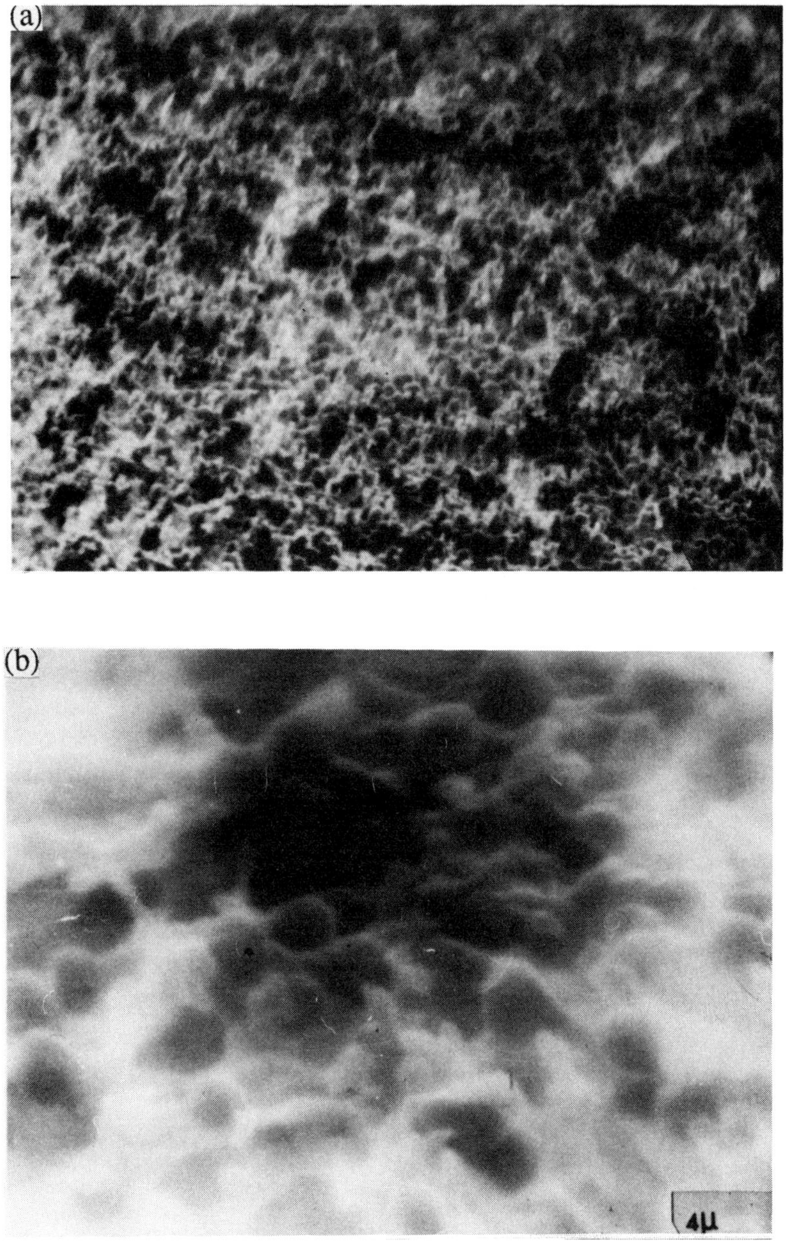


**Figure 8** Scanning electron micrographs of the surface view of chloroplasts trapped in the cavities of a cellulose nitrate membrane filter solid support (8 $\mu m$ nominal pore size). (a) Surface × 2000, (b) surface × 5000

chloroplast matrix; and (iv) engineering considerations such as ease of illumination and reactor design.

For intact chloroplasts capable of $CO_2$ fixation, these criteria cannot be easily met. The most popular technique, protein cross-linking, appears to be unsuitable while most other techniques result in disruption of the structural integrity of the chloroplast. The most nondestructive techniques use polyacrylamide (Karube *et al.*, 1979), polyvinyl alcohol (Ochiai *et al.*, 1978) and nitro cellulose filters (Cerovic *et al.*, 1987) for immobilization. Much more work is needed to establish a method which would allow successful application of intact chloroplasts to the production of carbon compounds or other metabolic products. Thylakoid systems, due to their robust nature, lend themselves much more to biotechnology and immobilization. The major problem appears to be inactivation of PSII or intersystem electron transport during immobilization. These problems are partially avoided by the use of agar and alginates as a matrix (see Gisby *et al.*, 1987) or by using polyvinyl alcohol (Ochiai *et al.*, 1978). Thylakoids have already been successfully used in bioreactors as a source of reducing power following immobilization (Gisby *et al.*, 1987). Biotechnological application of the systems will be dealt with in the following section.

## 15.4 BIOTECHNOLOGICAL APPLICATIONS OF CHLOROPLAST SYSTEMS

### 15.4.1 Efficiency of Energy Conversion

In its entirety photosynthesis is a photochemical energy transducer, converting light energy to soluble sugars and starch. Since isolated chloroplasts can synthesize starch but do not contain the enzymes required for the synthesis of soluble carbohydrates such as sucrose they are not ideal for application to the industrial production of carbon compounds. However, the chloroplast thylakoid is an efficient system for trapping solar energy and can be successfully used as a 'solar cell' to run a variety of redox reactions, since the redox potential of the primary acceptor of photosystem I is around $-600$ mV (Figure 3).

In this context, it is useful to examine the efficiency of light utilization in photosynthesis. Due to the absorption spectrum of chlorophyll molecules, only light in the wavelength region of 400–700 nm can be used in photosynthesis resulting in an immediate loss of approximately 50% of the available light energy. In addition to this, 20% of incident light is lost due to reflection, absorption and transmission of radiation by leaf tissue (see Hall, 1976). A further 77% loss occurs due to the quantum efficiency of photosynthesis (assuming 10 quanta per $CO_2$ fixed), resulting in a final efficiency of approximately 10%. If this calculation is restricted to red light alone the theoretical efficiency of photosynthetic energy conversion would be approximately 33% (based on the energy content of $CO_2$ of 0.47 MJ $mol^{-1}$, the energy of a mole quantum of 680 nm light as 0.176 MJ and a minimum quantum requirement of eight for the transfer of four electrons from water to NADP). Thus in the absence of energy dissipating processes occurring in whole leaves (such as respiration and photorespiration) a photochemical conversion efficiency of up to 33% might be achieved in a thylakoid or chloroplast system depending on the light regime applied.

### 15.4.2 Biocatalytic Hydrogen Production

Biological hydrogen-generating systems can be made in several ways. This may be simply by combination of individual enzymes and catalysts, or by the assembly of catalysts in organized units such as biological membranes or cells, or by coupling microorganisms with enzymes or organelles (Figure 9). A well-characterized chloroplast system for producing hydrogen consists of a combination of bacterial hydrogenase (Cammack *et al.*, 1985) with thylakoids and a soluble intermediate electron acceptor (to photosystem 1) such as methyl viologen or ferredoxin (Gisby and Hall, 1980; Packer, 1976, 1980). In such a system (as described by Packer) which is summarized in Figure 9, the four electrons generated from water oxidation reduce ferredoxin. This, in turn, is used by the added hydrogenase to reduce protons to hydrogen gas. The only reaction which can compete with the hydrogenase for reducing power is the oxidation of reduced ferredoxin by molecular oxygen. Thus, it is desirable to include an oxygen trap in the form of the enzyme glucose oxidase together with its substrate glucose in order to lower the oxygen content of the reaction medium.

Using the hydrogenase/thylakoid/ferredoxin system it is possible to generate hydrogen at rates of over 50 $\mu$mol $H_2$ per mg chlorophyll per hour (or up to 72% of the endogenous rate of photosynthetic oxygen evolution (Packer, 1980). However, more typical rates are in the region of 10–25 $\mu$mol $H_2$ per mg chlorophyll per hour (Gisby and Hall, 1980; Packer, 1980). Inorganic or synthetic catalysts can also be used in this system with similar results, as shown by Gisby and Hall (1980) who used platinum–polyvinyl alcohol and methyl viologen in place of hydrogenase to produce hydrogen in a thylakoid preparation.

An alternative system for biophotolytic hydrogen production has also been used where intact bacterial cells (such as *Chlostridium butyricum*) are combined with thylakoids and an electron acceptor such as benzyl viologen or ferredoxin (Karube *et al.*, 1981). There are two advantages to this system in relation to the isolated hydrogenase plus thylakoid coupled reaction. Firstly, it appears that in the reaction mixture the isolated hydrogenase enzyme is inactivated by oxygen (Karube *et al.*, 1981) a problem which is overcome when intact bacteria or immobilized enzymes are used. Secondly, the time-consuming isolation of pure hydrogenase is avoided with intact bacteria.

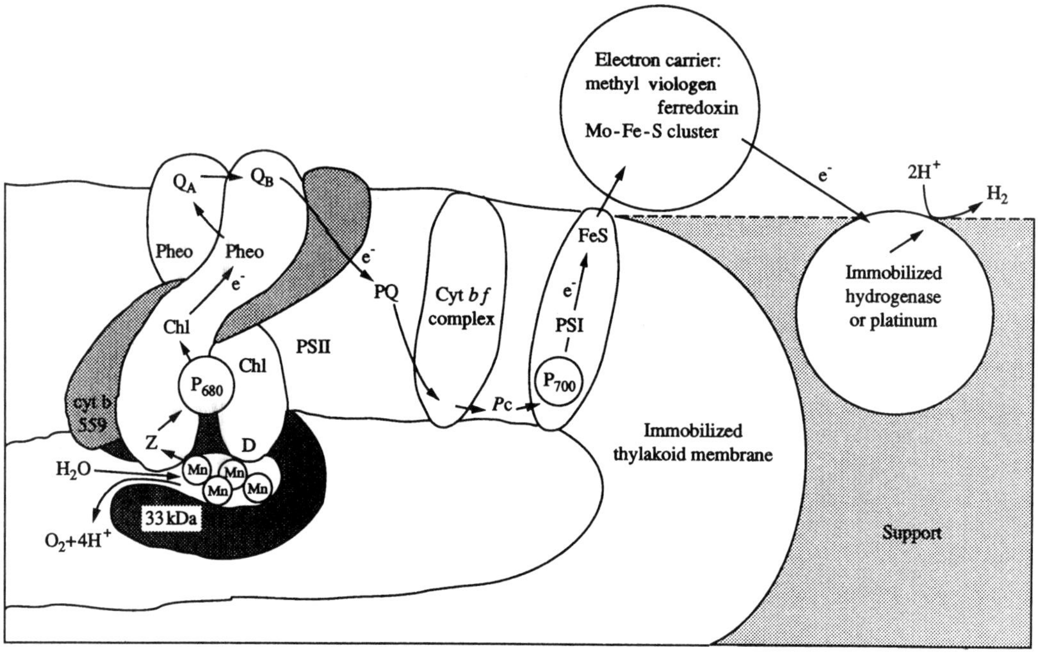


**Figure 9** Diagrammatic representation of a biocatalytic system for $H_2$ production

The conversion efficiency of biophotolytic systems is somewhat variable but efficiencies of up to 12.6% have been reported for the thylakoid/ferredoxin/hydrogenase reaction (Gisby and Hall, 1980) used in conjunction with an oxygen trap. These efficiencies compare very favourably with other solar energy systems in industrial use. The efficient utilization of hydrogen produced biologically also seems feasible. Biocatalytic hydrogen production has been successfully coupled to a hydrogen–oxygen fuel cell which produced 0.4–1.5 mA continuously for 4 h (Kayano *et al.*, 1981) with a conversion ratio from hydrogen to current of 80–100%. The possible economic feasibility of such systems was assessed by Hall (1976) who calculated that assuming a 10% conversion efficiency 130 km$^2$ of biophotolytic reactors could produce 2500 tonnes of hydrogen gas per day. Such a system is extremely attractive due to the low cost of reactants (sunlight and water) and the nonpolluting nature of the product.

### 15.4.3 Stability of Hydrogen-producing Systems

The major barrier to the large-scale application of biophotolytic hydrogen production is the inherent instability of the system. The major problem is the loss of thylakoid electron transport as described in Section 15.3. Immobilization procedures for the hydrogenase/electron acceptor/thylakoid system, the platinum/electron acceptor/thylakoid system and the bacterial reactor have been investigated by a number of laboratories (Gisby and Hall, 1980; Karube *et al.*, 1981; Kayano *et al.*, 1981; Packer, 1976). Agar and alginate have been the most popular materials used, resulting in a doubling of the half-life for loss of hydrogen evolution (Gisby and Hall, 1980). However, an optimistic estimate of the lifetime of any of these systems is 6–8 h in the light (see Gisby *et al.*, 1987). Unless this value can be increased to several days, it is unlikely that biophotolytic hydrogen production using thylakoids would be economically viable.

### 15.4.4 Industrial Applications

Immobilized thylakoids could potentially be used to reduce a wide variety of chemical substrates. Energy-rich redox potentials of $-600$ mV are produced in photosystem 1 and continuous flow through reactors of the type described by Shioi and Sasa (1979) could be employed industrially. In the system described by these authors, 2,6-dichloroindophenol (DCIP) was photoreduced by passing it through a column of thylakoids immobilized on cellulose. At flow rates of 18 mL h$^{-1}$, a stable rate of DCIP reduction was maintained for several hours.

An alternative application of chloroplast thylakoids is the utilization of the electrochemical potential difference produced across the membrane rather than the redox potential between the photosystems. Theoretically, this photogalvanic system could be used directly to produce electricity although the efficiency of such systems may be quite low. There has recently been considerable interest in using charged biological membranes in biosensors and as components in microprocessors for construction of 'biological computers'. In this context one could envisage the use of a chlorophyll containing membrane in conjunction with laser technology to process a binary signal by virtue of charge separation in the membrane, or the formation of a transthylakoid potential difference.

## 15.5 PROSPECTS FOR FUTURE DEVELOPMENT

The chloroplast clearly has unexploited potential for use in biotechnological development of synthetic processes. We are faced with the dilemma that biological material is inherently unstable yet the chloroplast is the only system that can effectively split water to $O_2$ and reducing power. The use of chloroplast systems in biotechnology (mainly to continuously produce hydrogen and oxygen) has been explored but thus far it has not provided the basis for process to harness the available energy or provide a functional production system for carbohydrates and other compounds. The reasons for this are readily appreciated in view of the features of chloroplast metabolism discussed above. There are, at present, major drawbacks to the use of chloroplast systems for biotechnological exploitation. These are: (i) the inherent instability of the chloroplast and accompanying loss of function; (ii) the regulatory processes that reduce the effective use of the light and $CO_2$ available; and (iii) the low activities or absence of enzymes other than for carbohydrate synthesis that can be used for biosynthesis of products of economical value. However, this situation may soon change with an improved understanding of the regulatory mechanisms that impose metabolic constraints that limit photosynthesis, coupled to the application of new molecular biological techniques. The fact that most of the protein complement of the chloroplasts that carry out the synthetic processes are products of nuclear genes and must be imported across the chloroplast envelope might successfully be exploited. It is now possible through controlled mutagenesis to modify genes with precision to meet the biochemical or structural requirements of the genetic engineers. With increasing success, genes for specific proteins can be isolated, inserted into bacteria and muliplied to large amounts and then nucleotide sequences taken out and replaced as required. This means that it has now become increasingly possible to alter directly the protein structure, an ability that opens up many possibilities. Firstly it is now possible to relate protein structure to function and secondly to adapt or improve an enzyme to meet with precise requirements. The exciting probability then exists to redesign an enzyme such as ribulose-1,5-bisphosphate carboxylase to function better. The engineered protein must then be expressed in or introduced into the chloroplast. The latter process requires that engineered proteins or foreign proteins are efficiently delivered and incorporated at the correct subcompartmental location within the chloroplast. The mechanism of transport of proteins into the chloroplast is known but not fully characterized (Della-Cioppa *et al.*, 1987; Ellis, 1981). Posttranslational targetting into the chloroplast and transport across the envelope is facilitated by the presence of amino terminal topogenic sequences known as 'transit peptides' (Wasmann *et al.*, 1986). The presence of a transit peptide is sufficient to target a foreign protein into the chloroplast and also contains sufficient information to define the final destination of the protein (stroma, thylakoid lumen, surface-exposed thylakoid) in the chloroplast. The transit peptides are recognized by sites on the outer membrane of the chloroplast envelope and mediate translocation across the envelope. They are proteolytically removed after import into the organelle. Chimaeric proteins made with the transit peptide of the small subunit of ribulose-1,5-bisphosphate carboxylase have been successfully imported into chloroplast and processed, for example the transit peptide for the small subunit of ribulose-1,5-bisphosphate carboxylase will most efficiently target the bacterial protein NPT II into chloroplasts *in vitro* (Wasmann *et al.*, 1986). We therefore have the foundations of a system by which the biological competence of the chloroplasts may be replenished or stabilized and numerous possibilities for manipulating carbon assimilation and other pathways in isolated chloroplasts.

It is of interest to note that some of the plant species that exhibit the $C_4$ photosynthetic pathway have dimorphic chloroplasts with differences in function and synthetic capabilities. Such differences might be used to advantage for biotechnological purposes (Furbank and Foyer, 1988). The mesophyll chloroplasts from *Zea mays* and other $C_4$ species, for example, produce ATP and NADPH in a similar fashion to chloroplasts that display the normal $C_3$ photosynthetic metabolism. However, they do not contain all the enzymes of the Calvin cycle but function instead primarily to produce phosphoenolpyruvate (when supplied with pyruvate) and triose phosphate (when supplied with glycerate

3-phosphate). Conversely, the thylakoids in the bundle sheath chloroplasts from *Z. mays* are capable of producing substantial amounts of ATP but have a reduced capacity for NADPH production. These variations and complexities of structure and metabolism between chloroplasts from different sources might be successfully used to meet the precise requirements of the biotechnologist.

The exploitation and emulation of photosystem design and even exploitation of functional components of thylakoid network provide considerable scope for future development. Various types of photoelectrochemical cells using chloroplast components have been described. At present the major catalytic systems capable of generating hydrogen by catalytic cleavage of water may be summarized in three categories; (a) the essentially complete biological system of thylakoids coupled to hydrogenase; (b) the semibiological systems where thylakoids are coupled to synthetic electron donors and hydrogenase or platinum for hydrogen production; and (c) the completely synthetic systems using semiconductors such as $TiO_2$ or $RuO_2$ coupled to platinum. The immobilization of functional photosynthetic membranes on semiconducting colloidal particles incorporating titanium and ruthenium has proved successful. Thylakoid fragments have been deposited on an $SnO_2$ electrode (Ochai *et al.*, 1982) and photosystem I and photosystem II particles have been immobilized on $TiO_2$ (Cuendet and Gratzel, 1984; Rao *et al.*, 1984). Such systems are fraught with problems and the process of electrode sensitization is only poorly understood; nevertheless, such systems have the capability to split water and produce hydrogen and oxygen. In most cases the biological material can be deposited onto the surface of an electrode and illumination results in a photoeffect through the charge separation generated in the reaction centres and the subsequent charge transfer to the supporting electrode and the surrounding medium. Electrical coupling between the thylakoid particles and artificial or natural catalysts localized on the support also shows promise.

## 15.6 ABBREVIATIONS

$C_3$ = 3 carbon compound; $C_4$ = 4 carbon compound; Chl = chlorophyll; DHAP = dihydroxyacetone phosphate; E4P = erythrose 4-phosphate; Fd = ferredoxin; FeS = iron-sulphur centre; F6P = fructose 6-phosphate; F-1,6-$P_2$ = fructose 1,6-bisphosphate; GAP = glyceraldehyde phosphate; GBP = glycerate 1,3-bisphosphate; G1P = glucose 1-phosphate; G6P = glucose 6-phosphate; GSH = reduced glutathione; GSSG = oxidized glutathione; Hepes = *N*-2-hydroxyethylpiperazine-*N'*-2-ethanesulfonic acid; Pc = plastocyanin; PGA = phosphoglycerate, Ph(eo) = phaeophytin; Pi = orthophosphate, PPi = pyrophosphate; PQ (Q) = plastoquinone; PSI = photosystem I; PSII = photosystem II; $Q_A$ = the primary electron acceptor of PSII; $q_Q$ = photochemical quenching of chlorophyll *a* fluorescence associated with the oxidation of $Q_A$; $q_{NP}$ = nonphotochemical quenching of chlorophyll *a* fluorescence; R5P = ribose 5-phosphate; RuBP = ribulose 1,5-bisphosphate; S7P = sedoheptulose 7-phosphate; SBP = sedoheptulose 1,7-bisphosphate; Sucrose-P = sucrose 6-phosphate; TP = triose phosphate, Tris = 2-amino-2-(hydroxymethyl)propane-1,3-diol, Xu5P = xylulose 5-phosphate.

## ACKNOWLEDGEMENT

We are grateful to Dr Deborah Rees for the data comprising Figure 1.

## 15.7 REFERENCES

Allen, J. F. (1983). Protein phosphorylation—carburetter of photosynthesis. *Trends Biochem. Sci.*, **8**, 369–373.

Anderson, J. W., C. H. Foyer and D. A. Walker (1983). Light-dependent reduction of hydrogen peroxide by intact spinach chloroplasts. *Biochim. Biophys. Acta*, **724**, 69–74.

Asada, K. and M. Takahashi (1987). Production and scavenging of active oxygen in photosynthesis. In *Photoinhibition; Topics in Photosynthesis*, ed. D. J. Kyle, C. B. Osmond and C. J. Arntzen, vol. 9, pp. 227–288, Elsevier, Amsterdam.

Barber, J. (1987). Photosynthetic reaction centres: a common link. *Trends Biochem. Sci.*, **12**, 321–326.

Barber, J. (1988). Signal from the reaction centre. *Nature* (*London*), **332**, 111–112.

Barbotin, J.-N., M.-F. Cocquempot, V. Larreta-Garde, B. Thomasset, G. Gellf, J.-D. Clement-Metral and D. Thomas (1987). Immobilized organelles in cross-linked proteins. *Methods Enzymol.*, **135**, 454–472.

Barbotin, J.-N., and B. Thomasset (1980). Immobilized organelles and whole cells into protein foam structures: scanning and transmission electron microscope observations. *Biochimie*, **62**, 359–365.

Broun, G., D. Thomas, G. Gellf, D. Domurado, A. M. Berjonneau and C. Guillon (1973). New methods for binding enzyme molecules into a water insoluble matrix: properties after immobilization. *Biotechnol. Bioeng.*, **15**, 359–375.

Brudvig, G. W. and R. H. Crabtree (1986). Mechanism of photosynthetic $O_2$ evolution. *Proc. Natl. Acad. Sci. USA*, **83**, 4586–4588.

Cammack, R., D. O. Hall and K. K. Rao (1985). Hydrogenases: structure and applications in hydrogen production. In *Microbial Gas Production: Mechanistic, Metabolic and Biotechnological Aspects*, pp. 75–102. Society for General Microbiology, London.

Cerovic, Z. G., J. K. Cheesbrough and D. A. Walker (1987). Photosynthesis by intact isolated chloroplasts on solid support. *Plant Physiol.*, **84**, 1249–1251.

Coleman, W. and Govindjee (1987). A model for the mechanism of chloride activation of oxygen evolution in photosystem II. *Photosynth. Res.*, **13**, 199–223.

Cuendet, P. and M. Gratzel (1984). Immobilization of PSI and PSII on semiconducting particles. *Adv. Photosynth. Res.*, **II**, 813–816.

Deisenhofer J., O. Epp, K. Miki, R. Huber and H. Michel (1985). Structure of the protein subunit in the photosynthetic reaction centre of *Rhodopseudomonas viridis* at 3 Å resolution. *Nature (London)*, **318**, 618–624.

Della-Cioppa, G., G. M. Kishore, R. N. Beachy and R. T. Fraley (1987). Protein trafficking in plant cells. *Plant Physiol.*, **84**, 965–968.

Diner, B. A. (1986). Photosystems I and II; structure, proteins and cofactors. In *Encyclopedia of Plant Physiology. Photosynthesis III; Photosynthetic membranes and light harvesting systems*, ed. L. A. Staehelin and C. J. Arntzen, vol. 19, pp. 422–436. Springer-Verlag, Berlin.

Douce, R. and J. Joyard (1979). Structure and function of the plastid envelope. *Adv. Bot. Res.*, **7**, 1–116.

Ellis, J. R. (1979). The most abundant protein in the world. *Trends Biochem. Sci.*, **4**, 241–244.

Ellis, J. R. (1981). Chloroplast proteins: synthesis, transport and assembly. *Annu. Rev. Plant Physiol.*, **32**, 111–137.

Farquhar, G. D., S. Von Caemmerer and J. A. Berry (1980). A biochemical model of photosynthetic $CO_2$ assimilation in leaves of $C_3$ species. *Planta*, **149**, 78–90.

Foyer, C. H. (1986). The regulation of carbon assimilation in photosynthesis. *Chem. Br.*, **22**, 723–726.

Foyer, C. H. and B. Halliwell (1976). The presence of glutathione and glutathione reductase in chloroplasts: a proposed role in ascorbic acid metabolism. *Planta*, **133**, 21–25.

Foyer, C. H., R. T. Furbank and D. A. Walker (1987). Interactions between ribulose-1,5-bishosphate carboxylase and stromal metabolites. I. Modulation of enzyme activity by Benson–Calvin cycle intermediates. *Biochim. Biophys. Acta*, **894**, 157–164.

Foyer, C. H., R. T. Furbank, J. Harbinson and P. Horton (1990). The mechanisms contributing to photosynthetic control of electron transport by carbon assimilation in leaves. *Photosynth. Res.*, **25**, 83–100.

Fujimura, T., F. Yoshii and I. Kaetsu (1981). Stabilization of photosystem II ($O_2$ evolution) of spinach chloroplasts by radiation-induced immobilization. *Plant Physiol.*, **67**, 351–354.

Furbank, R. T., C. H. Foyer and D. A. Walker (1986). Inhibition of photophosphorylation by ribulose-1,5-bisphosphate carboxylase. *Biochim. Biophys. Acta*, **852**, 46–54.

Furbank, R. T., C. H. Foyer and D. A. Walker (1987a). Interactions between ribulose-1,5-bisphosphate carboxylase and stromal metabolites. II. Corroboration of the rôle of this enzyme as a metabolite buffer. *Biochim. Biophys. Acta*, **894**, 165–173.

Furbank, R. T., C. H. Foyer and D. A. Walker (1987b). Regulation of photosynthesis in isolated spinach chloroplasts during orthophosphate limitation. *Biochim. Biophys. Acta*, **894**, 552–561.

Furbank, R. T. and C. H. Foyer (1988). $C_4$ plants as valuable model experimental systems for the study of photosynthesis. *New Phytol.*, **109**, 265–272.

Gisby, P. E. and D. O. Hall (1980). Biophotolytic $H_2$ production using alginate-immobilized chloroplasts, enzymes and synthetic catalysts. *Nature (London)*, **287**, 251–253.

Gisby, P. E., K. K. Rao and D. O. Hall (1987). Entrapment techniques for chloroplasts, cyanobacteria and hydrogenases. *Methods Enzymol.*, **135**, 440–471.

Glazer, A. N. and A. Melis (1987). Photochemical reaction centers: structure, organization and function. *Annu. Rev. Plant Physiol.*, **381**, 11–45.

Govindjee, T. Kambara and W. Coleman (1985). The electron donor side of photosystem II: the oxygen evolving complex. *Photochem. Photobiol.*, **42**, 187–210.

Hall, D. O. (1972). Nomenclature for isolated chloroplasts. *Nature (London)*, **235**, 125–126.

Hall, D. O. (1976). Photobiological energy conversion. *FEBS Lett.*, **64**, 6–16.

Halliwell, B. (1984). A classification of chloroplasts. In *Chloroplast Metabolism, the Structure and Function of Chloroplasts in Green Leaf Cells*. pp. 18–21. Clarendon Press, Oxford.

Hardt, H. and B. Kok (1976). Stabilization by glutaraldehyde of high-rate electron transport in isolated chloroplasts. *Biochim. Biophys. Acta*, **449**, 125–135.

Heber, U. and D. A. Walker (1979). The chloroplast envelope, barrier or bridge. *Trends Biochem. Sci.*, **4**, 252–256.

Heber, U., S. Neimanis and K.-J. Dietz (1988). Fractional control of photosynthesis by the $Q_B$ protein, the cytochrome $f/b_6$ complex and other components of the photosynthetic apparatus. *Planta*, **173**, 267–274.

Heldt, H. W. (1976). Metabolite transport in intact spinach chloroplasts. In *The Intact Chloroplasts*, ed. J. Barber, chap. 6, pp. 135–170, Elsevier, Amsterdam.

Heldt, H. W. and L. Rapley (1970). Specific transport of inorganic phosphate, 3-phosphoglycerate and dihydroxyacetone phosphate, and of dicarboxylates across the inner membrane of spinach chloroplasts. *FEBS Lett.*, **10**, 143–148.

Horton, P. (1983). Control of chloroplast electron transport by photophosphorylation of thylakoid proteins. *FEBS Lett.*, **152**, 47–52.

Horton, P. (1985). Regulation of photochemistry and its interaction with carbon metabolism. In *Regulation of Sources and Sinks in Crop Plants*, ed. B. Jeffcoat A. F. Hawkins and A. D. Stead. British Plant Growth Regulator Group Monograph 12, pp. 19–33. Parchments, Oxford.

Horton, P. and C. H. Foyer (1983). Relationships between protein phosphorylation and electron transport in the reconstituted chloroplast system. *Biochem. J.*, **210**, 517–521.

Ikeuchi, M. and Inoue (1987). Specific $^{125}$I labeling of D1 (herbicide-binding protein): an indication that D1 functions on both the donor and acceptor sites of photosystem II. *FEBS Lett.*, **210**, 71–76.

Ikeuchi, M., F. G. Plumley, Y. Inoue and G. W. Schmidt (1987). Phosphorylation of photosystem II components, CP43 apoprotein, D1, D2, and 10 and 11 kDa protein in chloroplast thylakoids of higher plants. *Plant Physiol.*, **85**, 638–642.

Jensen, R. G. and J. A. Bassham (1966). Photosynthesis by isolated chloroplasts. *Proc. Natl. Acad. Sci. USA.*, **56**, 1095–1101.

Kacser, H. and J. A. Burns (1979). Molecular democracy; who shares the controls? *Biochem. Soc. Trans.*, **7**, 1149–1160.
Kaetsu, I., F. Yoshii and T. Fujimura (1980). Effect of viscous solvents and manomer on conservation of intact and immobilized chloroplasts. *Z. Naturforsch., Teil C.*, **35**, 1052–1056.
Karube, I., K. Aizawa, S. Ideda and S. Suzuki (1979). Carbon dioxide fixation by immobilized chloroplasts. *Biotechnol. Bioeng.*, **21**, 253–260.
Karube, I., T. Matsunaga, T. Otsuka, H. Kayano and S. Suzuki (1981). Hydrogen evolution by coimmobilized chloroplasts and *Clostridium butyricum*. *Biochim. Biophys. Acta*, **637**, 490–495.
Kayano, H., T. Matsunaga, I. Karube and S. Suzuki (1981). Photochemical energy conversion system using immobilized chloroplasts. *Biotechnol. Bioeng.*, **23**, 2283–2291.
Kitajima, M. and W. L. Butler (1976). Microencapsulation of chloroplast particles. *Plant Physiol.*, **57**, 746–750.
Kuhlemeier, C., P. Green and N.-H. Chua (1987). Regulation of gene expression in higher plants. *Annu. Rev. Plant Physiol.*, **38**, 221–257.
Leegood, R. C. and R. Malkin (1986). Isolation of subcellular photosynthetic systems. In *Photosynthesis Energy Transduction: A Practical Approach*, ed. M. F. Hipkins and N. R. Baker, pp. 9–26, IRL Press, Oxford.
Leegood, R. C. and D. A. Walker (1985). Chloroplasts and protoplasts. In *Techniques in Bioproductivity and Photosynthesis*, ed. J. Coombs, D. O. Hall, S. P. Long and J. M. O. Skurlock, 2nd edn., pp. 118–132. Pergamon Press, Oxford.
Leegood, R. C., D. A. Walker and C. H. Foyer (1985). Regulation of the Benson–Calvin cycle. In *Photosynthetic Mechanisms and the Environment* ed. J. Barber and N. R. Baker, pp. 191–258. Elsevier, Amsterdam.
Lilley, R. McC., M. P. Fitzgerald, K. G. Rienits and D. A. Walker (1975). Criteria of intactness and the photosynthetic activity of spinach chloroplast preparations. *New Phytol.*, **75**, 1–10.
Michel, H. and J. Deisenhofer (1986). X-ray diffraction studies on a crystalline bacterial reaction centre: a progress report and conclusions on the structure of photosystem II reaction centres. In *Encyclopedia of Plant Physiology Photosynthesis III. Photosynthetic membranes and light harvesting systems*, ed. L. A. Staehelin and C. J. Arntzen, vol. 19, pp. 371–381. Springer-Verlag, Berlin.
Mourioux, G. and R. Douce (1981). Slow passive diffusion of orthophosphate between intact isolated chloroplasts and suspending medium. *Plant Physiol.*, **67**, 470–473.
Ochiai, H., H. Shibata, T. Matsuo, K. Hashinokuchi and I. Inamura (1978). Immobilization of chloroplast photosystems with polyvinyl alcohols. *Agric. Biol. Chem.*, **42**, 683–685.
Ochiai, H., H. Shibata, Y. Sawa and T. Katoh (1982). Properties of the chloroplast film electrode immobilized on an Sn-coated glass plate. *Photochem. Photobiol.*, **35**, 149–155.
Packer, L. (1976). Problems in the stabilization of the *in vitro* photochemical activity of chloroplasts used for $H_2$ production. *FEBS Lett*, **64**, 17–19.
Packer, L. (1980). $H_2$ production *in vitro* by chloroplast/ferredoxin/hydrogenase reconstituted system. *Methods Enzymol.*, **69**, 625–630.
Papageorgiou, G. C. (1980). Stabilization of chloroplasts and subchloroplast particles. *Methods Enzymol.*, **69**, 613–25.
Papageorgiou, G. C. (1983). Immobilization of photosynthetically active intact chloroplasts in a cross-linked albumin matrix. *Biotechnol Lett.*, **5**, 819–824.
Parry, M., A. Keys, C. H. Foyer, R. T. Furbank and D. A. Walker (1988). The mechanism of regulation of ribulose-1,5-bisphosphate carboxylase activity by the activase system in lysed spinach chloroplasts. *Plant Physiol.*, **87**, 558–561.
Portis, Jr., A. R. (1990). RuBisCo activase. *Biochim. Biophys. Acta*, **1015**, 15–28.
Powles, S. B. (1984). Photoinhibition of photosynthesis induced by visible light. *Annu. Rev. Plant Physiol.*, **35**, 15–44.
Prince, R. C. (1986). Manganese at the active site of the chloroplast oxygen-evolving complex. *Trends Biochem. Sci.*, **11**, 491–492.
Rao, K. K., D. O. Hall, P. Cuendet and M. Gratzel (1984). Photoproduction of hydrogen from water using immobilized biological and synthetic catalysts. *Adv. Photosynth. Res.* **II**, 777–780.
Robinson, S. P., Z. G. Cerovic and D. A. Walker (1986). Isolation of intact chloroplasts; general principles and criteria of integrity. *Methods Enzymol.*, **148**, 145–157.
Robinson, S. P. and A. R. Portis, Jr. (1988). Involvement of stromal ATP in the light activation of ribulose-1,5-bisphosphate carboxylase/oxygenase in intact isolated chloroplasts. *Plant Physiol.*, **86**, 293–298.
Robinson, S. P. and D. A. Walker (1981). Photosynthetic carbon reduction cycle. *The Biochemistry of Plants*, ed. M. D. Hatch and N. K. Boardman, vol. 8, pp. 193–236, Academic Press, New York.
Salvucci, M. E., A. R. Portis and W. Ogren (1985). A soluble chloroplast protein catalyzes activation of ribulosebisphosphate carboxylase/oxygenase *in vitro*. *Photosynth. Res.*, **7**, 193–201.
Sane, P. V., D. I. Goodchild and R. B. Park (1970). Characterization of chloroplast photosystems I and II separated by a nondetergent method. *Biochim. Biophys. Acta*, **216**, 162–170.
Seftor, R. E. B. and R. G. Jenson (1986). Causes for the disappearance of photosynthetic $CO_2$ fixation with isolated spinach chloroplasts. *Plant Physiol.*, **81**, 81–85.
Shioi, Y. and T. Sasa (1979). Immobilization of photochemically active chloroplasts onto diethyl aminoethyl–cellulose. *FEBS Lett*, **101**, 311–315.
Staehelin, L. A. and C. J. Arntzen (1983). Regulation of chloroplasts membrane function: protein phosphorylation changes the spatial organization of membrane components. *J. Cell Biol.*, **97**, 1327–1337.
Stitt, M. (1987). Limitation of photosynthesis by sucrose synthesis. In *Progress in Photosynthesis Research*, ed. J. Biggins, vol. 3, pp. 685–692. Nijhoff, Dordrecht.
Trebst, A. (1986). The topology of the plastoquinone and herbicide binding peptides of photosystem II. *Z. Naturforsch., Teil C.*, **41**, 240–245.
Von Caemmerer, S. and G. D. Farquhar (1981). Some relationships between the biochemistry of photosynthesis and the gas exchange of leaves. *Planta*, **153**, 376–387.
Walker, D. A. (1971). Chloroplasts (and grana)-aqueous (including high carbon fixation ability). *Methods Enzymol.*, **23**, 211–220.
Walker, D. A. (1976). $CO_2$ fixation by intact chloroplasts: photosynthetic induction and its relationship to transport phenomena and control mechanisms. In *Intact Chloroplast*, ed. J. Barber, pp. 235–238. Elsevier, Amsterdam.
Walker, D. A. (1980a). Preparation of higher plant chloroplasts. *Methods Enzymol.*, **69**, 94–104. Academic Press, New York.

Walker, D. A. (1980b). Regulation of starch synthesis in leaves—the role of orthophosphate. In *Physiological Aspects of Crop Productivity. Proc. 15th Colloq. Int. Potash Inst. Bern*, 195–207.
Walker, D. A. (1987). In *The use of Oxygen Electrode and Fluorescence Probes in Simple Measurements of Photosynthesis*, pp. 1–145. Oxygraphics, Sheffield.
Walker, D. A. and A. R. Herold (1977). Can the chloroplast support photosynthesis unaided? In *Photosynthetic Organelles: Structure and Function*, ed. Y. Fujita, S. Fatoh, K. Shibata and S. Miyachi, pp. 295–310. Japanese Society of Plant Physiologists and Centre for Academic Publications, Tokyo.
Walker, D. A. and S. P. Robinson (1978). Chloroplast and cell. A contemporary view of photsynthetic carbon assimilation. *Ber. Deutsch. Bot. Ges.*, **91**, 513–526.
Walker, D. A. and M. N. Sivak (1986). Photosynthesis and phosphate: a cellular affair? *Trends Biochem. Sci.*, **11**, 176–179.
Warman, J. M. (1987). A quantum jump for chemistry. *Nature (London)*, **327**, 462–464.
Wasmann, C. C., B. Reiss, S. G. Bertlett and H. J. Bohnert (1986). The importance of the transit peptide and transported protein for protein import into chloroplasts. *Mol. Gen. Genet.*, **205**, 446–453.
Woodrow, I. E. (1986). Control of the rate of photosynthetic carbon dioxide fixation. *Biochim. Biophys. Acta*, **851**, 181–192.

# SUBJECT INDEX